CLYMER
YAMAHA
WATER VEHICLES SHOP MANUAL
1993-1996

The World's Finest Publisher of Mechanical How-to Manuals

PRIMEDIA
Business Directories & Books

P.O. Box 12901, Overland Park, KS 66282-2901

Copyright ©1997 PRIMEDIA Business Magazines & Media Inc.

FIRST EDITION
First Printing July, 1997
Second Printing June, 2000
Third Printing May, 2002

Printed in U.S.A.

CLYMER and colophon are registered trademarks of PRIMEDIA Business Magazines & Media Inc.

ISBN: 0-89287-686-7

Library of Congress: 97-61086

TECHNICAL PHOTOGRAPHY: Ron Wright and Mark Jacobs.

TECHNICAL ILLUSTRATIONS: Steve Amos and Michael St. Clair.

COVER: Photo provided by Yamaha Motor Corporation

PRODUCTION: Dylan Goodwin.

All rights reserved. Reproduction or use, without express permission, of editorial or pictorial content, in any manner, is prohibited. No patent liability is assumed with respect to the use of the information contained herein. While every precaution has been taken in the preparation of this book, the publisher assumes no responsibility for errors or omissions. Neither is any liability assumed for damages resulting from use of the information contained herein. Publication of the servicing information in this manual does not imply approval of the manufacturers of the products covered.

All instructions and diagrams have been checked for accuracy and ease of application; however, success and safety in working with tools depend to a great extent upon individual accuracy, skill and caution. For this reason, the publishers are not able to guarantee the result of any procedure contained herein. Nor can they assume responsibility for any damage to property or injury to persons occasioned from the procedures. Persons engaging in the procedure do so entirely at their own risk.

General Information	1
Troubleshooting	2
Lubrication, Maintenance and Tune-up	3
Engine (500 cc)	4
Engine (650, 700 and 760 cc)	5
Engine (1100 cc)	6
Drive Train	7
Fuel and Water Box Systems	8
Electrical System	9
Oil Injection System	10
Bilge System	11
Steering System	12
Index	13
Wiring Diagrams	14

CLYMER PUBLICATIONS
PRIMEDIA Business Magazines & Media

Chief Executive Officer Timothy M. Andrews
President Ron Wall

EDITORIAL

Editors
Mike Hall
James Grooms

Technical Writers
Ron Wright
Ed Scott
George Parise
Mark Rolling
Michael Morlan
Jay Bogart
Rodney J. Rom

Production Supervisor
Dylan Goodwin

Lead Editorial Production Coordinator
Shirley Renicker

Editorial Production Coordinators
Greg Araujo
Shara Pierceall

Editorial Production Assistants
Susan Hartington
Holly Messinger
Darin Watson

Technical Illustrators
Steve Amos
Robert Caldwell
Mitzi McCarthy
Bob Meyer
Michael St. Clair
Mike Rose

MARKETING/SALES AND ADMINISTRATION

Vice President, PRIMEDIA Business Directories & Books
Rich Hathaway

Marketing Manager
Elda Starke

Advertising & Promotions Coordinator
Melissa Abbott

Associate Art Directors
Chris Paxton
Tony Barmann

Sales Manager/Marine
Dutch Sadler

Sales Manager/Manuals
Ted Metzger

Sales Manager/Motorcycles
Matt Tusken

Operations Manager
Patricia Kowalczewski

Customer Service Manager
Terri Cannon

Customer Service Supervisor
Ed McCarty

Customer Service Representatives
Susan Kohlmeyer
April LeBlond
Courtney Hollars
Jennifer Lassiter
Ernesto Suarez

Warehouse & Inventory Manager
Leah Hicks

The following product lines are published by PRIMEDIA Business Directories & Books.

More information available at *primediabooks.com*

Contents

QUICK REFERENCE DATA IX

CHAPTER ONE
GENERAL INFORMATION 1
Manual organization 2
Notes, cautions and warnings 2
Safety first 2
Service hints 3
Engine operation 5
Parts replacement 6
Torque specifications 6
Fasteners 7
Gasket sealant 12
Threadlocking compound 13
Basic hand tools 14
Precision measuring tools 18
Special tools 24
Mechanic's tips 25

CHAPTER TWO
TROUBLESHOOTING 39
Operating requirements 40
Emergency troubleshooting 41
Engine starting system 43
Charging system 47
Ignition system 49
Fuel system 57
Engine 61
Two-stroke pressure testing 76
Steering 77
Jet pump 77

CHAPTER THREE
LUBRICATION, MAINTENANCE AND TUNE-UP 80
Operational checklist 80	Cooling system flushing 108
Break-in procedure 85	Engine tune-up 108
10 hour inspection 85	Carburetor synchronization 118
Lubrication 86	Choke synchronization 119
Oil injection service 90	Storage 119
50 and 100 hour maintenance schedule 94	

CHAPTER FOUR
ENGINE (500 CC) 126
Engine lubrication 126	Special tools 129
Service precautions 127	Engine removal 129
Serial numbers 128	Muffler housing and exhaust guide 133
Servicing engine in hull 128	Engine 139

CHAPTER FIVE
ENGINE (650, 700 AND 760 CC) 168
Engine lubrication 168	Engine removal 171
Service precautions 169	Muffler housing 177
Serial numbers 170	Engine top end 183
Servicing engine in hull 170	Cylinder 189
Special tools 170	Piston, piston pin, and piston rings 192
Precautions 171	Crankcase and crankshaft 200

CHAPTER SIX
ENGINE (1100 CC) 219
Engine lubrication 219	Engine removal 222
Service precautions 220	Muffler housing 226
Serial numbers 221	Engine top end 233
Servicing engine in hull 221	Cylinder 236
Special tools 221	Piston, piston pin, and piston rings 239
Precautions 221	Crankcase and crankshaft 246

CHAPTER SEVEN
DRIVE TRAIN 261
Intermediate shaft and housing (WR500 and WR650) 261	Jet pump (WR500 and WR650) 285
Intermediate shaft and housing (WRA650, WRA650A and WRA700) 267	Jet pump (RA700, RA700A, RA700B, RA760, RA1100, WRA650, WRA650A, WRA700, WVT700 and WVT1100) 300
Intermediate shaft and housing (FX700, SJ650, SJ700, SJ700A, WB700, WB700A, WB760, WRB650, WRB650A and WRB700) 273	Jet pump (FX700, SJ650, SJ700, SJ700A, WB700, WB700A, WB760, WRB650, WRB650A and WRB700) 319
Intermediate shaft and housing (RA700, RA700A, RA700B, RA760, RA1100, WVT700 and WVT1100) 279	Ride plate adjustment 333

CHAPTER EIGHT
FUEL AND WATER BOX SYSTEMS 336
- Precautions 336
- Flame arrestor 336
- Intake manifold and reed valve 346
- Carburetor 351
- Fuel tank 362
- Fuel petcock 370
- Vent trap system (FX700, RA700, RA700A, RA700B, RA760, RA1100, WB700, WB700A and WB760) 383
- Water box 383

CHAPTER NINE
ELECTRICAL SYSTEM 399
- Battery 399
- Charging system 405
- Electric starting system 405
- Flywheel and stator plate (500 cc models) 415
- Flywheel, idler gear and stator plate (650, 700 and 760 cc models) 419
- Flywheel, idler gear and stator plate (1100 cc) 423
- Electric box (500 cc models) ... 427
- Electric box (650, 700 and 760 cc models) 431
- Electric box (1100 cc models) .. 436
- Switches 438
- Multifunction meters 441
- Fuel meter (RA700, RA700A, RA760, RA1100, WB760, WRA650, WRA650A, WRA700, WRB650, WRB650A, WRB700, WVT700 and WVT1100) 443
- Oil meter (RA700A, RA760, RA1100, WVT700 and WVT1100) 446
- Oil warning indicator (RA700, RA700B and WB760) 449
- Overheat warning indicator (RA700, RA700A, RA760, RA1100, WB760, WRB650, WRB650A, WRB700, WVT700 and WVT1100) 451
- Fuse 452
- Wiring diagrams 452

CHAPTER TEN
OIL INJECTION SYSTEM 454
- Oil tank 454
- Sub oil tank 465
- Check valves 465
- Oil tank breather valve 466
- Oil pump 466

CHAPTER ELEVEN
BILGE SYSTEM 472

CHAPTER TWELVE
STEERING SYSTEMS 481
- Steering assembly (RA700, RA700A, RA700B, RA760, RA1100, WB700, WB700A, WB760, WR500, WR650, WRA650, WRA650A, WR700, WRB650, WRB650A, WRB700, WVT700 and WVT1100) 481
- Steering assembly (SJ650, SJ700, SJ700A and FX700) 495
- Cables 510

INDEX 542

WIRING DIAGRAMS 545

Quick Reference Data

ENGINE SPECIFICATION (500 CC)

Bore × stroke	72 × 61 mm (2.83 × 2.40 in.)
Displacement	496 cc (30.27 cu. in.)
Compression ratio	7.0:1

ENGINE SPECIFICATIONS (650 CC)

Bore × stroke	77 × 68 mm (3.03 × 2.68 in.)
Displacement	633 cc (38.61 cu. in.)
Compression ratio	
WR650	7.0:1
WRA650, WRA650A	7.2:1
WRB650, WRB650A	7.2:1
SJ650	7.2:1

ENGINE SPECIFICATIONS (700 CC)

Bore × stroke	81 × 68 mm (3.19 × 2.68 in.)
Displacement	701 cc (42.78 cu. in.)
Compression ratio	7.2:1

ENGINE SPECIFICATIONS (760 CC)

Bore × stroke	84 × 68 mm (3.31 × 2.68 in.)
Displacement	754 cc (46.02 cu. in.)
Compression ratio	
RA760	F-7.2:1, R-6.8:1
WB760	7.2:1

ENGINE SPECIFICATION (1100CC)

Bore × stroke	81 × 68 mm (3.19 × 2.68 in.)
Displacement	1051 cc (64.14 cu. in.)
Compression ratio	5.8:1

ENGINE LUBRICATION SYSTEM

WR500, WR650, SJ650, SJ700, SJ700A, FX700	Premix 50:1
All other models	Oil injection

RECOMMENDED LUBRICANTS AND FUELS

Grease	Yamaha marine grease*
Engine lubrication	Yamalube Two-Cycle Outboard Oil
Fuel	Regular unleaded, minimum octane: 86

*Or equivalent water-resistant marine grease.

APPROXIMATE REFILL CAPACITIES

Fuel tank	Full liter (gal.)	Reserve liter (gal.)
Fuel tank		
SJ650, SJ700, SJ700A	18 (4.8)	5.5 (1.45)
FX700	14 (3.7)	3.6 (0.95)
WR500, WR650	22 (5.8)	4 (4.2)
WRA650, WRA650A, WRA700	40 (10.6)	7 (1.9)
WRB650, WRB650A, WRB700	30 (7.9)	5 (1.32)
WB700, WB700A	25 (6.6)	5 (1.32)
RA700, WB760	40 (10.6)	11.6 (3.06)
RA700A, RA700B, RA760, RA1100, WVT700	50 (13.2)	8.8 (2.32)
WVT1100	50 (13.2)	12 (3.2)
Oil tank		
WRA650, WRA650A, WRA700, RA700, WB760	4 (1.1)	–
WRB650, WRB650A, WRB700, WB700, WB700A	3.6 (0.95)	–
RA700A, RA700B, RA760, RA1100, WVT700, WVT1100	3.8 (1.0)	–

SPARK PLUG SPECIFICATIONS

Model	Type	Gap
WR500	NGK B7HS	0.5-0.6 mm (0.20-0.24 in.)
WR650, SJ650, 1994 SJ700	NGK B8HS	0.5-0.6 mm (0.20-0.24 in.)
	NGK B8HS/BR8HS	0.5-0.6 mm (0.20-0.24 in.)
All other models	NGK BR8HS	0.5-0.6 mm (0.20-0.24 in.)

CARBURETOR TUNING SPECIFICATIONS

Model	Idle speed (rpm)	Low speed screw (turns out)	High speed screw (turns out)
WR500	1500	1 1/4 ± 1/4	3/4 ± 1/4
WR650	1250	2 1/16 ± 1/4	1.0 ± 1/4
WRA650, WRA650A	1250	1 1/8 ± 1/4	1 1/8 ± 1/4
WRB650, WRB650A			
Carb ID: 61L00	1250	1 1/16 ± 1/4	1 1/4 ± 1/4
Carb ID: 61L01	1250	1 1/4 ± 1/4	1 5/8 ± 1/4
Carb ID: 61L02	1250	1.0 ± 1/4	1 5/8 ± 1/4

(continued)

CARBURETOR TUNING SPECIFICATIONS (continued)

Model	Idle speed (rpm)	Low speed screw (turns out)	High speed screw (turns out)
SJ650			
Carb ID: 6R700	1250	1 1/8 ± 1/4	1 1/8 ± 1/4
Carb ID: 6R701-03	1250	1.0 ± 1/4	1 3/8 ± 1/4
SJ700, FX700, RA700B, WRA700, WRB700, WB700, WB700	1250	1 7/8 ± 1/4	1 5/8 ± 1/4
SJ700A	1250	7/8 ± 1/4	F: 1 1/8 ± 1/4 R: 1 1/2 ± 1/4
RA700, RA700A, WVT700	1250	5/8 ± 1/4	F: 5/8 ± 1/4 R: 1 1/8 ± 1/4
RA1100, WVT1100	1250	1 1/8 ± 1/4	7/8 ± 1/4
RA760	1300	1 3/4 ± 1/4	1/2 ± 1/4

ELECTRICAL SPECIFICATIONS

Fuse capacity	10 Amp
Battery capacity	12 volt/19 amp hours

JET PUMP SERVICE SPECIFICATIONS

	New mm (in.)	Wear limit mm (in.)
Impeller clearance		
WR500, WR650	0.3 (0.012)	0.6 (0.024)
SJ650	0.2 (0.008)	0.6 (0.024)
SJ700, SJ700A, RA700, RA700A, RA700B RA760, RA1100, WB760, WVT700, WVT1100	0.3-0.4 (0.012-0.016)	0.6 (0.024)
FX700	0.25-0.35 (0.010-0.014)	0.6 (0.024)
WRA650, WRA650A, WRA700, WRB650, WRB650A, WRB700, WB700, WB700A	0.2-0.6 (0.008-0.024)	0.6 (0.024)
Drive shaft runout		
WR500, WR650, SJ650	–	0.5 (0.020 in.)
All other models	–	0.3 (0.011 in.)

MAINTENANCE TIGHTENING TORQUE

	N·m	ft.-lb.
Cylinder head		
WR500, WR650		
1st step	15	11
2nd step	28	20
(continued)		

MAINTENANCE TIGHTENING TORQUE (continued)

	N·m	ft.-lb.
WRA650, WRA650A, WRA700, WRB650, WRB650A, WRB700, WB700, WB700A, SJ650, SJ700, FX700		
1st step	15	11
2nd step	30	22
RA700, RA700A, RA700B, RA760, RA1100, SJ700A, WB760, WVT700, WVT1100		
1st step	15	11
2nd step	36	25
Spark plugs	20	14
Flywheel		
WR500	140	100
All other models	70	50

Chapter One

General Information

This Clymer shop manual covers Yamaha personal water vehicles manufactured from 1993-1996.

Troubleshooting, tune-up, maintenance and repair are not difficult, if you know what tools and equipment to use and what to do. Step-by-step instructions guide you through jobs ranging from simple maintenance to complete engine and jet pump overhaul.

This manual can be used by anyone from a do-it-yourselfer to a professional mechanic. Detailed drawings, clear photographs and detailed, easy-to-read text provide you with all the information you need to do the work right, the first time.

Some of the procedures in this manual require the use of special tools. The resourceful mechanic can, in some cases, design an acceptable substitute for a given special tool—there is always another way. This can be as simple as using a piece of threaded rod, washers and nuts to remove or install a bearing. However, using a substitute for a special tool is generally not recommended because it can be dangerous to the user and may damage the part or assembly. If, however, you determine that a tool can be designed and safely fabricated, you may want to search out a local community college or high school that has a machine shop curriculum. Shop teachers sometimes welcome outside work that can be used as practical shop applications for advanced students.

Table 1, at the end of this chapter, lists model coverage with hull and engine serial numbers.

Metric and U.S. dimensions are used throughout this manual. U.S. to metric equivalents are given in **Table 2**.

Critical torque specifications are located in table form at the end of the appropriate chapter (as required). The general torque specifications listed in **Table 3** can be used if a torque specification is not listed for a specific component or assembly.

Metric tap drill sizes can be found in **Table 4**.

A list of technical abbreviations are provided in **Table 5**.

General specifications for all Yamaha personal watercraft covered in this manual are listed in **Tables 6-11**.

Tables 1-11 are at the end of the chapter.

MANUAL ORGANIZATION

This chapter provides general information useful to watercraft owners and mechanics. Information in this chapter discusses the tools and techniques for performing preventive maintenance, troubleshooting and repair.

Chapter Two provides methods and suggestions for quick and accurate diagnosis and repair. Troubleshooting procedures discuss typical symptoms and logical methods to locate the trouble.

Chapter Three explains all periodic lubrication and routine maintenance necessary to keep your watercraft operating at peak performance. Chapter Three also includes recommended tune-up procedures, eliminating the need to consult other chapters on the various assemblies.

Subsequent chapters describe specific systems, providing disassembly, repair, assembly and adjustment procedures in simple step-by-step form. If a repair is impractical for the home mechanic, it is so indicated. It may be faster and less expensive to deliver such repairs to a Yamaha dealership or other qualified repair shop.

Specifications for a specific system are included at the end of the appropriate chapter.

NOTES, CAUTIONS AND WARNINGS

The terms NOTE, CAUTION and WARNING have specific meanings in this manual. A NOTE provides additional information to make a step or procedure easier or more clear. Disregarding a NOTE may cause inconvenience but would not cause damage or personal injury.

A CAUTION emphasizes areas where equipment damage could occur. Disregarding a CAUTION can cause permanent damage to the unit; however, personal injury is unlikely.

A WARNING emphasizes areas where personal injury or even death can result from negligence. Damage to the unit is also possible. WARNINGS *must be taken seriously*.

SAFETY FIRST

Professional mechanics can work for years and never sustain a serious injury. If you observe a few rules of common sense and safety, you can enjoy many safe hours servicing your own ma-

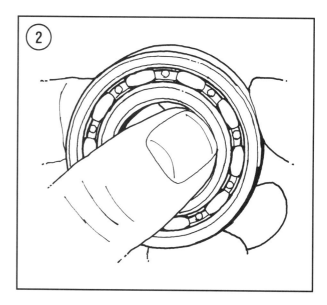

GENERAL INFORMATION

chine. If you ignore these rules, personal injury or damage to your machine is possible.

1. Never use gasoline as a cleaning solvent.

2. Never smoke or use a torch in the vicinity of flammable liquids, such as cleaning solvent or gasoline.

3. If welding or brazing is required, remove the fuel tank from the watercraft. Place the tank at least 50 ft. (15.2 m) away from the watercraft.

4. Use the proper sized wrenches to avoid damage to fasteners and injury to yourself.

5. When loosening a tight or stuck fastener, be guided by what will happen if the wrench slips. Be careful and protect yourself accordingly.

6. If replacing a fastener, always use one with the same measurements and strength as the original. Incorrect or mismatched fasteners can result in damage to the watercraft and possible personal injury. Refer to *Fasteners* in this chapter for additional information.

7. Maintain all hand and power tools in good working condition. Wipe grease and oil from tools after using them. Greasy or oily tools are difficult to hold and can result in injury. Replace or repair worn or damaged tools.

8. Maintain a clean and uncluttered work area.

9. Wear eye protection during all operations involving drilling, grinding, the use of a cold chisel or *anytime* you are unsure about the safety of your eyes. Also wear suitable eye protection if using solvent and compressed air to clean parts.

10. Keep an approved fire extinguisher close by. See **Figure 1**. Make sure the fire extinguisher is rated for gasoline (Class B) and electrical (Class C) fires.

11. If using compressed air to dry bearings or other rotating components, never allow the air jet to rotate the bearing or part. The air jet is capable of rotating the part at speeds far in excess of those for which they were designed. The bearing or rotating part is very likely to disintegrate and cause serious injury. Hold the bearing inner race as shown in **Figure 2** if using compressed air to dry the bearing.

SERVICE HINTS

Most of the service procedures covered in this manual are straight-forward and can be performed by anyone reasonable handy with tools. It is suggested, however, that you consider your own capabilities carefully before attempting any operation involving major disassembly of the unit.

1. Front, as used in this manual, refers to the front or bow of the watercraft. The front of any component is the end closest to the front of the vehicle. Left and right sides refer to the position of the parts as viewed by a rider sitting or standing on the watercraft, facing forward. For example, the throttle control is located on the right side of the vehicle. See **Figure 3**. These rules are simple, but confusion can cause a major inconvenience during service.

2. During engine or drive system disassembly, mark the location and direction of the parts as they are removed. Be sure to mark parts that mate

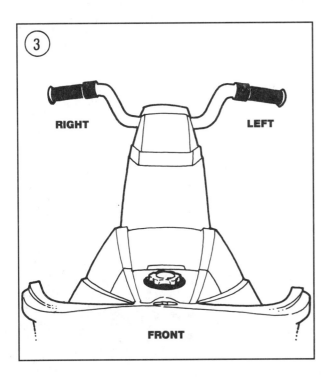

together. Small parts, such as bolts, can be identified by placing them in plastic bags such as those shown in **Figure 4**. Seal the bags and label them with masking tape and marking pen. Another accepted practice for keeping small parts organized is to place them in a cupcake tin or egg carton in the order of disassembly.

3. Protect finished surfaces from physical damage or corrosion. Keep gasoline off painted surfaces.

4. Apply penetrating oil to frozen or corroded fasteners, then strike the head of the fastener with a hammer and punch (use a screwdriver on screws). Avoid using heat where possible, as heat can warp, melt or affect the temper (hardness) of metal parts. Heat also ruins finishes, especially paint and plastics.

5. Except when removing and installing bushings and bearings, unusual force is not necessary to remove most parts. If a part is especially difficult to remove or install, determine why before proceeding.

6. Cover all openings after removing parts or assemblies to prevent dirt, small tools or contamination from entering.

7. Read each procedure *completely* while looking at the actual unit before starting a job. Make sure you thoroughly understand what is to be done, then carefully follow the procedure, step by step.

NOTE
Some of the procedures or service specifications listed in this manual may not be accurate if your watercraft is modified or if it is equipped with aftermarket equipment. If installing aftermarket equipment, or if someone else has modified your watercraft, file all printed instructions or technical information regarding the new equipment for future reference. If your watercraft was purchased used, the previous owner may have modified it or installed aftermarket parts. If necessary, consult your dealer or accessory manufacturer regarding service related changes.

8. Recommendations are occasionally made in this manual to refer a specific service or procedure to a watercraft dealership or specialist in a particular field. If so, it is possible that this work may be completed more quickly and economically than if you performed the job yourself.

9. In the procedural steps contained in this manual, the term *replace* means to discard a defective part and replace it with a new or exchange unit. *Overhaul* means to remove, disassemble, inspect, measure, repair or replace defective parts, then reassemble and reinstall the assembly.

10. Some operations require the use of a hydraulic or arbor press. If a press is not available, have these operations performed by a shop equipped for such work. Trying to do the work with makeshift equipment may damage the component.

11. Repairs are easier and proceed much faster if your watercraft is clean before beginning work. Many special cleaners, such as Bel-Ray Degreaser, are available to wash the engine and related components. Follow the cleaner manufacturer's instructions for best results.

WARNING
Never use gasoline as a cleaning solvent. Gasoline presents an extreme fire hazard. Only use solvent in a well-ventilated area. Keep a fire extinguisher rated for gasoline fires nearby.

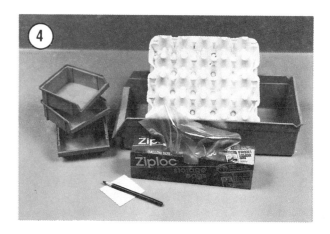

GENERAL INFORMATION

12. Often, much of the labor charge for a repair job at a dealership is for time involved during removal, disassembly, assembly and installation of other parts just to access the defective part. It may be possible to perform the preliminary operations yourself and then take the defective unit to the dealership for repair, often at substantial savings.

13. If special tools are required for a specific procedure, arrange to get them before starting the work. It is frustrating and time consuming to get partly into a repair procedure and not be able to complete it.

14. Make a diagram, or take an instant photograph, before disassembling an assembly with similar—appearing parts. You may think you can remember where everything came from, but mistakes are costly. It is also possible that you may be sidetracked and not return to the work for days or even weeks, in which time carefully arranged parts may be disturbed.

15. If servicing an assembly containing shims or washers, be certain all shims/washers are reinstalled in exactly the same location and facing the same direction as removed.

16. Whenever a rotating parts abuts a stationary part, look for a shim or washer separating the parts.

17. Always use new gaskets, seals and O-rings during reassembly. Unless specified otherwise, install new gaskets dry (without sealant).

18. Use grease to hold small parts in place if they tend to fall from position during assembly. However, make sure any grease used inside the engine crankcase is gasoline soluble. In addition, keep grease and oil away from electrical components.

19. Never use wire or drill bits to clean carburetor jets or air passages. If the jets or passages are enlarged, the carburetor's calibration will be changed.

20. Remember that a newly rebuilt engine must be broken in just like a new engine.

21. Most of all, take your time and do the job right.

ENGINE OPERATION

NOTE
During this discussion, assume the crankshaft is turning counterclockwise in **Figure 5**.

All Yamaha watercraft are equipped with 2-stroke marine engines. As the piston travels downward, a transfer port (A, **Figure 5**) located between the crankcase and cylinder is uncovered. The exhaust gases leave the cylinder through the exhaust port (B, **Figure 5**), which is also opened by the downward movement of the piston. A fresh air/fuel charge, which has been compressed slightly, moves from the crankcase (C, **Figure 5**) to the cylinder through the transfer port (A **Figure 5**) as the port opens. Since the incoming charge is under pressure, it rushes into the cylinder quickly and helps expel the exhaust gases from the previous combustion event.

Figure 6 illustrates the next phase of the cycle. As the crankshaft continues to rotate, the piston moves upward, closing the exhaust and transfer ports. As the piston continues upward, the air/fuel mixture in the cylinder is compressed. Notice also that a vacuum is created in the crankcase at the same time. Further upward movement of the piston uncovers the intake port (D, **Figure 6**). A fresh air/fuel charge is then drawn into the

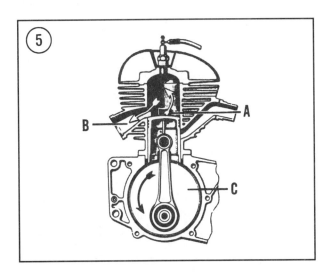

crankcase through the intake port because of the vacuum created by the upward piston movement.

The third phase of the cycle is shown in **Figure 7**. As the piston approaches top dead center (TDC), the spark plug fires, igniting the compressed air/fuel mixture. The piston is then driven downward by the expanding gases.

When the top of the piston uncovers the exhaust port, the fourth phase begins, as shown in **Figure 8**. The exhaust gases leave the cylinder through the exhaust port. As the piston continues downward, the intake port is closed and the mixture in the crankcase is compressed in preparation for the next cycle.

It can be seen from this discussion that every downward stroke of the piston is a power stroke.

PARTS REPLACEMENT

The manufacturer makes frequent changes to the watercraft during a model year; some are minor, some are relatively major. Therefore, when you order replacement parts, always supply the hull and engine serial numbers. The hull number is stamped on the outside of the hull. The engine number is stamped on the crankcase. Record the numbers for later reference. Also, compare new parts to old before purchasing them. If they are not exactly alike, have the parts manager explain the difference. **Table 1** lists engine and hull serial numbers for all watercraft covered in this manual. Refer to *Serial Numbers* in Chapter Four, Five and Six for the location of serial numbers on an individual watercraft.

TORQUE SPECIFICATIONS

Torque specifications throughout this manual are given in Newton-meters (N•m) and foot-pounds (ft.-lb.)

Table 3 lists general torque specifications for fasteners not listed in their respective chapters. To use the table, first determine the size of the fastener by measuring it with a vernier caliper.

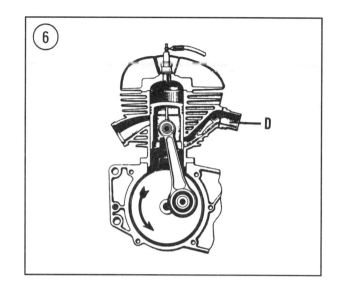

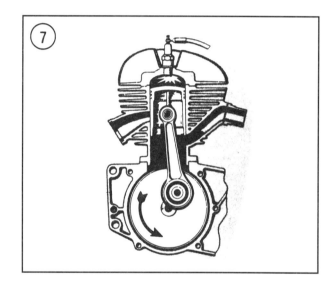

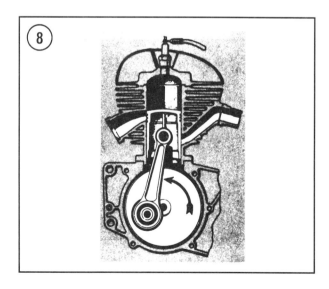

GENERAL INFORMATION

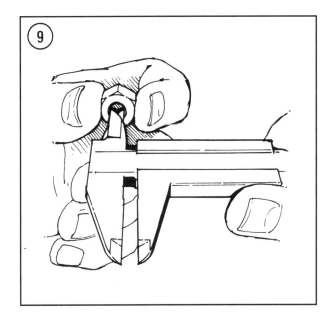

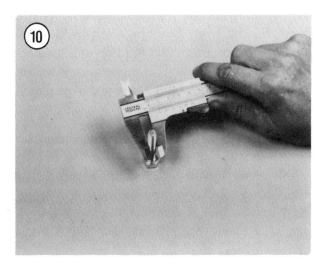

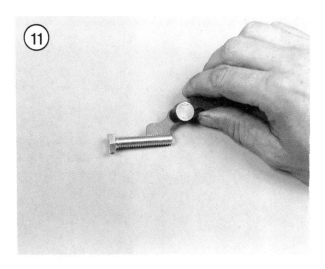

Figure 9 (nut) and **Figure 10** (bolt) shows how to measure fasteners.

FASTENERS

Fastener design and the materials used to make them are carefully considered during the manufacturing process. Fastener design determines the type of tool required to turn the fastener. Fastener material is carefully selected to decrease the possibility of physical failure from breakage, corrosion or other factors.

Nuts, bolts and screws are manufactured in a wide range of thread patterns. To join a nut and bolt, the diameter of the bolt and the diameter of the hole in the nut must be the same. It is just as important that the threads on both are properly matched.

The best way to determine if the threads on 2 fasteners are matched is to turn the nut on the bolt (or the bolt into a threaded hole) using only your fingers. Be sure the threads on both fasteners are clean and undamaged. If much force is required, check the thread condition on each fastener. If the thread condition is good but the fasteners jam, the threads are not compatible. A thread pitch gauge can also be used to determine thread pitch. See **Figure 11**. Yamaha personal watercraft are manufactured with ISO metric fasteners.

Most threads are cut so the fastener must be turned clockwise to tighten it. These are called right-hand threads. Some fasteners have left-hand threads; they must be turned counterclockwise to tighten. Left-hand threads are used in locations where normal rotation of the equipment would tend to loosen a right-hand threaded fastener.

ISO Metric Screw Threads

ISO (International Organization for Standardization) metric threads are available in 3 standard thread sizes: coarse, fine and constant pitch. The ISO coarse pitch is used for most common fastener applications. The fine pitch thread is used on certain precision tools and instruments. The constant pitch thread is used mainly on machine parts and is not generally used on fasteners. The constant pitch thread is, however, used on all metric thread spark plugs.

ISO metric threads are specified by the capital letter M followed by the diameter in millimeters and the pitch (or distance between each thread) in millimeters. For example a M8 × 1.25 bolt has a diameter of 8 millimeters with a distance of 1.25 millimeters between the crest of each thread. **Figure 10** shows how to measure bolt diameter. The measurement across 2 flats on the head of the bolt indicates the proper wrench size used to turn the fastener.

NOTE
*When purchasing a bolt from a dealer or parts store, it is important to know how to specify bolt length. To correctly measure bolt length, measure from the bottom of the bolt head to the end of the bolt. See **Figure 12**. Always measure bolt length in this manner to avoid purchasing bolts that are too long.*

Machine Screws

There are many different types of machine screws. **Figure 13** shows a number of screw heads which require different types of tools to turn them. The screw heads are also designed to either protrude above the metal (round) or to be slightly recessed (flat) in the metal. See **Figure 14**.

Bolts

Commonly called a bolt, the correct technical name for this type of fastener is cap screw. ISO metric bolts are described by the length and diameter in millimeters and the pitch (or distance between each thread) in millimeters. For example a M8 × 1.25 bolt has a diameter of 8 millimeters with a distance of 1.25 millimeters between each thread.

Nuts

Nuts are manufactured in a variety of types and sizes. Most are hexagonal (6 sides) and fit on bolts, screws and studs having the same diameter and pitch.

Figure 15 shows several types of nuts. The common nut is generally used with a lockwasher. Self-locking nuts have a nylon insert which prevents the nut from loosening. Self-locking fasteners do not normally require lockwashers. Wing nuts are designed for fast removal by hand and are used for convenience in noncritical locations.

To indicate the size of a metric nut, manufacturers specify the diameter of the opening (in millimeters) and the thread pitch. This is similar to bolt specifications, but without the length designation. The measurement across 2 flats on the nut indicates the proper wrench size to turn the nut.

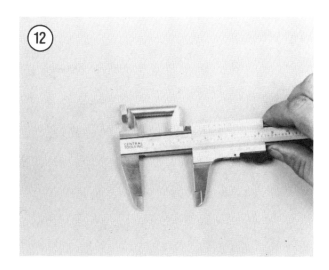

GENERAL INFORMATION

Prevailing Torque Fasteners

Several types of bolts, screws and nuts incorporate a system that develops an interference between the bolt, screw, nut or threaded hole. Interference is achieved in various ways: by slightly distorting the threads, coating the threads with dry adhesive or nylon, or distorting the top of an all-metal nut.

Self-locking fasteners offer greater holding strength and better vibration resistance than a plain nut and lockwasher. Some self-locking fasteners can be reused if still in good condition. Others, like the nylon insert nut, form an initial locking condition when the nut is first installed. Once installed, however, the nylon insert conforms closely to the bolt thread pattern. Therefore, once the fastener is removed, it should be replaced to offer the greatest security.

Washers

There are 2 basic types of washers: flat washers and lockwashers. Flat washers are simple discs with a hole to fit a screw, bolt or stud. Lockwashers are designed to prevent a fastener from working loose due to vibration, expansion or contraction. **Figure 16** shows several types of washers. Washers are also used in the following functions:

a. As spacers.

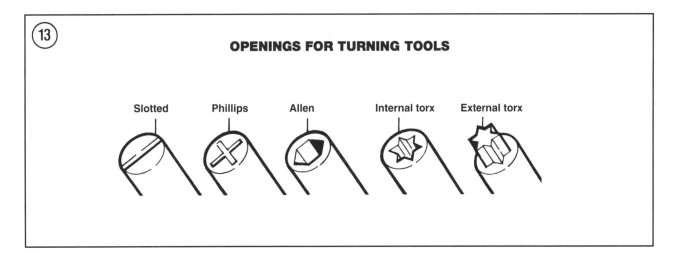

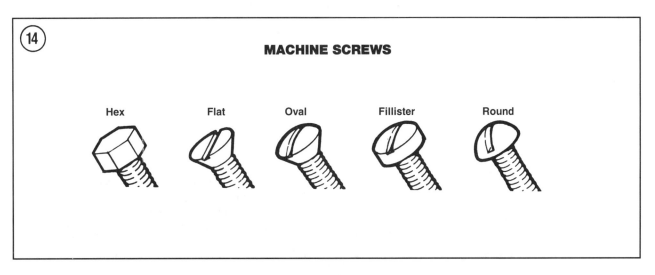

b. To prevent galling or damage of the equipment by the fastener.
c. To help distribute fastener load during tightening.
d. As seals.

Note that flat washers are often used between a lockwasher and a fastener to provide a smooth bearing surface.

Cotter pins

Cotter pins (**Figure 17**) are used to secure special kinds of fasteners. The stud or bolt must have a hole in it, and the nut or nut lock piece must have slots or castellations through which the cotter pin is inserted. Cotter pins should not be reused after removal.

Snap Rings

Snap rings can be an internal or external design. They are used to retain items on shafts (external) or within housings (internal). In some applications, snap rings of varying thicknesses are used to control the thrust or end play of shafts or assemblies. This type of snap ring is called a selective snap ring. Snap rings should normally be replaced during installation, as removal often weakens or deforms them.

Two basic types of snap rings are available: machined and stamped. Machined snap rings (**Figure 18**) can be installed in either direction (shaft or housing) because both faces are machined, thus creating 2 sharp edges. Stamped snap rings (**Figure 19**) are manufactured with one sharp edge and one rounded edge. When installing stamped snap rings in a thrust situation, the sharp edge must face away from the part producing the thrust. When installing snap rings, observe the following:
a. Compress or expand the snap ring only enough to permit installation.
b. After the snap ring is installed, make sure it is completely seated in its groove.

LUBRICANTS

Periodic lubrication helps ensure a long service life for any type of equipment. Regular lubrication is especially important to marine equipment, because it is exposed to salt or brackish water and other harsh environments. The *type* of lubricant used is just as important as the lubrication service itself, although in an emer-

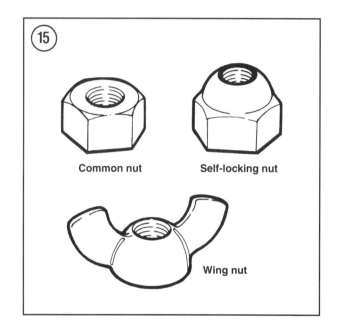

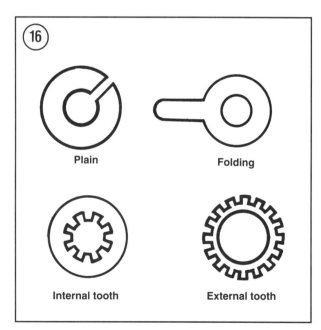

GENERAL INFORMATION

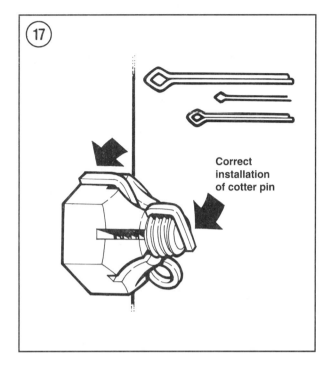

Correct installation of cotter pin

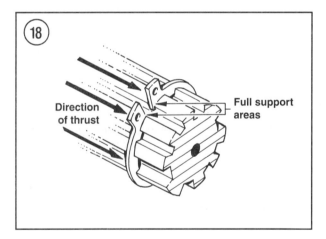

Direction of thrust / Full support areas

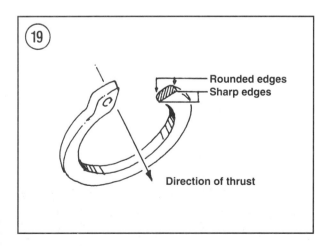

Rounded edges / Sharp edges / Direction of thrust

gency the wrong type of lubricant is much better than none at all. The following paragraphs describe the types of lubricants most often used on mechanical equipment. Be sure to follow the manufacturer's recommendations regarding lubricant use.

Generally, all liquid lubricants are called *oil*. Oil may be mineral-based (petroleum base), natural-based (vegetable and animal base), synthetic-based or emulsions (mixtures). *Grease* is an oil to which a thickening base has been added so the end product is semi-solid. The resulting material is then usually enhanced with anticorrosion, antioxidant and extreme pressure (EP) additives. Grease is often classified by the type of thickener added. Lithium and calcium soap are commonly used thickeners.

4-Stroke Engine Oil

Oil designed for use in 4-stroke engines is graded by the American Petroleum Institute (API) and the Society of Automotive Engineers (SAE) in several categories. Oil containers display these ratings on the top or label.

API oil grade is indicated by letters. Oil for gasoline engines is identified by an *S* and oil for diesel engines is identified by a *C*. Most modern 4-stroke gasoline engines require SF or SG graded oil. Automotive and marine diesel engines use CC or CD graded oil.

Viscosity is an indication of the oil's thickness or resistance to flow. The SAE uses numbers to indicate viscosity; thin oil has a low number, and thick oil has a high number. A *W* after the number indicates that the viscosity testing was done at low temperature to simulate cold wear operation. Engine oil falls into the 5W-20 to 20W-50 range.

Multigrade oil (for example, 10W-40) is less viscous (thinner) at low temperature and more viscous (thicker) at high temperature. This allows the oil to perform efficiently across a wide range of engine operating temperatures.

CAUTION
Four-stroke oil is only discussed here to provide a comparison. All Yamaha personal watercraft are equipped with 2-stroke engines. Never use 4-stroke oil in a 2-stroke engine.

2-Stroke Engine Oil

Lubrication for 2-stroke engines is provided by oil mixed with the incoming air/fuel mixture. Some of the oil mixture settles out in the crankcase, lubricating the crankshaft and lower end of the connecting rods. The rest of the oil enters the combustion chamber to lubricate the piston, rings and cylinder wall. This oil is then burned along with the air/fuel mixture during the combustion process.

Engine oil must have several special qualities to work well in a 2-stroke engine. It must mix easily and stay suspended in gasoline. When burned, it cannot leave behind excessive deposits. It must also be able to withstand the high temperatures associated with 2-stroke engines.

The National Marine Manufacturer's Association (NMMA) has set standards for oil used in 2-stroke, water-cooled engines. This is the NMMA TC-W (two-cycle, water-cooled) grade. The oil's performance in the following areas is evaluated:

   a. Lubrication (prevention of wear and scuffing).
   b. Spark plug fouling.
   c. Preignition.
   d. Piston ring sticking.
   e. Piston carbon and varnish (piston coking).
   f. General engine condition (including deposits).
   g. Exhaust port blockage.
   h. Rust prevention.
   i. Mixing ability with gasoline.

In addition to oil grade, manufacturers specify the ratio of gasoline to oil required during break-in and normal operation.

Grease

Grease is graded by the National Lubricating Grease Institute (NLGI). Grease is graded by number according to the consistency of the grease. These classifications range from No. 000 to No. 6, with No. 6 being the most solid. A typical multipurpose grease is NLGI No. 2. For specific applications, equipment manufacturers may require grease with an additive such as molybdenum disulfide (MOS2).

GASKET SEALANT

Gasket sealant is used instead of preformed gaskets on some applications or as a gasket dressing on others. Two types of gasket sealant are commonly used: room temperature vulcanizing (RTV) and anaerobic. Because these 2 materials have different sealing properties, they cannot be used interchangeably.

RTV Sealant

Room temperature vulcanizing (RTV) sealant is used on some preformed gaskets and to seal some components, such as the jet pump, during watercraft service. RTV is a silicone gel supplied in tubes. RTV sealant can generally fill gaps up to 1/4 in. (6.3 mm) and works well on slightly flexible surfaces.

Moisture in the air causes RTV to cure. Therefore, always place the cap on the tube as soon as possible after using RTV. RTV has a shelf life of approximately 1 year and will not cure properly after the shelf life has expired. Check the expiration date on RTV tubes before use. Keep partially used tubes tightly sealed.

Applying RTV Sealant

Clean all gasket material from mating surfaces. If scraping is necessary, use a broad, flat scraper or a somewhat dull putty knife to avoid nicks, scratches or other damage to mating surfaces. Before applying RTV sealant, the mating surfaces must be absolutely free of gasket material, sealant, dirt, oil, grease or any other contamination. Lacquer thinner, acetone, isopropyl alcohol or similar solvents work well for cleaning mating surfaces. Avoid using solvents with an oil, wax or petroleum base as they may not be compatible with some sealants. Remove all RTV gasket material from blind attaching holes because it can create hydraulic pressure in the hole and affect fastener torque.

Apply RTV sealant in a continuous bead 1/8 to 3/16 in. (3-5 mm) thick. Avoid excess application as some of the sealant could be forced into bearings or seals. Circle all mounting holes unless otherwise specified. Tighten the fasteners within 10 minutes after application.

Anaerobic Sealant

Anaerobic sealant is a gel supplied in tubes and is used to seal rigid case assemblies instead of a gasket. It cures only in the absence of air, as when squeezed tightly between two machined mating surfaces. For this reason, it will not spoil if the cap is left off the tube. It should not be used if one mating surface is flexible. Anaerobic sealant is able to fill gaps up to 0.030 in. (0.8 mm) and generally works best on rigid, machined flanges or surfaces.

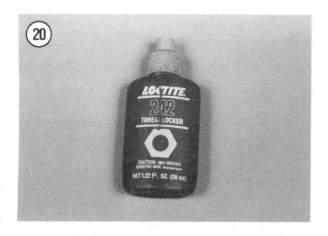

Applying Anaerobic Sealant

Clean all gasket material from mating surfaces. If scraping is necessary, use a broad, flat scraper or a somewhat dull putty knife to avoid nicks, scratches or other damage to mating surfaces. Before applying anaerobic sealant, the mating surfaces must be absolutely free of gasket material, sealant, dirt, oil, grease or any other contamination. Lacquer thinner, acetone, isopropyl alcohol or similar solvents work well for cleaning mating surfaces. Avoid using solvents with an oil, wax or petroleum base as they may not be compatible with some sealants. Remove all anaerobic sealant from blind attaching holes because it can create hydraulic pressure in the hole and affect fastener torque.

Apply anaerobic sealant in a 1/8 in. (3 mm) or less bead to one sealing surface. Circle all mounting holes. Tighten mating parts with 15 minutes after application.

THREADLOCKING COMPOUND

A threadlocking compound is used to help secure many of the fasteners used throughout the Yamaha watercraft. A threadlocking compound secures fasteners against vibration loosening and also seals against leaks. Loctite and ThreeBond products are recommended for many threadlocking requirements described in this manual. See **Figure 20**.

Loctite 242 (blue) is a medium-strength threadlocking compound and component disassembly can be performed with normal hand tools. Loctite 271 (red) is a high-strength threadlocking compound. Heat or special tools, such as a press or puller, are sometimes necessary for component disassembly. Loctite 572 is a thread sealing compound that blocks the flow of fluids through the threads.

Applying Threadlocking Compound

Fastener threads must be clean and dry before applying threadlocking compound. If necessary, remove old threadlocking compound using a wire brush or similar tool. Remove oil, grease or other contamination using lacquer thinner, acetone or isopropyl alcohol. Apply the compound sparingly to the fastener, following the instructions on the container.

BASIC HAND TOOLS

Many of the procedures in this manual can be performed with simple hand tools and test equipment familiar to the average home mechanic. Keep your tools clean and well organized in a tool box. After using a tool, wipe off any dirt, oil or grease, and return the tool to its correct place.

Top quality tools are essential. They are also more economical in the long run. If you are just now starting to build your tool collection, avoid inexpensive tools that are thick, heavy and clumsy or are made of inferior material. Quality tools are made of alloy steel and are heat treated for greater strength. Good-quality tools may cost more initially, but they are less expensive over time because good quality tools generally last forever.

The following tools are required to perform virtually any repair job. Each tool is described and the recommended sizes given for starting a tool collection. Additional tools and some duplicates may be added as necessary. Yamaha personal watercraft are built with metric fasteners, so if you are just starting your tool collection, buy metric.

Screwdrivers

The screwdriver is a very basic tool, but if used improperly, can do more damage than good. The slot on a screw has a definite dimension and shape. A screwdriver must be selected to conform with this shape. Use a small screwdriver for small screws and a large one for large screws or the screw head will be damaged.

Two basic types of screwdrivers are required: common (flat-blade) screwdriver (**Figure 21**) and Phillips screwdriver (**Figure 22**).

Screwdrivers are available in sets which often include an assortment of common and Phillips blades. If you purchase them individually, buy at least the following:

a. Common screwdriver—5/16 × 6 in. blade.

b. Common screwdriver—3/8 × 12 in. blade.

c. Phillips screwdriver—size 2 tip, 6 in. blade.

Use screwdrivers only for removing and installing screws. Never use a screwdriver for prying or chiseling metal. Do not try to remove a Phillips or Allen head screw with a common screwdriver. The screw head can be damaged so that even the correct tool will be unable to remove it.

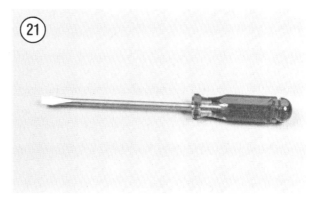

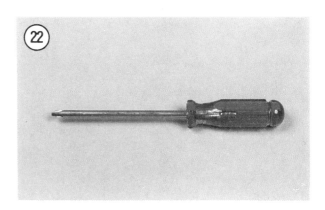

GENERAL INFORMATION

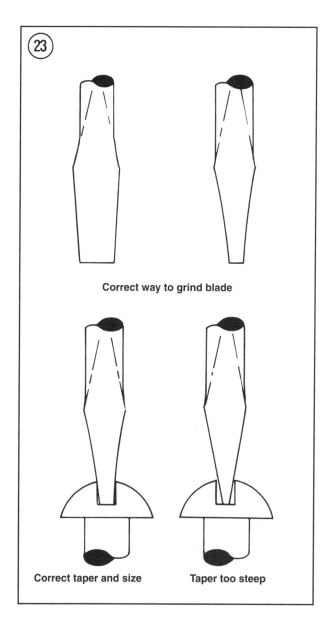

Correct way to grind blade

Correct taper and size · Taper too steep

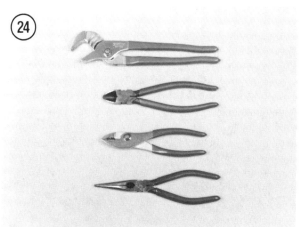

Keep screwdrivers in the proper condition. They will last longer and perform better. Always keep the tip of a common screwdriver in good condition. **Figure 23** shows how to dress the tip to the proper shape if it becomes damaged. Note the symmetrical sides of the tip.

Pliers

Pliers come in a wide range of types and sizes. Pliers are useful for cutting, bending and crimping. They should never be used to cut hardened objects or to turn bolts or nuts. **Figure 24** shows several pliers useful in watercraft service.

Each type of pliers has a specialized function. Combination or slip-joint pliers are general purpose pliers and are used mainly for holding and bending. Needlenose pliers are used to grasp or bend small objects, or objects in a difficult to reach area. Adjustable pliers (commonly called Channel Locks) can be adjusted to hold various sizes of objects while the jaws remain parallel to grip round objects such as pipe or tubing. There are many more types of specialized pliers.

Locking pliers

Locking pliers (commonly called vise-grips) are used as pliers or to hold objects very tightly while another task is performed on the object. See **Figure 25**. Locking pliers can sometimes be used to remove a rounded-off fastener. Locking

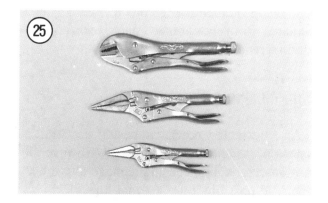

pliers are available in many types and sizes for more specific tasks.

Snap Ring Pliers

Snap ring pliers (**Figure 26**) are special in that they are only used to remove snap rings from shafts or within a housing. When purchasing snap ring pliers, there are 2 basic types to consider. External pliers (spreading) are used to remove snap rings that fit on the outside of a shaft. Internal pliers (squeezing) are used to remove snap rings which fit inside a housing.

Box-end and Open-end Wrenches

Box-end and open-end wrenches are available in sets or separately in a variety of sizes. The size number stamped near the end refers to the distance between 2 parallel flats on the hex-head bolt or nut.

Box-end wrenches usually have superior fastener turning ability compared to open-end wrenches. Open-end wrenches grip the fastener only on 2 flats. Unless a wrench fits well, it may slip and round off the fastener. A box-end wrench contacts the fastener on 6 flats, providing a much more secure grip. Both 6-point and 12-point openings on box-end wrenches are available. The 6-point provides better holding power, but the 12-point allows a shorter swinging radius when working in a confined area.

Combination wrenches (**Figure 27**) which are open on one end and boxed on the other are also available. Both ends are the same size on combination wrenches.

Adjustable Wrenches

An adjustable wrench can be adjusted to fit a variety of fasteners. See **Figure 28**. However, it can loosen and slip, causing damage to the fastener and possible injury to your knuckles. Use an adjustable wrench only when other wrenches are not available.

Adjustable wrenches are available in various sizes.

Socket Wrenches

Socket wrenches are undoubtedly the fastest, safest and most convenient to use. Sockets which

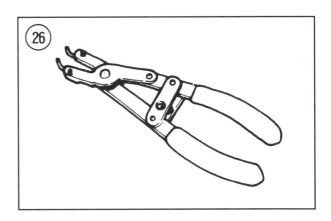

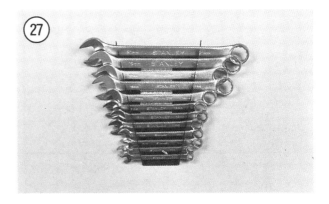

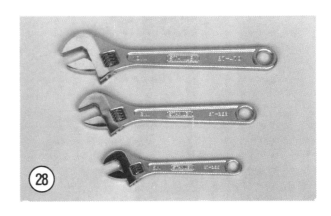

GENERAL INFORMATION

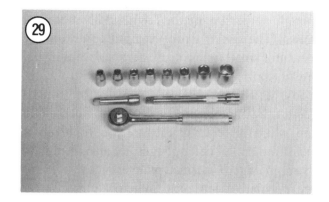

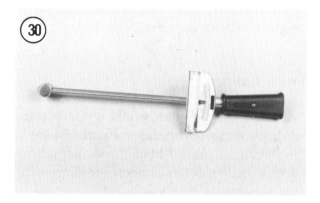

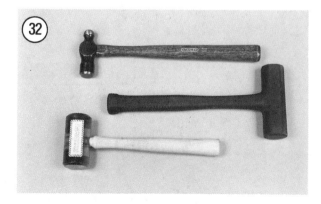

attach to a ratchet handle are available with 6- or 12-point openings and 1/4, 3/8, 1/2 and 3/4 in. drives. The drive size indicates the size of the square hole which mates with the ratchet handle. See **Figure 29**.

Torque Wrench

A torque wrench is used with a socket to measure how tightly a fastener is installed. Torque wrenches are available in a wide range of price ranges, with either a 3/8 in. (in.-lb.) or 1/2 in. (ft.-lb.) drive size. A beam-type torque wrench is shown in **Figure 30**. A beam torque wrench has a pointer which is read against a scale near the wrench handle. Other common types of torque wrench are dial type and click type. A click type torque wrench makes an audible click when the desired torque is obtained. A click type torque wrench is especially useful in hard to reach areas because the user does not have to closely watch the wrench for a reading.

Impact Driver

An impact driver makes removal of tight screws easy and also helps eliminate damage to screw heads (especially screws secured with threadlocking compound). Impact drivers with interchangeable bits (**Figure 31**) are available from tool vendors, most hardware stores and automotive parts stores. Sockets can also be used with a hand impact driver. However, make sure the socket is designed for impact use. Do not use regular hand type sockets, as they can shatter during use.

Hammers

The correct hammer (**Figure 32**) is necessary for many procedures during repair and maintenance. Always use a soft-face mallet (the type that is filled with lead shot) or a hammer with a rubber or plastic face (or head) whenever a steel hammer could damage or mar a component. *Never* use a metal-face hammer on engine or drive train parts as severe damage will likely result. You can generally produce the same amount of force with a soft-face hammer as a metal-face hammer. A metal-face hammer is necessary, however, when using a hand impact driver or a punch or cold chisel.

PRECISION MEASURING TOOLS

Precision measurement is an important part of watercraft service. When performing many of the procedures in this manual, you will be required to make a number of measurements. These include such basic checks as engine compression and spark plug gap. During major engine disassembly and service, however, highly accurate, precision measurements are required to determine the condition of internal engine components such as the pistons, cylinders and crankshaft. When performing these measurements, the degree of accuracy required will dictate which tool to use.

Precision measuring tools are relatively expensive. If this is your first experience at engine service, it may be worthwhile to have the measurements taken at a dealership or other qualified repair shop. Then, as your skills increase, you may want to purchase some of these specialized measuring tools. The following is a description of the measuring tools required to perform engine service.

Because precision measuring instruments are somewhat fragile, they must be stored, handled and used carefully to ensure continued accuracy.

Feeler Gauge

The feeler gauge is made of either a flat or round piece of hardened steel of a specified thickness. See **Figure 33**. Feeler gauges are available in U.S. standard or metric sizes, although some are marked with both measurements. The feeler gauge is used to accurately measure small gaps such as spark plug gap, breaker point gap, valve clearance (4-stroke engines) and crankshaft end play. Wire (round) gauges are mostly used to measure spark plug gap, while flat feeler gauges are used for all other measurements. Feeler gauges are also used with a straightedge or surface plate to measure warpage of mating surfaces. When using a feeler gauge, insert feelers of various thicknesses into the gap being measured until a slight drag is felt as the gauge is moved through the gap.

Vernier Caliper

A vernier caliper is used to take inside, outside and depth measurements. See **Figure 34** for a typical vernier caliper. Although a vernier caliper is not as precise as a micrometer, it allows rea-

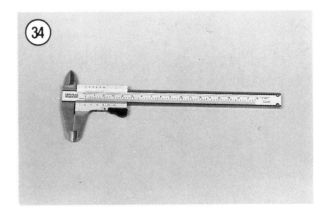

GENERAL INFORMATION

sonably accurate measurements, typically to within 0.001 in. (0.05 mm).

Outside Micrometer

An outside micrometer is one of the most reliable instruments for precision measurement. Most outside micrometers are accurate to within 0.0001 in. (0.0025 mm). See **Figure 35**. Outside micrometers are required to precisely measure piston diameter, piston pin diameter, crankshaft journal and crankpin diameter. Used with a telescopic gauge, an outside micrometer can be used to measure cylinder bore size and to determine cylinder taper and out-of-round. Outside micrometers are delicate instruments; if dropped on the floor, they most certainly will be knocked out of calibration. Always handle and use micrometers carefully to ensure accuracy. Store micrometers in their padded case when not in use to prevent damage.

Dial Indicator

Dial indicators (**Figure 36**) are precision tools used to check ignition timing, gear lash and runout of shafts. Dial indicators take precision measurements to within 0.001 in. (0.01 mm). A dial indicator is typically mounted on a holder or bracket (**Figure 37**) and the indicator plunger is positioned against the shaft or component being checked. The dial should then be rotated to align zero with the indicator needle. Always make sure the indicator plunger is set at a right angle to the shaft, gear or other component for accurate results.

Select a dial indicator with a continuous dial (**Figure 38**). A continuous dial is required to

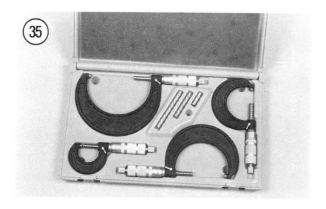

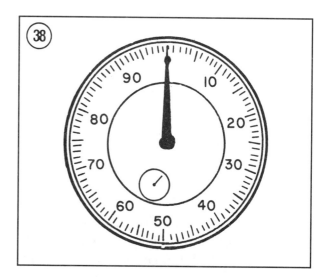

accurately measure piston position during timing adjustments

Cylinder Bore Gauge

The cylinder bore dial gauge set shown in **Figure 39** is comprised of a dial indicator, handle and a number of length adapters (anvils) and shims to adapt the gauge to different bore sizes. The bore gauge is used to make cylinder bore measurements such as bore size, taper and out-of-round. An outside micrometer of the correct size must be used to calibrate the bore gauge to a particular cylinder bore size.

Small Hole Gauges

A small hole gauge (**Figure 40**) allows inside measurement of a hole, groove or slot. An outside micrometer must be used together with the hole gauge to perform the measurement. The hole gauges are equipped with a thimble on one end, that when turned, causes the hole gauge to expand to fit the size of the hole, groove or slot being measured. Then, measure the end of the hole gauge with an outside micrometer to obtain an accurate inside measurement.

Telescopic Gauges

Telescopic gauges (**Figure 41**) are similar to small hole gauges, except they are designed for inside measurements of larger holes (approximately 8 mm [5/16 in.] and larger). Also like small hole gauges, telescopic gauges do not have a scale for direct readings. Therefore, an outside micrometer is required to measure the telescopic gauge after it is carefully fitted to a hole or bore.

To use a telescopic gauge, insert it into the bore being measured. Then, loosen the thimble and allow the gauge anvils to expand into the bore. Next, tighten the thimble to lock the anvils in place. Remove the gauge and carefully measure the width of the anvils with an outside micrometer. A micrometer stand is very useful when performing this procedure.

A fairly high degree of skill is necessary to obtain consistent, accurate measurements using a telescopic gauge. Be sure to hold the gauge squarely inside the bore. The handle of the gauge must be parallel to the walls of the bore. Also, take the measurement several times to ensure

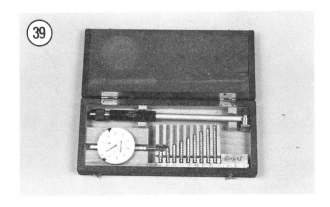

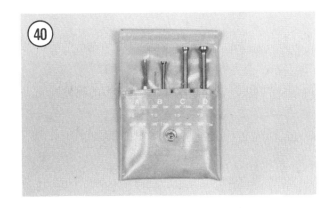

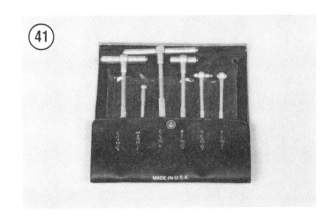

GENERAL INFORMATION

consistent results. Make sure the gauge thimble is tightened securely when measuring it with the outside micrometer. If not, the micrometer can compress the anvils slightly, resulting in an inaccurate measurement.

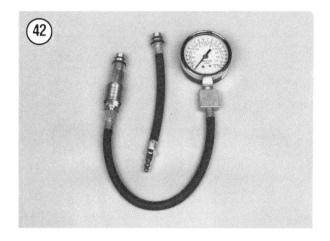

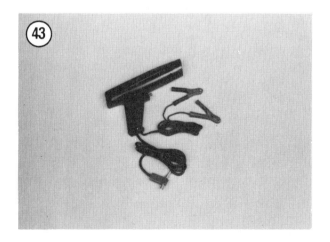

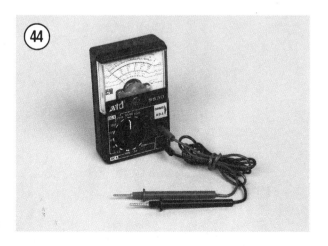

Compression Gauge

An engine with low compression cannot be properly tuned and will not develop full power. A compression gauge measures the amount of pressure present in the engine's combustion chamber during the compression/power stroke. Compression readings can be interpreted to pinpoint specific engine mechanical problems. The gauge shown (**Figure 42**) has a flexible stem with an extension that screws into the spark plug hole. Although compression gauges with press-in rubber tips are available, a gauge with screw-in adapters are generally more accurate and easier to use. See Chapter Three for instructions regarding the use of a compression gauge and interpretation of test results.

Strobe Timing Light

A timing light is used to accurately check ignition timing. By flashing light at the precise instant the spark plug fires, the position of the timing mark can be seen. The flashing light makes a moving mark appear to stand still opposite a stationary pointer or mark.

Suitable timing lights range from inexpensive neon bulb types to powerful xenon strobe lights. **Figure 43** shows a typical timing light. A light with an inductive pickup is recommended to eliminate any possible damage to ignition components or wiring.

Multimeter

A multimeter (**Figure 44**) is invaluable for electrical system troubleshooting and service. It combines a voltmeter, an ohmmeter and an ammeter into one unit, so it is often called VOM.

Two types of multimeter are commonly available, analog and digital. Analog meters have a moving needle with marked bands indicating the volt, ohm and amperage scales. The digital meter (DVOM) is ideally suited for troubleshooting because it is easy to read, more accurate than analog, contains internal overload protection, is auto-ranging (analog meters must be calibrated each time the scale is changed) and has automatic polarity compensation.

Tachometer

A portable tachometer is necessary for tuning and carburetor adjustments. Ignition timing and carburetor adjustments must be performed at specified engine speeds. The best instrument for this purpose is one with a low range of 0-1000 or 0-2000 rpm and a high range of up to 6000 rpm. Extended range instruments sometimes lack accuracy at lower speeds. The tachometer should be capable of detecting speed changes as low as 25 rpm.

Battery Hydrometer

A hydrometer (**Figure 45**) is used to determine a battery's state of charge. A hydrometer measures the specific gravity of the electrolyte in each battery cell. Specific gravity is the weight or density of the electrolyte as compared to pure water. As a battery is charged, the specific gravity goes up; when a battery is discharged, the specific gravity goes down. See Chapter Nine for hydrometer testing instructions.

Screw Thread Gauge

A screw thread gauge (**Figure 46**) determines the thread pitch of bolts, screws, studs and other threaded fasteners. The thread gauge used for U.S. standard fasteners measures in threads per inch. The gauge used for metric fasteners measures thread pitch. The gauge is made up of a number of thin plates. Each plate has a thread shape cut on one edge to match one thread size. When using a thread gauge to determine a thread size, try to fit different thread sizes onto the fastener until a perfect match is obtained.

Magnetic Stand

A magnetic stand (**Figure 47**) is used to securely hold a dial indicator when checking the runout of a round object or when checking the end play of a shaft. See **Figure 37**.

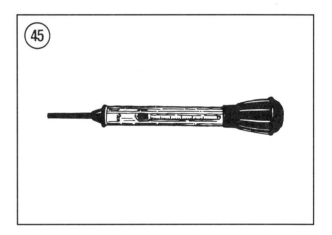

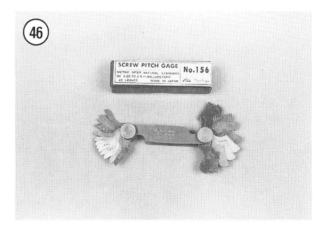

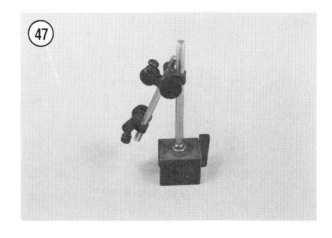

GENERAL INFORMATION

V-blocks

V-blocks (**Figure 48**) are precision ground blocks used to hold a round object (shaft) when checking its runout or condition. The shaft is placed into the V-blocks and rotated to check for excessive runout using a dial indicator.

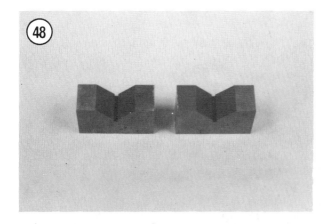

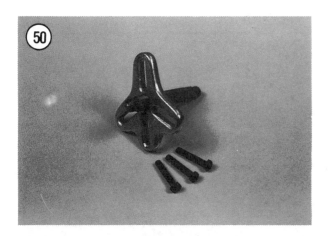

Expendable Supplies

Certain expendable supplies are also required for watercraft service and maintenance. These include the recommended greases, oils and gasket sealants. In addition, shop towels and a suitable cleaning solvent are necessary. Ask your watercraft dealer for the special threadlocking compounds, silicone sealants and lubrication products that are required for maintaining your watercraft. Cleaning solvent is generally available at automotive parts stores.

WARNING
Having a stack of clean shop towels readily available is important when performing engine work. However, to prevent the possibility of a fire from spontaneous combustion, store oil or solvent soaked towels inside a sealed container until they can be washed.

SPECIAL TOOLS

This section describes special tools unique to Yamaha personal watercraft service and repair.

Flywheel Puller

A flywheel puller is required to remove the flywheel from the engine. Flywheel removal is necessary to service the stator or trigger assemblies or anytime major engine service is necessary. **Figure 49** shows a Yamaha flywheel puller. **Figure 50** shows a universal puller. Because flywheel design varies with different models, a universal puller can often be used on several flywheels.

There is no suitable substitute for a flywheel puller. Because the flywheel is a tapered fit on the crankshaft, makeshift removal often results in crankshaft and flywheel damage. Do not attempt to remove the flywheel without the correct flywheel puller. Refer to Chapter Four, Five, or Six for flywheel removal procedures.

Flywheel Holder

A flywheel holder is used to hold the flywheel and keep it from turning during flywheel or coupler removal. **Figure 51** shows a number of holders that can be used when servicing Yamaha personal watercraft.

Coupler Holder

A coupler holder (A, **Figure 52**) is required to secure the engine coupler when removing or installing the engine or drive shaft coupler half. Because the coupler half arms are somewhat brittle, the use of improper tools can damage the coupler half.

Drive Shaft Holder

The drive shaft holder (B, **Figure 52**) is used to secure the drive shaft during impeller removal and installation.

Shaft Holder

The shaft holder (C, **Figure 52**) is used to secure the intermediate shaft during intermediate coupler removal and installation.

MECHANIC'S TIPS

Removing Frozen Fasteners

If a fastener cannot be removed due to corrosion or damaged threads, several methods may be used to loosen it. First, apply penetrating oil such as Liquid Wrench or WD-40 (available at hardware or automotive parts stores). Apply the oil liberally and allow it to penetrate for 10-15 minutes. Tapping the fastener several times with a small hammer helps break loose rust and corrosion and may help loosen the fastener. But, do not strike the fastener hard enough to cause damage. Reapply the penetrating oil as necessary.

For frozen screws, apply penetrating oil as previously described, then insert a screwdriver in the slot and tap the screwdriver with a hammer. If the screw head becomes damaged, apply a small amount of valve grinding compound to the screw head. The compound helps the screwdriver

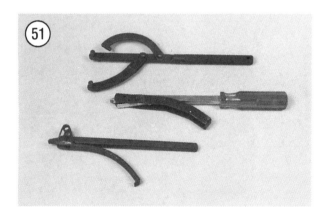

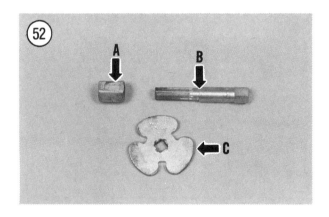

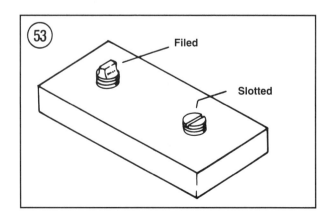

GENERAL INFORMATION

grip the damaged screw head. If the screw head is excessively damaged, it may be necessary to grip the head with locking pliers and attempt to remove it.

Avoid applying heat unless specifically instructed, as it may melt, warp or remove the hardness from the components.

Removing Broken Fasteners

If the head of a screw or bolt breaks, several methods are available to remove the remaining portion.

If a large piece of the broken fastener projects above the part, try gripping it with locking pliers. If the projecting piece is too small, file it to fit a wrench or cut a slot in the piece to fit a screwdriver. See **Figure 53**.

If the head breaks off flush, use a screw extractor to remove the broken fastener. To do this, centerpunch the exact center of the remaining piece of the screw or bolt. Placing the centerpunch exactly in the center of the fastener is not an easy task; take your time and be precise. Drill an appropriately sized hole into the fastener at the centerpunch, then tap the screw extractor into the hole. Place a wrench on the screw extractor and back out the broken fastener. See **Figure 54**.

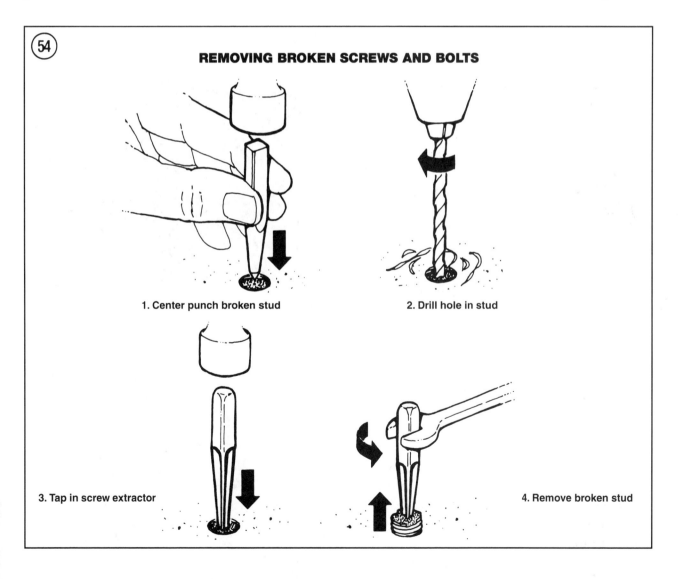

Repairing Damaged Threads

Occasionally, threads are damaged through carelessness or impact damage. Often, the threads can be cleaned up by running a tap (internal threads) or die (external threads) through the threads. See **Figure 55**. Use a spark plug tap to clean or repair spark plug threads.

NOTE
Taps and dies can be purchased individually or in a set as shown in ***Figure 56***.

If an internal thread is damaged, it may be necessary to install a Helicoil (**Figure 57**) or other type of thread insert. Follow the insert manufacturer's instructions when installing the insert.

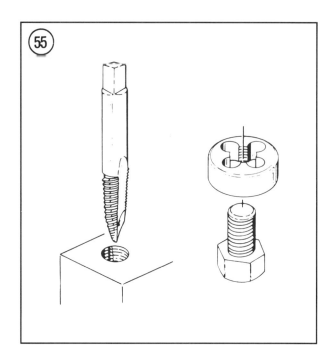

Removing Broken or Damaged Studs

If a stud is broken or the threads severely damaged, perform the following. Two nuts that fit the stud and 2 wrenches are required for this procedure.

1. Thread the 2 nuts onto the damaged stud. Then, securely tighten the 2 nuts against each other so that they are locked.

NOTE
If the threads on the damaged stud do not allow installation of the 2 nuts, remove the stud with a pair of locking pliers or a stud remover.

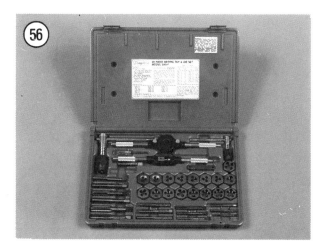

2. Turn the bottom nut counterclockwise to unscrew and remove the broken stud.
3. Clean the threads with clean solvent and allow to dry thoroughly.
4. Install 2 nuts on the top half of the new stud as in Step 1. Make sure they are securely locked.
5. Coat the bottom threads of a new stud with Loctite 271 (red), ThreeBond 1303 (green), or equivalent.
6. Install the stud and tighten securely by turning the top nut.

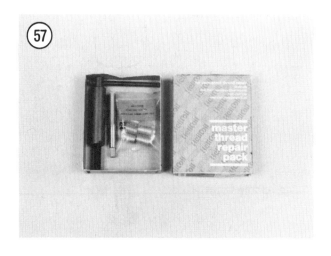

GENERAL INFORMATION

7. Remove the nuts and repeat for each stud to be replaced.

8. Allow sufficient time for the threadlocking compound to cure before returning the unit to service.

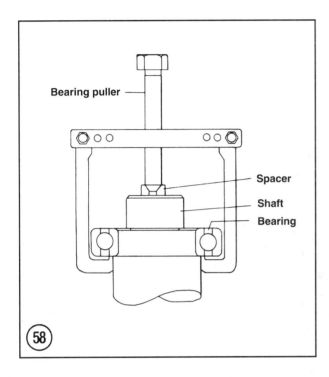

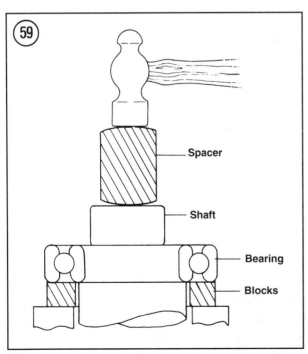

BEARING REPLACEMENT

Bearings are precision made components and therefore, require regular lubrication and maintenance. However, even with the proper maintenance, a bearing can eventually fail. If a bearing is found to be excessively worn or damaged, replace immediately to prevent additional damage.

When installing a new bearing, care must be exercised to prevent damage to the new bearing. While bearing replacement is described in the individual chapters where applicable, use the following as a general guide to bearing service.

NOTE
Unless otherwise specified, install bearings with their manufacturer's mark or part number facing outward.

Bearing Removal

While bearings are normally removed only if damaged, there may be times when it is necessary to remove a bearing that is in good condition. However, bearing removal often damages an otherwise good bearing, and improper bearing removal almost certainly will damage the bearing, shaft or case half. Note the following points when removing a bearing.

1. When using a puller to remove a bearing from a shaft, care must be taken to prevent shaft damage. Always place a metal protector between the end of the shaft and the puller screw. In addition, place the puller arms next to the inner bearing race. See **Figure 58**.

2. When using a hammer to remove a bearing from a shaft, do not strike the hammer directly against the shaft. Instead, use a brass or aluminum rod between the hammer and shaft. In addition, make sure to support both bearing races with wooden blocks as shown in **Figure 59**.

3. The best method to remove bearings is using a hydraulic or arbor press. However, certain procedures must be closely followed or damage

can occur to the bearing, shaft or bearing housing. Note the following when using a press:

a. Always support the inner and outer bearing races with a wooden or aluminum spacer (**Figure 60**). If you only support the outer race, pressure applied against the inner bearing race will damage the bearing.
b. Always make sure the press ram (**Figure 60**) aligns with the center of the shaft. If the ram is not centered, it may damage the bearing and/or shaft.
c. The moment the shaft is free of the bearing, it will drop to the floor. Secure or hold the shaft to prevent it from falling.

Bearing Installation

1. When installing a bearing into a housing, pressure must be applied to the *outer* bearing race (**Figure 61**). When installing a bearing on a shaft, pressure must be applied to the *inner* bearing race (**Figure 62**).
2. When installing a bearing as described in Step 1, some type of driver is required. Never strike the bearing directly with a hammer or the bearing will be damaged. When installing a bearing, a piece of pipe or a socket with an outer diameter that matches the bearing race is required. **Figure**

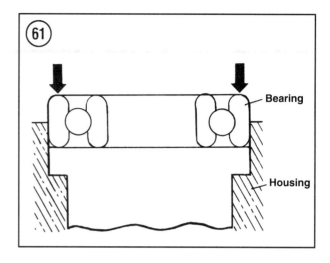
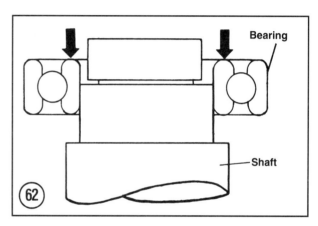
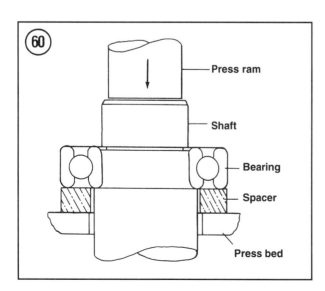
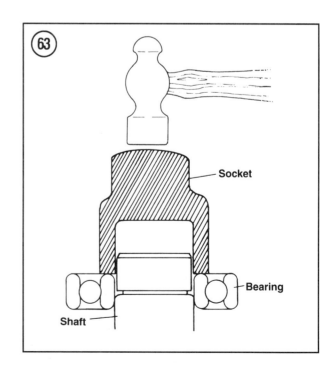

GENERAL INFORMATION

63 shows the correct way to use a socket and hammer when installing a bearing.

3. Step 1 described how to install a bearing in a case half or housing and on a shaft. However, when installing a bearing over a shaft and into a housing at the same time, a snug fit on both the inner and outer races is necessary. In this situation, a spacer must be installed under the driver tool so that pressure is applied evenly across both races. See **Figure 64**. If the outer race is not supported as shown in **Figure 64**, the bearing balls or rollers will push against the outer bearing track and damage it.

Shrink Fit

1. *Installing a bearing over a shaft*: When a tight (interference) fit is required, the bearing inside diameter is slightly smaller than the shaft. In this case, simply driving the bearing on the shaft may cause bearing or shaft damage. Instead, heat the bearing before installation. Note the following:

   a. Secure the shaft so it is ready for bearing installation.

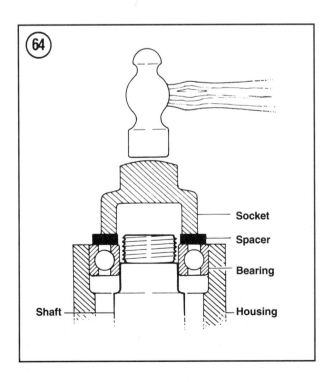

   b. Clean the bearing surface on the shaft of all residue. Remove burrs with a file or sandpaper.

   c. Fill a suitable pot or beaker with clean mineral oil. Place a thermometer (rated higher than 248°F [120°C]) in the oil. Support the thermometer so that it does not rest on the bottom or side of the pot.

   d. Secure the bearing with a piece of heavy wire bent to hold it in the pot. Hang the bearing in the pot so it does not touch the bottom or sides of the pot.

   e. Turn the heat on and monitor the thermometer. When the oil temperature rises to approximately 248°F (120°C), remove the bearing from the pot and quickly install it. If necessary, place a socket on the inner bearing race and tap the bearing into place. As the bearing cools, it will tighten on the shaft so you must work quickly during installation. Make sure the bearing is installed all the way.

2. *Installing a bearing in a housing*: Bearings are generally installed in a housing with a slight interference fit. Driving the bearing into the housing may damage the housing or bearing. Instead, heat the housing as follows before the bearing is installed:

   a. Heat the housing to a temperature of about 212°F (100°C) in an oven, on a hot plate or using a heat gun. A quick way to check for the proper temperature is to place small drops of water on the housing; if they sizzle and evaporate immediately, the temperature is about correct. Heat only one housing at a time.

   CAUTION
   Do not heat housings with open flame. The direct heat will destroy the bearing and could warp the housing.

   b. Remove the housing from the oven or hot plate and hold onto the housing with heavy

gloves or shop cloths—remember the part is *hot*.
   c. Hold the housing with the bearing side down and tap the bearing out. Repeat for all bearings in the housing.
   d. Prior to heating the housing for bearing installation, place the bearing in a freezer, if possible. Chilling a bearing will shrink it slightly making installation easier.
   e. To install a new bearing, reheat the housing as previously described. While the housing is still hot, install the new bearing(s). Install the bearings by hand, if possible. If necessary, lightly tap the bearing(s) into the housing with a suitable driver placed on the bearing outer race. Do not install new bearings into a housing by driving on the outer race. Make sure the bearing(s) are fully seated.

SEALS

Seals (**Figure 65**) are used to contain oil, coolant, grease or combustion gasses. Improper removal of a seal can damage the housing or shaft. Improper installation of the seal can damage the seal. Note the following:
   a. Prying is generally the easiest and most effective method to remove a seal from a housing. However, prying out a seal is often a difficult task. Use caution not to damage the housing or shaft. Place a shop towel under the pry tool to prevent damage.
   b. A suitable grease should be packed in the seal lips before the seal is installed.
   c. Unless otherwise specified, install seals so their manufacturer's numbers or marks face outward.
   d. Install seals with a suitable driver placed on the outer edge of the seal as shown in **Figure 66**. Make sure the seal is driven squarely into the housing. Never install a seal by striking directly on the seal with a hammer.

CLEARING A SUBMERGED WATER VEHICLE

If the water vehicle is submerged, water may get into the engine and fuel tank. To prevent corrosion and serious damage to the engine, the following steps must be taken before trying to restart the engine.

CAUTION
To prevent corrosion damage inside the engine, accomplish this procedure as soon as possible after submerging the water vehicle.

1. Beach the craft and remove the engine cover.
2. If your craft is equipped with a hull drain plug, remove the plug and raise the front of the hull. If your model does not have a hull drain

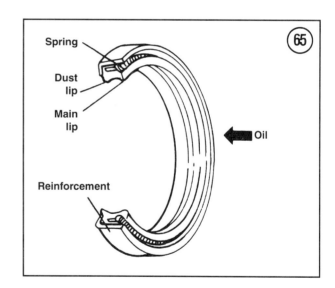

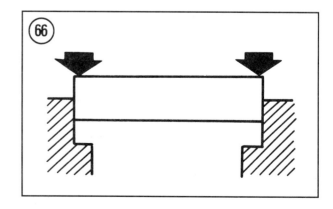

GENERAL INFORMATION

plug, turn the craft onto its *left* side. Precautions should be taken to avoid damaging the hull, engine hood or steering pole when turning the craft on its side. When all of the water has drained, reinstall the drain plug or turn the hull upright.

NOTE
Pay particular attention to the oil injection system when cleaning out a submerged 1993-1994 WRA650, WRA650A, WRB650, WRB650A, WRB700, WB700, WB700A or 1994 WRA700, RA700 or RA700A. These models are equipped with a self-bleeding oil injection system. If the craft was capsized for 5 minutes or more, air may have entered the oil injection system. You must perform one of the following 2 steps to ensure that the self-bleeding system is allowed to function properly.

1. After your craft is right side up, do not start the engine for a minimum of 10 minutes.

2. If it is necessary to start your craft after it is right side up, allow the engine to idle for 10 minutes. Do not exceed idle speed or engine damage may result.

3. Remove the spark plugs and ground their leads to the engine to prevent damage to the CDI ignition system. If the spark plugs are wet, spray them with electrical contact cleaner (if available) and then dry them with compressed air before performing Step 4.

CAUTION
If there is a possibility that sand may have entered the engine, do not try to start the engine or severe internal damage may occur.

4. Turn the fuel valve to OFF.

5. Crank the engine with the starter button to force water in the engine out of the spark plug holes. Continue until water stops exiting from the plug holes.

WARNING
Keep your hands away from the jet pump while cranking the engine.

CAUTION
Do not force the engine if it does not turn over. This may be an indication of internal damage.

6. Reinstall the spark plugs and connect their leads, then install the engine cover. Turn the fuel valve to ON and launch the craft. Operate the engine for approximately 10 minutes at trolling speed. If the engine will not start:
   a. Remove the spark plugs again. If the firing tips are covered with water, repeat Step 5, making sure to ground the spark plugs when cranking the engine with the start button.
   b. Reinstall the spark plugs.
   c. Repeat until engine starts. If water fouling continues, there may be water in the fuel.
   d. If there is water in the fuel tank, use a pump or siphon to empty the tank. Refill with fresh fuel.

7. After performing Step 6, beach the craft once again and remove the engine cover and spark arrestor (**Figure 67**). Spray a rust preventive, such as Yamaha Stor-Rite Engine Fogging Oil, into the carburetor while operating the starter button.

8. To dry out the electrical components and to keep them from rusting, remove the flywheel cover and spray the components with electrical contact cleaner.

Table 1 HULL AND ENGINE SERIAL NUMBERS

Year and model	Hull*	Engine*
1994 FX700S	FX1-900101	62L-000101
1995 FX700T	FX1-901710	62L-001490
1993 SJ650R	EW2-813101	6R7-C18911
1994 SJ700S	GH7-800101	62N-000101
1995 SJ700T	GH7-801601	62N-002301
1996 SJ700AU	GM6-900101	64V-000101
1994 RA700S	GH1-900101	62T-000101
1995 RA700T	GH1-920461	62T-004464
1995 RA700AT	GH1-800101	62T-004464
1996 RA700BU	GJ8-800101	64R-000101
1996 RA760U	GP2-800101	64X-000101
1995 RA1100T	GJ1-800101	63M-000101
1996 RA1100U	GJ1-809101	63M-006201
1993 WB700R	GA7-900101	62E-000101
1994 WB700S	GA7-904345	62E-001801
1995 WB700T	GA7-940001	62E-011113
1996 WB700AU	GR7-900101	64U-000101
1996 WB760U	GK5-900101	64Y-000101
1993 WR500R	EU0-819101	6K8-075820
1993 WR650R	FK7-815401	6M6-026717
1993 WRA650R	FJ0-809501	6R8-036494
1993 WRA650RA	FJ0-812301	6R8-036494
1995 WRA650T	FJ0-819271	6R8-050234
1996 WRA650U	FJ0-826673	6R8-058292
1994 WRA700S	GE2-800101	62G-000101
1995 WRA700T	GE2-815396	62G-016586
1993 WRB650R	FN8-807001	61L-022701
1993 WRB650RA	FN8-820601	61L-020601
1994 WRB650S	FN8-822901	61L-026801
1995 WRB650T	FN8-825801	61L-029701
1993 WRB700R	GD0-800101	61X-000101
1994 WRB700S	GD0-808001	61X-009001
1995 WVT700T	GJ3-800101	63N-000101
1996 WVT700U	GJ3-814819	63N-013801
1996 WVT1100U	GH3-800101	64T-000101
*Starting serial number		

GENERAL INFORMATION

Table 2 DECIMAL AND METRIC EQUIVALENTS

Fractions	Decimal in.	Metric mm	Fractions	Decimal in.	Metric mm
1/64	0.015625	0.39688	33/64	0.515625	13.09687
1/32	0.03125	0.79375	17/32	0.53125	13.49375
3/64	0.046875	1.19062	35/64	0.546875	13.89062
1/16	0.0625	1.58750	9/16	0.5625	14.28750
5/64	0.078125	1.98437	37/64	0.578125	14.68437
3/32	0.09375	2.38125	19/32	0.59375	15.08125
7/64	0.109375	2.77812	39/64	0.609375	15.47812
1/8	0.125	3.1750	5/8	0.625	15.87500
9/64	0.140625	3.57187	41/64	0.640625	16.27187
5/32	0.15625	3.96875	21/32	0.65625	16.66875
11/64	0.171875	4.36562	43/64	0.671875	17.06562
3/16	0.1875	4.76250	11/16	0.6875	17.46250
13/64	0.203125	5.15937	45/64	0.703125	17.85937
7/32	0.21875	5.55625	23/32	0.71875	18.25625
15/64	0.234375	5.95312	47/64	0.734375	18.65312
1/4	0.250	6.35000	3/4	0.750	19.05000
17/64	0.265625	6.74687	49/64	0.765625	19.44687
9/32	0.28125	7.14375	25/32	0.78125	19.84375
19/64	0.296875	7.54062	51/64	0.796875	20.24062
5/16	0.3125	7.93750	13/16	0.8125	20.63750
21/64	0.328125	8.33437	53/64	0.828125	21.03437
11/32	0.34375	8.73125	27/32	0.84375	21.43125
23/64	0.359375	9.12812	55/64	0.859375	21.82812
3/8	0.375	9.52500	7/8	0.875	22.22500
25/64	0.390625	9.92187	57/64	0.890625	22.62187
13/32	0.40625	10.31875	29/32	0.90625	23.01875
27/64	0.421875	10.71562	59/64	0.921875	23.41562
7/16	0.4375	11.11250	15/16	0.9375	23.81250
29/64	0.453125	11.50937	61/64	0.953125	24.20937
15/32	0.46875	11.90625	31/32	0.96875	24.60625
31/64	0.484375	12.30312	63/64	0.984375	25.00312
1/2	0.500	12.70000	1	1.00	25.40000

Table 3 GENERAL TORQUE SPECIFICATIONS*

	N·m	in.-lb.	ft.-lb.
Nut			
8mm	5.0	44	—
10mm	8.0	71	—
12mm	18	—	13
14mm	36	—	26
17mm	43	—	31
Bolt			
M5	5.0	44	—
M6	8.0	71	—
M8	18	—	13
M10	36	—	26
M12	43	—	31

* Use these torque values for all fasteners not individually listed.

Table 4 METRIC TAP DRILL SIZES

Metric tap (mm)	Drill size	Decimal equivalent	Nearest fraction
3 × 0.50	No. 39	0.0995	3/32
3 × 0.60	3/32	0.0937	3/32
4 × 0.70	No. 30	0.1285	1/8
4 × 0.75	1/8	0.125	1/8
5 × 0.80	No. 19	0.166	11/64
5 × 0.90	No. 20	0.161	5/32
6 × 1.00	No. 9	0.196	13/64
7 × 1.00	16/64	0.234	15/64
8 × 1.00	J	0.277	9/32
8 × 1.25	17/64	0.265	17/64
9 × 1.00	5/16	0.3125	5/16
9 × 1.25	5/16	0.3125	5/16
10 × 1.25	11/32	0.3437	11/32
10 × 1.50	R	0.339	11/32
11 × 1.50	3/8	0.375	3/8
12 × 1.50	13/32	0.406	13/32
12 × 1.75	13/32	0.406	13/32

Table 5 TECHNICAL ABBREVIATIONS

ABDC	After bottom dead center
ATDC	After top dead center
BBDC	Before bottom dead center
BDC	Bottom dead center
BTDC	Before top dead center
C	Celsius (Centigrade)
cc	Cubic centimeters
CDI	Capacitor discharge ignition
cu. in.	Cubic inches
F	Fahrenheit
ft.-lb.	Foot-pounds
g	Gram
gal.	Gallons
hp	Horsepower
in.	Inches
kg	Kilogram
kg/cm^2	Kilograms per square centimeter
kgm	Kilogram meters
km	Kilometer
L	Liter
m	Meter
MAG	Magneto
mm	Millimeter
N.A.	Not available
N•m	Newton-meters
oz.	Ounce
psi	Pounds per square inch
pto	Power take off
pts.	Pints
qt.	Quarts
rpm	Revolutions per minute
WOT	Wide-open throttle

GENERAL INFORMATION

Table 6 GENERAL SPECIFICATIONS (WR500, WR650)

Dimensions	
Length	2,770 mm (109.1 in.)
Width	1,020 mm (40.2 in.)
Height	870 mm (34.3 in.)
Dry weight	
WR500	168 kg (348 lb.)
WR650	172 kg (379 lb.)
Performance	
Maximum speed	58 km/h (36 mph)
Maximum rpm	5,500 rpm
Horsepower	
WR500	32 hp
WR650	42 hp
Fuel consumption	
Maximum	
WR500	14 l/h (3.7 gal./h)
WR650	18 l/h (4.7 gal./h)
Cruising range @ full throttle	
WR500	1.5 hours
WR650	1.2 hours

Table 7 GENERAL SPECIFICATIONS (WRA650, WRA650A, WRA700)

Dimensions	
Length	2,990 mm (117.7 in.)
Width	1,110 mm (43.7 in.)
Height	950 mm (37.4 in.)
Dry weight	
WRA650, WRA650A	206 kg (454.1 lb.)
WRA700	214 kg (472 lb.)
Performance	
Maximum speed	
WRA650, WRA650A	60.5 km/h (38 mph)
WRA700	67.0 km/h (41.6 mph)
Maximum rpm	
WRA650, WRA650A	6,000 rpm
WRA700	6,250 rpm
Horsepower	
WRA650, WRA650A	50 hp
WRA700	63 hp
Fuel consumption	
Maximum	
WRA650, WRA650A	21 l/h (5.5 gal./h)
WRA700	26 l/h (6.9 gal./h)
Cruising range @ full throttle	
WRA650, WRA650A	1.9 hours
WRA700	1.5 hours

Table 8 GENERAL SPECIFICATIONS (WRB650, WRB650A, WRB700)

Dimensions	
Length	2,770 mm (109.1 in.)
Width	1,020 mm (40.2 in.)
Height	900 mm (35.4 in.)
Dry weight	
WRB650, WRB650A	170 kg (374 lb.)
WRB700	185 kg (407 lb.)
Performance	
Maximum speed	
WRB650, WRB650A	60.5 km/h (38 mph)
WRB700	70.5 km/h (44 mph)
Maximum rpm	
WRB650, WRB650A	6,000 rpm
WRB700	6,250 rpm
Horsepower	
WRB650, WRB650A	50 hp
WRB700	64 hp
Fuel consumption	
Maximum	
WRB650, WRB650A	21 l/h (5.5 gal./h)
WRB700	26 l/h (6.9 gal./h)
Cruising range @ full throttle	
WRB650, WRB650A	1.4 hours
WRB700	1.2 hours

Table 9 GENERAL SPECIFICATIONS (WB700, WB700A, WB760)

Dimensions	
Length	
WB700, WB700A	2,430 mm (95.7 in.)
WB760	2,720 (107.1 in.)
Width	
WB700, WB700A	880 mm (34.6 in.)
WB760	1,030 mm (40.6 in.)
Height	
WB700, WB700A	910 mm (35.8 in.)
WB760	970 mm (38.2 in.)
Dry weight	
WB700, WB700A	145 kg (320 lb.)
WB760	180 kg (397 lb.)
Performance	
Maximum speed	
WB700, WB700A	69 km/h (43 mph)
WB760	75 km/h (46.6 mph)
Maximum rpm	
WB700, WB700A	6,250 rpm
WB760	6.350 rpm
Horsepower	
WB700, WB700A	64 hp
WB760	90 hp
Fuel consumption	
Maximum	
WB700, WB700A	26 l/h (6.9 gal./h)
WB760	38 l/h (10.0 gal./h)
Cruising range @ full throttle	
WB700, WB700A, WB760	1.0 hours

GENERAL INFORMATION

Table 10 GENERAL SPECIFICATIONS (RA700, RA700A, RA700B, RA760, RA1100, WVT700, WVT1100)

Dimensions	
Length	
RA700, RA700A, RA700B,	
RA760, RA1100	2,860 mm (112.6 in.)
WVT700, WVT1100	3,150mm (124.0 in.)
Width	
RA700, RA700A, RA700B,	
RA760, RA1100	1,120 mm (44.1 in.)
WVT700, WVT1100	1,250 mm (49.2 in.)
Height	
RA700, RA700A, RA700B,	
RA760, RA1100	970 mm (38.2 in.)
WVT700, WVT1100	1,050 mm (41.3 in.)
Dry weight	
RA700	176 kg (388 lb.)
RA700A	219 kg (483 lb.)
RA700B	214 kg (472 lb.)
RA760	211 kg (465 lb.)
WVT700, RA1100	245 kg (540 lb.)
WVT700	271 kg (597 lb.)
Performance	
Maximum speed	
RA700, WVT1100	83 km/h (51.6 mph)
RA700A	81 km/h (50.3 mph)
RA760	84 km/h (52.2 mph)
RA1100	91 km/h (56.5 mph)
RA700B, WVT700	73 km/h (45.4 mph)
Maximum rpm	
RA700, RA700A, WVT700	6,250 rpm
RA700B	6,300 rpm
RA760	6,350 rpm
RA1100, WVT1100	6,500 rpm
Horsepower	
RA700, RA700A, RA700B,	
WVT700	80 hp
RA760	90 hp
RA1100, WVT1100	110 hp
Fuel consumption	
Maximum	
RA700B	27 l/h (7.1 gal./h)
RA700, RA700A, WVT700	34 l/h (9.0 gal./h)
RA1100, WVT1100	46 l/h (12.2 gal./h)
Cruising range @ full throttle	
RA700	1.2 hours
RA700A, WVT700	1.5 hours
RA700B	1.9 hours
RA760	1.3 hours
RA1100, WVT1100	1.1 hours

Table 11 GENERAL SPECIFICATIONS (FX700, SJ650, SJ700, SJ700A)

Dimensions	
Length	
SJ650, SJ700, SJ700A	2,240 mm (88.2 in.)
FX700	2,130 mm (83.9 in.)
Width	
SJ650, SJ700, SJ700A	680 mm (26.8 in.)
FX700	630 mm (24.8 in.)
Height	
SJ650, SJ700, SJ700A	660 mm (26.0 in.)
FX700	680 mm (26.8 in)
Dry weight	
SJ650	130 kg (287 lb.)
SJ700, SJ700A	132 kg (291 lb.)
FX700	121 kg (267 lb.)
Performance	
Maximum speed	
SJ700	70 km/h (43.5 mph)
SJ700A	73 km/h (45.4 mph)
FX700	75 km/h (46.6 mph)
Maximum rpm	
SJ650	6,000 rpm
SJ700, FX700	6,250 rpm
SJ700A	6300 rpm
Horsepower	
SJ650	50 hp
SJ700	62.5 hp
FX700	62 hp
SJ700A	73 hp
Fuel consumption	
Maximum	
SJ650	23 l/h (6.1 gal./h)
SJ700, FX700	26 l/h (6.9 gal./h)
SJ700A	29 l/h (7.7 gal./h)
Cruising range at full throttle	
SJ650	0.8 hours
SJ700	0.7 hours
SJ700A	0.6 hours
FX700	0.5 hours

Chapter Two

Troubleshooting

Diagnosing mechanical and electrical problems is relatively simple if you use orderly procedures and keep a few basic principles in mind. The first step in any troubleshooting procedure is to define the symptoms as closely as possible and then localize the problem. Subsequent steps involve testing and analyzing those areas which could cause the symptoms. A haphazard approach may eventually solve the problem, but it can be very costly in terms of wasted time and unnecessary parts replacement.

Proper lubrication, maintenance and periodic tune-up as described in Chapter Three will reduce the necessity for troubleshooting. Even with the best of care, however, all watercraft are prone to problems which will require troubleshooting.

Never assume anything. Do not overlook the obvious. If the engine will not start, is the engine stop switch lock plate properly inserted into the engine stop switch? Is the engine cranking slowly because the battery is discharged?

If the engine suddenly quits, check the easiest, most accessible problem first. Is there gasoline in the tank? Has a spark plug wire fallen off? Is the fuel vent check valve allowing air to enter the tank as it should?

If nothing obvious turns up in a quick check, look a little further. Learning to recognize and describe symptoms will make repairs easier for you or a mechanic at the shop. Describe problems accurately and fully.

Gather as many symptoms as possible to aid in diagnosis. Note whether the engine lost power gradually or all at once, what color smoke came from the exhaust and so on. Remember that the more complicated a machine is, the easier it is to troubleshoot because symptoms point to specific problems.

After the symptoms are defined, areas which could cause problems should be tested and analyzed. Guessing at the cause of a problem may provide the solution, but it can easily lead to frustration, wasted time and a series of expensive, unnecessary parts replacements.

You do not need expensive equipment or complicated test gear to determine whether repairs can be attempted at home. A few simple checks could save a large repair bill and lost time while your water vehicle sits in a dealer's service department. On the other hand, be realistic. Do not attempt repairs beyond your abilities. Service departments tend to charge heavily for putting together a disassembled engine that may have been abused. Some will not even take on such a job-so use common sense and do not get in over your head.

This chapter is divided into 2 parts. The first part covers emergency and cause related troubleshooting procedures. The second part describes troubleshooting tests and checks that can be performed within the individual engine operating assemblies.

Table 1 and **Table 2** are at the end of the chapter.

NOTE
There is one kind of trouble you can expect to run into occasionally if you like to push your water vehicle to its performance and handling limits. If the engine cover has come loose and the engine compartment is full of water, refer to **Cleaning a Submerged Water Vehicle** *in Chapter One.*

OPERATING REQUIREMENTS

An engine needs 3 basics to run properly: correct air/fuel mixture, compression and a spark at the right time (**Figure 1**). If one basic requirement is missing, the engine will not run. When troubleshooting your water vehicle, it is important to follow a process of elimination, making sure you do not overlook the obvious.

Two-stroke engine operating principles are described in Chapter One under *Engine Operation*. The ignition system is the weakest link of the 3 basics. More problems result from ignition breakdowns than from any other source. Keep that in mind before you begin tampering with carburetor adjustments.

If the water vehicle has been sitting for any length of time and refuses to start, check and clean the spark plugs. Then check the condition of the battery to make sure it has an adequate charge. If these are okay, then look to the gasoline delivery system. This includes the tank, fuel shutoff valve, inline fuel filter and fuel line to the carburetor. Gasoline deposits may have gummed

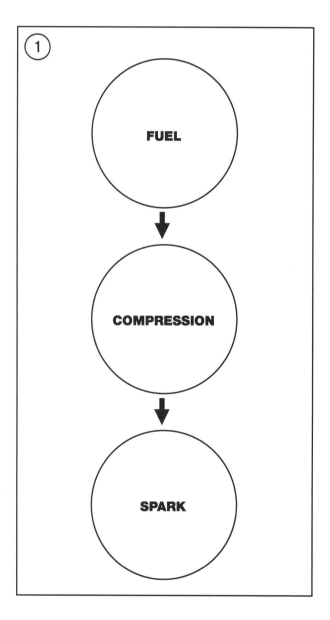

TROUBLESHOOTING

up carburetor jets and air passages. Gasoline tends to lose its potency after standing for long periods. Condensation may contaminate it with water. Drain the old gas and try starting with a fresh tankful.

EMERGENCY TROUBLESHOOTING

If the engine is difficult to start or will not start at all, it does not help to wear down the battery. Check for obvious problems even before getting out your tools. Go down the following list step by step. You may be embarrassed to find that the stop switch lock plate has pulled out of the switch, but that is better than wearing down the battery. If the starter does not operate or if the engine still will not start after performing the following steps, refer to the appropriate troubleshooting procedure which follows in this chapter.

NOTE
The carburetor was initially set at the factory for sea level operation. If your engine started and ran okay at a lower altitude, but you are experiencing starting problems after a change in altitude, carburetor adjustment is in order. When changing altitudes, note that a higher altitude will make your engine run richer. Readjust your carburetor as described in Chapter Three.

1. Make sure the stop switch lock plate is properly installed in the engine stop switch (A, **Figure 2**).

WARNING
*Do **not** use an open flame when looking inside the tank. A serious explosion is certain to result.*

2. Is there fuel in the tank? Open the hatch and remove the filler cap. You should be able to see the fuel level in the tank. If the fuel level is low, refill the tank.

NOTE
If the engine has not been run for some time, gasoline deposits may have gummed up carburetor jets and air passages. In addition, gasoline tends to loose its potency after standing for long periods or you may find water in the tank. Drain the old gas and try starting with a fresh tankful.

3. Is the choke (**Figure 3**) in the correct position? The choke knob should be pulled out when starting a cold engine and pushed in when restarting a warm or hot engine.

NOTE
The condition of your engine's spark plugs is a deciding factor in its performance and an important reference point during troubleshooting and general maintenance. To avoid mixups when re-

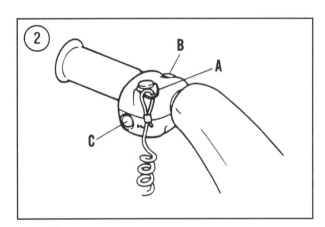

moving the spark plugs in Step 4, identify each plug so you know which cylinder it came from.

4. After attempting to start the engine, immediately remove the spark plugs and check their firing tips. Refer to Chapter Three for information on reading spark plugs. Fuel should be present on both plugs' firing tips. This indicates that fuel is being pumped from the fuel tank to the engine. If there is no sign of fuel on the plugs, suspect a fuel delivery problem; refer to *Fuel System* in this chapter. If it appears that there is water on the plugs, water has probably entered the engine from contaminated fuel or there is water in the crankcase. Remove water from the engine as follows:

NOTE
If water has contaminated the fuel, remove the fuel tank as described in Chapter Eight and drain the tank thoroughly. Refill the tank with fresh fuel before starting the engine.

NOTE
If there is no sign of water on the plugs and the plugs do not appear to be fouled, proceed to Step 5.

  a. Turn the fuel petcock OFF.
  b. Disconnect the stop switch lock plate from the engine stop switch to turn the ignition system OFF.
  c. Remove and ground both spark plugs against the cylinder head. If you plan on turning the watercraft over, tie the spark plugs in place so they are grounded against the cylinder head, then check the plugs after turning the craft over to make sure they are properly grounded.

CAUTION
Whenever the engine is cranked with the spark plugs removed, the plugs must be grounded to protect the ignition system from permanent damage. While substep b turns the ignition system OFF, substep c guarantees that the ignition system is properly grounded should the engine stop switch malfunction.

  d. Open the throttle fully and operate the starter switch for 10-15 seconds to force water out of the engine.
  e. Release the starter switch and throttle.

5. Perform a spark test as described under *Engine Fails to Start (Spark Test)* in this chapter. If there is a strong spark, perform Step 6.

6. Check cylinder compression as follows:
  a. Turn the fuel petcock OFF.
  b. Insert the stop switch lock plate into the engine stop switch to turn the ignition system ON.
  c. Remove and ground the spark plugs as described in Step 4, substep c. Also review the CAUTION following substep c. The spark plugs must be grounded when performing the following steps or the ignition system will be permanently damaged.

WARNING
When grounding the spark plugs in substep c, make sure the plugs are placed away from the spark plug holes in the cylinder head. Because you will be placing your fingers over the spark plug holes, you could receive quite a shock if you accidentally touch a plug while cranking the engine.

  d. Put your finger over one of the spark plug holes.
  e. Crank the engine with the starter button. Rising pressure in the cylinder should force your finger off of the spark plug hole. This indicates that the cylinder probably has sufficient cylinder compression to start the engine.
  f. Repeat for the opposite cylinder.
  g. Lack of cylinder compression indicates a problem with that cylinder. This could be worn or damaged piston rings or a damaged piston. Refer to *Engine* in this chapter.

TROUBLESHOOTING

NOTE
Engine compression can be checked more accurately with a compression gauge as described in Chapter Three.

NOTE
If cylinder compression is sufficient for both cylinders, your engine may be suffering from a loss of crankcase pressure. During 2-stroke operation, the air/fuel mixture is compressed twice, first in the crankcase and then in the combustion chamber. Crankcase pressure forces the air/fuel mixture to flow from the crankcase chamber through the transfer ports and into the combustion chamber. Before continuing, you should perform a crankcase pressure check as described in this chapter.

ENGINE STARTING SYSTEM

An engine that refuses to start or is difficult to start is very frustrating. More often than not, the problem is very minor and can be found with a simple and logical troubleshooting approach.

The following items show a beginning point from which to isolate engine starting problems.

Description

An electric starter motor (**Figure 4**) is used on all Yamaha water vehicles. The motor is mounted horizontally on the side of the engine. When battery current is supplied to the starter motor, its pinion gear is thrust forward to engage the teeth on the engine flywheel. Once the engine starts, the pinion gear disengages from the flywheel.

The electric starting system requires a fully charged battery to provide the large amount of current required to operate the starter motor. Because there are no lights on the watercraft, the battery is only used to provide power to the starter motor. A lighting coil (mounted on the stator plate) and a voltage regulator, connected in circuit with the battery, keeps the battery charged while the engine is running. The battery can also be charged externally.

The starting circuit consists of the battery, start switch, engine stop switch, starter solenoid, starter motor and connecting wiring.

The starter relay carries the heavy electrical current to the motor. Depressing the starter switch (B, **Figure 2**) allows current to flow through the relay coil. The relay contacts close and allow current to flow from the battery through the relay to the starter motor. The starter relay is mounted in the electric box.

CAUTION
Do not operate an electric starter motor continuously for more than 5 seconds. Allow the motor to cool for at least 15 seconds between attempts to start the engine.

Troubleshooting

Before troubleshooting the starting circuit, make sure that:
a. The battery is fully charged.
b. Battery cables are the proper size and length. Replace cables that are undersize or damaged.
c. All electrical connections are clean and tight.
d. The wiring harness is in good condition, with no worn or frayed insulation or loose harness sockets.

e. The electrical circuit fuse is in good condition.
f. The fuel system is filled with an adequate supply of fresh gasoline. On WR500, WR650, SJ650, SJ700, SJ700A and FX700 models, be sure that the fuel and oil have been properly mixed. See Chapter Three.
g. The spark plugs are in good condition and properly gapped.
h. The ignition system is properly timed. See Chapter Three.

Troubleshooting is intended only to isolate a malfunction to a certain component. Refer to Chapter Nine for electrical component removal and installation procedures.

Engine Stop Switch and Lanyard

The engine stop switch on all models is designed to operate with a lanyard. The lanyard is a safety device commonly used in marine applications. The lanyard assembly is comprised of a lock plate that fits into the engine stop switch, a wrist band which attaches to the operator's wrist and a lanyard connecting the lock plate and wrist band (A, **Figure 2**).

Before starting the engine, the lock plate must be inserted into the engine stop switch. The operator then attaches the lanyard hook to his or her wrist. If the operator and craft should part company when the engine is running, the lock plate pulls out of the engine stop switch and the engine stops.

When troubleshooting a starting problem, make sure the lanyard lock plate is inserted into the ignition stop switch properly.

One-Touch Stop Switch

The engine start/stop switch on all models also includes a one-touch stop switch (C, **Figure 2**) that immediately kills the engine when pressed. When the one-touch stop switch is pressed, the circuit from the charge coil is grounded, so there is no spark at the spark plug. The circuit remains grounded for 2 seconds to prevent ignition, then resets itself.

This is particularly important during troubleshooting. The engine will not restart if you press the start switch immediately after pressing the stop switch. Wait 2 to 3 seconds before attempting to restart the engine.

Starting System Check

Troubleshoot the starting system through a process of elimination.

1. Connect a remote starter to the starter relay in the electric box.
2. If the starter motor functions when you press the remote switch, the battery, battery cables, and starter motor are in good operating condition.

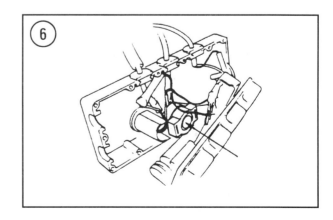

TROUBLESHOOTING

3. Connect a jumper from the red fuse lead to the brown lead on the starter relay.
   a. If the starter motor operates, the starter switch or switch wiring is defective.
   b. If the starter relay does not click when the jumper wire is connected, replace the relay.
4. If the battery, cables, switch, and starter relay are good, check the starter motor. Refer to *Starter Motor* in Chapter Nine.

Starter Relay Test

An ohmmeter, a fully charged 12-volt battery and 2 jumper cables are required for this procedure.

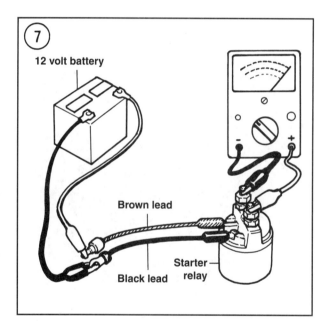

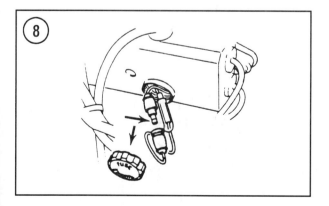

NOTE
If you plan to use the battery from your water vehicle when making this test, remove the battery from the hull and make sure it is fully charged.

1. Remove the starter relay (**Figure 5** or **Figure 6**) from the electric box as described in Chapter Nine.
2. Connect an ohmmeter between the 2 starter terminals shown in **Figure 7**. Switch the ohmmeter to the R × 1 scale. The ohmmeter must indicate no continuity.
3. Use jumper cables to connect the brown relay lead to the positive battery terminal and the black relay lead to the negative battery terminal. See **Figure 7**.
4. The relay should click and the ohmmeter show continuity when the relay leads are connected to the battery in Step 3. If both of these do not occur, replace the relay.
5. Disconnect all test equipment.

Starter Switch Continuity Test

Refer to Chapter Nine.

Starter Motor Does Not Turn

Check for one or more of the following possible malfunctions:

1. *Fuse*—Pull the fuse out of the electric box (**Figure 8**) and check the metal element. Replace the fuse if blown. If the fuse is okay, check the fuse box clips for corrosion and clean with sandpaper, if required. Reinstall the fuse into its holder and then install the fuse and holder into the electric box opening. Install the electric box fuse cover.
2. *Discharged battery*—Check the battery with a hydrometer as described in Chapter Nine. If the reading is below 1.230, recharge or replace the battery. Also check the charging system output as described in this chapter.

3. *Wire harness*—Check the battery and starter cables for loose, corroded or damaged wiring. Check all related wiring and connectors for the same conditions. Repair wiring as required.
4. *Start switch*—Check as described in Chapter Nine.
5. *Engine stop switch*—Check as described in Chapter Nine.
6. *Starter relay*—Test relay as described in this chapter.
7. *Starter motor*—If checks in Steps 1-6 do not locate the problem, remove the starter motor and perform the bench checks as described in Chapter Nine.

Starter Turns Too Slowly

1. Check for a discharged battery. See Step 2 under *Starter Motor Does Not Turn*.
2. Check for poor battery cable connections. Then check for poor contact at the starter solenoid and starter motor. Check terminals for looseness or corrosion. Clean and tighten as required.
3. Remove starter as described in Chapter Nine. Turn the starter pinion or shaft by hand. See **Figure 9** (WR500 models) or **Figure 10** (all other models). The pinion gear and motor should turn freely. If the motor does not turn easily, clean the starter and replace all defective parts.
4. If the pinion gear and starter motor operate freely when checked in Step 3, disassemble the starter motor and overhaul or replace it as required.

Starter Does Not Engage Freely

1. Inspect the starter pinion and flywheel gears for excessive wear. Replace all defective parts.
2. If the pinion gear interferes with the flywheel gear after the engine starts, remove the starter and inspect the anti-drift spring located under the pinion gear. The spring may be broken. Replace all worn or damaged parts.

3. Inspect the pinion gear for excessive wear or damage. Remove the flywheel and pinion gear as described in Chapter Nine. Inspect the pinion gear assembly.

Engine Fails to Start (Spark Test)

Perform the following test to determine if the ignition system is operating properly.
1. Remove the engine hood.
2. Remove each spark plug from the cylinder head as described in Chapter Three.
3. Connect the spark plug wire and cap to the spark plug, and insert the plug into a spark checker.

NOTE
If a spark checker is not available, touch the spark plug's base to a good ground such as the engine cylinder head. Make sure the spark plug is away from the

TROUBLESHOOTING

spark plug hole and that it is against bare metal, not a painted surface. Position the plug so it will stay in contact with the metal during cranking, and be sure you can see the electrode while you crank the engine.

WARNING
*If it is necessary to hold the high voltage lead, do so with an **insulated** pair of pliers. The high voltage generated by the CDI could produce a serious or fatal shock.*

4. Crank the engine with the starter. A fat blue spark should be evident across the plug's electrodes. Repeat the test for each plug.

5. If the spark is good, check for one or more of the following possible malfunctions:
   a. Check the fuel tank for a low fuel level or for water contamination. Add fuel or drain the fuel tank as required.
   b. Check for a contaminated fuel filter or fuel line. If it appears that fuel is entering the carburetor, check for clogged carburetor jets and passages or a defective fuel pump diaphragm.
   c. Check for an incorrectly adjusted or defective choke system. Check choke cable adjustment as described in Chapter Three.
   d. Low engine compression. Check secondary (this chapter) and primary (Chapter Three) engine compression.
   e. Throttle not operating properly.

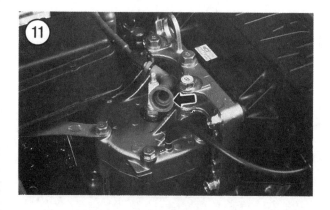

6. If there is no spark, check for one or more of the following:
   a. Loose or damaged spark plug cap(s). See **Figure 11**.
   b. Loose or damaged ignition coil high tension wire.
   c. Loose or corroded electrical connectors or wiring within the ignition system.
   d. Defective ignition coil.
   e. Defective pulse or charge coil(s).
   f. Weak or faulty CDI unit.

Engine is Difficult to Start

Check for one or more of the following possible malfunctions:
   a. Fouled spark plug(s).
   b. Improperly adjusted choke.
   c. Contaminated fuel system.
   d. Weak ignition coil.
   e. Weak or faulty CDI.
   f. Poor engine or crankcase compression.

CHARGING SYSTEM

The charging system consists of a lighting coil mounted on the magneto stator plate assembly, permanent magnets located within the flywheel rim, a rectifier to change alternating current (AC) to direct current (DC), the battery and connecting wiring. The rectifier is a solid-state device that also contains a regulator to prevent overcharging of the battery.

A malfunction in the charging system generally causes the battery to remain undercharged.

Troubleshooting

Before performing any charging circuits tests, visually check the following.
1. Make sure the battery cables are properly connected. If polarity is reversed, check for a damaged rectifier.

2. Carefully inspect all wiring between the magneto base and battery for worn or cracked insulation and for corroded or loose connections. Replace wiring or clean and tighten connections as required.

3. Check battery condition. Clean and recharge as required. See Chapter Nine.

Lighting Coil Output Test

WARNING
*To prevent an electrical shock, do not perform this test with the water vehicle in the water. Instead, use an auxiliary water supply during the test. Do not exceed the recommended rpm during this test, and only operate the engine for a short period of time. Refer to **Cooling System Flushing** in Chapter Three.*

CAUTION
Never operate the engine for more than 15 seconds without a water supply or engine damage will result.

1. Remove the engine cover.

2. Connect an ammeter in series with the red fuse wire.

3. Start the engine. Then turn on the auxiliary water supply.

4. Watch the ammeter gauge and gradually increase engine speed to approximately 5,000 rpm and note the gauge reading. If it is not 1-1.5 amps, replace the lighting coil as described in Chapter Nine.

5. After checking the lighting coil, bring the engine rpm back to idle. Then turn off the auxiliary water supply and turn the engine off.

6. Remove all test equipment and reinstall all components.

Lighting Coil Resistance Test

1A. *Most models*—Remove the electric box from its mounting position as described in Chapter Nine. Then open the box to gain access to the electrical connectors.

1B. *RA1100 and WVT1100*—Remove the electric box cover from the electric box to gain access to the electrical connectors.

2. Disconnect the 2 green lighting coil connectors from the stator plate-to-CDI wire harness inside the electric box.

3. Connect an ohmmeter between the 2 disconnected leads (**Figure 12**). With the ohmmeter set on the R × 1 scale, note the reading and compare to **Table 1**. If not within specification, replace the lighting coil as described in Chapter Nine.

Rectifier/Regulator Test

1A. *Most models*—Remove the electric box from its mounting position as described in Chapter Nine. Then open the box to gain access to the electrical connectors.

1B. *RA1100 and WVT1100*—Remove the electric box cover from the electric box to gain access to the electrical connectors.

NOTE
On WR500 cc models, the rectifier/regulator housing is mounted on the outside of the electric box. On other models, the rectifier/regulator is mounted inside the

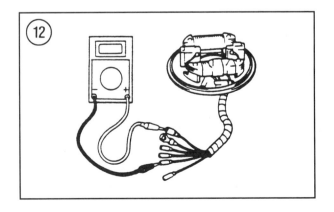

TROUBLESHOOTING

electric box. If the rectifier/regulator is mounted outside the electric box, remove it together with the electric box.

2. Disconnect the red, black and 2 green rectifier/regulator connectors from inside the electric box.
3. Set the ohmmeter to the R × 1000 scale.
4. Refer to **Figure 13** for test connections and values. If any of the meter readings differ from the stated values, replace the rectifier/regulator assembly.

IGNITION SYSTEM

All Yamaha water vehicles are equipped with a capacitor discharge ignition (CDI) system. This solid state system uses no contact breaker point or other moving parts. Because of the solid state design, problems with the capacitor discharge system are relatively few. However, if a malfunction occurs, it generally causes the ignition system to have a weak spark or no spark at all. It is relatively easy to test an ignition system that has weak or no spark output. It is difficult, however, to test an ignition system that has an intermittent problem that only occurs when the engine is hot and/or under a load. General troubleshooting procedures are provided in **Figure 14**.

Precautions

Certain measures must be taken to protect the capacitor discharge system from secondary damage. Instantaneous damage to the semiconductors in the system will occur if the following is not observed.

1. Do not reverse the battery connections. This reverses polarity and can damage the regulator/rectifier or CDI unit.
2. Do not disconnect the battery while the engine is running. A voltage surge will occur and damage the voltage regulator and CDI unit.
3. Do not arc the battery terminals with the battery cable connections to check polarity.
4. Do not crank the engine if the spark plugs are not grounded to the engine.
5. Do not crank the engine if the CDI unit is not grounded to the engine.
6. Do not touch or disconnect any ignition components when the engine is running or while the battery cables are connected.
7. Keep all connections between the various units clean and tight. Be sure that the wiring connectors are pushed together firmly.

Troubleshooting Preparation

NOTE
To test the wiring harness for poor connections in Step 1, bend the molded rubber connector while checking each wire for resistance.

1. Check the wiring harness and all plug-in connections to make sure that all terminals are free of corrosion, all connectors are tight and the wiring insulation is in good condition.
2. Check all electrical components that are grounded to the engine for a good ground connection.
3. Make sure all ground wires are properly connected and the connections are clean and tight.
4. Check the remainder of the wiring for disconnected wires, short or open circuits.
5. Make sure there is an adequate supply of fresh fuel available to the engine and that the oil tank

⑬ REGULATOR/RECTIFIER TESTING

		Positive test lead			
		R	B	G_1	G_2
Negative test lead	R		∞	∞	∞
	B	2-20		1-10	1-10
	G_1	1-10	2-15		3-30
	G_2	1-10	2-15	3-30	

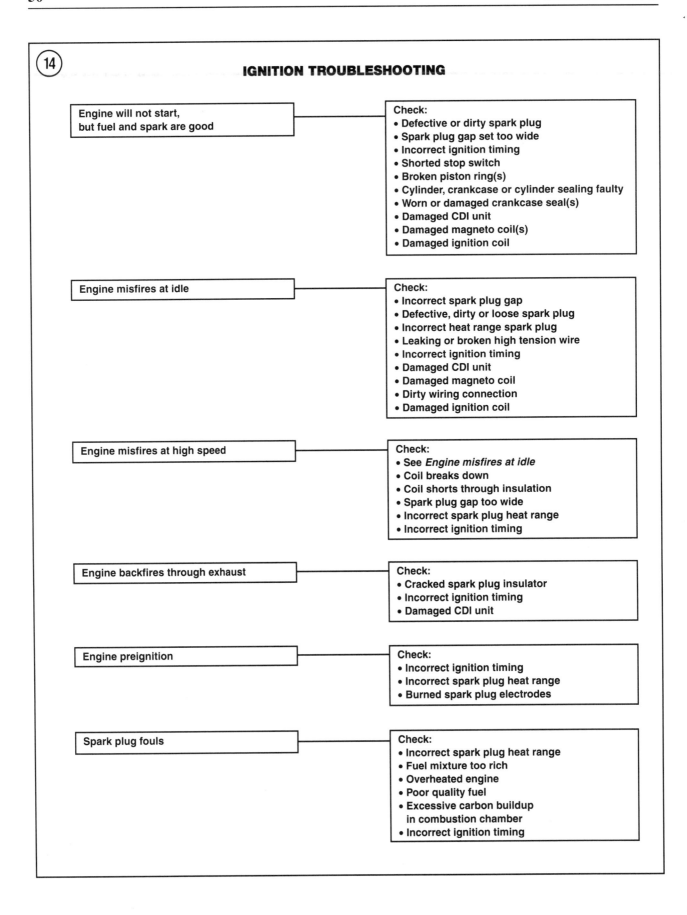

TROUBLESHOOTING

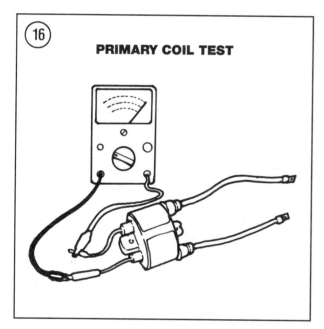

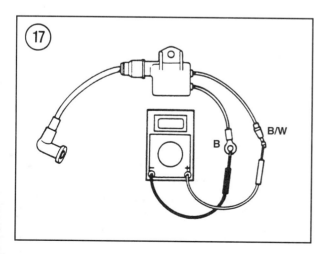

is properly filled. On WR500, SJ650, SJ700, SJ700A and FX700 models, be sure an adequate supply of properly mixed fuel is available to the engine.

6. Check the battery condition. Clean the terminals and recharge battery, if necessary. See Chapter Nine.

7. Check spark plug cable routing. Make sure the cables are properly connected to their respective spark plugs.

Ignition Coil Test

The ignition coil is mounted inside the electric box. See **Figure 15**, typical. The ignition coil is a step-up transformer which increases the low voltage produced by the magneto to the high voltage required to jump the spark plug gap. This test checks the primary and secondary windings of the coil for circuit continuity.

1A. *Except RA1100 and WVT1100*—Remove the electric box from its mounting position as described in Chapter Nine. Then open the box to gain access to the electrical connectors.

1B. *RA1100 and WVT1100*—Remove the electric box cover from the electric box to gain access to the electrical connectors.

2A. *Except RA1100 and WVT1100*—Disconnect the orange and black bullet connectors from the ignition coil.

2B. *RA1100 and WVT1100*—Disconnect the black/white and black bullet connectors from the ignition coil.

3. Check ignition coil primary resistance as follows:

   a. Switch a low-reading ohmmeter to the R × 1 scale.
   b. Measure the resistance between the ignition coil orange and black leads (**Figure 16**) and compare to the specifications listed in **Table 2**. For RA1100 and WVT1100 models, measure the resistance between the black/white and black wires (**Figure 17**).

c. Disconnect the ohmmeter leads.

4. Check ignition coil secondary resistance as follows:

   a. Switch an ohmmeter to the R × 1000 scale.

   b. Measure the resistance between the 2 high-tension leads at the spark plug connections (**Figure 18**). On *RA1100 and WVT1100* models, measure the resistance between the black/white wire and the spark plug connector. Refer to **Table 2** for specifications.

5. If the ohmmeter indicates an open circuit (no continuity) in Step 4, unplug both high-tension leads from the coil and test the coil again with the meter leads connected directly to the contact pins in the coil caps. On *RA1100 and WVT1100* models, remove the high-tension lead and measure the resistance between the coil's black/white wire and the contact pin in the coil cap. If there is continuity now at the coil terminals, the trouble is in the high-tension leads. It may be a bad connection at the spark plug or an internal break in the wire. Make sure the connections are good and check the leads themselves for continuity. If an open circuit is still indicated, replace the high-tension leads. If high tension leads have continuity, the coil is faulty and must be replaced.

NOTE
Normal resistance in both the primary and secondary windings in the coil is not a guarantee that the unit is working properly. Only an operational spark test can determine if a coil is producing an adequate spark from the input voltage. A Yamaha dealership or auto electrical repair shop may have the equipment to test the coil's output. If not, substitute a known good coil to see if the problem goes away.

6. If the resistance values are not as specified, replace the ignition coil.

Pulser Coil Resistance Check

1A. *Except RA1100 and WVT1100*—Remove the electric box from its mounting position as described in Chapter Nine. Then open the box to gain access to the electrical connectors.

1B. *RA1100 and WVT1100*—Remove the electric box cover from the electric box.

2A. *Except 760 and 1100 cc models*—Disconnect the white/red and black pulser coil bullet connectors at the stator plate-to-CDI wire harness.

2B. *RA1100 and WVT1100*—Disconnect the white/red, white/black, white/green and black pulser coil bullet connectors at the stator plate-to-CDI wire harness.

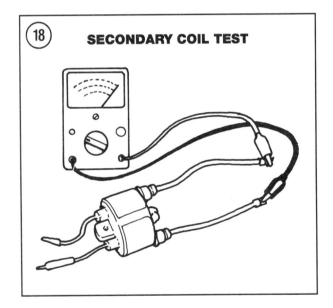

SECONDARY COIL TEST

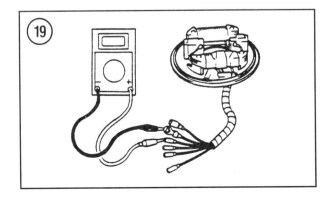

TROUBLESHOOTING

2C. *RA760 and WB760*—Disconnect the stator to CDI harness connector.

3A. *Except 760 and 1100 cc models*—With the ohmmeter set on the R × 10 or R × 100 scale (depending upon your particular model), measure the resistance between the 2 disconnected leads (**Figure 19**). Compare the reading to the specification in **Table 1**. If not within specification, replace the pulser coil as described in Chapter Nine.

3B. *RA1100 and WVT1100*—With the ohmmeter set on the R × 100 scale, check the resistance between the wires (**Figure 19**) indicated in **Table 1**. If the readings are not within specification, replace the pulser coil as described in Chapter Nine.

3C. *RA760 and WB760*—With an ohmmeter set on the R × 100 scale, check the resistance between the white/red and white/black leads on the coil side of the connector (**Figure 20**). Compare the reading to the specification given in **Table 1**. If not within specification, replace the pulser coil as described in Chapter Nine.

Charge Coil Resistance Check

1A. *Except 1100 cc models*—Remove the electric box from its mounting position as described in Chapter Nine. Then open the box to gain access to the electrical connectors.

1B. *RA1100 and WVT1100*—Remove the electric box cover from the electric box.

2A. *Except 760 and 1100 cc models*—Refer to **Table 1**, and disconnect the indicated charge coil connectors at the stator plate-to-CDI wire harness.

2B. *RA1100 and WVT1100*—Disconnect the black/red, brown/red, and blue connectors at the stator plate-to-CDI wire harness.

2C. *RA760 and WB760*—Disconnect the stator to CDI harness connector.

3A. *Except 760 and 1100 cc models*—With the ohmmeter set on the R × 100 scale (R × 10 for WR500 models), measure the resistance between the 2 disconnected leads (**Figure 21**). Compare the reading to the specification in **Table 1**. If not within specification, replace the charge coil as described in Chapter Nine.

3B. *RA1100 and WVT1100*—With the ohmmeter set on the R × 100 scale, check the resistance between the wires indicated in **Table 1** (**Figure 21**). If the readings are not within specification, replace the charge coil as described in Chapter Nine.

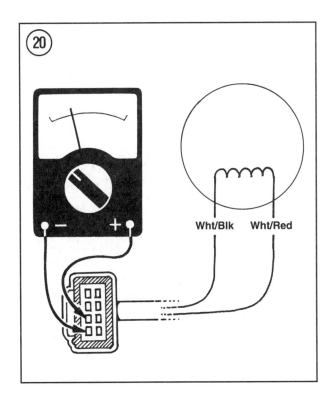

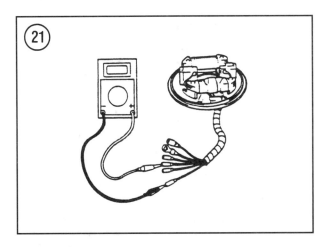

3C. *RA760 and WB760*: With an ohmmeter set on the R × 100 scale, check the resistance between the brown and blue leads on the coil side of the connector (**Figure 22**). Compare the reading to the specification given in **Table 1**. If not within specification, replace the pulser coil as described in Chapter Nine.

CDI Unit Resistance Test

The resistance values provided by Yamaha are based on the use of its multimeter (part No. YU-34899-A [SJ650, WRA650 and WRB650] or part No. YU-3112-A [all other models]). If another ohmmeter is used, the readings obtained may not agree with those specified due to the internal resistance of the ohmmeter. When switching between ohmmeter scales with an analog meter, always cross the test leads and zero the needle to ensure a precise reading.

1A. *Except RA1100 and WVT1100*—Remove the electric box from its mounting position as described in Chapter Nine. Then open the box to gain access to the electrical connectors.

1B. *RA1100 and WVT1100*: Remove the electric box cover from the electric box.

2. Disconnect the CDI bullet connectors.

3. Refer to the following figures for test connections and values for your model. Make

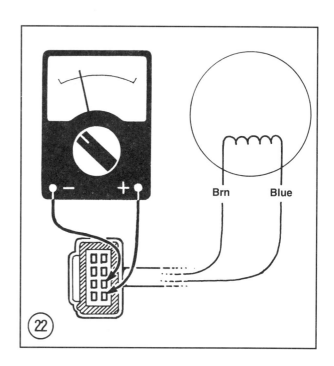

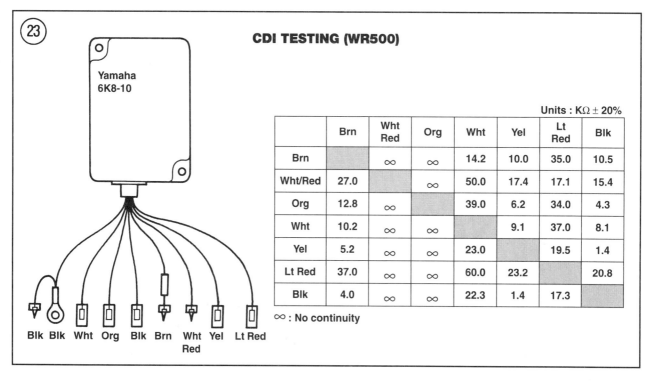

CDI TESTING (WR500)

Units : KΩ ± 20%

	Brn	Wht Red	Org	Wht	Yel	Lt Red	Blk
Brn		∞	∞	14.2	10.0	35.0	10.5
Wht/Red	27.0		∞	50.0	17.4	17.1	15.4
Org	12.8	∞		39.0	6.2	34.0	4.3
Wht	10.2	∞	∞		9.1	37.0	8.1
Yel	5.2	∞	∞	23.0		19.5	1.4
Lt Red	37.0	∞	∞	60.0	23.2		20.8
Blk	4.0	∞	∞	22.3	1.4	17.3	

∞ : No continuity

TROUBLESHOOTING

each connection and compare the meter reading to the stated value. If any of the meter readings differ from the stated values, replace the CDI unit.

a. **Figure 23**: WR500.
b. **Figure 24**: WR650.
c. **Figure 25**: FX700, SJ650, SJ700, WB700, WB700A, WRA650, WRA650A, WRA-700, WRB650, WRB650A and WRB700.
d. **Figure 26**: RA700, RA700A, RA700B, SJ700A and WVT700.
e. **Figure 27**: RA760 and WB760.

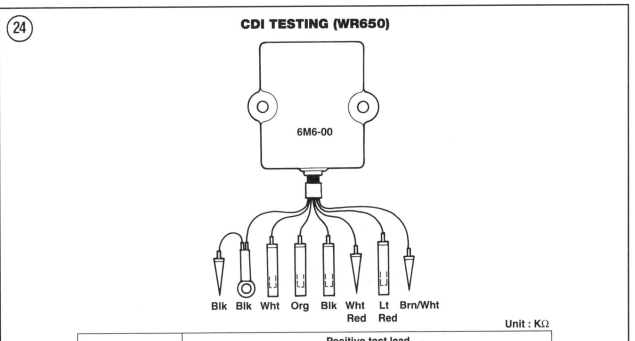

CHAPTER TWO

㉕ **CDI TESTING
(FX700, SJ650, SJ700, WB700, WB700A, WRA650,
WRA650A, WRA700, WRB650, WRB650A AND WRB700)**

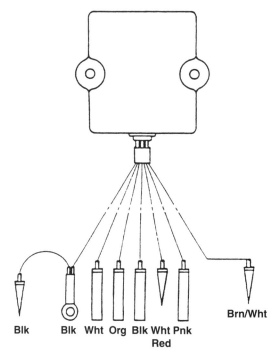

Blk   Blk Wht Org Blk Wht Pnk   Brn/Wht
              Red

Unit : KΩ

		Positive test lead						
		Blk	Wht	Org	Blk	Wht Red	Pnk	Brn Wht
Negative test lead	Blk		20 +20/−10	●	0	300 +300/−150	7 ± 4	4 ± 2
	Wht	∞		∞	∞	∞	∞	∞
	Org	●	●		●	●	●	●
	Blk	0	20 +20/−10	●		300 +300/−150	7 ± 4	4 ± 2
	Wht/Red	18 +18/−9	100 +100/−50	●	18 +18/−9		20 +20/−10	35 +35/−18
	Pnk	∞	∞	∞	∞	∞		∞
	Brn/Wht	40 +40/−20	30 +30/−15	●	40 +40/−20	500 +500/−250	100 +100/−50	

∞ : No continuity.

● : Indicates that the pointer deflects and then returns to the home position.

TROUBLESHOOTING

f. **Figure 28**: RA1100 and WVT1100.

FUEL SYSTEM

Many watercraft owners automatically assume that the carburetor is at fault if the engine does not run properly. While fuel system problems are not uncommon, carburetor adjustment is seldom the answer. In many cases, adjusting the carburetor only compounds the problem by making the engine run worse.

Fuel system troubleshooting should start at the fuel tank and work through the system, reserving the carburetor as the final point. Most fuel system problems result from an empty fuel tank, a

㉖

CDI TESTING (RA700, RA700A, RA700B, SJ700A, AND WVT700)

Unit : KΩ

		Negative test lead					
		Blk	Brn/Wht	Org	Pnk	Wht	Wht/Red
Positive test lead	Blk		2~6	●	3~11	10~40	150~600
	Brn/Wht	20~80		●	50~200	15~60	500
	Org	●	●		●	●	●
	Pnk	∞	∞	∞		∞	∞
	Wht	∞	∞	∞	∞		∞
	Wht/Red	9~36	17~70	●	10~40	50~200	

∞ : No continuity.

● : Indicates that the pointer deflects and then returns to the home position.

㉗

CDI TESTING (RA760 AND WB760)

Unit : KΩ

		Negative test lead							
		Wht	Pnk	Wht/Blk	Wht/Red	Org	Brn	Blu	Blk
Positive test lead	Wht		∞	3.8~16	9.5~4.0	11~45	80~400	3.4~14	3.8~16
	Pnk	7.5~35		17~70	22~100	40~300	70~1000	16~70	17~80
	Wht/Blk	10~45	∞		4.4~18	2~9	70~400	6~26	0~0.6
	Wht/Red	16~70	∞	4~17		8~35	70~400	13~60	4~17
	Org	∞	∞	∞	∞		∞	∞	∞
	Brn	26~150	∞	2.4~11	9~40	7.5~35		16~70	2.4~11
	Blu	26~150	∞	2.4~11	9~40	7.5~35	80~500		2.4~11
	Blk	10~45	∞	0~0.6	4.4~19	2~8.5	70~400	6~26	

∞ : No continuity.

plugged fuel filter or check valve, malfunctioning fuel pump or sour fuel. **Figure 29** provides a series of symptoms and causes that can be useful in localizing fuel system problems.

Carburetor chokes (**Figure 30**) can also present problems. A choke stuck in the off position will cause hard starting when cold; one that sticks on will result in a flooding condition. Check choke operation and adjustment as described in Chapter Three.

Refer to the following figure for your model for an outline of the fuel system:

a. **Figure 31**: WR500.
b. **Figure 32**: WR650.
c. **Figure 33**: SJ650.
d. **Figure 34**: SJ700 and SJ700A.
e. **Figure 35**: FX700.
f. **Figure 36**: WRA650, WRA650A and WRA700.
g. **Figure 37**: WRB650, WRB650A and WRB700.
h. **Figure 38**: WB700, WB700A and WB760.
i. **Figure 39**: RA700.
j. **Figure 40**: RA700A, RA700B, RA760 and RA1100.
k. **Figure 41**: WVT700 and WVT1100.

CDI TESTING (RA1100 AND WVT1100)

Unit : KΩ

Positive test lead \ Negative test lead	Blk	Blk/Org	Blk Red	Blk Wht	Blk/Yel	Brn	Blu	Pnk	Wht	Wht Blk	Wht Grn	Wht Red
Blk		280~420	14.4~21.6	280~420	280~420	∞	2.9~4.3	280~420	280~420	60~90	60~90	60~90
Blk/Org	∞		∞	∞	∞	∞	∞	∞	∞	∞	∞	∞
Blk Red	∞	∞		∞	∞	∞	∞	∞	∞	∞	∞	∞
Blk Wht	∞	∞	∞		∞	∞	∞	∞	∞	∞	∞	∞
Blk/Yel	∞	∞	∞	∞		∞	∞	∞	∞	∞	∞	∞
Brn	76~114	120~180	200~300	120~180	120~180		144~216	120~180	120~180	184~276	184~276	184~276
Blu	19.2~28.8	48~72	240~360	48~72	48~72	∞		56~84	45.6~68.4	168~252	168~252	168~252
Pnk	∞	∞	∞	∞	∞	∞	∞		∞	∞	∞	∞
Wht	∞	∞	∞	∞	∞	∞	∞	∞		∞	∞	∞
Wht Blk	200~300	280~420	400~600	280~420	280~420	∞	280~420	280~420	280~420		320~480	320~480
Wht Grn	200~300	280~420	400~600	280~420	280~420	∞	280~420	280~420	280~420	320~480		320~480
Wht Red	200~300	280~420	400~600	280~420	280~420	∞	280~420	280~420	280~420	320~480	320~480	

∞ : No continuity.

TROUBLESHOOTING

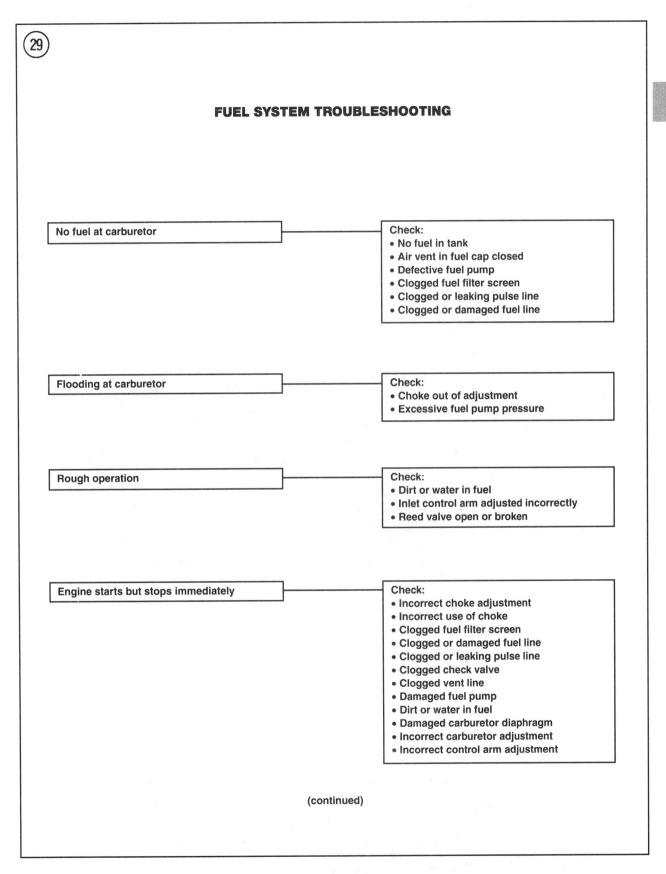

FUEL SYSTEM TROUBLESHOOTING

No fuel at carburetor → Check:
- No fuel in tank
- Air vent in fuel cap closed
- Defective fuel pump
- Clogged fuel filter screen
- Clogged or leaking pulse line
- Clogged or damaged fuel line

Flooding at carburetor → Check:
- Choke out of adjustment
- Excessive fuel pump pressure

Rough operation → Check:
- Dirt or water in fuel
- Inlet control arm adjusted incorrectly
- Reed valve open or broken

Engine starts but stops immediately → Check:
- Incorrect choke adjustment
- Incorrect use of choke
- Clogged fuel filter screen
- Clogged or damaged fuel line
- Clogged or leaking pulse line
- Clogged check valve
- Clogged vent line
- Damaged fuel pump
- Dirt or water in fuel
- Damaged carburetor diaphragm
- Incorrect carburetor adjustment
- Incorrect control arm adjustment

(continued)

CHAPTER TWO

㉙ (continued)

FUEL SYSTEM TROUBLESHOOTING (continued)

Engine misfires →
Check:
- Dirty carburetor
- Dirty or defective inlet seat or needle
- Choke out of adjustment
- Incorrect carburetor adjustment
- Clogged flame arrestor

Engine backfires →
Check:
- Poor quality fuel
- Air/fuel mixture too rich or too lean
- Incorrect carburetor adjustment

Engine preignition →
Check:
- Excessive oil in fuel
- Poor quality in fuel
- Lean carburetor mixture

Spark plug burns or fouls →
Check:
- Incorrect spark plug heat range
- Fuel mixture too rich
- Incorrect carburetor adjustment
- Poor quality fuel

High fuel consumption →
Check:
- Incorrect carburetor adjustment
- Clogged flame arrestor
- Clogged exhaust system
- Loose inlet seat and needle
- Defective inlet seat gasket
- Worn inlet seat and needle
- Foreign matter clogging inlet seat
- Leaks at fuel line connections

TROUBLESHOOTING

Troubleshooting

As a first step, visually check all of the fuel and vacuum hoses. Refer to the fuel system diagram for your model. Check all of the hoses for kinks, cracks or other damage. A leaking fuel hose will allow fuel to leak into the engine compartment and may result in a fire. Loose or damaged vacuum hoses will prevent fuel flow, thus causing fuel starvation and engine seizure. Replace all questionable or visibly damaged hoses.

If the previous inspection does not locate the problem, check fuel flow next. Remove the fuel tank cap and look into the tank. If there is fuel present, reinstall the cap securely. Locate the fuel hose connecting the fuel filter to the carburetor. Remove this hose and replace it with a clear hose with the same inside and outside diameter as the original hose. Connect the clear hose to the fuel filter and carburetor, making sure the hose bottoms out on both fittings. Operate the starter and watch for fuel flow through the clear hose.

If there is no fuel flow through the hose:
a. The fuel petcock may be shut off or blocked by foreign matter.
b. The fuel line may be stopped up or kinked.
c. One or more fuel filters may be blocked.
d. The fuel pump-to-engine pulse hose is blocked, damaged or disconnected.
e. The fuel pump may be defective.
f. Internal fuel tank cap or hose leakage.

If a good fuel flow is present:
a. Check the choke valve for proper operation.

b. The fuel may be contaminated with water.
c. Carburetor passages are blocked by dirt or other contamination.
d. On premix models, the fuel mixture in the tank is stale.
e. The reed valve is open or broken.
f. Check for a contaminated, defective or improperly adjusted inlet needle and seat.

Remove the clear fuel hose and install the original fuel hose. Make sure both hose ends bottom out on their fittings. Store the clear hose in your tool kit so you will have access to it for use later.

ENGINE

Engine problems are generally symptoms of something wrong in another system, such as ignition, fuel or starting system.

Overheating and Lack of Lubrication

Overheating and lack of lubrication cause the majority of engine mechanical problems. Engines used in water vehicles create a great deal of heat and are not designed to operate at a standstill; the cooling system does not circulate water at idle speed. Make sure the cooling system isn't clogged with sand or that the jet intake isn't clogged with weeds or other debris. See *Cooling System Flushing* in Chapter Three or *Jet Pump* in this chapter. Using a spark plug of the wrong heat range can burn a piston. Incorrect ignition timing, a faulty cooling system or an excessively lean fuel mixture can also cause engine overheating.

NOTE
Ignition timing is fixed (not adjustable) on all models. However, the CDI unit contains a high-speed spark retard circuit. The high-speed spark retard circuit retards ignition timing at high speed to prevent overheating and engine damage

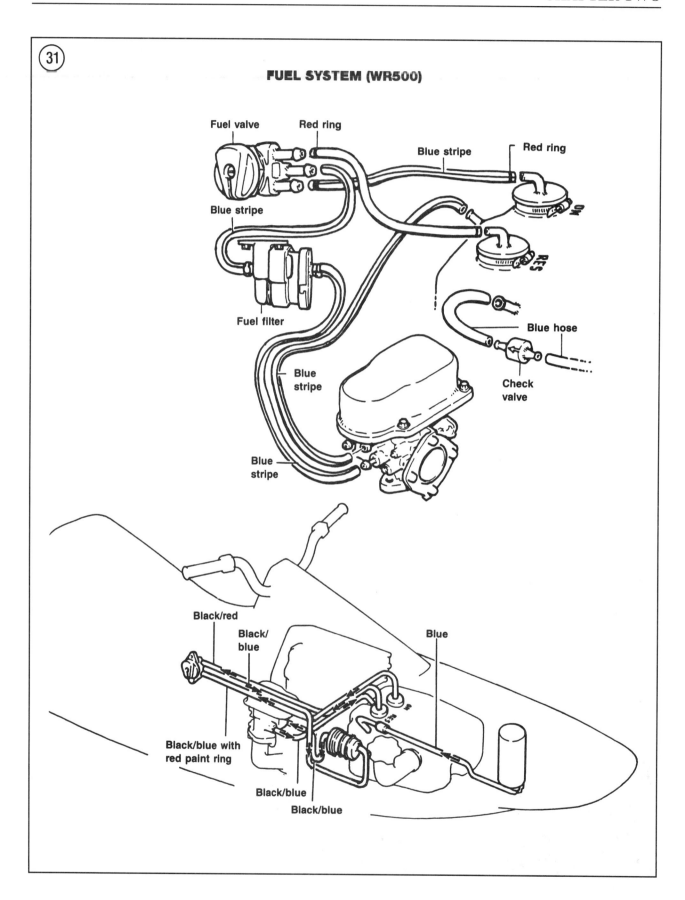

TROUBLESHOOTING

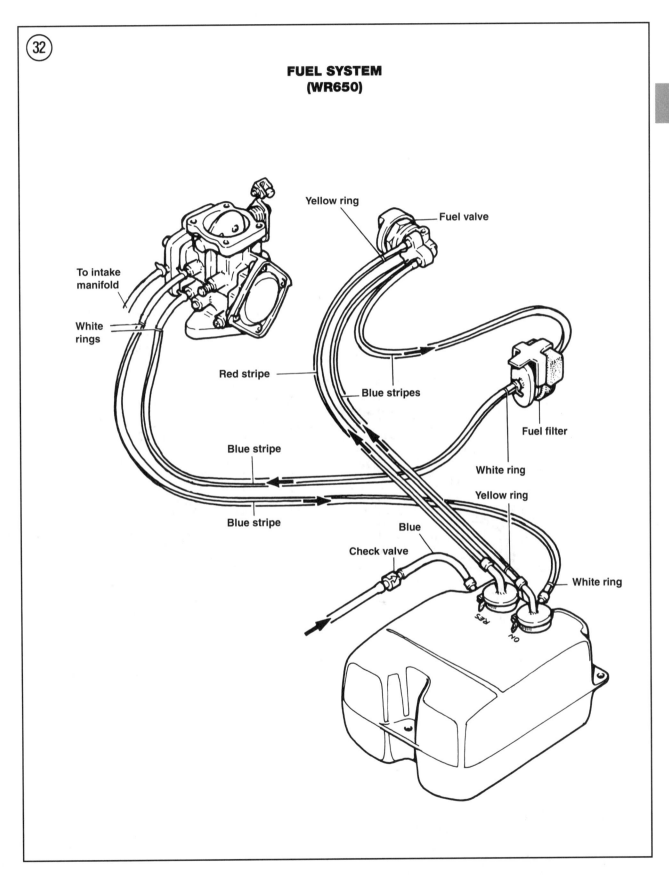

㉜

**FUEL SYSTEM
(WR650)**

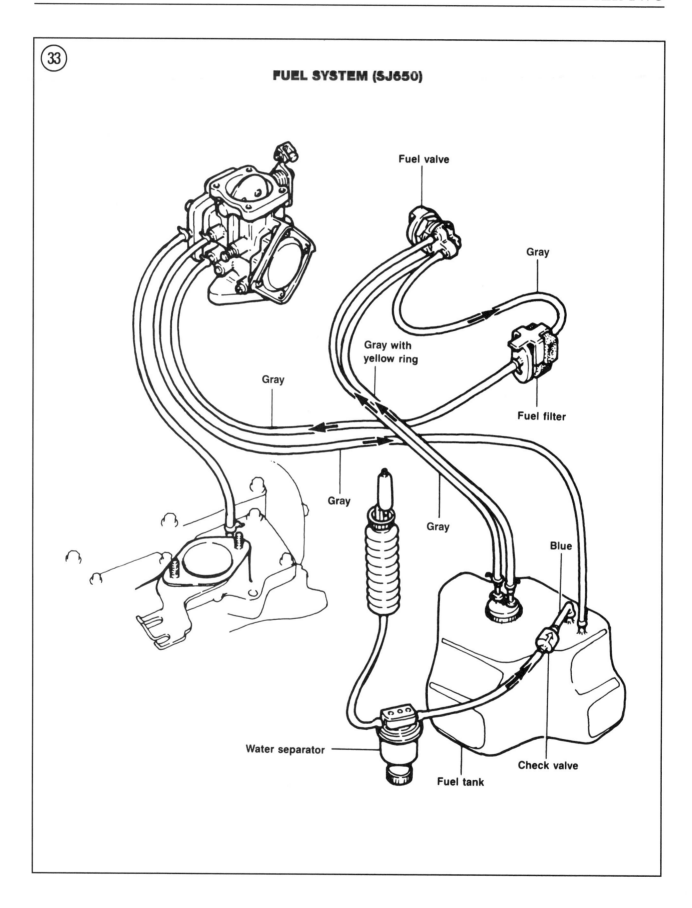

TROUBLESHOOTING

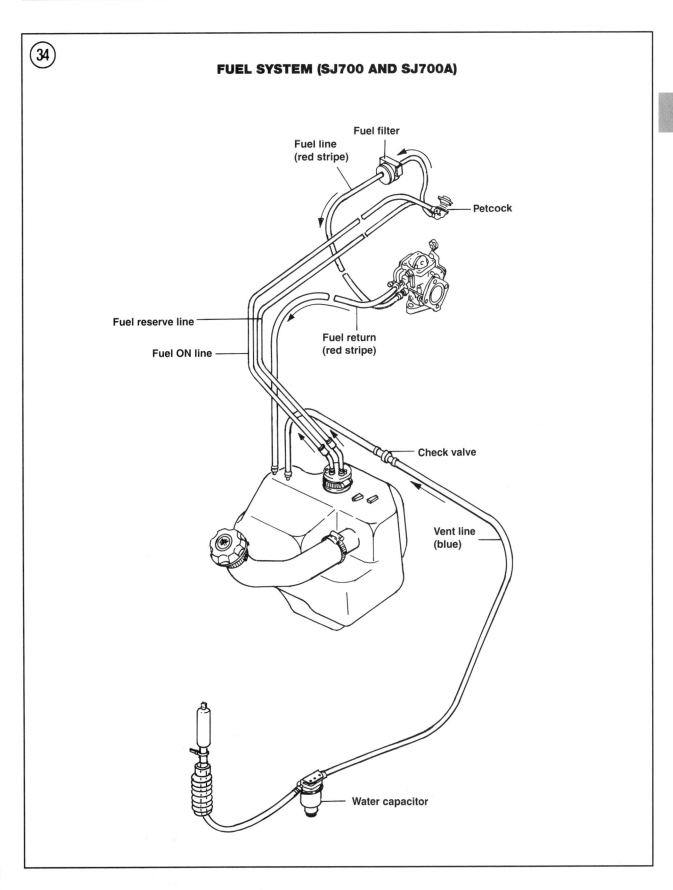

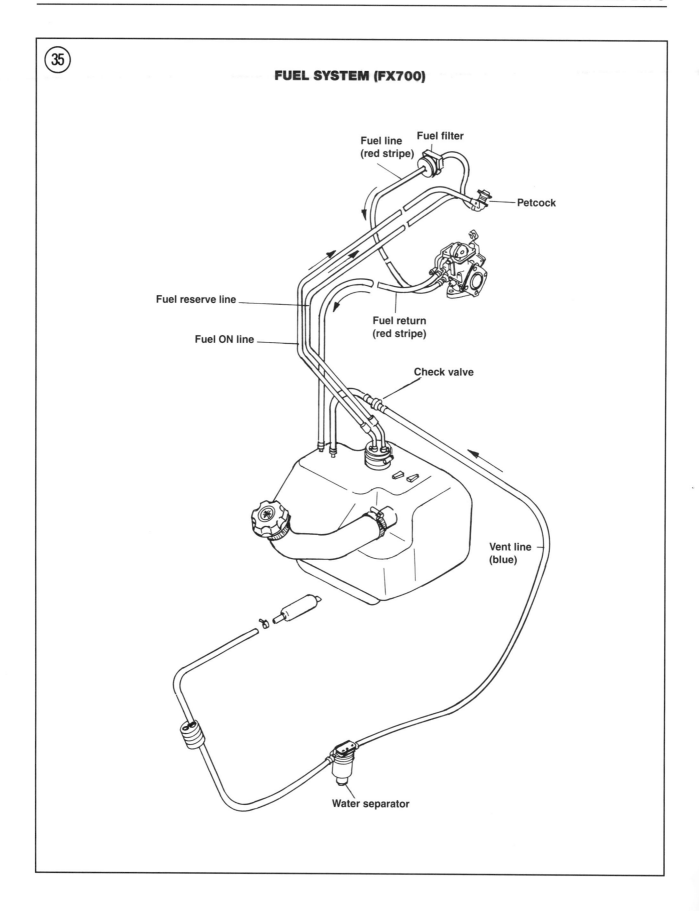

TROUBLESHOOTING

㊱

**FUEL SYSTEM
(WRA650, WRA650A, AND WRA700)**

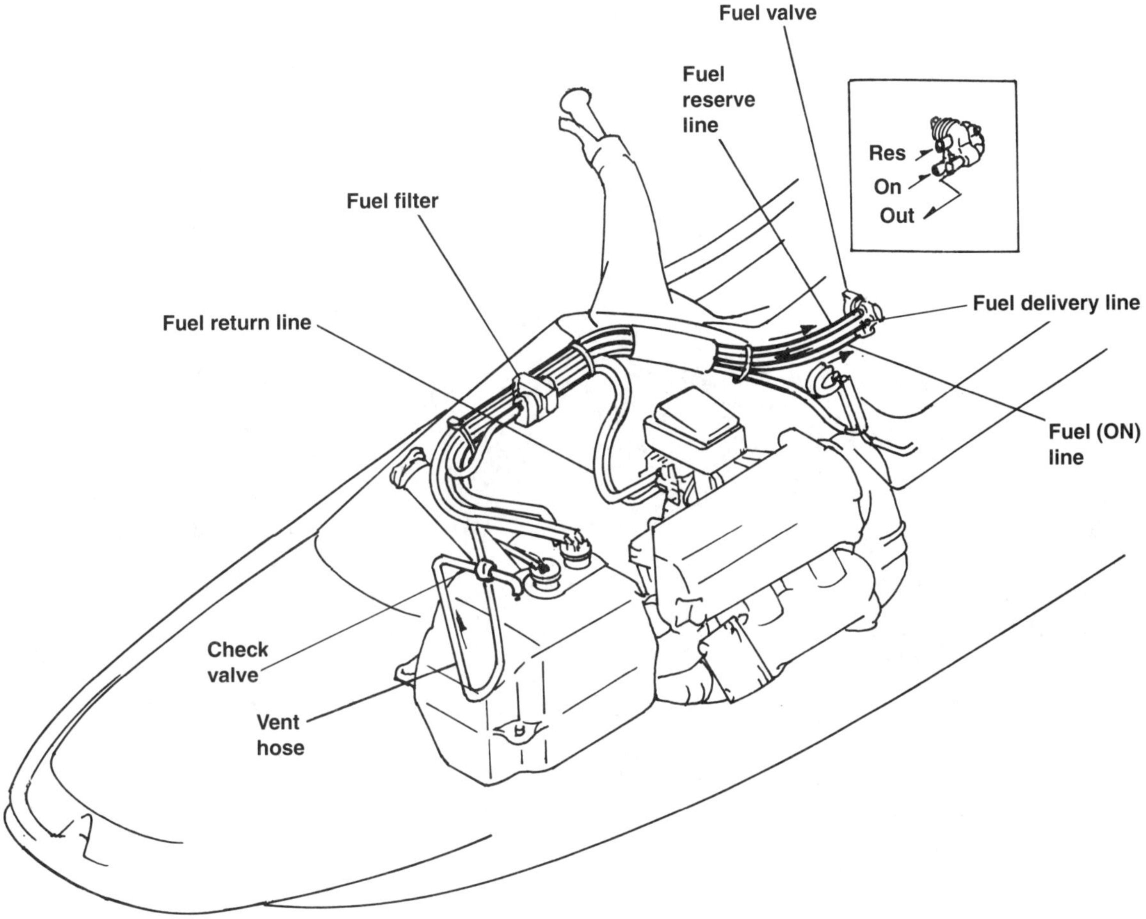

- Fuel valve
- Fuel reserve line
- Res / On / Out
- Fuel filter
- Fuel delivery line
- Fuel return line
- Fuel (ON) line
- Check valve
- Vent hose

CHAPTER TWO

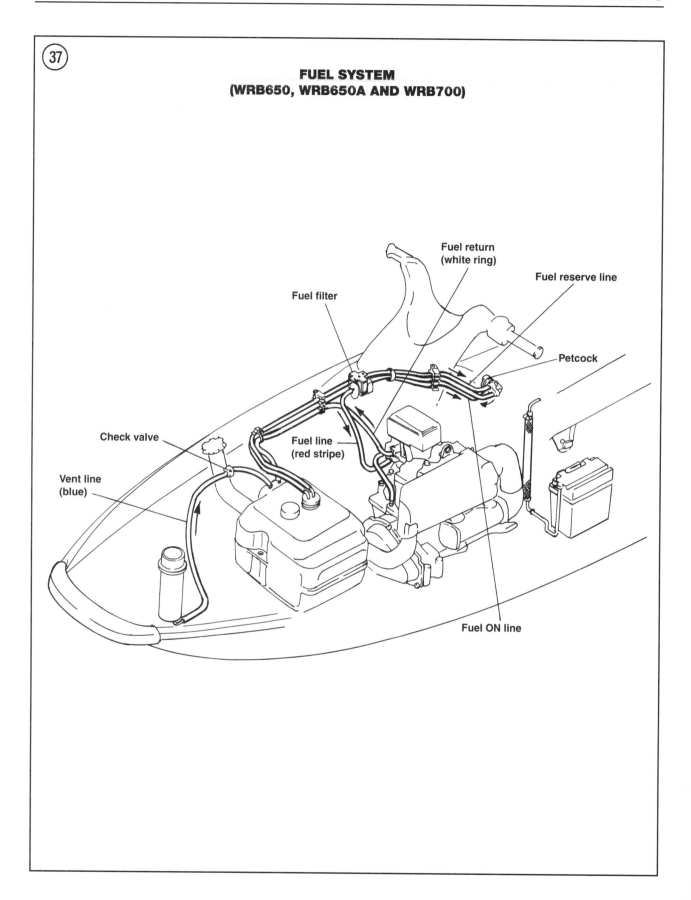

③⑦ FUEL SYSTEM (WRB650, WRB650A AND WRB700)

TROUBLESHOOTING

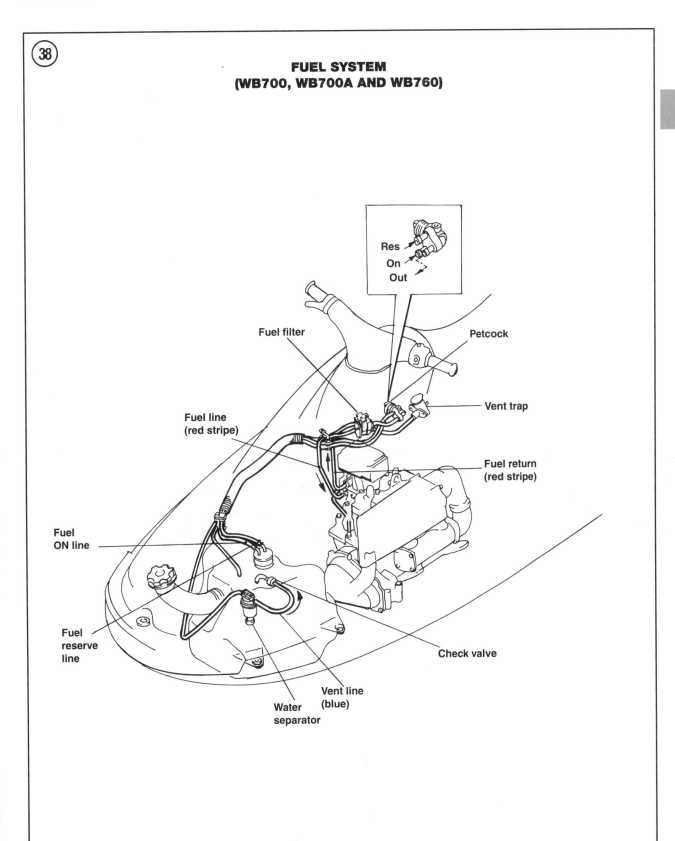

㊳ **FUEL SYSTEM (WB700, WB700A AND WB760)**

CHAPTER TWO

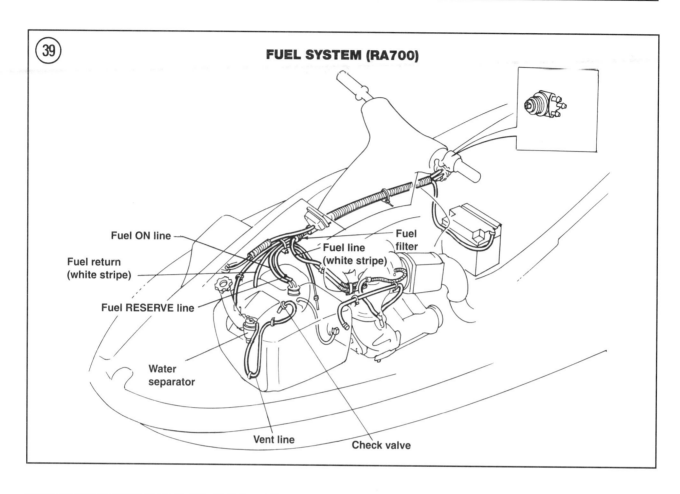

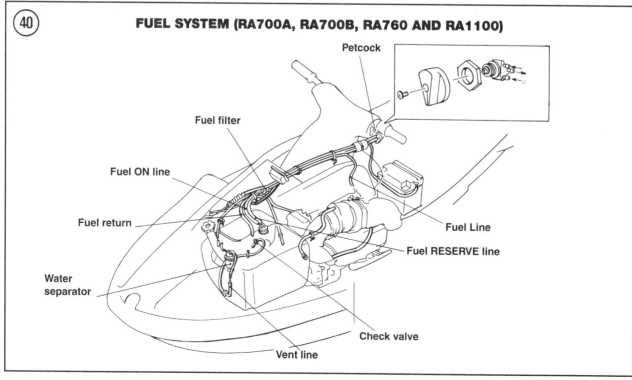

TROUBLESHOOTING

71

**FUEL SYSTEM
(WVT700 AND WVT1100)**

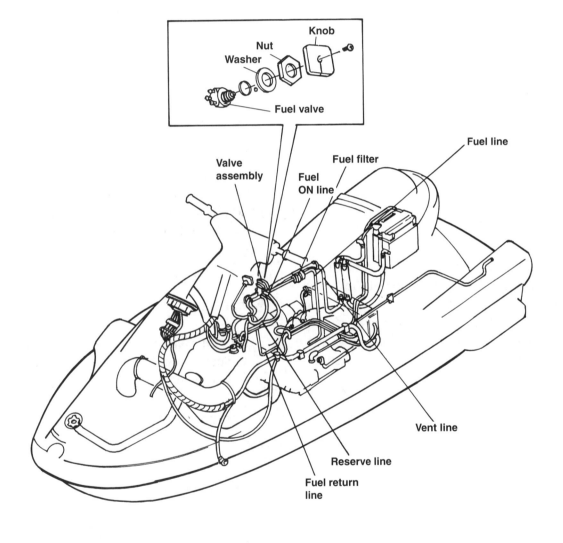

due to detonation. If the spark retard circuit malfunctions, ignition timing can continue to advance in proportion to engine speed causing overheating and possible engine damage.

Preignition

Preignition is the premature burning of fuel and is caused by hot spots in the combustion chamber (**Figure 42**). The fuel actually ignites before it is supposed to. Glowing deposits in the combustion chamber, inadequate cooling or overheated spark plugs can all cause preignition. This is first noticed in the form of a power loss. It will, however, eventually result in extended damage to the internal parts of the engine because of excessive combustion chamber temperature and pressure.

Detonation

Commonly called spark knock or fuel knock, detonation is the violent explosion of fuel in the combustion chamber instead of the controlled burn that occurs during normal combustion (**Figure 43**). Severe damage can result. Use of low octane gasoline is a common cause of detonation.

Even when high octane gasoline is used, detonation can still occur. Common other causes are over-advanced ignition timing, lean fuel mixture at or near full throttle, inadequate engine cooling, cross-firing of spark plugs, or the excessive accumulation of deposits on piston and combustion chamber.

Since the engine is covered, engine knock or detonation is likely to go unnoticed, especially at high engine speed when wind noise is also present. Such inaudible detonation, as it is called, is usually the cause when engine damage occurs for no apparent reason.

Poor Idling

A poor idle can be caused by improper carburetor adjustment or ignition system malfunction. Check the carburetor pulse and vent lines for an obstruction. Also check for loose carburetor mounting bolts or a faulty carburetor flange gasket.

Misfiring

Misfiring can result from a weak spark or a dirty spark plug. Check for fuel contamination. If misfiring occurs only under heavy load, as

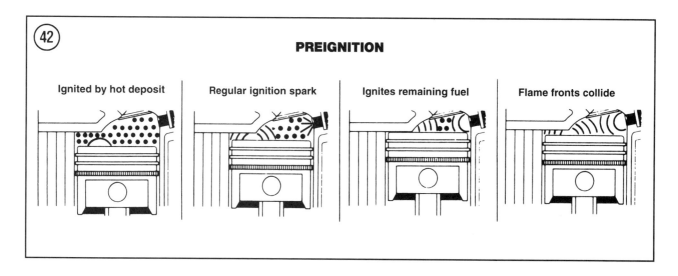

TROUBLESHOOTING

when accelerating, it is usually caused by a defective spark plug or spark plug wire.

Water Leakage in Cylinder

The fastest and easiest way to check for water leakage into a cylinder is to check the spark plugs. Water inside the combustion chamber during combustion will turn to steam and thoroughly clean the spark plug and combustion chamber. If one spark plug is very clean and the other plug(s) have normal deposits, water ingestion is possibly taking place on the cylinder with the clean plug.

Water ingestion can be verified by installing used spark plugs with normal deposits into each cylinder. Run the engine for 5-10 minutes or until warmed to normal operating temperature. Stop the engine, allow it to cool and remove the spark plugs. If one or more plugs are thoroughly clean, water leakage is probably occurring.

Flat Spots

If the engine seems to die momentarily when the throttle is opened and then recovers, check for a dirty or contaminated carburetor, water in the fuel or an excessively lean or rich low speed mixture.

Power Loss

Several factors can cause a lack of power and speed. Look for air leaks in the fuel line or fuel pump, a clogged fuel filter or a choke/throttle valve that does not operate properly. A piston or cylinder that is galling, incorrect piston clearance, or worn or sticky piston rings may be responsible. Look for loose bolts, defective gaskets or leaking machined mating surfaces on the cylinder head, cylinder or crankcase. Also check the outer crankcase seals and the crankshaft seals. Refer to *Two-Stroke Pressure Testing* in this chapter.

Exhaust fumes leaking within the engine compartment can slow and even stop the engine.

If the engine seems to operate correctly but you are experiencing performance related problems in the water, check the jet pump for pump case or impeller damage, weeds or other contamination in the jet pump. Also check for excessive impeller-to-pump case clearance.

Refer to **Figure 44** for a general listing of engine troubles.

NOTE
If the engine starts, idles and runs properly until engine speed reaches 2,500 rpm but then misfires and will not accelerate higher, stop the engine and disconnect the thermosensor wire bullet connectors in the electric box. Restart

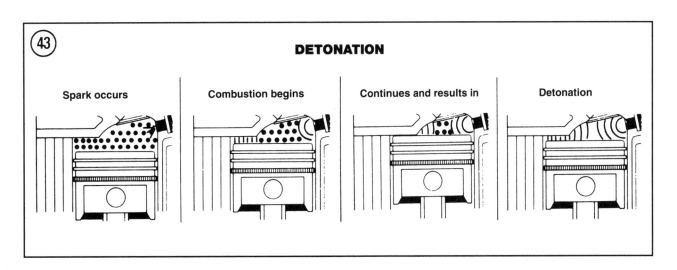

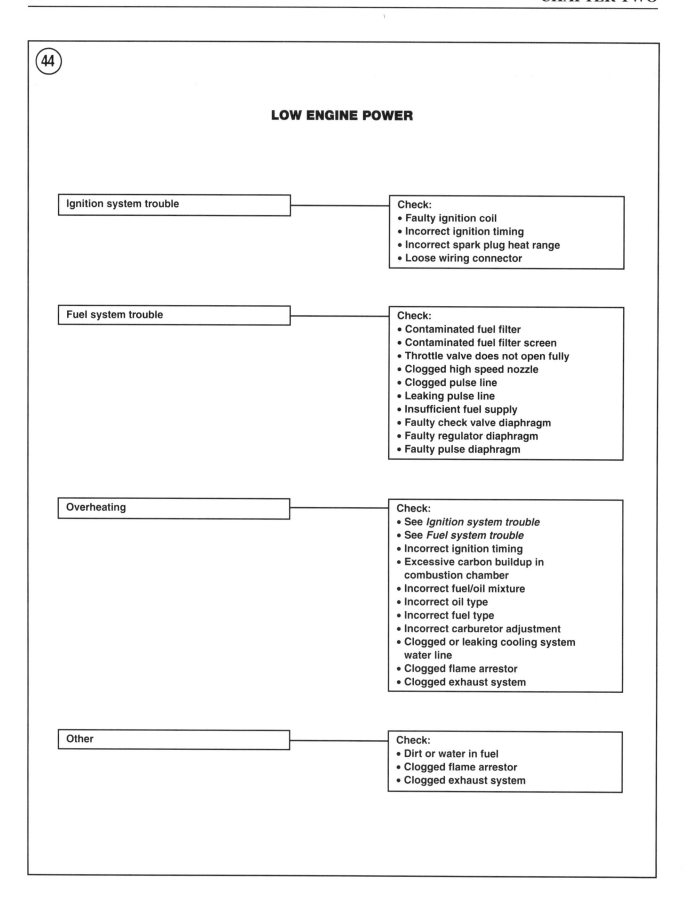

TROUBLESHOOTING

the engine and test ride it again. If the engine speed now climbs above 2,500 rpm with no sign of falling off, check the thermosensor (Chapter Nine) and CDI unit.

Piston Seizure

This is caused by one or more pistons with incorrect bore clearances, piston rings with an improper end gap, compression leak, incorrect type of oil, spark plug of the wrong heat range, incorrect ignition timing, the use of an incorrect fuel/oil mixture on premix models or an incorrectly operating oil injection pump on models so equipped. Overheating from any cause may result in piston seizure.

Excessive Vibration

Excessive vibration may be caused by loose engine mount bolts, worn engine or drive shaft bearings, a generally poor running engine, incorrect drive shaft alignment, loose jet pump bolts, a bent drive shaft or a damaged jet pump or impeller.

Engine Noises

Experience is needed to diagnose accurately in this area (**Figure 45**). Noises are difficult to differentiate and even harder to describe.

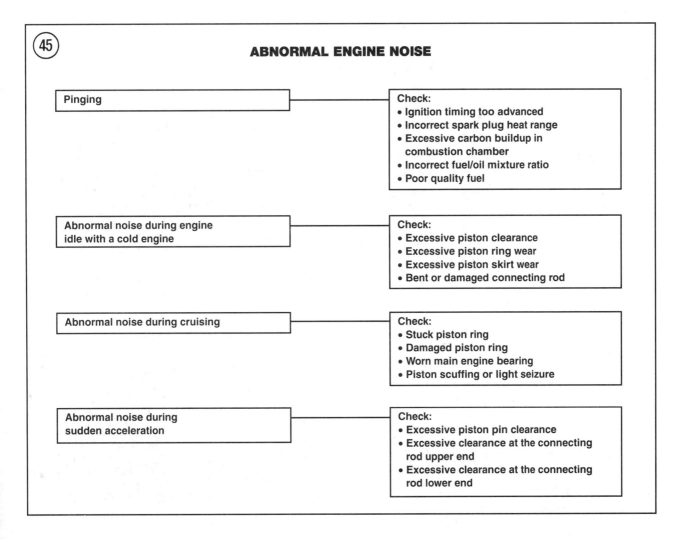

TWO-STROKE PRESSURE TESTING

Some 2-stroke engines are plagued by hard starting and generally poor running, for which there seems to be no cause. Carburetion and ignition may be good, and a compression test may show that all is well in the engine's upper end.

What a compression test does *not* show is lack of primary compression. In a 2-stroke engine, the crankcase must be alternately under pressure and vacuum. After the piston closes the intake port, further downward movement of the piston pressurizes the entrapped mixture so that it can rush quickly into the cylinder when the scavenging ports are opened. Upward piston movement creates a lower vacuum in the crankcase, enabling fuel/air mixture to pass in from the carburetor.

NOTE
The operational sequence of a 2-stroke engine is illustrated in Chapter One under Engine Principles.

If crankcase seals or cylinder gaskets leak, the crankcase cannot hold pressure or vacuum and proper engine operation becomes impossible. Any other source of leakage such as a defective cylinder base gasket or porous or cracked crankcase castings will result in the same conditions.

It is possible, however, to test for and isolate engine pressure leaks. The test is simple but requires special equipment. A typical 2-stroke pressure test kit is shown in **Figure 46**. Briefly, you seal off all natural engine openings, and then apply air pressure with the hand pump. If the engine does not hold air, a leak or leaks is indicated. Then it is only necessary to locate and repair all leaks.

The following procedure describes a typical pressure test.

1. Remove the carburetor as described in Chapter Eight but do not remove the intake manifold; the manifold could be the source of the leak.
2. Take a rubber plug and insert it tightly into the intake manifold.
3. Remove the exhaust assembly and block off the exhaust ports, using suitable adapters and fittings.
4. Disconnect the crankcase pulse hose and plug the pulse fitting.
5. Remove one spark plug and install the pressure gauge adapter into the spark plug hole. Connect the pressurizing lever and gauge to the pressure fitting installed where the spark plug was, then continue to squeeze the lever until the gauge indicates approximately 9 psi (62 kPa).

NOTE
The cylinders cannot be checked individually because of the seal installed on the crankshaft that fits between the crankcase chambers. When you pump up one cylinder you will also be pumping up the opposite cylinder.

6. Observe the pressure gauge. If the engine is in good condition, the pressure should not drop more than 1 psi (7 kPa) in several minutes. Any

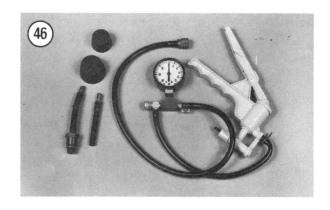

TROUBLESHOOTING

pressure loss of 1 psi (7 kPa) in one minute indicates serious sealing problems.

Before condemning the engine, first be sure that there are no leaks in the test equipment or sealing plugs. If the equipment shows no signs of leakage, go over the entire engine carefully. Large leaks can be heard; smaller ones can be found by going over every possible leakage source with a small brush and soap suds solution. Possible leakage points are listed below:

 a. Crankshaft seals (**Figure 47**).
 b. Spark plug(s).
 c. Cylinder head joint.
 d. Cylinder base joint.
 e. Intake manifold.
 f. Crankcase joint.

STEERING

The steering system should operate smoothly. **Figure 48** provides a series of causes that can be useful in localizing steering problems.

JET PUMP

Reduced jet thrust can occur gradually or all of a sudden and will cause cavitation, power loss or engine damage.

A gradual reduction of jet thrust can be hard to detect, and is usually caused by worn jet pump components. To detect damaged parts before they cause secondary engine damage, inspect the impeller and impeller housing for scuffing and damage. At the same time, impeller-to-housing clearance should be checked (Chapter Seven) as this clearance is critical to jet pump operation and overall vehicle performance.

The sudden loss of jet thrust is usually caused by weeds or other debris entering and clogging the pump's water intake area. Because engine cooling water is first picked up at the jet pump, a clogged hose can cause engine damage from overheating.

NOTE
All Yamaha water vehicles are equipped with an overheat warning system. This

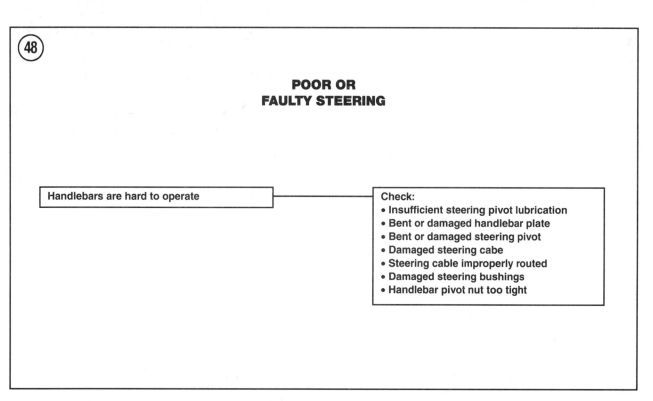

system consists of a thermosensor, related circuitry in the CDI unit and wiring connecting the thermosensor to the CDI unit. On most models, the thermosensor is mounted in the cylinder head (**Figure 49**). On RA760, RA1100, WB760 and WVT1100 models, the thermosensor is mounted on the rear boss of the exhaust housing (**Figure 50**). When the thermosensor detects an overheat situation, a signal is sent from the thermosensor to the CDI unit, grounding charge coil output. To prevent a sudden drop in engine speed, a delay circuit in the CDI unit allows engine speed to drop gradually until it is running at approximately 2,500-3,000 rpm. If you notice a drop in engine speed, beach the craft and check for a clogged jet intake or engine cooling hose.*

If you suspect weeds or other debris have clogged the pump intake area or if engine speed has started to drop, perform the following:

1. Beach the craft.
2. Remove the lanyard lock plate as a safety precaution to prevent accidental starting of the engine. See A, **Figure 2**.
3. Place a towel or piece of plastic next to the water vehicle and turn it onto its port (left) side. Support the handlebar to avoid damaging it.

CAUTION
Do not turn the water vehicle on its right (starboard) side or water in the exhaust system may drain into the engine's exhaust port and cause serious internal damage.

4. Remove the ride plate (Chapter Seven) to gain access to the intake area. Then remove weeds or other debris from the impeller/drive shaft area. If weeds, ski rope or other debris are tightly trapped around the drive shaft or impeller, partial disassembly of the jet pump may be required. Refer to Chapter Seven for jet pump service.

5. Reinstall the ride plate (Chapter Seven) and turn the vehicle right side up.

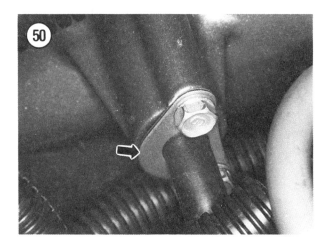

TROUBLESHOOTING

Table 1 STATOR PLATE COIL SPECIFICATIONS*

Model	Pulser Coil (ohms)	Charge Coil (ohms)	Lighting Coil (ohms)
WR500	White/red to black: 102 ± 10%	Brown to black: 90 ± 10%	Green to green: 0.10 ± 10%
RA700, RA700A, RA700B, SJ700A, WVT700	White/red to black: 14 ± 10%	Black/white to black: 553 ± 10%	Green to green: 1.27 ± 10%
RA1100	White/red to black: 310 ± 20% White/black to black: 310 ± 20% White/green to black: 310 ± 20%	Brown/red to brown: 215 ± 20% Brown/red to blue: 820 ± 20%	Green to green: 0.70 ± 10%
RA760, WB760	White/red to white/black: 495 ± 10%	Brown to blue: 352 ± 10%	Green to green: 1.27 ± 10%
All other models	White/red to black: 14 ± 10%	Brown/white to black: 365 ± 10%	Green to green: 0.90 ± 10%

*Perform tests with coil at 68° F (20° C).

Table 2 IGNITION COIL SPECIFICATIONS*

	Primary winding (ohms)	Secondary winding (ohms)
1993 WR500	0.15	3,500
All other 1993 models	0.092	4,100
All 1994 models	0.092	4,100
1995 RA700, RA700A, WVT700	0.092	22,500
RA1100	0.21	3,400
All other 1995 models	0.092	21,400
1996 RA700, RA700A, RA760, WB760, WVT700	0.092	22,500
RA1100, WVT1100	0.21	3,400
SJ700	0.92	4,100
All other 1996 models	0.092	21,400

*Perform tests with coil at 68° F (20° C).

Chapter Three

Lubrication, Maintenance and Tune-up

This chapter covers all of the regular maintenance required to keep your Yamaha water vehicle in top shape. Regular maintenance is the best guarantee of a trouble-free, long lasting vehicle. Because all of the Yamaha water vehicles are high-performance machines, proper lubrication, maintenance and tune-ups are important ways in which you can maintain a high level of performance, extend engine life and extract the maximum economy of operation.

You can do your own lubrication, maintenance and tune-ups if you follow the correct procedures and use common sense. Always remember that engine damage can result from improper tuning and adjustment. In addition, where special tools or testers are called for during a particular maintenance or adjustment procedure, the tool should be used or you should refer service to a qualified dealer or repair shop.

The following information is based on recommendations from Yamaha that will help you keep your watercraft operating at its peak level.

Table 1 and **Table 2** list maintenance schedules for models covered in this manual. **Tables 1-11** are at the end of the chapter.

NOTE
Due to the number of models and years covered in this book, be sure to follow the correct procedure and specifications for your specific model and year. Also use the correct quantity and type of fluid as indicated in the tables.

OPERATIONAL CHECKLIST

An important part of water vehicle maintenance and operation is the preparation given to the vehicle before and after riding it.

Pre-ride

Before starting and riding your water vehicle, remove the engine cover and check the following:

LUBRICATION, MAINTENANCE AND TUNE-UP

NOTE
Removing the engine cover serves a dual purpose. Besides allowing you access to the engine components, removing the cover ventilates the engine compartment of all fuel vapor. This step should always be performed prior to starting and riding your water vehicle.

1. Release vapor pressure from the fuel tank by loosening the fuel filler cap (**Figure 1**). When the pressure has been released (sound diminishes), tighten the cap securely.

CAUTION
Do not turn the water vehicle on its right side when performing Step 2 (if necessary) or water in the exhaust system may drain into the engine's exhaust ports and cause serious engine damage.

2. Excessive water in the engine compartment can splash onto engine components and may cause operating problems. If your craft is equipped with a hull drain plug, remove the plug and lift the front of the hull and allow the water to run out the drain plug port. If your model does not have a hull drain plug, turn the craft onto its *left* side. Precautions should be taken to avoid damaging the hull, engine hood or steering pole when turning the craft on its side. When all of the water has drained out, reinstall the drain plug or turn the hull upright.

3. If your craft is equipped with a drain plug and you did not remove it in Step 2, check it now for tightness.

4. Check the electrolyte level in the battery (**Figure 2**). Refill with distilled water as described in Chapter Nine.

CAUTION
*If the battery was serviced, check the battery vent tube (**Figure 3**) for proper routing after repositioning the battery.*

5. Check all fuel hoses for leakage, loose or damaged conditions. Do not operate the water vehicle until all leaks are repaired and until any spilled fuel inside the hull is thoroughly wiped up. Refer to Chapter Eight for a schematic diagram of your model's fuel system.

6. Tighten any loose bolts, nuts or hose clamps. Pay special attention to the following:

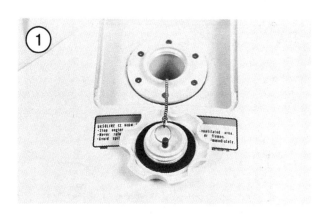

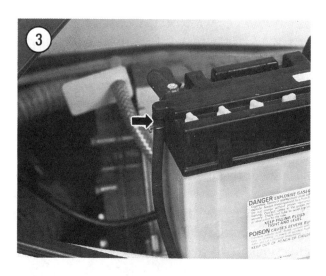

a. Fuel hoses and tank outlet retainer nut. See **Figure 4**.
b. Fuel (A, **Figure 5**) and oil tank hold down straps.
c. Cooling system and bilge hoses.
d. Exhaust system clamps and hoses (B, **Figure 5**).
e. Oil injection hoses and clamps (on oil injection models). See Chapter Ten.

7. *Oil injection models*—Check that the engine oil tank is full (**Figure 6**). If necessary, refill as described in this chapter.
8. Check the fuel level and refill as required.
9. Operate the throttle lever, and make sure it returns to the fully closed position when it is released (**Figure 7**). Check throttle movement at the carburetor also.
10. Check the handlebar (**Figure 8**) for tightness. Grasp each of the hand grips tightly and try to twist them on the bar; both grips must be tight. Then check the washer and screw (**Figure 9**) on the end of each grip for tightness. If a screw is loose, remove it and the washer. Then reinstall the washer and screw, first applying Loctite 242 (blue) to the screw threads. Tighten the screw securely.
11. Turn the handlebars from side-to-side and check the operation of the jet pump steering nozzle (**Figure 10**); make sure it moves from side-to-side corresponding to handlebar movement.
12. *WRA650, WRA650A, WRA700, WVT700 and WVT1100*—Check for proper shift operation as described in this chapter.
13. Remove any weeds or foreign objects from the water intake, jet pump and drive shaft. Check the ride plate and intake grate (**Figure 11**) bolts for tightness. If loose, tighten bolts securely.
14. Inspect the hull for damage.
15. Check that the seat is positioned and secured properly.
16. Remove the fire extinguisher from its compartment. See **Figure 12**, typical. Follow the directions listed by the fire extinguisher manu-

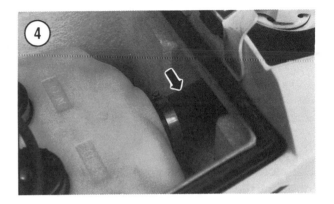

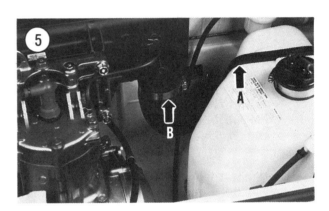

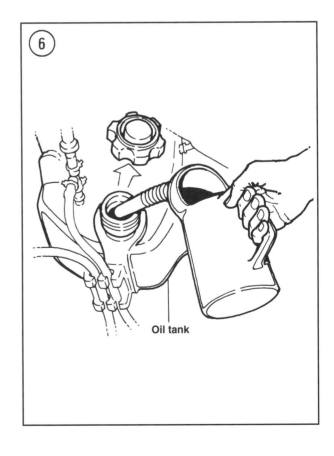

LUBRICATION, MAINTENANCE AND TUNE-UP

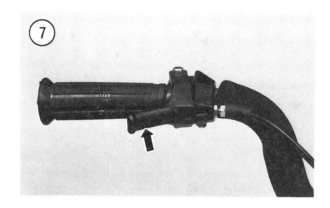

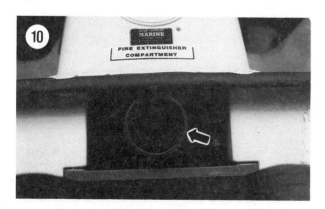

facturer to determine the charge or condition of the extinguisher. Replace or replenish the extinguisher as required. Check the O-ring or seal on the fire extinguisher compartment lid for deterioration or damage; replace if necessary.

NOTE
A fire extinguisher is not standard equipment on any Yamaha water vehicle. Before operating the vehicle, purchase a fire extinguisher that meets U.S. Coast Guard classification B-1 with a minimum charge capacity of 2 lb. (0.9 kg). This type and size of extinguisher is required to be onboard your craft whenever you operate in waters under U.S. Coast Guard jurisdiction. In addition, most local and state boating laws require that a U.S. Coast Guard approved fire extinguisher be onboard your craft when operating in their waters.

17. Check the condition of the engine cover seal and replace it if worn or damaged.

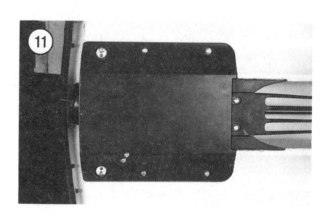

18. Check the water separator for accumulated water (**Figure 13**), on models so equipped. If there is water in the separator, place a clean container underneath the separator. Then remove the drain plug from the bottom of the separator and allow the water to drain into the container. When the separator is empty of water, reinstall the drain plug and tighten securely. Dispose of the water in the container.
19. Reinstall the engine cover, making sure that it is securely fastened.
20. After launching the craft, check the starter switch and stop switch (**Figure 14**) for proper operation. Do not operate the craft if either switch is operating incorrectly.
21. Be sure that water is flowing from the cooling water pilot outlet.

End of Day Checklist

Before putting your water vehicle away for the day, complete the following procedures:
1. After taking the water vehicle out of the water, lift the rear end 10 in. (25 cm) or more to allow water in the expansion chamber to drain away from the engine.
2. Flush the cooling system as described in this chapter.
3. With the watercraft out of the water, clear excess water out of the exhaust system by starting the engine and running it for *no more than 15 seconds*. During this short period, the throttle should be operated in quick bursts ranging from idle to 3/4 throttle. Stop the engine and turn the fuel valve OFF (**Figure 15**, typical).

WARNING
The exhaust gases are poisonous. Do not run the engine in a closed area. Make sure there is plenty of ventilation.

CAUTION
Do not run the engine for more than 15 seconds without a supply of cooling water or the rubber parts of the exhaust system will be damaged. Prolonged running without coolant will cause serious engine damage. Do not operate the engine at maximum speed out of the water.

4. Rinse the water vehicle off with clean freshwater.
5. Remove the engine cover.

NOTE
Cover the carburetor with a plastic cover when performing Step 6.

6. Lightly rinse the engine compartment with clean freshwater. Do not use a powerful water flow.
7. While all models are equipped with an automatic bilge draining assembly that removes water from the engine compartment during operation, some water will still collect in the bilge. Note the following to remove water in the bilge:
   a. Models equipped with a hull drain plug: Remove the hull drain plug and then raise the front of the hull to drain all water in the engine compartment. Check the drain plug threads for contamination and then reinstall the drain plug, making sure to tighten it securely.

CAUTION
Do not turn the water vehicle on its right side (substep b) or water in the exhaust system will drain into the engine's exhaust port and cause serious engine damage.

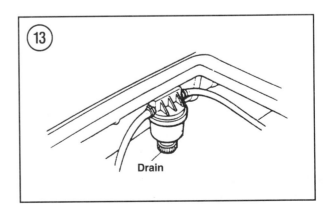

LUBRICATION, MAINTENANCE AND TUNE-UP

b. Models without a hull drain plug: Turn the craft onto its *left* side. Precautions should be taken to avoid damaging the hull, engine hood or steering pole when turning the craft on its side. When all of the water has drained out, turn the hull upright.

8. Wipe the engine and bilge areas with a clean-dry rag to absorb as much moisture on these components as possible.

9. Spray Yamaha Silicone Protectant and Lubricant or a similar rust inhibitor onto all metallic components to help reduce corrosion buildup.

10. Reinstall the engine cover.

BREAK-IN PROCEDURE

Following cylinder service (boring, honing and new rings) and major lower end work, the engine should be broken in just as if it were new. The performance and service life of the engine depends greatly on a careful and sensible break-in.

For the first 2 tanks of fuel, no more than 1/2 throttle should be used and the speed should be varied as much as possible. Avoid prolonged steady running at one speed, no matter how moderate. Also avoid hard acceleration.

To ensure adequate lubrication on non-oil injection models, a 25:1 gas/oil ratio should be used. After 2 tanks of fuel have been used and the break-in completed, you can resume engine operation at the standard 50:1 ratio.

On oil injection models, use a 50:1 gas/oil mixture *together* with the oil supplied by the injection system. Throughout the break-in period, check the oil injection reservoir tank to make sure the injection system is working properly (oil level diminishing). After using 2 tanks of fuel, resume engine operation with unmixed gasoline (used together with the oil injection system).

After engine break-in is complete, perform the *10-Hour Inspection* as described in the following section. In addition, retighten the cylinder head nuts as described in this chapter.

NOTE
After the break-in is complete, install new spark plugs as described in this chapter.

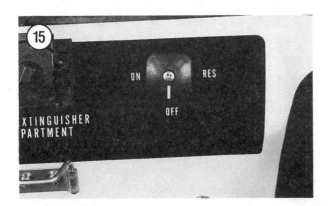

10-HOUR INSPECTION

Yamaha lists an initial 10-hour inspection that is to be performed on a new water vehicle after the first 10 hours of operation. The 10-hour checks are listed in **Table 1**. Repeat these checks whenever the engine top-end or bottom-end is overhauled or the engine removed from the hull. Likewise, perform steering inspection procedures after major service is performed to these components.

LUBRICATION

Lubrication intervals are listed in **Table 2**.

> *WARNING*
> *A serious fire hazard always exists around gasoline. Do not allow any smoking in areas where fuel is being mixed or while refueling your machine. Always have a fire extinguisher, rated for gasoline and electrical fires, within reach.*

Proper Fuel Selection

Two-stroke engines are lubricated by mixing oil with the fuel. The various components of the engine are thus lubricated as the fuel/oil mixture passes through the crankcase and cylinders. Since the fuel/oil mixture serves the dual function of producing combustion and distributing the lubrication, the use of low octane marine white gasoline should be avoided. Such gasoline has a tendency to cause ring sticking and port plugging.

Yamaha recommends the use of unleaded gasoline with a minimum posted pump rating of 86 to prevent engine knock and ensure proper operation.

Sour Fuel

Do not store fuel for more than 2-3 weeks (under ideal conditions). Gasoline forms gum and varnish deposits as it ages. Such fuel will

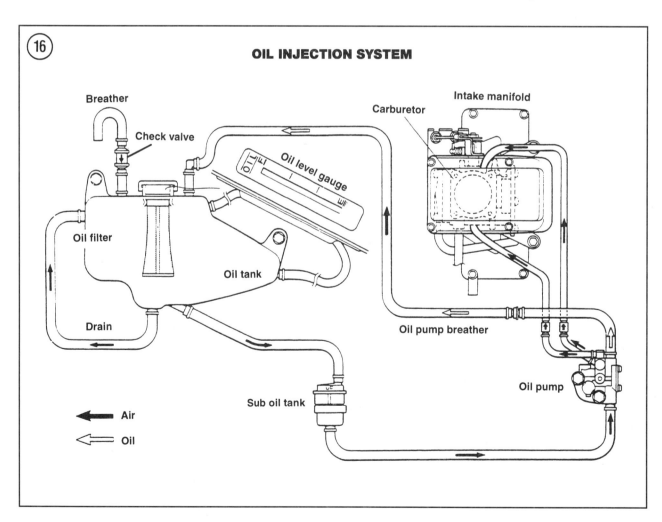

LUBRICATION, MAINTENANCE AND TUNE-UP

cause carburetion and engine deposit trouble that will result in engine starting problems. On premix models, make sure to keep the fuel tank cap tight; otherwise, some of the fuel will evaporate and result in a too rich oil mixture.

Lubrication System Identification

Refer to **Table 3** for engine lubrication system identification for your model.

Recommended Fuel Mixture (Oil Injection Models)

An oil injection system is used on many Yamaha models. A mechanical pump driven by the crankshaft is mounted on the flywheel cover located at the front of the engine. The oil pump hose routing diagram is shown in **Figure 16**. Oil added to the oil tank first flows through an internal filter and then settles into the oil tank. The main oil line connects the oil tank to the suboil tank and then to the oil pump. At the oil pump, 2 hoses (3 on the RA1100 and WVT1000) connect the pump to a nozzle on each intake manifold, thereby providing each cylinder with oil. During engine operation, oil is automatically injected into the engine at a variable ratio depending on engine rpm.

Oil capacity in the tank can be monitored by the oil level gauge on the starboard deck step or in the multifunction meter. Be sure to locate the oil gauge for your particular model.

To refill the oil tank, perform the following:

a. Remove the engine cover.

b. Open and remove the oil tank cap (**Figure 6**).

c. Pour in the required amount of Yamalube Two-Cycle Outboard Oil. Fill the oil tank until the oil level is approximately 25 mm (1 in.) from the top of the tank.

CAUTION
*For the oil filter to work properly, it must be installed in the oil tank when the tank is being filled. Do not add oil to the tank when the oil filter is removed. Debris or contamination may clog up the oil pump passages and cause engine damage. Service the oil filter at specified intervals (**Table 1**) and replace it if it is torn or otherwise damaged. Refer to **Oil Filter Inspection/Replacement** in this chapter.*

d. Reinstall the oil tank cap, first making sure the seal in the cap is in position. Tighten the cap securely.

e. Close and secure the engine cover.

Recommended Fuel Mixture (All Models Without Oil Injection)

WR500, WR650, SJ650, SJ700, SJ700A and FX700 models are lubricated by oil mixed with gasoline. **Table 4** lists the recommended 2-stroke oil. **Table 5** lists premix ratios for all models. Fuel tank capacity for each model is listed in **Table 6**.

CAUTION
Do not, under any circumstances, use multigrade oil or any oil designed for use in 4-stroke engines. Four-stroke engine oil will not mix well with gasoline and will not burn in the combustion chamber like 2-stroke oil will. The use of 4-stroke oil will result in piston scoring, ring sticking and bearing failure.

Correct Fuel Mixing

Mix the fuel and oil outdoors or in a well-ventilated indoor location. Combine the fluids in a separate container and pour the mixture into the fuel tank after it is properly mixed.

WARNING
Gasoline is an extreme fire hazard. Never use gasoline near sparks, heat or flame. Do not smoke while mixing fuel.

Using less than the specified amount of oil can result in insufficient lubrication and serious engine damage. Using more oil than specified causes spark plug fouling, erratic carburetion, excessive smoking and rapid carbon accumulation which can cause preignition and detonation.

Cleanliness is of prime importance. Even a very small particle of dirt can cause carburetion problems. Always use fresh unleaded gasoline with an octane rating of 86 or higher. Gum and varnish deposits tend to form in gasoline stored in a tank for any length of time. Use of sour fuel can result in carburetor problems and spark plug fouling.

Mix the oil and gasoline thoroughly in a separate clean, sealable container larger than the quantity being mixed to allow room for agitation. Always measure the quantities exactly. See **Table 4** and **Table 6**.

Use a discarded baby bottle with graduations in cubic centimeters (cc) or fluid ounces (oz.) on the side. Pour the required amount of oil into the mixing container and add approximately 1/2 the required amount of gasoline. Agitate the mixture thoroughly, then add the remaining fuel and agitate again until well mixed.

To keep contaminants out of the fuel system, use a funnel with a filter when pouring the fuel into the tank.

Consistent Fuel Mixtures

The carburetor idle adjustment is sensitive to fuel mixture variations which result from the use of different oils and gasolines or from inaccurate measuring and mixing. This may require carburetor adjustment. To prevent the necessity for constant readjustment of the carburetor from one batch of fuel to the next, always be consistent. Prepare each batch of fuel exactly the same as previous ones.

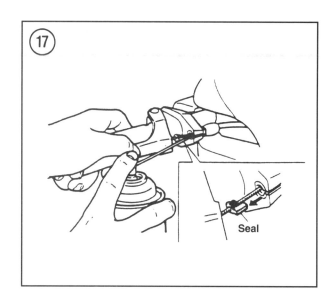

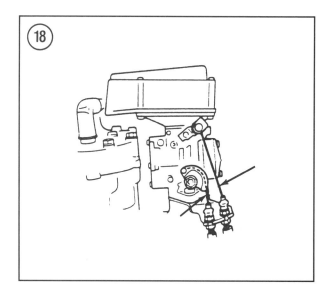

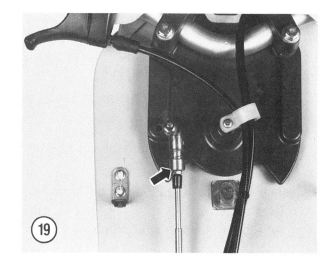

LUBRICATION, MAINTENANCE AND TUNE-UP

General Lubrication

Lubricate the following components at the intervals specified in **Table 2**. Use Yamaha Marine Grease (or equivalent) where grease is called for. Use WD-40, LPS or an equivalent penetrating rust inhibitor where a rust inhibitor is called for.

1. Remove the engine cover.

2. Push the throttle lever (**Figure 17**) and remove the cable seal from the housing, if so equipped. Spray rust inhibitor along the cable and into the cable housing. Refit the seal into the housing (if used) and release the throttle lever.

3. Wipe marine grease along the exposed inner cable on both carburetor cables (**Figure 18**). Then operate the throttle lever to distribute the grease evenly.

4. Lubricate the steering cable as follows:

   a. Locate the steering cable ball joint on the handlebar bracket (**Figure 19**). Push the cable sleeve rearward then pull the cable sleeve off the ball joint; release the cable sleeve after it is free of the ball joint. Repeat to disconnect the cable from the steering nozzle (**Figure 20**).

   b. At the front of the cable, pull the inner cable out of the outer cable to expose as much of the inner cable as possible. Wipe the exposed inner cable with marine grease. Repeat for opposite cable end.

   c. Lubricate the cable sleeve (portion which mates with ball) and the ball joint with marine grease. Repeat for both cable ends.

   d. Reconnect the cable by opening the cable sleeve and then inserting the sleeve opening over the ball joint. Release the sleeve and check that the cable connection is secure. Repeat for opposite cable end.

5. Pull the choke shaft (**Figure 21**) out and lightly wipe the exposed portion of the shaft with marine grease, then work the shaft back and forth to distribute the grease.

6. Remove the steering nozzle pivot shafts (**Figure 22**) and lubricate them with marine grease. Reinstall the shafts and tighten securely.

7. The steering column requires partial disassembly to gain access to the steering shaft for lubrication. Refer to Chapter Twelve. Lubricate

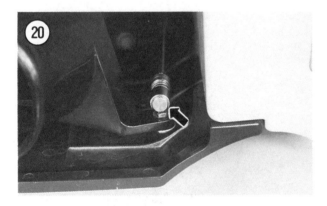

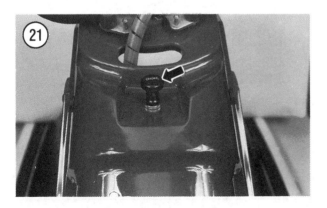

the steering column pivot shaft and nylon bearings (**Figure 23**) with marine grease.

8. *WRA650, WRA650A, WRA700, WVT700 and WVT1100*—Lubricate the reverse shift cable as follows. Refer to **Figure 24**.
   a. Disconnect the reverse cable at the shift lever.
   b. Disconnect the spring at the pin and loosen the pin.
   c. Apply marine grease to the reverse cable end and at the pin and collar.
   d. Retighten the pin and reconnect the spring.
   e. Reconnect and secure the reverse cable at the shift lever.

9. *RA700, RA700A, RA760 and RA1100*—Lubricate the trim control cables as follows. Refer to **Figure 25**.
   a. Spray rust inhibitor along both trim adjust cables and into the cable housings at the handlebar grip.
   b. Apply marine grease along the exposed portions of both trim adjust cables at the trim wheel under the engine cover.
   c. Locate the trim control cable ball joint on the steering nozzle (**Figure 26**). Push the cable sleeve rearward and then pull the cable sleeve off the ball joint. Release the cable sleeve after it is free of the ball joint.
   d. At the trim wheel, pull the inner cable out of the outer cable to expose as much of the trim control inner cable as possible. Wipe the exposed inner cable with marine grease. Repeat for the opposite cable end.
   e. At the steering nozzle, lubricate the cable sleeve (portion which mates with ball) and the ball joint with marine grease. Repeat for both cable ends.
   f. Reconnect the cable by opening the cable sleeve and then inserting the sleeve opening over the ball joint. Release the sleeve and check that the cable connection is secure.

10. Fill the intermediate shaft bearing housing with marine grease applied with a grease gun through the housing nipple. To avoid damage, fill the housing slowly and carefully.

OIL INJECTION SERVICE

Many Yamaha water vehicles are equipped with an oil injection system that provides proper engine lubrication during all engine operating conditions see **Table 3**. Premixing fuel is not

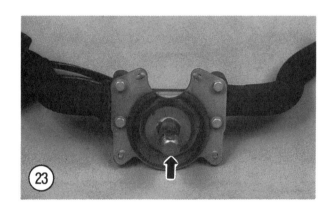

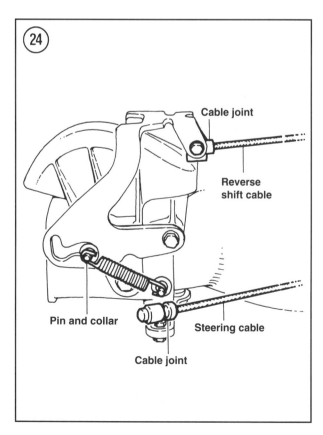

LUBRICATION, MAINTENANCE AND TUNE-UP

required (except during engine break-in). This section describes basic maintenance required to keep the oil injection system working correctly. Maintenance intervals are listed in **Table 1**.

Oil Pump Bleeding

Automatic air-bleeding systems

The oil pump used on 1993-1994 WRA650, WRA650A, WRA700, WRB650, WRB650A, WRB700 and WB700 models is automatic air-bleeding type. This type of injection pump differs from most in that if the oil tank runs dry or if the oil hoses are disconnected or replaced, it is unnecessary to bleed the system manually after reconnecting the hoses or refilling the oil tank before starting the engine. When air enters an automatic air-bleeding oil injection system, it is automatically purged (when the hoses are reconnected or the oil tank refilled). When servicing an automatic air-bleeding oil injection system (hose replacement, pump removal/installation), you can consider your system bled once you have reconnected all of the hoses and refilled the oil tank.

Another cause of air entering the oil injection system is when the craft capsizes. The efficiency of the pump bleeding the system will depend on how quickly you can put the craft right side up. For example, if your craft is capsized for less than 5 minutes, the system should bleed itself automatically. However, if the craft is capsized for more than 5 minutes, you should perform *one* of the following two procedures to make sure that all of the air is bled from the system:

a. After your craft is right side up, do not start the engine for a minimum of 10 minutes.
b. If it is necessary to start your craft after it is right side up, allow the engine to idle for 10 minutes. Do not exceed idle speed or engine damage may result.

NOTE
For the automatic air-bleeding system to work properly, the oil injection system must be kept CLOSED. A CLOSED system is one in which the oil tank is kept full and all of the hoses are in good condition and properly connected. Consider the system OPEN if a hose is disconnected for whatever purpose or if the oil tank is allowed to run empty. An OPEN system cannot automatically air-bleed itself.

Manual air-bleeding systems

In 1995, Yamaha stopped using automatic air-bleeding oil injection systems. Models produced

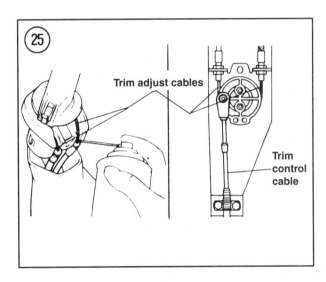

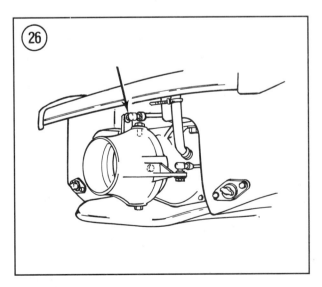

in 1995 and later, use manual air-bleeding systems. To bleed the air from these systems:
1. Be sure all hoses are connected.
2. Refill the oil tank.
3. Place a rag under the oil pump to prevent oil from spilling into the bilge.
4. Open the bleed screw on top of the oil pump so oil can flow from the pump (**Figure 27**). Watch for bubbles in the oil. Let the oil flow from the opened bleed screw until no bubbles are present in the oil.
5. Tighten the bleed screw, and wipe up any spilled oil.

Oil Tank Drain Hose
Inspection/Draining

The oil tank is equipped with a drain hose that connects from a high point on the tank to a lower point (**Figure 16**). Periodically inspect the hose for damage or contamination. If the hose is contaminated, drain it as follows:

> *WARNING*
> *Spilled oil in the bilge can be a serious fire hazard when the craft is operating on the water. When draining the oil tank drain hose in the following steps, take the appropriate precautions to prevent oil from spilling into the bilge.*

1. Place a clean container next to the oil tank.
2. Disconnect the drain hose from its upper position (**Figure 16**).
3. Insert the open end of the hose into the container and allow hose to drain completely. Reconnect the hose by pushing it all the way onto the oil tank fitting. Secure the hose with its clamp.

> *NOTE*
> *If the hose is plugged, remove the oil tank as described in Chapter Ten and remove the hose from the tank; flush the hose to clean it. Replace the hose if it cannot be properly cleaned. Reinstall the hose and oil tank.*

Oil Filter
Inspection and Cleaning

A serviceable oil filter is installed in the oil tank (**Figure 28**). At the intervals specified in **Table 1**, the filter should be removed and inspected, cleaned (if not damaged) or replaced.
1. Remove the engine cover.
2. Open and remove the oil tank cap. Then remove the oil filter from the oil tank, placing a rag underneath the filter as it is removed to prevent oil from dripping into the bilge. Reinstall the oil fill cap or cover the oil tank opening with a clean rag to prevent debris from being accidentally dropped into the tank.
3. Hold the filter up to a light and check the filter for tearing, fraying or other damage. Replace the filter if it is damaged-do not attempt to repair it.
4. If the filter is serviceable, immerse it in *clean* solvent, working it back and forth to remove all

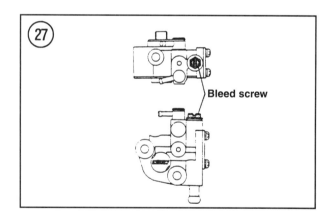

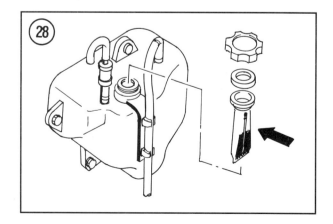

LUBRICATION, MAINTENANCE AND TUNE-UP

dirt and other residue trapped inside the filter. Repeat this cleaning action until the filter is thoroughly clean, then place it on a clean lint-free cloth and allow the filter to air dry thoroughly. Do not install the filter into the tank until it has dried completely. If the filter cannot be properly cleaned, replace it.

5. Uncover the oil tank opening and reinstall the oil filter. If the oil level in the tank is low, refill it with the correct engine oil *after* installing the filter. See **Table 4**.

6. Check the rubber seal in the oil tank cap and replace it if worn or damaged. Then reinstall the cap and tighten securely.

CAUTION
For the oil filter to work properly, it must be installed in the oil tank as the tank is being filled. Do not add oil to the tank if the oil filter has been removed. Debris or contamination may clog up the oil pump passages and cause engine damage.

7. Close and secure the engine cover.

Oil Filter Replacement

Replace the oil filter at the intervals listed in **Table 1**.

Oil Hose Inspection

A number of hoses are used to carry the oil from the reservoir to the engine. Inspect all hoses at the intervals listed in **Table 1**. Look for loose or damaged connections or damaged hoses. Tighten all connections and replace any hose that is damaged or cracked, otherwise, engine damage may occur.

When inspecting the oil hoses, also inspect the hose clamps for weakness or damage and replace as required. A worn, loose or damaged hose clamp may allow a hose to disconnect.

Check Valve Inspection

A check valve is connected to the oil tank vent hose (**Figure 16**). The check valve prevents oil from spilling out the vent hose. During operation, air can flow through the check valve and into the tank and oil pump, but oil cannot flow out of the check valve.

The check valve is color coded for proper installation (**Figure 29**). Install the check valve so its green end faces *toward* the oil tank, not toward the open hose.

Inspect the check valve at the service intervals in **Table 1**.

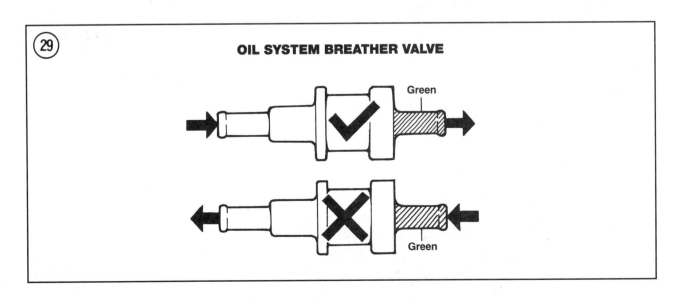

1. Disconnect the check valve from the 2 hoses.
2. Wipe the ends of the check valve with a clean towel.
3. Test the check valve by blowing through both ends. For the check valve to work properly, you should only be able to blow through the valve one way. See **Figure 29**. If you can blow air through the valve both ways or not at all, replace the check valve.
4. Reinstall the check valve, making sure the green end faces toward the oil tank. Push the hoses onto the check valve until they bottom out, then secure them with cable ties.

Oil Tank
Cleaning and Inspection

At the specified intervals (**Table 1**), remove the oil tank and drain all oil. Then clean the tank in solvent and allow to dry thoroughly before filling with oil. Inspect the oil tank for cracks, wear spots or other damage. Replace the tank if it is leaking or damaged in anyway. Refer to Chapter Ten.

50 AND 100 HOUR MAINTENANCE SCHEDULE

Perform the following periodic maintenance items at the intervals specified in **Table 1**.

Nuts, Bolts and Other Fasteners

Visually check for loose or missing fasteners in the engine compartment, on the steering assembly and at the jet pump assembly. Replace missing fasteners with the same type of material as the original. For example, many of the fasteners used on the water vehicle are made of stainless steel. Replacing a stainless steel bolt with a carbon steel bolt will allow corrosion buildup that will eventually result in broken fasteners and various service related problems. After checking for missing fasteners, check all of the exposed fasteners for looseness with a wrench or socket. Use a torque wrench when tightening critical fasteners. Refer to the appropriate chapters in this manual for tightening torques. General torque specifications are listed in Chapter One.

Fuel System Service

Before servicing the fuel system components in the following sections, observe the following information:

WARNING
Gasoline is an extreme fire hazard. Keep gasoline away from all sparks, heat or

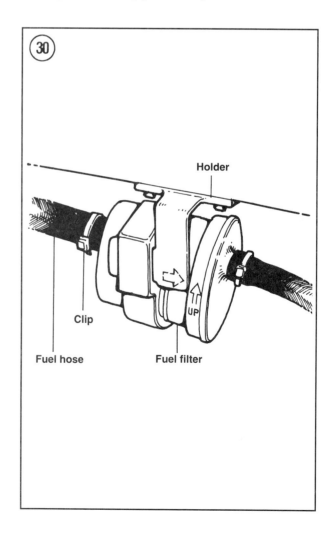

LUBRICATION, MAINTENANCE AND TUNE-UP

flame. Never smoke while working near gasoline.

Fuel Filter

The fuel filter is mounted in the bilge area next to the engine.

1. Locate the filter in the bilge. It is held in a rubber housing which is secured to the hull by a special mounting bracket. Note the position of the filter relative to the holder. You must reinstall the filter is the same position during reassembly (**Figure 30**).

2. Label the inlet and outlet hoses at the filter so you do not mix them up during reassembly.

3. Pull the filter and its rubber housing off of the mounting bracket and cut the 2 plastic ties. Carefully twist and pull the 2 hoses off of the filter, then plug the hoses to prevent leakage. Handle the filter carefully as it will be full of gasoline. Pour the filter contents into a clean glass jar.

4. Pull the filter out of the rubber housing.

5. Examine the plastic filter housing for cracks or damage. Inspect the filter element for debris or water contamination. Replace the filter if necessary.

6. While the filter is removed, check the 2 fuel hoses for damage and replace if necessary.

7. Insert the filter into the rubber housing, making sure the arrow marked UP points upward.

8. Reconnect the 2 hoses to the filter, following the reference marks made prior to their removal. If you did not mark the hoses, refer to the hose routing diagram in Chapter Eight for your model. Push the hoses onto the filter until they bottom out. Secure the hoses with 2 new plastic ties; cut the ties to length after tightening.

9. Install the rubber housing into its mounting bracket. Be sure the UP arrow on the filter points up, and that it is next to the long, flat side of the holder as shown in **Figure 30**.

10. Start the engine and check each of the fuel filter hoses for leaks and make repairs as required.

Fuel Vent Check Valve Inspection/Replacement

All models are equipped with a fuel vent check valve that allows air to enter the fuel tank when the engine is running, but keeps fuel from spilling out of the tank if the water vehicle is tipped over. See **Figure 31**, typical. The arrow stamped on the check valve must always point toward the fuel tank, not toward the open hose.

Review the *WARNING* listed under *Fuel System Service* in this chapter, then proceed with the following.

1. Locate the fuel vent check valve in the engine compartment and label the hoses for reassembly reference.

2. Cut the 2 plastic ties and carefully pull the hoses off of the check valve.

3. Wipe the ends of the check valve with a clean rag.

4. Test the check valve by blowing through both ends. For the check valve to work properly, you should only be able to blow through the valve

one way (**Figure 32**). If you can blow air through the valve both ways or not at all, replace the check valve.

5. Reconnect the 2 hoses to the check valve so the arrow on the valve faces *toward* the fuel tank, then push the hoses onto the check valve fittings until they bottom out. Secure the hoses with 2 new plastic ties. Cut ties to length after tightening.

**Water Separator
(SJ650, SJ700, SJ700A, FX700, WB700, WB700A, WB760, RA700, RA700A, RA700B, RA760, RA1100, WVT700 and WVT1100)**

A water separator (**Figure 13**) is installed in the fuel tank air vent line to prevent water which enters the air vent line from entering the fuel tank. Drain the water separator periodically of all trapped water.

Review the *WARNING* listed under *Fuel System Service* in this chapter, then proceed with the following.

1. Visually examine the water separator housing for cracks or damage. If the housing is cracked, replace it.

2. Check the water separator for the presence of water. If there is water in the separator, place a clean container underneath the separator. Then remove the drain plug (**Figure 13**) from the bottom of the separator and allow the water to drain into the container. When the separator is empty of water, reinstall the drain plug and tighten securely. Dispose of the water in the container.

**Fuel Filler Cap
and Seal Inspection**

The fuel filler cap (**Figure 1**) is equipped with a seal to prevent fuel leakage during operation. Unscrew and remove the cap and check the seal

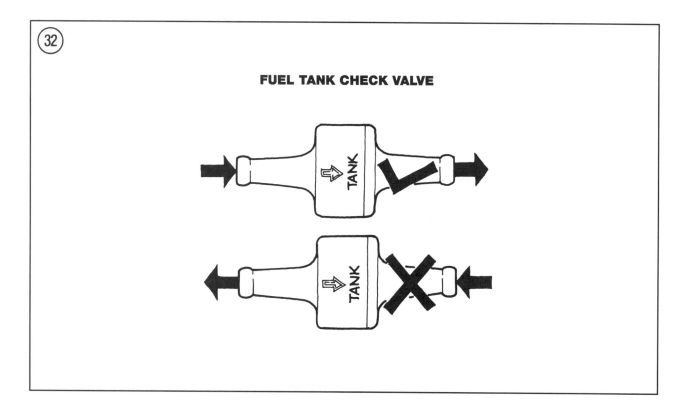

LUBRICATION, MAINTENANCE AND TUNE-UP

for cracks or damage and replace if necessary. Check the cap for cracks or damage and replace if necessary.

Fuel Tank Pickup

Periodically remove the fuel tank pickup and inspect the hoses for contamination, damage or loose hose clamps.

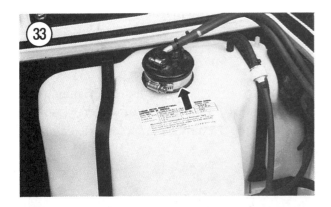

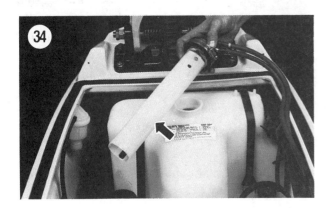

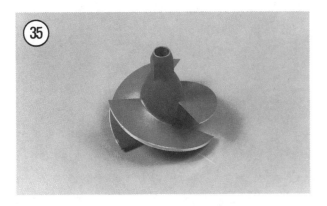

1. If you are going to disconnect the 2 fuel hoses from the pickup cap (**Figure 33**), first label each hose so you do not mix them up during reassembly. Then cut the plastic ties and pull the hoses off of the cap fittings.

2. Loosen and remove the steel hose clamp securing the cap to the fuel tank. Carefully twist the cap from side to side and pull it off of the fuel tank by hand. Do not use any type of tool to remove the cap or you may damage it. For the fuel system to work properly, this cap must provide an air-tight seal on the fuel tank.

3. Cover the fuel tank opening with a clean shop rag.

4. Remove the cotter pin from the top of the guide and remove the washer, pin and guide (**Figure 34**). Inspect the 2 fuel hoses for cracks or other damage. Check the hose clamps for tightness. Replace the hoses as required. Reverse to install the guide. Secure the pin with a new cotter pin.

5. Push the cap down onto the fuel tank neck until it bottoms out (**Figure 33**). Secure the cap with the hose clamp. Tighten the clamp securely, but not so tight that the clamp tears the cap. Following the marks made prior to disassembly, reconnect the fuel hoses to their respective cap fitting. Push the hoses onto the fittings until they bottom out. Secure each hose with a new plastic tie. Cut the ties to length after tightening.

Spark Plugs

Inspect the spark plugs as described under *Engine Tune-up* in this chapter.

Impeller Inspection

Inspect the impeller blades (**Figure 35**) for nicks, deep scratches or gouges at the specified service intervals (**Table 1**) or whenever performance deteriorates while the engine seems to be running well. Refer to Chapter Seven.

Impeller Clearance

Excessive impeller housing clearance will reduce jet thrust. At the intervals listed in **Table 1**, check the impeller housing clearance. Refer to *Impeller Clearance Check and Adjustment* for your model in Chapter Seven.

Battery

Check the battery electrolyte level on a periodic schedule. Refill with distilled water as required. Refer to Chapter Nine for complete battery service information.

Electrical Wiring and Connectors

Remove the electrical box (Chapter Nine) and check all wiring connectors for corrosion or loose connectors. On RA1100 and WVT1100 remove the cover from the electric box and check the wiring. Disconnect the connectors and spray them with electrical contact cleaner. Reconnect the connectors, making sure they snap together.

Inspect the high-tension leads to the spark plugs (**Figure 36**) for cracks and breaks in the insulation and replace the leads if they are less than perfect; breaks in the insulation allow the spark plug to arc to ground and will impair engine performance.

Check primary ignition wiring for damaged insulation. Minor damage can be repaired by wrapping the damaged area with electrical insulating tape. If insulation damage is extensive, replace the damaged wire.

Check that you have a spare fuse of the correct amperage placed in the fuse holder. See *Fuses* in Chapter Nine.

When reinstalling the electrical box, make sure it is secured and that all exposed wiring harnesses are properly routed.

Engine Stop Switch Operation

If the engine stop switch fails to operate properly, test it as described in Chapter Nine and replace if damaged.

Engine Start Switch Operation

If the engine start switch fails to operate properly and the battery is properly charged, test the start switch as described in Chapter Nine. Replace the switch if damaged.

Hull Inspection

Inspect the hull for scratches or punctures. Punctures or other types of severe damage should be repaired by a competent mechanic.

Engine Cover Seal

A special seal is used between the engine cover and hull to prevent water from leaking into the engine compartment. Inspect the seal for cracks, deterioration or other damage. Replace the seal if it is visibly damaged or if its condition is questionable.

Hose Clamps

Check all coolant and exhaust hoses for loose or missing clamps.

LUBRICATION, MAINTENANCE AND TUNE-UP

Carburetor Adjustment

Refer to *Engine Tune-up* in this chapter.

Steering Inspection and Adjustment

Move the handlebars from side to side and then up and down. They should turn smoothly with no sticking or roughness. If the steering is not smooth, check for a damaged steering cable or loose or damaged steering components. If the steering play feels too tight or too loose, perform the following procedure for your model.

WR500, WR650, WRA650, WRA650A, WRA700, WB760, RA700, RA700A, RA700B, RA760, RA1100, WVT700 and WVT1100

The steering column on these models includes an adjuster band to allow minor steering play adjustment.

1. First check the steering shaft bracket bolts (**Figure 37**). If these bolts are loose, retighten and then recheck steering play. If play is still abnormal, perform Step 2.
2. Tighten the adjuster band (**Figure 37**) until the steering play becomes acceptable. If tightening the adjuster band does not change the steering play, inspect the steering stem and bearings as described in Chapter Twelve.

WRB650, WRB650A, WRB700, WB700 and WB700A

The steering column on these models includes adjuster nuts to allow minor steering play adjustment.

1. First check the steering shaft bracket bolts (**Figure 38**). If these bolts are loose, retighten and recheck steering play. If play is still abnormal, perform Step 2.
2. Tighten the adjuster nuts (**Figure 38**) until the steering play becomes acceptable. If tightening the adjuster nuts do not change the steering play, inspect the steering stem and bearings as described in Chapter Eleven.

SJ650, SJ700, SJ700A and FX700

Two adjustments are provided on these models—steering column and steering pole. Adjust the steering column as follows:

1. On FX700 models, remove the handle cover.

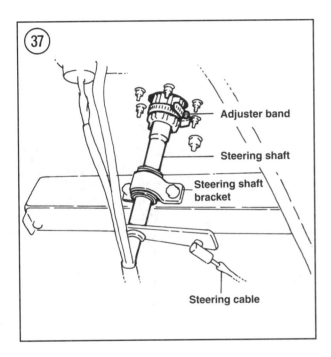

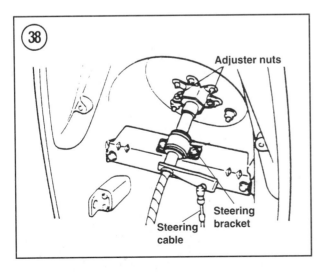

2. Loosen the steering column locknut. See **Figure 39** (SJ650 and SJ700) or **Figure 40** (FX700 and SJ700A).

3. Next, loosen or tighten the steering column adjustment nut (second nut) until the steering play is acceptable.

4. When steering play is correct, hold the adjustment nub with a wrench and tighten the locknut to 29 N•m (21 ft.-lb.).

5. Recheck the steering play adjustment after tightening the locknut. The steering pole should pivot smoothly without excessive play. In addition, the steering play must not have excessive side play.

6A. *SJ650 and SJ700*—Adjust the steering pole as follows:

   a. Remove the bow cover.

   b. Loosen the pivot shaft locknut (A, **Figure 41**).

   c. Tighten the adjustment nut (B, **Figure 41**) to 15 N•m (11 ft.-lb.). Then, while holding the adjustment nut from turning, tighten the locknut to 75 N•m (55 ft.-lb.) on SJ650 models or 70 N•m (52 ft.-lb.) on SJ700 models.

   d. Recheck steering pole side play. If play is still excessive, the steering pole bushings

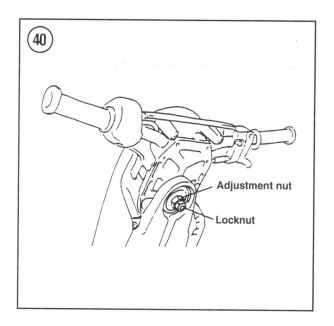

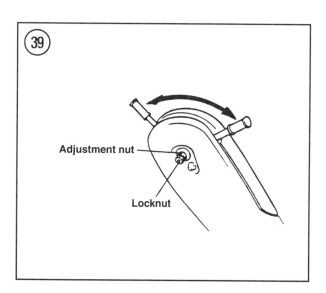

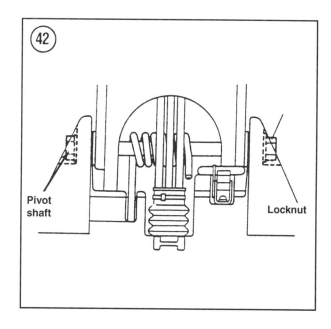

LUBRICATION, MAINTENANCE AND TUNE-UP

may be excessively worn. See Chapter Twelve.

6B. *FX700 and SJ700A*—Adjust the steering pole as follows:

   a. Loosen the locknut (**Figure 42**).
   b. Tighten the pivot shaft to 15 N•m (15 ft.-lb.).
   c. Hold the shaft bolt and tighten the locknut to 70 N•m (52 ft.-lb.).

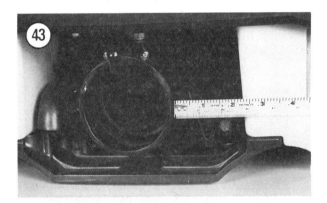

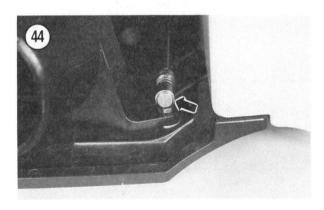

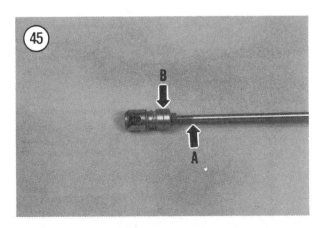

Steering Cable Adjustment

For proper steering, the handlebars must be aligned with the jet pump steering nozzle.

1. Turn the handlebars all the way to the right and measure the distance from the steering nozzle to a point on the hull with a tape measure (**Figure 43**). Record the distance.
2. Turn the handlebars all the way to the left and repeat Step 1.
3. For the steering adjustment to be correct, the distances measured in Step 1 and Step 2 must be the same. If they are different, proceed to Step 4.
4. Adjust the steering cable as follows:

   a. Disconnect the steering cable from the steering nozzle or steering arm by pushing the cable sleeve toward the cable and then lifting the sleeve off of the ball joint. See **Figure 44**, typical.

> *WARNING*
> *The steering cable (A, **Figure 45**) must be threaded into the steering sleeve (B, **Figure 45**) so the minimum cable-to-sleeve engagement distance is maintained as shown in **Figure 46**. Proper engagement will prevent the cable and sleeve from separating when steering the craft. If you cannot properly adjust the steering cable while maintaining the minimum cable-to-sleeve engagement, thread the sleeve into the cable so the*

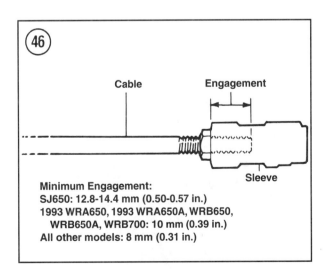

minimum engagement is obtained, then tighten the cable locknut against the sleeve and reconnect the cable to the steering ball joint. Go to the opposite cable end and make your adjustment there. Failure to maintain this critical adjustment can cause loss of control if the cable separates from the sleeve.

NOTE
The steering cable on WRA650 models is marked with a black band or ring on the cable threads. If this ring is visible, the cable engagement is less than its minimum length.

b. Loosen the cable locknut and turn the sleeve IN to increase the steering distance or OUT to decrease it. Tighten the locknut securely against the sleeve. Make sure to read the previous *WARNING* regarding proper cable engagement.

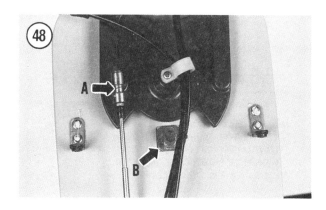

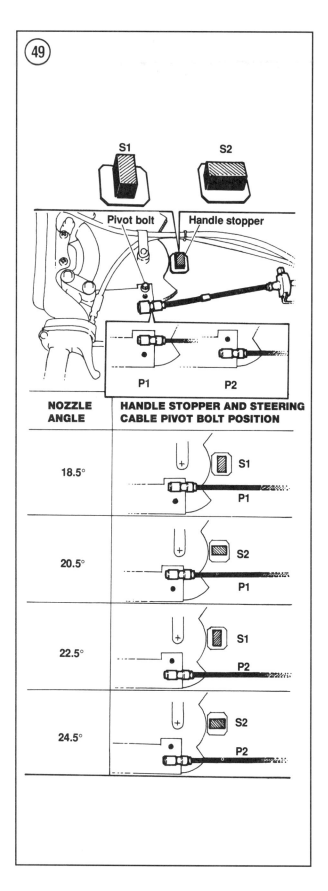

LUBRICATION, MAINTENANCE AND TUNE-UP

c. Repeat Steps 1-4 until the distance is the same.

5. Make sure the sleeve engages the ball joint securely.

Steering Nozzle Angle Adjustment (SJ650, SJ700, SJ700A and FX700)

The steering nozzle angle (**Figure 47**) on these models can be changed by repositioning the steering cable ball joint (A, **Figure 48**) and/or the steering column stop (B, **Figure 48**).

1. Remove the 2 steering pad bolts and remove the steering pad.

2. Select the desired steering angle by referring to the chart in **Figure 49** (SJ650 and SJ700) or **Figure 50** (FX700 and SJ700A). After selecting the steering angle, determine the changes required to obtain adjustment.

3. For example, suppose you wish to set the steering nozzle angle at 20.5°. First, disconnect the steering cable from the ball joint and move the ball joint to position P1 (**Figure 49**). Apply Loctite 242 (blue) to the ball joint stud threads and tighten to 7 N•m (62 in.-lb.).

Then remove the steering column stop nut (B, **Figure 49**) and move the column stop to position S2 (**Figure 49**). Wipe the nut threads with Loctite 242 (blue) and tighten to 32 N•m (23 ft.-lb.).

4. Reinstall the steering pad and its 2 mounting bolts.

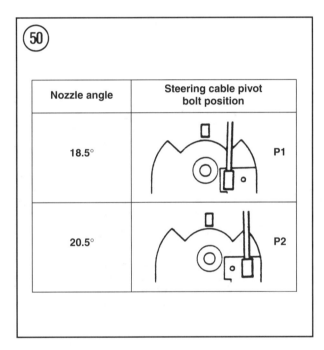

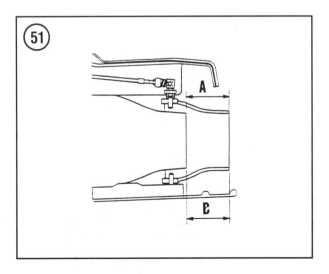

Trim Control Adjustment (RA700, RA700A, RA760 and RA1100)

The RA700, RA700A, RA760 and RA1100 models have an operator adjustable trim system. With this system, the operator controls the angle of the steering nozzle by turning the left-hand grip up or down.

1. With the hand grip in the neutral position, measure the distance from the end of the nozzle deflector to the hull. See A, **Figure 51**. If dimension B, **Figure 51** is not 68 mm (2.7 in.), adjust the trim system as described below.

2. Disconnect the trim control cable from the trim wheel at the back side of the hood.

3. Remove trim adjust cables 1 and 2 from the trim wheel.

4. With the hand grip in the neutral position (centered), pull the slack from the inner wire on each cable.

5. Turn the adjuster nut on each cable until dimension B, **Figure 52** equals 75 mm (2.95 in.). Measure this distance from the outer side of the adjuster nut to the cable end.

6. Reattach both pull cables to the trim wheel. Tighten the locknut on each cable to 16 N•m (11 ft.-lbs.).

7. Loosen the locknut on the trim control cable (**Figure 53**) and remove the sleeve from the trim wheel.

8. Turn the sleeve in or out to adjust dimension A, **Figure 51** at the steering nozzle to 68 1 mm (2.68 0.04 in.). Turn the sleeve in to decrease dimension A, **Figure 51**.

WARNING
The control sleeve must be threaded onto the nozzle control cable so at lease 8 mm (0.31) of cable thread engages the sleeve. Proper engagement will prevent the cable and sleeve from separating during operation. If you cannot adjust nozzle deflection while maintaining this minimum sleeve-to-cable engagement, thread the sleeve into the cable so the minimum engagement is obtained, then tighten the cable locknut against the sleeve and reconnect the cable to the trim wheel. Go to the opposite cable end and make your adjustment at the nozzle deflector. Failure to maintain this critical adjustment can cause loss of control if the cable separates from the sleeve.

9. Reattach the sleeve to the trim wheel (**Figure 53**) and finger tighten the sleeve nut.

10. Torque the locknut to 3 N•m (26 in.-lb.) and then torque the sleeve nut to 3 N•m (26 in.-lb.).

Throttle Cable Adjustment

Throttle cable adjustment is maintained by monitoring throttle lever free play. The throttle cable is provided with a cable adjuster mounted on the throttle housing assembly.

1. Adjust the low speed screw as described under *Engine Tune-up* in this chapter.

2. Squeeze the throttle lever to make sure it moves smoothly. Perform this check as you move the handlebar from side to side. If the throttle lever is tight or sluggish, check the lever and throttle cable before proceeding to Step 3.

3. Depress the throttle lever until resistance is felt and then measure the gap between the throttle lever and its housing as shown in **Figure**

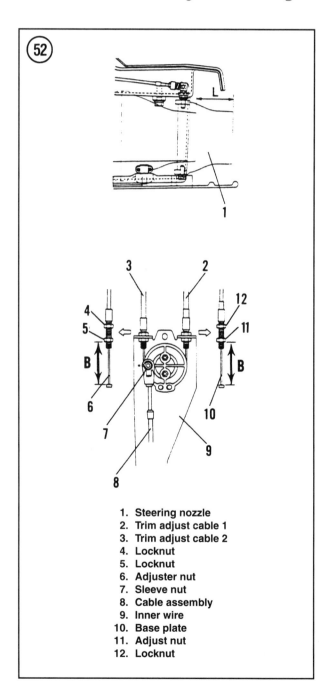

1. Steering nozzle
2. Trim adjust cable 1
3. Trim adjust cable 2
4. Locknut
5. Locknut
6. Adjuster nut
7. Sleeve nut
8. Cable assembly
9. Inner wire
10. Base plate
11. Adjust nut
12. Locknut

LUBRICATION, MAINTENANCE AND TUNE-UP

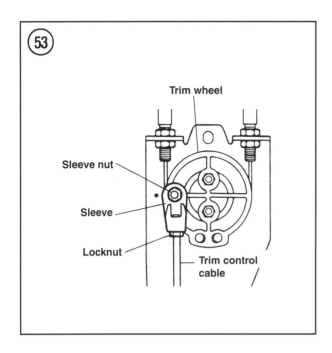

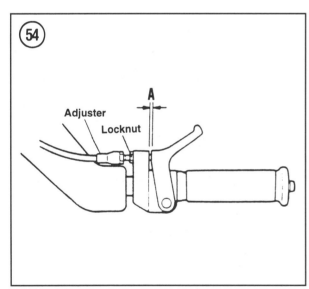

54. This gap (A, **Figure 54**) is throttle lever free play. The correct throttle lever free play for all models is listed in **Table 7**. If the free play on your craft is incorrect, proceed to Step 4.

4. To adjust throttle lever free play, loosen the throttle cable adjuster locknut and turn the adjuster (**Figure 54**) in or out until free play is correct. Tighten the locknut and recheck.

5. After adjusting the throttle lever free play in Step 4, swing the handlebar from side to side and repeat Step 2. Make sure that the throttle lever moves smoothly and that the cable does not bind or kink in any position.

Choke Cable Adjustment

An incorrectly adjusted choke cable will cause hard starting when the engine is cold.

1. Remove the flame arrestor as described in Chapter Eight.

2. Operate the choke knob (**Figure 55**, typical) while watching the choke valve in the carburetor (**Figure 56**). The choke valve should be fully CLOSED when the knob is pulled out and fully OPEN when the knob is pushed in.

3. If adjustment is necessary, loosen the choke cable at the choke lever on the carburetor. Reposition the cable in the lever and retighten. Recheck choke operation. Adjust the choke free play as necessary.

4. Reinstall the flame arrestor (Chapter Eight).

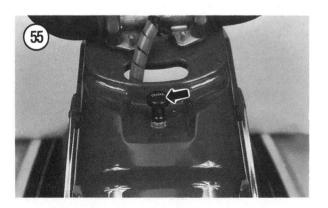

Choke Cable Freeplay

1. Operate the choke knob until resistance is felt, then measure the gap between the choke knob and its housing as shown in **Figure 57**. This gap is choke lever free play. The correct choke lever free play for all models is 1-6 mm (0.04-0.24 in.). If the free play on your craft is incorrect, proceed to Step 2.

2. To adjust choke cable free play, loosen the choke cable adjuster locknut and turn the adjuster (**Figure 58**) at the carburetor. Turn the adjuster in or out until free play is correct. Tighten the locknut and recheck the freeplay.

Reverse Shift Cable Adjustment (WRA650, WRA650A, WRA700, WVT700, and WVT1100)

1. Move the shift lever into REVERSE.

2. Measure the reverse cable length at the point indicated in **Figure 59**. The length should be 114 mm (4.5 in.). If the cable length is incorrect, adjust the cable length to obtain the correct overall length.

3. With the shift lever still in REVERSE, turn the handlebars all the way to port (left). Make sure the pin (A, **Figure 60**) is between the 2 lines on the shift lever (B, **Figure 60**). If it not, adjust by turning the shift lever pivot.

4. Move the shift lever into and out of REVERSE while checking reverse gate movement. Recheck adjustments.

Engine Compartment

Inspect all components mounted in the engine compartment for loose, missing or damaged parts. Tighten, replace or repair components as required.

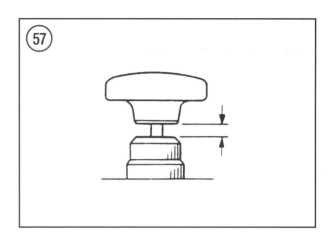

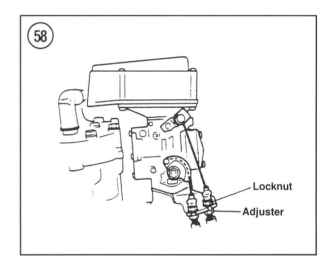

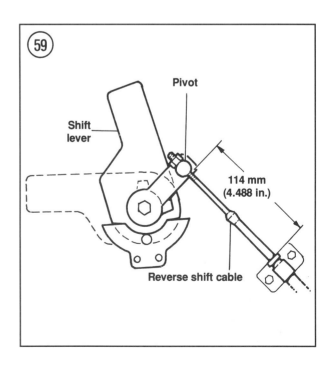

LUBRICATION, MAINTENANCE AND TUNE-UP

Starter Operation

If the starter fails to operate properly, troubleshoot the starting system as described in Chapter Two. If the starter motor is at fault, service it as described in Chapter Nine.

Engine Cover Lock

Periodically check the engine cover lock for looseness or damage. Anytime the cover becomes difficult to lock, or if it should open suddenly, repair or replace the lock mechanism as required.

Rubber Coupler

Inspect the rubber coupler (**Figure 61**) at the specified service intervals (**Table 1**). Refer to the engine chapter for your particular model.

Engine Alignment

Check engine alignment at the specified intervals (**Table 1**) or whenever the engine is removed and installed in the hull. Refer to the engine chapter for your particular model.

Hull Drain Plug

Most late model Yamaha water vehicles are equipped with a hull drain plug that simplifies draining water from inside the hull. Periodically inspect the drain plug for cracks or damage that could allow water to enter the hull. When removing the drain plug, check its O-ring for wear or damage. Replace the O-ring as required. Tighten the drain plug securely.

COOLING SYSTEM FLUSHING

The cooling system can become blocked by sand and salt deposits. Therefore, always flush the cooling system after each use, especially if your craft is operated in saltwater. Read this procedure through before flushing your engine's cooling system.

The following procedure can be used to flush the cooling system and to provide cooling water to run the engine on shore for tuning or adjustment.

NOTE
A flush kit must be installed to flush your water vehicle's engine properly on all models except for RA1100 and

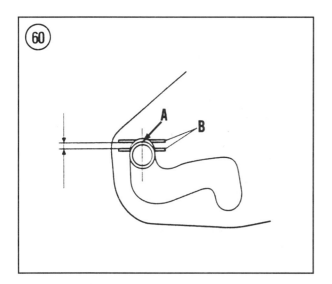

WVT1100 models. Flush kits are standard equipment on these models. Flush kits are manufactured by Yamaha and a number of accessory manufacturers. Purchase and install a kit according to its manufacturer's directions. See **Figure 62**, typical.

This procedure describes cooling system flushing with the use of a flush kit.

1A. *Except RA1100 and WVT1100 models:* Attach a garden hose to the flush inlet following the kit manufacturer's instructions.

1B. *RA1100 and WVT1100:* Attach a garden hose to the flush inlet near the battery box (**Figure 63**).

2. Attach the garden hose to a faucet.

> *WARNING*
> *Exhaust gases are poisonous. Do not run the engine in a closed area. Make sure there is plenty of ventilation.*

3. Turn the fuel valve ON. Start the engine then, turn the water on. Flush the cooling system for approximately 10-15 minutes while running the engine at a fast idle.

> *CAUTION*
> *Too slow a coolant flow will cause exhaust system and engine damage. Too fast a coolant flow may kill the engine and flood the cylinders; this may cause hydraulic locking of the engine and severe damage. If the engine dies while flushing the engine, shut the water off immediately.*

4. When you are finished flushing the cooling system, turn the water supply OFF and clean excess water out of the exhaust system by running the engine for an additional 10-15 *seconds* while operating the throttle in quick bursts from idle to 3/4 throttle. Then stop the engine and turn the fuel valve OFF.

> *CAUTION*
> *Do not run the engine for more than 15 seconds after the water has been turned off. The rubber parts of the exhaust system will be damaged. Prolonged running without coolant will cause serious engine damage. Never operate the engine at maximum speed out of water.*

ENGINE TUNE-UP

The number of definitions of the term tune-up is probably equal to the number of people defining it. For the purposes of this book, a tune-up is general adjustment and maintenance to ensure peak engine performance.

The following paragraphs discuss each phase of a proper tune-up. Be sure to perform the tune-up steps in the order given.

Have the new parts on hand before you begin.

To perform a tune-up on your vehicle, you need the following tools and equipment:

  a. 14 mm spark plug wrench.
  b. Socket wrench and assorted sockets.
  c. Phillips head screwdriver.
  d. Spark plug feeler gauge and gap adjusting tool.
  e. Compression gauge.

Cylinder and Cylinder Head Nuts

Retighten the cylinder head bolts as described in the engine chapter for your particular model.

LUBRICATION, MAINTENANCE AND TUNE-UP

Cylinder Compression

A cylinder compression test is one of the quickest ways to check the condition of the pistons, rings, head gasket and cylinder. It is a good idea to check compression at each tune-up, and compare it to the reading obtained at the next tune-up. This will help you spot any developing problems.

1. Warm the engine to normal operating temperature. If the craft is beached or in your shop, a temporary cooling system must be used. Refer to *Cooling System Flushing* in this chapter.

CAUTION
Do not run the engine for more than 15 seconds without a fresh supply of cooling water. The rubber parts of the exhaust system will be damaged. Prolonged running without coolant will cause serious engine damage. Never operate the engine at maximum speed with the watercraft out of the water.

2. Remove the spark plugs. Insert the plugs in the caps and ground both plugs against the cylinder head.

CAUTION
If the plugs are not grounded during the compression test, the CDI unit could be damaged.

3. Screw a compression gauge into one spark plug hole or, if you have a press-in type gauge, hold it firmly in position.

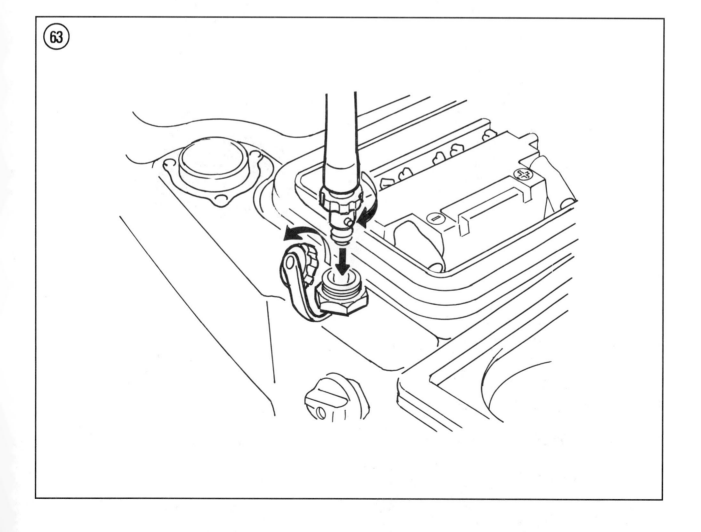

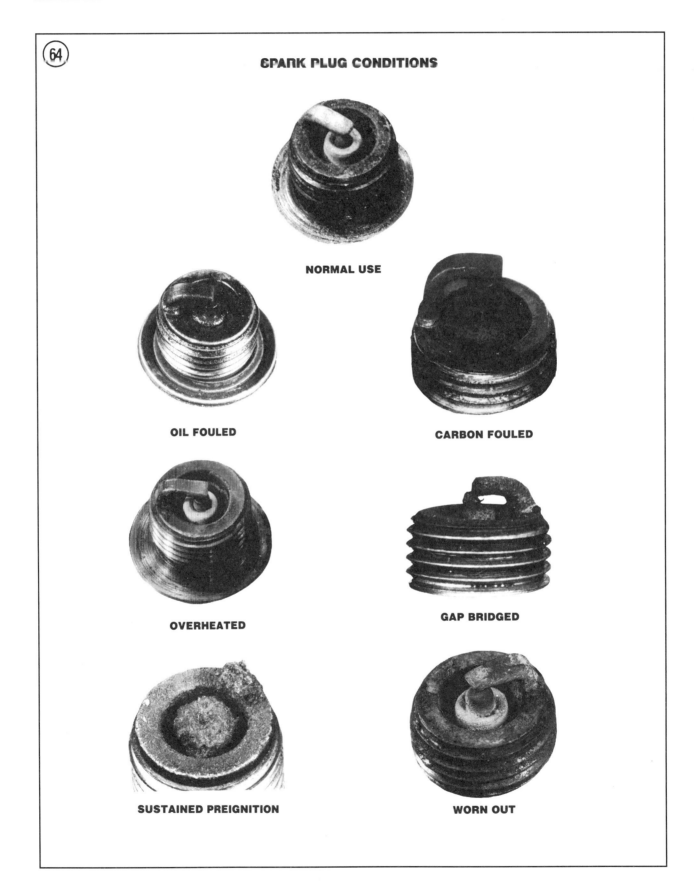

LUBRICATION, MAINTENANCE AND TUNE-UP

4. Check that the stop switch lock plate is removed.

5. Hold the throttle wide open and crank the engine several revolutions until the gauge gives its highest reading. Record the pressure reading.

6. Repeat for the opposite cylinder.

7. There should be no more than a 10% difference in compression between cylinders.

8. Refer to **Table 9** for compression readings. If the compression is very low, it is likely that a ring is stuck or broken or there is a hole in the piston.

Correct Spark Plug Heat Range

The proper spark plug is very important in obtaining maximum performance and reliability. The condition of a used spark plug can tell a trained mechanic a lot about engine condition and carburetion.

Select plugs of the heat range designed for the loads and conditions under which the water vehicle will be run. Using an incorrect heat range can cause a seized piston, scored cylinder wall or damaged piston crown.

In general, use a hot plug for low speeds and low temperatures. Use a cold plug for high speeds, high engine loads and high temperatures. The plugs should operate hot enough to burn off unwanted deposits, but not so hot that they burn themselves or cause preignition. A spark plug of the correct heat range will show a light tan color on the insulator. See **Figure 64**.

The reach (length) of a plug is also important. A longer than normal plug could interfere with the piston, causing severe damage. Refer to **Figure 65**.

The recommended spark plug for the various models is listed in **Table 10**.

Spark Plug Removal/Cleaning

1. Grasp the spark plug lead (**Figure 66**) as near the plug as possible and pull it off the plug. If it is stuck to the plug, twist it slightly to break it loose.

2. Blow away any dirt that has accumulated next to the spark plug base.

CAUTION
The dirt could fall into the cylinder when the plug is removed, causing serious engine damage.

3. Remove the spark plug with a 14 mm spark plug wrench.

NOTE
If the plug is difficult to remove, apply penetrating oil, like WD-40 or Liquid Wrench, around the base of the plug and let it soak in about 10-20 minutes.

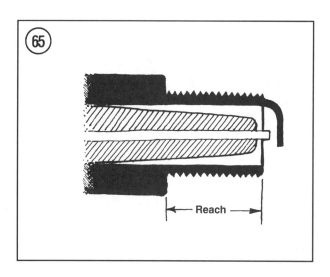

4. Inspect the plug carefully. Look for a broken center porcelain, excessively eroded electrodes, and excessive carbon or oil fouling. See **Figure 64**.

NOTE
The most common spark plug problem on a water vehicle is water fouling. Water or a water/oil emulsion on the plug electrodes indicates water in the fuel or inside the engine. A plug with this condition should be dried with electrical contact cleaner or replaced with a clean, dry plug.

Gapping and Installing the Plug

Carefully gap new spark plugs to ensure a reliable, consistent spark. You must use a special spark plug gapping tool and a wire feeler gauge.
1. Insert a wire feeler gauge between the center and side electrode (**Figure 67**). The correct gap is listed in **Table 10**. If the gap is correct, you will feel a slight drag as you pull the wire through. If there is no drag, or the gauge will not pass through, bend the side electrode with a gapping tool (**Figure 68**) as necessary.
2. Apply antiseize compound to the plug threads before installing the spark plug.

NOTE
Antiseize compound can be purchased at most automotive parts stores.

3. Screw the spark plug in by hand until it seats. Very little effort is required. If force is necessary, you may have the plug cross-threaded. Unscrew it and try again.
4. Use a torque wrench and tighten the plug to the torque specification listed in **Table 8**. If you do not have a torque wrench, tighten the plug an additional 1/4 to 1/2 turn after the gasket has made contact with the head.

NOTE
Do not overtighten. This will only crush the gasket and destroy its sealing ability.

5. Install the spark plug cap, making sure it fits tightly on the spark plug.

CAUTION
Make sure the spark plug wire is pulled away from the exhaust pipe.

Reading Spark Plugs

Because the firing end of a spark plug operates in the combustion chamber, it reflects the operating condition of the engine. Important information about engine and spark plug performance can be determined by careful examination of the spark plug. This information is only valid after performing the following steps.
1. Launch the water vehicle and warm it to normal operating temperature. Then ride the water vehicle a short distance at full throttle.
2. While running at full throttle, push on the stop switch before closing the throttle; coast back to shore.
3. Remove the spark plug and examine it. Compare it to **Figure 64** while noting the following information.

Normal condition

If the plug has a light tan- or gray-colored deposit and no abnormal gap wear or erosion,

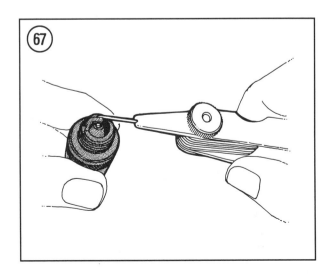

LUBRICATION, MAINTENANCE AND TUNE-UP

good engine, carburetion and ignition conditions are indicated. The plug in use is of the proper heat range and may be serviced and returned to use.

Carbon fouled

Soft, dry, sooty deposits covering the entire firing end of the plug are evidence of incomplete combustion. Even though the firing end of the plug is dry, the plug's insulation decreases. An electrical path is formed that lowers the voltage from the ignition system. Engine misfiring is a sign of carbon fouling. Carbon fouling can be caused by one or more of the following:
  a. Too rich fuel mixture (incorrect adjustment).
  b. Spark plug heat range too cold.
  c. Over-retarded ignition timing.
  d. Ignition component failure.
  e. Low engine compression.

Oil fouled

The tip of an oil fouled plug has a black insulator tip, a damp oily film over the firing end and a carbon layer over the entire nose. The electrodes will not be worn. Common causes for this condition are:
  a. Too much oil in the fuel (incorrect jetting or malfunctioning oil pump).
  b. Wrong type of oil.
  c. Ignition component failure.
  d. Spark plug heat range too cold.
  e. Engine still being broken in.

Oil fouled spark plugs may be cleaned in an emergency, but it is better to replace them. It is important to correct the cause of fouling before the engine is returned to service.

Gap bridging

Plugs with this condition exhibit gaps shorted out by combustion deposits between the electrodes. If this condition is encountered, check for an improper oil type, excessive carbon in the combustion chamber or a clogged exhaust port and pipe. Be sure to locate and correct the cause of this condition.

Overheating

Badly worn electrodes and premature gap wear are signs of overheating, along with a gray or white blistered porcelain insulator surface. The most common cause for this condition is using a spark plug of the wrong heat range (too hot). If you have not changed to a hotter spark plug and the plug is overheated, consider the following causes:
  a. Lean fuel mixture (incorrect carburetor adjustment or malfunctioning oil pump).
  b. Ignition timing too advanced.
  c. Cooling system malfunction.
  d. Engine air leak.

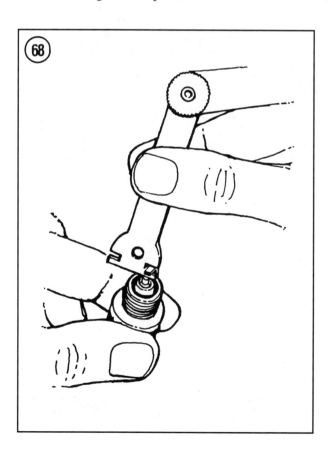

e. Improper spark plug installation (overtightening).
f. No spark plug gasket.

Worn out

Corrosive gases formed by combustion and high voltage sparks have eroded the electrodes. Spark plugs in this condition require more voltage to fire under hard acceleration. Replace with a new spark plugs.

Preignition

If the electrodes are melted, preignition is almost certainly the cause. Check for carburetor mounting or intake manifold leaks and overadvanced ignition timing. It is also possible that a plug of the wrong heat range (too hot) is being used. Find the cause of the preignition before returning the engine to service.

Ignition Timing

All models are equipped with a capacitor discharge ignition (CDI). This system uses no breaker points which makes the ignition system much less susceptible to failures caused by dirt, moisture and wear.

Yamaha does not specify ignition timing procedures for its engines.

a. On WR500, WB760, RA760, RA1100 and WVT1100 models, the stator plate is mounted in a fixed position (**Figure 69** shows WR500 stator). Repositioning the plate to change ignition timing is not possible in these models. When the stator is properly installed onto the engine, ignition timing is automatically set.

b. On all other models, the stator plate is not fixed; slots are provided in the plate that can be used to advance or retard ignition timing. Set the ignition timing by aligning the timing mark on the stator plate with the crankcase index mark (**Figure 70**).

Carburetor Adjustment

The following procedure describes carburetor adjustment. Initial carburetor settings are for sea level conditions. If operating your craft at a higher elevation, carburetor adjustment may be required. When operating at a higher altitude, use spark plug readings as the determining factor when making carburetor adjustments.

When making carburetor adjustments, keep the following points in mind:

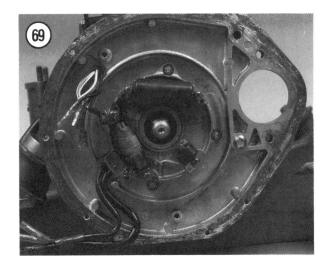

LUBRICATION, MAINTENANCE AND TUNE-UP

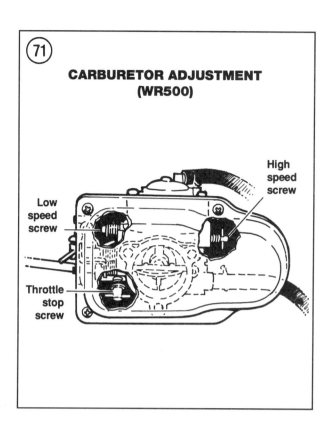

Figure 71 CARBURETOR ADJUSTMENT (WR500)

a. To increase fuel flow (enrichen mixture), turn the mixture needles *counterclockwise*.

b. To decrease fuel flow (lean mixture), turn the mixture needles *clockwise*.

Initial Carburetor Adjustment

To set the low- and high-speed screws, turn each screw in until it seats lightly, then back it out the number of turns specified in **Table 11**. Repeat to set the pilot screw on WR650 models. Refer to the figure for your model for screw identification:

a. **Figure 71**: WR500.
b. **Figure 72**: FX700, RA700B, RA760, SJ650, SJ700, SJ700A, WB700, WB700A, WB760, WRA650, WRA650A, WRA700, WRB650, WRB650A and WRB700.
c. **Figure 73**: WR650.

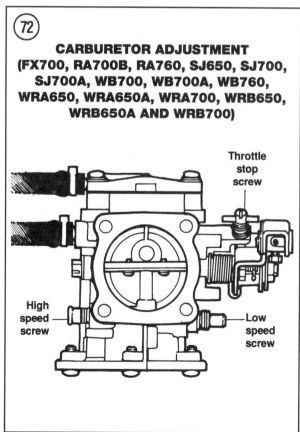

Figure 72 CARBURETOR ADJUSTMENT (FX700, RA700B, RA760, SJ650, SJ700, SJ700A, WB700, WB700A, WB760, WRA650, WRA650A, WRA700, WRB650, WRB650A AND WRB700)

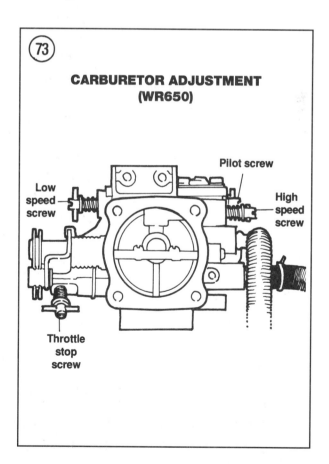

Figure 73 CARBURETOR ADJUSTMENT (WR650)

d. **Figure 74**: RA700, RA700A, RA1100, WVT700 and WVT1100.

CAUTION
Never force the screws into their seat. You will damage the screw or its seat in the carburetor.

CAUTION
Do not ride the water vehicle with the high-speed screw turned in (clockwise) more than specified or the engine may be damaged by an excessively lean fuel mixture.

Low-Speed Screw Adjustment (Adjustment Under Load)

The following adjustment must be made with the craft in the water.

WARNING
The engine produces a large thrust at high engine speeds. Keep a 100 ft. (30.5 m) clear area in front of your craft in case the anchor rope fails. Keep your clothing, hands and feet away from the jet pump intake and outlet.

1. Put the craft in at least 2 ft. (0.6 m) of water.
2. Secure the craft with a strong rope (minimum 500 lb. test) connected at one end to the rear of the craft and the other end to a stationary object (**Figure 75**). Be sure to choose an object that is strong enough to withstand the full thrust of the engine.
3. Remove the engine cover and set both the high- and low-speed screws to their normal initial settings. See **Table 11**.
4. Start the engine and warm it up. Check that the idle speed is normal (see **Table 11**).
5. Make sure the anchor rope has no slack and is secure at both ends, then set the idle speed to specification with the throttle stop screw. Give the throttle 2 or 3 quick applications to see if the engine accelerates without stalling or hesitation and returns to a smooth idle. If it hesitates, readjust the low-speed screw until the engine accelerates from idle smoothly.
6. Readjust idle speed to specifications in **Table 11**.

High-Speed Screw Adjustment (Adjustment Under Load)

The high-speed screw adjustment specifications listed in **Table 11** are for sea-level operation. When operating your craft at higher elevations, you can adjust the high-speed screw to a leaner setting by turning it in slightly. Take spark plug readings to ensure the high-speed mixture is correct.

CAUTION
A large volume of water is pumped at high engine speeds. Provide a 50 ft. (15 m) clear area behind the water vehicle when performing the following.

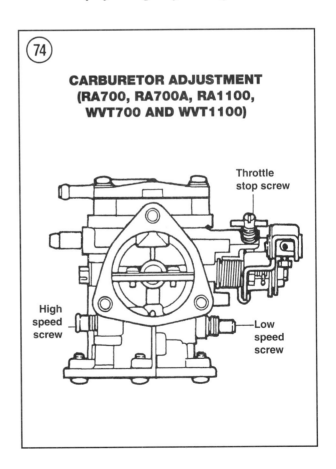

CARBURETOR ADJUSTMENT (RA700, RA700A, RA1100, WVT700 AND WVT1100)

LUBRICATION, MAINTENANCE AND TUNE-UP

1. Make sure the pump intake stays submerged at full throttle by putting a 50 lb. (25 kg) weight on the riding platform.

2. Make sure the anchor rope has no slack and is secure at both ends, then gradually give the engine full throttle.

3. Turn the high-speed screw in (clockwise) until the engine begins to misfire and lose speed.

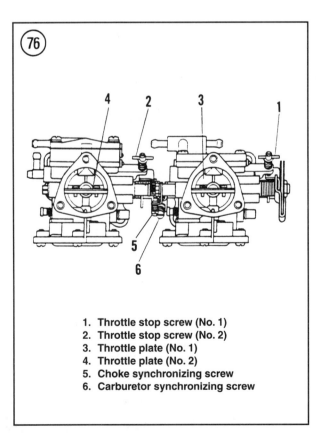

1. Throttle stop screw (No. 1)
2. Throttle stop screw (No. 2)
3. Throttle plate (No. 1)
4. Throttle plate (No. 2)
5. Choke synchronizing screw
6. Carburetor synchronizing screw

4. Slowly turn the high-speed screw back out until the highest engine speed is reached.

CAUTION
Do not ride the water vehicle with the high-speed screw turned in (clockwise) more than specified (at sea level) or the engine may be damaged by an excessively lean fuel mixture.

5. After adjusting the high-speed mixture under load, install new spark plugs and check the mixture to make sure you have not set it so lean that engine damage will result. Ride the water vehicle a short distance at full throttle then, without releasing the throttle, use the stop switch to stop the engine. Drift back to shore. Remove the spark plugs and examine them.

  a. If the insulator is white or burned, the fuel mixture is too lean.
  b. Black, sooty deposits indicate the mixture is too rich.
  c. If the insulator is light tan or gray colored, the mixture is correct.

NOTE
These plug readings are correct only for the standard heat range spark plugs.

CARBURETOR SYNCHRONIZATION

Multiple carburetor systems require carburetor synchronization to ensure optimal performance. Carburetor can be synchronized on a non-running engine. Yamaha recommends that the carburetors be removed for synchronization.

Dual Carburetors

1. Turn out throttle stop screw on each carburetor so the screw no longer contacts the throttle lever. See **Figure 76**. Note how many turns you backed out each screw so you can reset them later.

2. Make sure the throttle plate in the rear carburetor (No. 2) is fully closed.

3. Turn the synchronization screw in or out so the throttle plate in the front carburetor (No. 1) is also fully closed.

4. Reset each throttle stop screw back to its original setting.

Triple Carburetors

1. Turn out throttle stop screw so it no longer contacts the throttle lever. See **Figure 77**. Note how many turns you backed the screw so you can reset it later.

2. Back out synchronization screws No. 1 and No. 2 so they no longer contact the throttle lever.

3. Turn in synchronization screw No. 2 until the throttle plates in carburetor No. 2 and carburetor No. 3 are fully closed. Do not turn the screw too much or the No. 3 throttle plate will start to reopen.

4. Turn in synchronization screw No. 1 until the throttle plate in carburetor No. 1 is fully closed.

5. Check that all throttle plates are fully closed. Repeat the above procedures as necessary.

6. Reset the throttle stop screw to its original setting.

CHOKE SYNCHRONIZATION

Visually check to see if all throttle plates are fully closed when the choke is engaged. If not, synchronize the choke plates as described below.

Dual Carburetors

1. Be sure the choke is fully on.

2. Turn the choke synchronizing screw in or out until both choke plates are fully closed. (**Figure 76**).

3. Visually inspect the choke plates to be sure they are fully closed. If not, repeat the above steps.

Triple Carburetors

1. Be sure the choke is fully on.

2. Turn out choke synchronizing No. 1 and choke synchronizing No. 2 until they do not contact the synchronization lever. See **Figure 77**.

3. Turn in choke synchronizing No. 2 until the throttle plates in carburetors No. 2 and No. 3 are fully closed.

4. Turn in choke synchronizing screw No. 1 until the throttle plate in carburetor No. 1 is fully closed.

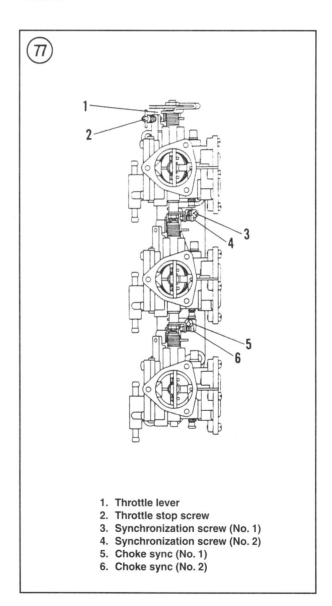

1. Throttle lever
2. Throttle stop screw
3. Synchronization screw (No. 1)
4. Synchronization screw (No. 2)
5. Choke sync (No. 1)
6. Choke sync (No. 2)

LUBRICATION, MAINTENANCE AND TUNE-UP

5. Visually inspect the throttle plates to be sure they are closed. If not, repeat the above steps.

STORAGE

Several months of inactivity can cause problems and a general deterioration of the water vehicle's condition if proper care is neglected. This is especially true in areas of weather extremes. You should prepare your craft carefully for storage.

Preparation for Storage

Careful preparation will minimize deterioration and make it easier to restore the water vehicle to service later. Use the following procedure.

WARNING
The exhaust gases are poisonous. Do not run the engine in a closed area. Make sure there is plenty of ventilation.

CAUTION
Do not run the engine for more than 15 seconds without a supply of cooling water or the rubber parts of the exhaust system will be damaged. Prolonged running without coolant will cause serious engine damage. Do not operate the engine at maximum speed out of the water.

1. Flush the cooling system as described in this chapter.
2. With no water running through the engine, clear excess water out of the exhaust system by starting the engine and running it for *no more than 15 seconds*. During this short period, operate the throttle in quick bursts ranging from idle to 3/4 throttle. Stop the engine and turn the fuel valve OFF.
3. Perform the lubrication procedures in this chapter.

WARNING
Some fuel may spill during these procedures. Work in a well-ventilated area at least 50 ft. (15 m) from any sparks or flames, including gas appliance pilot lights. Do not smoke in the area. Keep a B/C rated fire extinguisher handy.

4. Drain all gasoline from the fuel tank, fuel hoses, filter and carburetor. Run the engine in 15 second intervals to use up all the fuel in the carburetors.
5. Remove the carburetors' cover and clean the flame arrestor element. See Chapter Eight. Open the choke and throttle and spray a penetrating rust inhibitor such as WD-40 down the carburetor bore. Install the flame arrestor and cover.
6. Remove the spark plug(s) and add an ounce of 2-stroke oil to each cylinder. Crank the engine a few revolutions to distribute the oil, then reinstall the spark plugs.

CAUTION
Do not add more than a small amount of oil to the cylinder(s) or the crankshaft oil seals may be blown out when the engine is cranked.

7. Remove the battery and coat the cable terminals with petroleum jelly. See Chapter Nine. Check the electrolyte level and refill with distilled water if it is low. Store the battery in an area where it will not freeze and recharge it once a month. A trickle charger works best.
8. Spray the whole engine with a penetrating rust inhibitor such as WD-40. Wipe off the excess.
9. Wash the craft completely, dry it and thoroughly drain the water from the engine compartment. Wax all painted and polished surfaces.
10. Place the engine cover on the hull, but prop it up off the gasket so that air can circulate freely throughout the engine compartment.
11. Cover the craft with material that will allow air circulation-do not use plastic.

Removal From Storage

1. Perform the lubrication procedures as described in this chapter.
2. Install the battery as described in Chapter Nine.
3. Fill the fuel tank with fresh fuel and check for leaks.
4. Inspect or replace the fuel filter element(s) as described in this chapter.
5. Refill the oil injection tank on models so equipped. Refer to **Table 3**.
6A. *1993-1994 WRA650, WRA700, WRB650, WRB700 and WB700*—Allow the oil pump to self-bleed for at least 10 minutes before starting the engine.
6B. *All 1995 and later oil-injection models*—Manually bleed the oil injection system as described in this chapter.
7. Start the engine and run it for 15 seconds. Allow the engine to cool for 5 minutes. Then repeat a few times. Check for fuel or exhaust leaks.

WARNING
The exhaust gases are poisonous. Do not run the engine in a closed area. Make sure there is plenty of ventilation.

CAUTION
Do not run the engine for more than 15 seconds without a supply of cooling water or the rubber parts in the exhaust system will be damaged. Prolonged running without coolant will cause serious engine damage. Do not operate the engine at maximum speed out of the water.

8. Perform a tune-up as described in this chapter.
9. Check all rubber bilge and cooling hoses for cracking and weathering. Replace any faulty parts.
10. Check the steering for proper operation.
11. Check all controls for proper operation.
12. Make sure your fire extinguisher has a full charge.
13. Inspect your life jacket and make sure your boat registration is up to date.

LUBRICATION, MAINTENANCE AND TUNE-UP

Table 1 MAINTENANCE SCHEDULE*

Initial 10 hours	Check all nuts, bolts and fasteners for tightness.
	Check all hose clamps for tightness.
	Clean and check spark plug gap; replace plug if necessary.
	Check carburetor adjustment.
	Check fuel filter for contamination or damage; replace if necessary.
	Check all components in the engine compartment for looseness or damage; repair or replace components as required.
	Check battery electrolyte level; check more frequently in hot weather.
	Check all of the electrical wires for contamination, loose or damaged connectors and wiring; clean connectors and repair wiring as required.
	Check the engine stop switch operation; replace switch if faulty.
	Check the starter switch operation; replace switch if faulty.
	Check the hatch lock for wear of damage.
	Check the rubber seal in the hatch grooves for cracks or wear; replace seal as required.
	SJ650, SJ700, SJ700A, FX700: Check the steering pole operation for excessive sideplay; adjust as required.
	Check the steering column operation; adjust as required.
	Check the hull for damage; repair as required.
	Clean and inspect the oil filter on models with oil injection system.
	Inspect the oil line check valve on models with oil injection system.
	Check all of the oil lines on models with oil injection system.
	Grease the starter motor idle gear on RA700, RA700A, RA700B, RA760, RA1100 and WB760.
	Grease the intermediate shaft bearing housing.
Initial 50 hours of operation	Clean and check spark plug gap; replace plug if necessary.
	Check fuel filter for contamination or damage; replace if necessary.
	Clean and inspect the oil filter on models with oil injection system.
	Check the impeller for abnormal wear, damage or surface defects; repair or replace impeller as required.
	Check battery electrolyte level: check more frequently in hot weather.
	Check all of the electrical wires for contamination, loose or damaged connectors and wiring; clean connectors and repair wiring as required.
	Check the engine stop switch operation; replace switch if faulty.
	Check the starter switch operation; replace switch if faulty.
	Check the hull for damage; repair as required.
Initial 100 hours; thereafter every 100 hours of operation	Check all nuts, bolts and fasteners for tightness.
	Check all hose clamps for tightness.
	Clean and check spark plug gap; replace plug if necessary.
	Check carburetor adjustment.
	Check fuel filter for contamination or damage; replace if necessary.
	Check all of the fuel system components for any sign of wear, damage or fuel leakage; repair worn or damaged components as required.
	Remove and test the check valve for proper operation; replace as required.
	Clean and inspect the oil filter on models with oil injection system.
	Inspect the oil line check valve on models with oil injection system.
	Check all of the oil lines on models with oil injection system.
	Check steering cable adjustment; adjust if necessary.
	Check throttle lever adjustment; adjust if necessary.
	Check the impeller for abnormal wear, damage or surface defects; repair or replace the impeller as required.
	Check the impeller housing clearance; replace impeller if necessary.

(continued)

Table 1 MAINTENANCE SCHEDULE* (continued)

Initial 100 hours; thereafter every 100 hours of operation (continued)	Check all components in the engine compartment for looseness or damage; repair or replace components as required. Check battery electrolyte level; check more frequently in hot weather. Check all of the electrical wires for contamination, loose or damaged connectors and wiring; clean connectors and repair wiring as required. Check the engine stop switch operation; replace switch if faulty. Check the starter switch operation; replace switch if faulty. Check the hatch lock for wear of damage. Check the rubber seal in the hatch grooves for cracks or wear; replace seal as required. SJ650, SJ700, SJ700A, FX700: Check the steering pole operation for excessive sideplay; adjust as required. Check the steering column operation; adjust as required. Check the hull for damage; repair as required. Grease the intermediate shaft bearing housing. Check the trim cable on RA700, RA700A, RA1100, WVT700 and WVT1100 models; adjust as necessary.
Every 200 hours	Replace the oil filter on models with oil injection system. Remove and flush the oil tank on models with oil injection system. Check all components in the engine compartment for looseness or damage; repair or replace components as required. Remove and inspect the engine rubber coupling; replace if necessary. Check engine alignment; realign if necessary. Check the fuel tank for contamination or damage; clean or repair as required. Replace fuel filter. Check the hull drain plug for damage; replace if necessary. Check the reverse shift adjustment on WRA650, WRA650A, WRA700, WVT700 and WVT1100 models; adjust if necessary.

*Consider this maintenance schedule a guide to general maintenance and lubrication intervals. Harder than normal use will naturally dictate more frequent attention to most maintenance items.

Table 2 LUBRICATION INTERVALS

At every 100 hours of operation, lubricate the following:	Throttle cable Choke cable Choke knob shaft Trim control cable Steering cable Steering column pivot shaft Steering pole pivot shaft Steering nozzle pivot bolts Ball joint Reverse shift mechanism on WRA650, WRA650A, WRA700, WVT700 and WVT1100 models. Intermediate shaft bearing housing Starter idle gear

LUBRICATION, MAINTENANCE AND TUNE-UP

Table 3 ENGINE LUBRICATION SYSTEM

WR500, WR650, SJ650, SJ700, J700A, FX700	Premix 50:1
All other models	Oil injection

Table 4 RECOMMENDED LUBRICANTS AND FUELS

Grease	Yamaha marine grease*
Engine lubrication	Yamalube Two-Cycle Outboard Oil
Fuel	Regular unleaded, minimum octane: 86

*Or equivalent water-resistant marine grease.

Table 5 FUEL/OIL PREMIX RATIO—50:1

U.S. gal.	Oil (oz.)	Oil (cc)
1	2.6	77
2	5.1	151
3	7.7	228
4	10.2	302
5	12.8	378

Table 6 APPROXIMATE REFILL CAPACITIES

	Full—liter (gal.)	Reserve—liter (gal.)
Fuel tank		
SJ650, SJ700, SJ700A	18 (4.8)	5.5 (1.45)
FX700	14 (3.7)	3.6 (0.95)
WR500, WR650	22 (5.8)	4 (4.2)
WRA650, WRA650A, WRA700	40 (10.6)	7 (1.9)
WRB650, WRB650A, WRB700	30 (7.9)	5 (1.32)
WB700, WB700A	25 (6.6)	5 (1.32)
RA700, WB760	40 (10.6)	11.6 (3.06)
RA700A, RA700B, RA760, RA1100, WVT700	50 (13.2)	8.8 (2.32)
WVT1100	50 (13.2)	12 (3.2)
Oil tank		
WRA650, WRA650A, WRA700, RA700, WB760	4 (1.1)	—
WRB650, WRB650A, WRB700, WB700, WB700A	3.6 (0.95)	—
RA700A, RA700B, RA760, RA1100, WVT700, WVT1100	3.8 (1.0)	—

Table 7 THROTTLE LEVER FREEPLAY

	mm	in.
WR500, WR650	2-5	0.08-0.20
RA1100	4-7	0.16-0.28
All other models	7-10	0.28-0.39

Table 8 MAINTENANCE TIGHTENING TORQUE

	N.m	ft.-lb.
Cylinder head		
WR500, WR650		
1st step	15	11
2nd step	28	20
WRA650, WRA650A, WRA700, WRB650, WRB650A, WRB700, WB700, WB700A, SJ650, SJ700, FX700		
1st step	15	11
2nd step	30	22
RA700, RA700A, RA700B, RA760, RA1100, SJ700A, WB760, WVT700, WVT1100		
1st step	15	11
2nd step	36	26
Spark plugs	20	14
Flywheel		
WR500	140	103
All other models	70	51

Table 9 ENGINE COMPRESSION SPECIFICATIONS

	kg/cm^2	psi
WR500, WRA650, WRA650A, SJ650, 1995 WRB650, WRB650A	10	142
WR650	8.4	119
All other models	9.4	134

Table 10 SPARK PLUG SPECIFICATIONS

Model	Type	Gap
WR500	NGK B7HS	0.5-0.6 mm (0.20-0.24 in.)
WR650, SJ650,	NGK B8HS	0.5-0.6 mm (0.20-0.24 in.)
1994 SJ700	NGK B8HS / BR8HS	0.5-0.6 mm (0.20-0.24 in.)
All other models	NGK BR8HS	0.5-0.6 mm (0.20-0.24 in.)

Table 11 CARBURETOR TUNING SPECIFICATIONS

Model	Idle speed (rpm)	Low-speed screw (turns out)	High-speed screw (turns out)
WR500	1500	1 1/4 ± 1/4	3/4 ± 1/4
WR650	1250	2 1/16 ± 1/4	1.0 ± 1/4
WRA650, WRA650A	1250	1 1/8 ± 1/4	1 1/8 ± 1/4
WRB650, WRB650A			
Carb ID: 61L00	1250	1 1/16 ± 1/4	1 1/4 ± 1/4
Carb ID: 61L01	1250	1 1/4 ± 1/4	1 5/8 ± 1/4
Carb ID: 61L02	1250	1.0 ± 1/4	1 5/8 ± 1/4

(continued)

LUBRICATION, MAINTENANCE AND TUNE-UP

Table 11 CARBURETOR TUNING SPECIFICATIONS (continued)

Model	Idle speed (rpm)	Low-speed screw (turns out)	High-speed screw (turns out)
SJ650			
Carb ID: 6R700	1250	1 1/8 ± 1/4	1 1/8 ± 1/4
Carb ID: 6R701-03	1250	1.0 ± 1/4	1 3/8 ± 1/4
SJ700, FX700, RA700B, WRA700, WRB700, WB700, WB700A	1250	1 7/8 ± 1/4	1 5/8 ± 1/4
SJ700A	1250	7/8 ± 1/4	F: 1 1/8 ± 1/4 R: 1 1/2 ± 1/4
RA700, RA700A, WVT700	1250	5/8 ± 1/4	F: 5/8 ± 1/4 R: 1 1/8 ± 1/4
RA1100, WVT1100	1250	1 1/8 ± 1/4	7/8 ± 1/4
RA760, WB760	1300	1 3/4 ± 1/4	1/2 ± 1/4

Chapter Four

Engine (500 cc)

The 500 cc engine used on all 1993 WR500 models is based on Yamaha's 2-stroke 30 hp outboard marine engine. The engine is mounted inside the hull and is secured with 4 rubber engine mounts. An aluminum coupler half is mounted onto the end of the crankshaft and to the front of the drive shaft. A rubber coupler is used to engage the crankshaft and drive shaft couplings. The coupler cushions crankshaft and drive shaft engagement and absorbs small amounts engine vibration resulting from drive train misalignment and drive shaft runout.

Anticorrosion protection consists of a replaceable zinc anode mounted in the cylinder block water jacket and a 5-step baked on paint process. Stainless steel fasteners are used throughout the engine to assist in anticorrosion and ease of maintenance.

This chapter covers complete information on engine disassembly, inspection, cleaning and reassembly.

Work on the engine requires considerable mechanical ability. Carefully consider your own capabilities before attempting any operation involving major disassembly of the engine.

Much of the labor charge for repairs performed at a dealership involves the removal and disassembly of other parts to reach the defective component. Even if you decide not to tackle the entire engine overhaul after studying the text and illustrations in this chapter, it can be less expensive to perform the preliminary operations yourself then take the engine to your dealership. Since dealership service departments have lengthy waiting lists for service (especially during the spring and summer season), this practice can reduce the time your unit is in the shop.

General engine specifications are listed in **Table 1**. **Tables 1-3** are found at the end of the chapter.

ENGINE LUBRICATION

Engine lubrication is provided by the 50:1 fuel/oil mixture used to power the engine. Refer

ENGINE (500 CC)

to Chapter Three for recommended mixing capacities.

SERVICE PRECAUTIONS

Whenever you work on your water vehicle, there are several precautions that should be followed to help with disassembly, inspection, and reassembly.

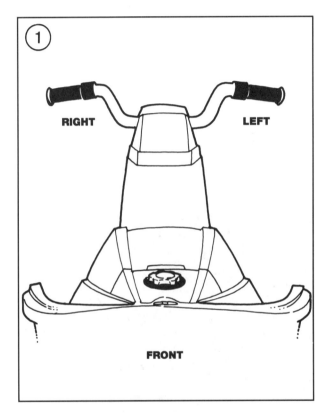

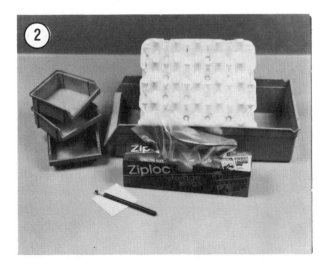

1. Before beginning the job, read Chapter One of this manual. You will do a better job with this information fresh in your mind.

2. In the text there is frequent mention of the left and right side of the engine. This refers to the engine as it is mounted in the hull, not as it sits on your workbench. Left and right refers to the the rider's point of view when seated on the watercraft and facing forward. See **Figure 1**.

3. Always replace a worn or damaged fastener with one of the same size, type and torque requirements. Stainless steel fasteners are used throughout the watercraft. Make sure to identify each fastener before replacing it with another. Lubricate fastener threads with engine oil, unless otherwise specified, before torque is applied. If a tightening torque is not listed in **Table 3** at the end of this chapter, refer to the torque and fastener information in Chapter One.

4. Use special tools where noted. In some cases, it may be possible to perform the procedure with makeshift tools, but this procedure is not recommended. The use of makeshift tools can damage the components and may cause serious personal injury. Where special tools are required, these may be purchased through any Yamaha water vehicle dealer. Other tools can be purchased through your dealership, or from a motorcycle or automotive accessory store. When purchasing tools from an automotive accessory dealer or store, remember that all threaded parts must have metric threads.

5. Before removing the first bolt and to prevent frustration during installation, get a number of boxes, plastic bags and containers and store the parts as they are removed (**Figure 2**). Also have on hand a roll of masking tape and a permanent, waterproof marking pen to label each part or assembly as required. If your craft was purchased second hand and it appears that some of the wiring may have been changed or replaced, it

will be helpful to label each electrical connection before disconnecting it.

6. Use a vise with protective jaws to hold parts. If protective jaws are not available, insert wooden blocks on each side of the part(s) before clamping them in the vise.

7. Remove and install pressed-on parts with an appropriate mandrel, support and hydraulic press. *Do not* try to pry, hammer or otherwise force them on or off.

8. Refer to **Table 3** at the end of this chapter for torque specifications. Proper torque is essential to ensure long life and satisfactory service from marine components.

9. Replace all O-rings and oil seals during assembly. Apply a small amount of marine grease to the inner lips of each oil seal to prevent damage when the engine is first started.

10. Keep a record of all shims and where they came from. As soon as the shims are removed, inspect them for damage and write down their thickness and location.

11. Work in an area where there is sufficient lighting and room for component storage.

SERIAL NUMBERS

Yamaha water vehicles are identified by hull and engine identification numbers. The hull number is stamped on a plate mounted on the rear of the footrest floor (**Figure 3**). The engine number is stamped on a plate mounted to the muffler box (**Figure 4**). The primary identification number is stamped on a plate mounted inside the engine compartment.

Because Yamaha may make a number of design changes during production or after the craft is sold, these numbers identifying your craft should always be used when ordering replacement parts.

At the front of this book, write down all of the identification numbers for your craft.

SERVICING ENGINE IN HULL

Some of the components can be serviced while the engine is mounted in the hull:
a. Carburetor.
b. Flywheel.
c. Starter motor.

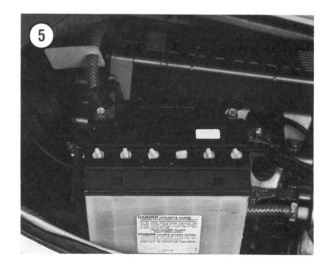

ENGINE (500 CC)

SPECIAL TOOLS

Where special tools are required or recommended for engine overhaul, the tool part numbers are provided. Yamaha tool part numbers have a YB, YU or YW prefix. These tools can be purchased from a Yamaha water vehicle dealer.

Precautions

Because of the explosive and flammable conditions that exist around gasoline, always observe the following precautions.

1. Immediately after removing the engine cover, check for the presence of raw gasoline fumes. If strong fumes can be smelled, determine the source and correct the problem.

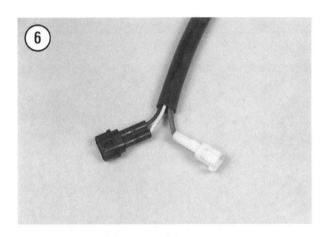

2. Allow the engine compartment to air out before beginning work.
3. Disconnect the negative battery cable. See **Figure 5**.
4. Gasoline dripping onto a hot engine component may cause a fire. Always allow the engine to cool completely before working on any fuel system component.
5. Wipe up spilled gasoline immediately with dry rags. Then store the rags in a suitable metal container until they can be cleaned or disposed of properly. Do not store gas or solvent soaked rags in the engine compartment.
6. Do not service any fuel system component while in the vicinity of open flames, sparks or while anyone is smoking.
7. Always have a Coast Guard-approved fire extinguisher close at hand when working on the engine.

ENGINE REMOVAL

Engine removal and disassembly is required for piston and crankshaft service. Engine removal is also required for service to the intermediate shaft and housing.

1. Support the watercraft on a stand or on wooden boxes so it is secure.
2. Remove the engine cover and allow the engine compartment to air out before continuing.
3. Remove the battery as described in Chapter Nine.
4. Disconnect the start switch and stop switch electrical connectors (**Figure 6**).
5. Remove the carburetor as described in Chapter Eight.
6. Disconnect the spark plug wires from the plugs (A, **Figure 7**), then loosen but do not remove the spark plugs.
7. Remove the electric box from its mounting position and set it aside in the hull. The electric box will be removed with the engine.

8. Disconnect the cooling water hose from the engine (**Figure 8**).

9. Disconnect the exhaust hose from the engine (**Figure 8**).

10. Remove the bolts holding the coupling cover to its mounting brackets and remove the cover (B, **Figure 7**).

NOTE
Shim packs are used underneath the engine mount bolts to achieve proper engine-to-drive shaft alignment. Use any appropriate means to mark each shim pack so it can be identified and reinstalled in its original position.

11. Loosen the 4 engine bed bolts. Remove the bolts and shims.

12. Check to make sure all wiring and hoses are disconnected from the engine. Place the electric box on top of the engine (**Figure 9**).

WARNING
A minimum of 2 people are required to remove the engine assembly in Step 13.

NOTE
Loop a chain around the center muffler-housing bracket for best engine balance.

13. Slide the engine forward to disengage it from the coupler, then lift it up and out of the engine compartment together with the electric box. Take it to a workbench for further disassembly.

14. Remove the rubber coupler (**Figure 10**).

Installation

1. Wash the engine compartment with clean water.

2. Spray all exposed electrical connectors with electrical contact cleaner.

3. Now is a good time to check components which are normally inaccessible for visual inspection or replacement. Before installing the

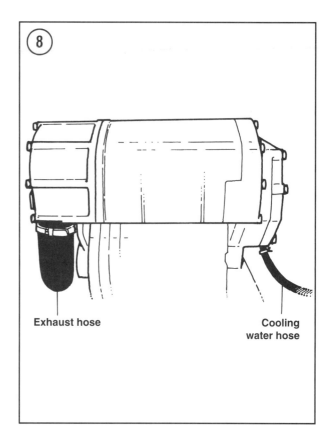
Exhaust hose — Cooling water hose

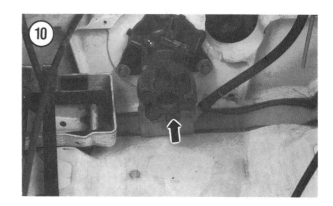

ENGINE (500 CC)

engine, check all components for damage and replace as required.

4. Remove and clean the bilge filter as described in Chapter Eleven.

5. Examine the engine mounts and supports as described in this chapter.

6. Check inside the hull for tools or other objects that may interfere with engine installation. Make sure all hoses, cables and wiring connectors are routed and secured properly.

7. Glue the rubber damper to the front engine bracket before installing the engine in the hull.

8. Install the rubber engine mounts into the hull, if previously removed.

NOTE
Before installing the engine in the hull, install a 6 × 25 mm bolt into each of the engine bracket threaded holes as shown in Figure 11. These bolts will be used to jack the engine into position when aligning the couplers and checking coupling alignment.

9. Check the rubber coupler (**Figure 12**) for wear grooves or deterioration. Especially check for wear on the round outer knobs. Replace the coupler if worn so much that any rotational play exists between the engine half and the drive shaft coupler half.

10. Install the rubber coupler into the forward bearing housing coupling (**Figure 10**).

11. With an assistant, place the engine in the hull and slide it backward. Be sure the engine and bearing housing couplings engage the rubber coupler.

12. If you have access to the Yamaha gauge block (part No. YW-6367), place it under the

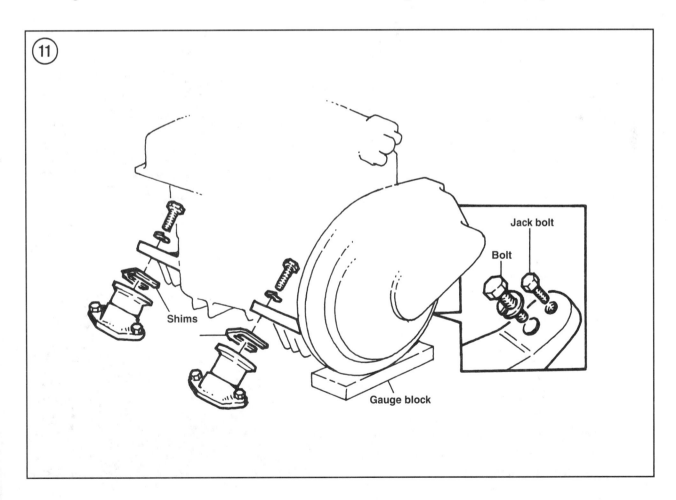

flywheel cover so the 19.5 mark on the block faces up. See **Figure 11**.

NOTE
You can make a gauge block with a piece of aluminum and pieces of shim stock to obtain a thickness of 19.5 mm (0.77 in.).

13. Temporarily install the 4 engine mount bolts (**Figure 13**) finger-tight. Do not tighten the bolts; they are installed to align the engine brackets with the rubber mounts.

14. Check coupling alignment as follows:
   a. Hold a straightedge against one side of the coupling halves (**Figure 14**). Push the straightedge against the flat sides of the couplings. If you can see a gap between one coupling flat and the straightedge or you can feel the straightedge rock as you push against one coupling flat and then the other, measure the clearance with a feeler gauge. See A, **Figure 14**. If the clearance exceeds 0.6 mm (0.024 in.), the coupling halves are misaligned. Repeat this check with the straightedge against the other side, then on top of the coupler halves.
   b. To correct misalignment, reposition the engine with the jack bolts previously installed in the engine brackets. Repeat substep a to check alignment; continue until alignment is correct.
   c. When alignment is correct, install the required thickness of shims in the gap between the engine mounts and brackets. Shims can be purchased through Yamaha water vehicle dealers in the following thicknesses: 0.1, 0.3, 0.5, 1.0 and 2.0 mm.
   d. After installing the shims, measure the gap between the front and rear coupling halves as shown in B, **Figure 14**. The correct clearance is 2-4 mm (0.079-0.157 in.). If the clearance is incorrect, reposition the engine by pushing it forward or rearward as required. Recheck clearance.
   e. Remove the gauge block from underneath the flywheel cover, and remove the 4 jack bolts from the engine brackets.
   f. Wipe the engine mount bolt threads with Loctite 242 (blue) and install the bolts and washers. Tighten the bolts to the torque specification listed in **Table 3**.
   g. Recheck engine alignment.

CAUTION
It is important to align the coupler halves as closely as possible. Any significant degree of misalignment will cause excessive vibration and result in damage to the drive train components.

15. Apply Loctite 242 (blue) to each of the coupling cover bolts. Then install the coupling cover with the bolts, flat washers and lockwash-

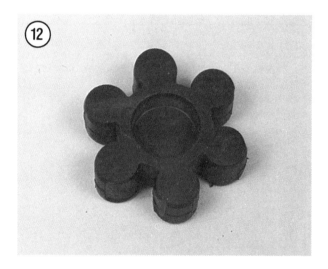

ENGINE (500 CC)

ers. Turn the drive shaft and check that the coupler assembly does not contact the coupler guard.

16. Reverse Steps 1-9 in *Engine Removal* to complete the installation. Note the following:
   a. Use new hoses or cable guides where required.
   b. Adjust the throttle cable and choke cable as described in Chapter Three.

MUFFLER HOUSING AND EXHAUST GUIDE

Refer to **Figure 15** when performing procedures in this section.

Removal

NOTE
Gaskets are used between all exhaust components and you may find that separating the individual components is difficult. Work carefully when separating components in the following steps to avoid damaging the mating surfaces.

1. Remove the long through bolts securing the cover (**Figure 16**) to the muffler housing. Then separate the cover and remove it and the muffler (**Figure 15**).
2. Remove the engine from the hull as described in this chapter.
3. Remove the muffler housing mounting bolts at the cylinder head (A, **Figure 17**) and at the exhaust guide (**Figure 18**). Then carefully break the gasket seal between the muffler housing and exhaust guide and remove the muffler housing (B, **Figure 17**).
4. Remove the exhaust guide assembly as follows:
   a. Unbolt and remove the outer exhaust guide cover (**Figure 19**).
   b. Carefully remove the inner cover (**Figure 20**) from the exhaust guide.
   c. Unbolt the exhaust guide (**Figure 21**) from the engine assembly and remove it.
   d. Remove the 2 exhaust guide dowel pins (**Figure 22**).

Cleaning/Inspection

1. Carefully remove all gasket and sealant residue (**Figure 23**) from all mating surfaces with lacquer thinner. Clean the aluminum surfaces carefully to avoid nicking them. A dull putty knife or gasket scraper can be used. Be careful not to scratch or gouge the surfaces while cleaning.
2. Remove all carbon residue from inside the muffler (**Figure 23**) and muffler housing (**Figure 24**) with a gasket scraper and solvent.
3. After the parts have been cleaned of all gasket and carbon residue, clean the parts in solvent and dry thoroughly.
4. Check the exhaust covers (**Figure 25**) for cracks or warping.

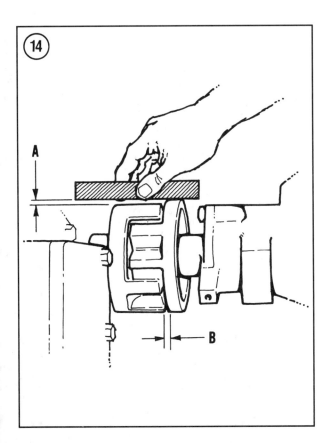

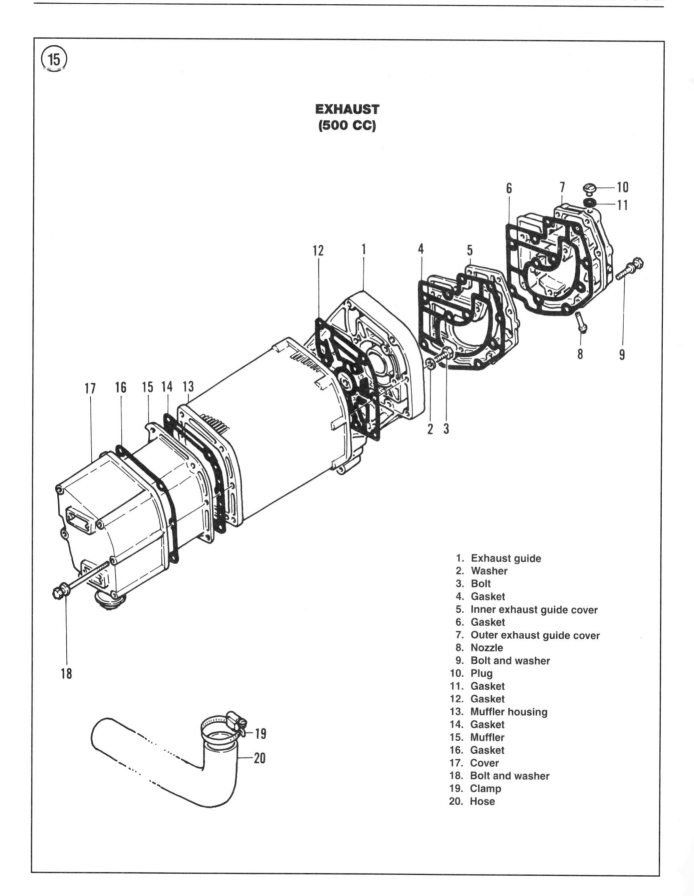

ENGINE (500 CC)

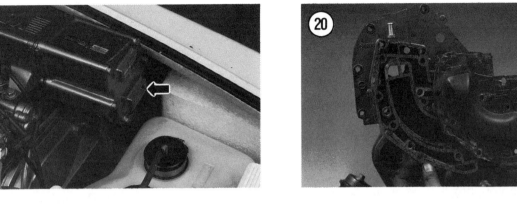

5. Chase all of the muffler housing threads (**Figure 26**) with a tap to remove carbon and corrosion residue. Repeat for engine housing threads.
6. Check the muffler assemblies for cracks or other damage.
7. Replace worn or damaged parts as required.

Coupling Removal/Installation

The engine coupling half is mounted to the engine drive shaft. When removing the coupling half, the Yamaha coupling holder (part No. YW-6365 [A, **Figure 27**]) and drive shaft holder (part No. YB-6079-A [B, **Figure 27**]) are required.

1. Slide the drive shaft holder onto the drive shaft and secure the drive shaft holder in a vise as shown in **Figure 28**.

CAUTION
Do not allow the exhaust guide to contact the top of the vise or its gasket surface may be damaged.

2. Attach the Yamaha coupling holder onto the end of a breaker bar or ratchet and fit the holder

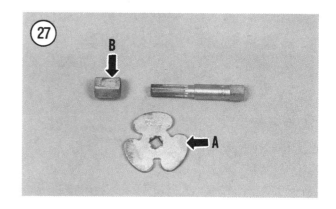

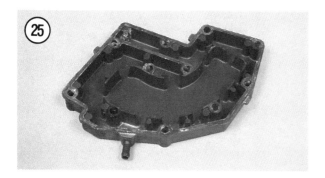

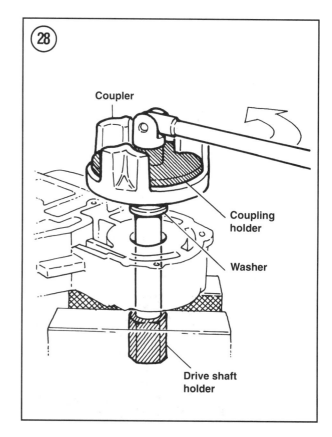

ENGINE (500 CC)

into the coupling half. Turn the breaker bar counterclockwise to loosen and remove the coupling half.

3. Installation is the reverse of these steps. Note the following:

   a. Wipe the drive shaft threads with Loctite 272 before installing the coupling half.

   b. Tighten the coupling half to the torque specification listed in **Table 3**.

Drive Shaft and Bearing Inspection/Disassembly/Reassembly

The crankshaft drive shaft and coupling half are mounted in the exhaust guide assembly. The drive shaft rides on a bearing pressed into the exhaust guide. A seal protects the bearing from contamination.

Refer to **Figure 29** for this procedure.

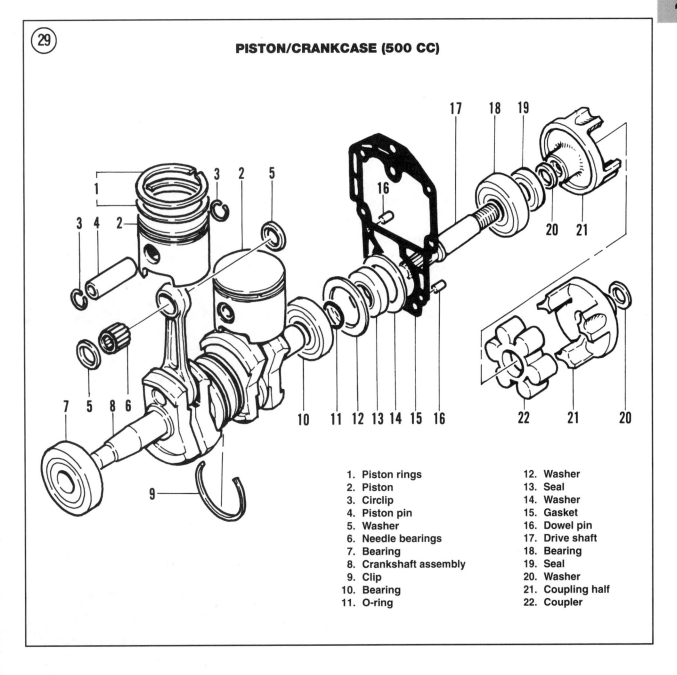

PISTON/CRANKCASE (500 CC)

1. Piston rings
2. Piston
3. Circlip
4. Piston pin
5. Washer
6. Needle bearings
7. Bearing
8. Crankshaft assembly
9. Clip
10. Bearing
11. O-ring
12. Washer
13. Seal
14. Washer
15. Gasket
16. Dowel pin
17. Drive shaft
18. Bearing
19. Seal
20. Washer
21. Coupling half
22. Coupler

1. Remove the coupling half as described under the previous *Coupling Removal/Installation* procedure.
2. Turn the drive shaft (**Figure 30**) by hand. The bearing should turn smoothly without excessive play. If the bearing or drive shaft is damaged, perform the following.
3. Support the exhaust guide in a press. Press the drive shaft and bearing out of the exhaust guide. See **Figure 31**.
4. Replace the seal as follows:
    a. Using a bearing remover (**Figure 32**), pull the seal out of the cover. If you do not have the special tool, carefully pry the seal out of the cover with a large wide-blade screwdriver; place a rag underneath the screwdriver to avoid damaging the cover gasket surface (**Figure 33**).
    b. Clean the bearing area with solvent and dry thoroughly.
    c. Fill the lips of a new seal with marine grease.

NOTE
Support the exhaust cover bearing area when installing the oil seal in substep d.

   d. Align the new seal so the closed end of the seal faces up as shown in **Figure 34**. Install the seal with a seal driver (**Figure 35**) or suitable size socket. Drive the seal into the housing until it bottoms. See **Figure 36**.
5. To separate the bearing and drive shaft (**Figure 31**), support the bearing in a press and press the drive shaft out of the bearing. Reverse to assemble the new bearing and drive shaft.
6. Install the bearing/drive shaft assembly as follows:
    a. Install the exhaust guide in a press bed, making sure the bearing area is supported properly.
    b. Align the bearing with the exhaust guide and press the drive shaft/bearing assembly into the bearing bore until the bearing seats. See **Figure 30**.

ENGINE (500 CC)

Installation

Remove all gasket residue from all muffler box component mating surfaces before performing the following procedure.

1. Install the 2 exhaust guide dowel pins into the crankcase (**Figure 22**).

2. Fit a new exhaust guide gasket onto the crankcase and install the exhaust guide (**Figure 21**), making sure it engages the 2 dowel pins. Install the 2 outer exhaust guide bolts finger-tight.

3. Match the inner and outer exhaust guide covers with their respective gaskets (**Figure 37**). Install the inner cover gasket and inner exhaust guide cover (**Figure 20**), then install the outer cover gasket and outer exhaust guide cover (**Figure 19**). Install the bolts finger-tight.

4. Install a new muffler housing gasket and install the muffler housing. Install the housing bolts finger-tight.

5. Tighten the exhaust guide bolts and the exhaust guide cover bolts in 2 steps to the torque listed in **Table 3**. Tighten the exhaust guide-to-muffler housing bolts to 10 N•m (88 in.-lb.). Tighten all of the bolts in the torque sequence shown in **Figure 38**.

6. Tighten the 2 top muffler housing bolts (A, **Figure 17**) to 15 N•m (11 ft.-lb.).

7. Assemble the muffler and front cover assembly, using new gaskets, in the order shown in **Figure 15**. Install the cover bolts (**Figure 16**) and tighten securely in a crisscross pattern.

8. Connect the exhaust hose the front cover nozzle and secure it with its hose clamp.

ENGINE

A large number of bolts and screws of different lengths are used to secure the various engine components. It is a good idea to use a cupcake tin or similarly compartmented container to hold the various fasteners removed from each component. This will make reassembly easier and faster.

Some mating surfaces can be easily separated by using the engine pry points. If no pry points are provided, break the gasket seal with a wide-blade putty knife and mallet, then carefully separate the components.

Engine Mounts

The engine mount assembly consists of 2 brackets bolted to the bottom of the engine, 4 rubber engine mounts connecting the brackets to the hull and shim packs placed between the rubber engine mounts and brackets. The engine bracket bolts are also used to secure the crankcase halves. See **Figure 39**.

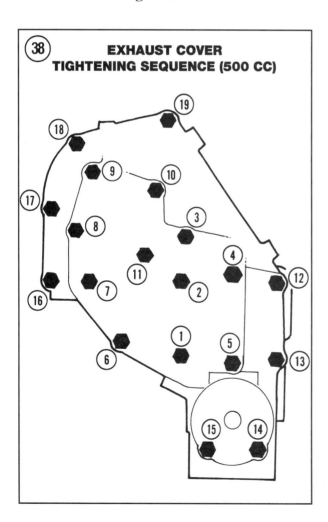

㊳ **EXHAUST COVER TIGHTENING SEQUENCE (500 CC)**

㊴ **CYLINDER/CRANKCASE (500 CC)**

1. Bolt
2. Washer
3. Cover
4. Gasket
5. Bolt and washer
6. Bolt and washer
7. Cylinder head
8. Cylinder head gasket
9. Screw
10. Anode
11. Plug
12. Crankcase/cylinder block assembly
13. Gasket
14. Inner exhaust cover
15. Gasket
16. Outer exhaust cover
17. Washer
18. Bolt
19. Washer
20. Bolt
21. Washer
22. Bolt
23. Bracket
24. Bolt
25. Washer
26. Bracket
27. Nipple
28. Clip
29. Hose
30. Crankshaft seal
31. Clamp
32. Hose
33. Nipple
34. Dowel pin
35. Check valve
36. Gasket
37. Housing
38. Bolt
39. Bolt
40. Bolt and washer
41. Bolt
42. Washer
43. Shim
44. Bolt
45. Engine mount
46. Bracket
47. Bracket
48. Dowel pin
49. Bolt
50. Damper
51. Damper

ENGINE (500 CC)

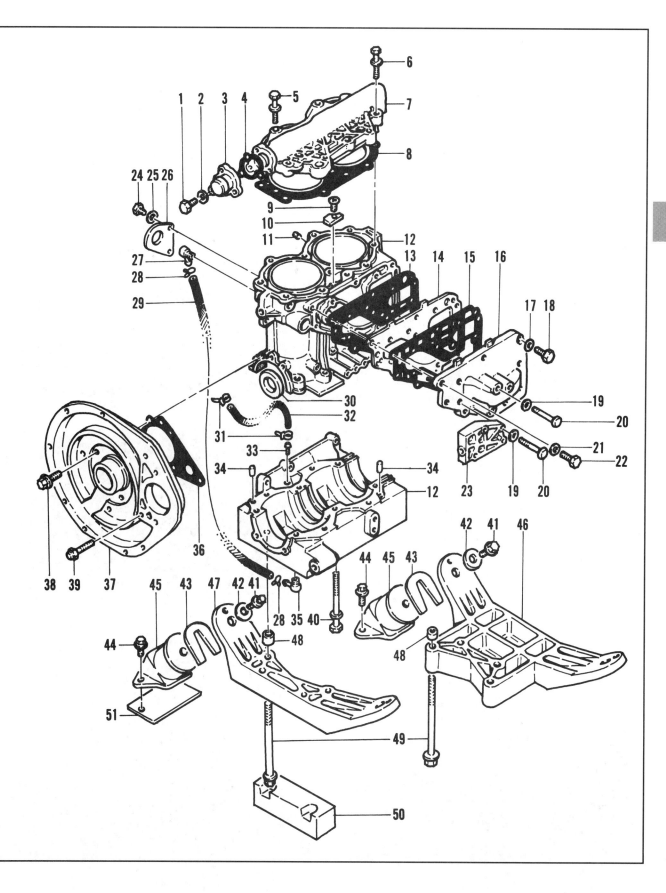

The drive system consists of 2 aluminum coupling flanges. One flange is mounted to the engine and the other is mounted to the forward bearing housing. A rubber coupler connects the 2 flanges together, thus coupling the engine crankshaft directly to the jet pump. Rubber is used as a coupling device to reduce drive train vibration and allow slight variances in engine-to-jet pump alignment.

The engine mounts and brackets are a critical part of the water vehicle drive train. Damaged or loose engine mounts will allow the engine to shift or pull out of alignment during operation. This condition will cause engine-to-jet pump misalignment that will reduce engine performance and accelerate wear to the drive train components. The engine mount and bracket assembly should be inspected carefully whenever the engine is removed from the water vehicle or if engine-to-jet pump alignment becomes a problem.

Engine Disassembly

Refer to **Figure 39** for this procedure.

1. Remove the engine from the hull as described in this chapter.
2. Remove the muffler housing and exhaust guide assembly as described in this chapter.
3. Unbolt and remove the starter motor (**Figure 40**).
4. Remove the flywheel and stator plate as described in Chapter Nine.
5. Remove the bolt securing the thermosensor to the cylinder head and pull the sensor (**Figure 41**) out of the cylinder head.
6. Pull the electrical box wire harness leads out of the oil seal housing and remove the electrical box.

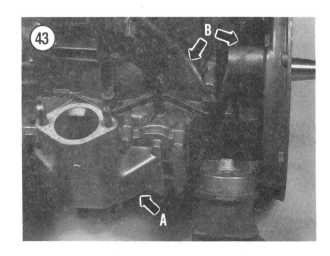

ENGINE (500 CC)

7. Unbolt and remove the oil seal housing (**Figure 42**).

8. Unbolt and remove the intake manifold (A, **Figure 43**) with the reed valve assembly attached. Set the manifold assembly aside so the reed valve assembly faces up (**Figure 44**).

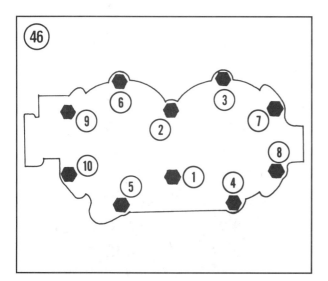

9. Label and disconnect the 2 engine hoses (B, **Figure 43**).

10. Unbolt and remove the outer (**Figure 45**) and inner exhaust covers.

11. Loosen the cylinder head bolts in a crisscross pattern (**Figure 46**), then remove the bolts and washers.

12. Loosen the cylinder head by tapping around its perimeter with a soft-faced mallet, then lift the cylinder head (**Figure 47**) off of the engine block.

13. Remove and discard the cylinder head gasket.

14. Remove the anode from the cylinder block water passage. Replace the anode if it is less than 60 percent of its original size. See **Figure 48**.

15. Turn the engine assembly over so the engine mounting brackets face up.

16. Tag and disconnect the 2 vacuum hoses from the bottom of the crankcase, if not previously done.

17. Mark each bracket with an arrow pointing forward.

18. Loosen the engine bracket and crankcase bolts in a crisscross pattern (**Figure 49**). Then remove the bolts and washers.

19. Remove the front and rear engine brackets (**Figure 50**) then remove the 4 dowel pins (A, **Figure 51**).

20. Loosen the lower crankcase by tapping around its perimeter with a soft-faced mallet. Remove the lower crankcase (B, **Figure 51**).

21. Grasp the crankshaft assembly on both sides and lift the crankshaft up and remove it from the cylinder block; you will have to steady the cylinder block at the same time to prevent it from rising with the pistons. Be careful not to damage the pistons when removing the crankshaft. See **Figure 52**.

22. Locate and remove the 2 crankcase dowel pins (**Figure 53**).

23. Remove the washers and seal from the crankshaft (**Figure 54**).

NOTE
*The 2 washers shown in **Figure 54** have the same thickness.*

Piston and Piston Ring Removal

1. Disassemble the engine as described in this chapter.

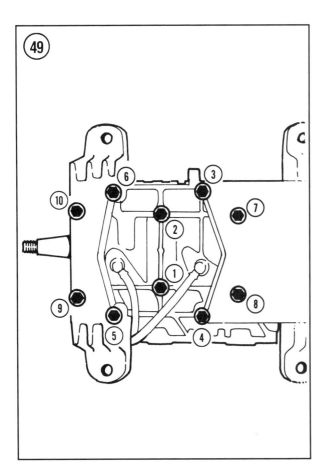

ENGINE (500 CC)

2. Mark each piston with a felt-tipped pen for reassembly on the same connecting rod. In addition, keep each piston together with its own pin, bearings and piston rings to avoid confusion during reassembly.

3. Before removing the piston, hold the rod tightly and rock the piston as shown in **Figure 55**. Any rocking motion (do not confuse with the normal sliding motion) may indicate wear on the piston pin, needle bearing, piston pin bore, or more likely a combination of all three.

CAUTION
The piston rings are a tight fit on the piston. When removing them in Step 4, use a piston ring expander tool to avoid breaking them.

4. Remove the rings from each piston using a suitable ring expander tool (**Figure 56**). If the

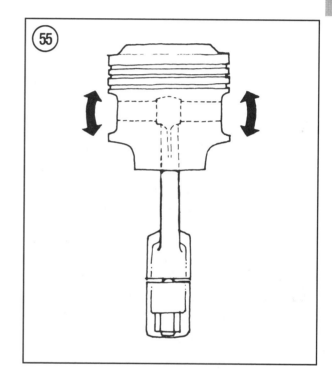

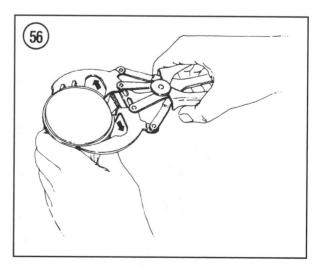

rings are to be reused, tie a parts tag around each set and identify the tag with the piston number.

WARNING
Wear protective eye glasses while performing Step 5.

5. Remove the clips from each side of the piston pin bore with needlenose pliers (**Figure 57**). Hold your thumb over one edge of the clip when removing it to prevent it from springing out. Discard the clips-piston pin clips should never be reused.

CAUTION
The piston pin is a snug sliding fit. You can usually remove it by pushing it out with a wooden dowel. However, if the engine ran hot or seized, the pin may be very tight in the piston. You will have to remove the pin with a suitable tool. Attempting to drive the pin out with force may damage the piston, needle bearing or connecting rod. If the pin is tight, remove it as described in Step 6.

6. If the piston pin is tight, fabricate the tool shown in **Figure 58**. Assemble the tool onto the piston and pull the piston pin out of the piston. Make sure to install a rubber or plastic pad between the piston and piece of pipe to avoid scoring the side of the piston.

NOTE
*Uncaged needle bearings (**Figure 59**) are used on the WR500. When removing the piston, the bearings and retaining washers will fall out. Do not drop or lose any uncaged bearings if reuse is intended.*

7. Use a proper size wooden dowel or socket extension and push the piston pin through the piston and connecting rod (**Figure 60**). Remove the piston from the rod and catch the bearings.

8. Repeat Steps 4-7 to remove the remaining piston.

Cylinder Head Cleaning and Inspection

1. Wipe away any soft deposits from the cylinder head gasket surface (**Figure 61**). Remove hard deposits from the combustion chambers using a wire brush mounted in a drill or drill press or with a soft-metal scraper and lacquer thinner. Be careful not to gouge the aluminum

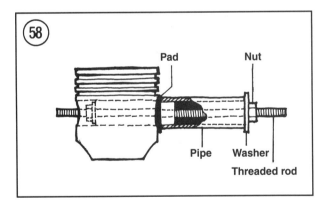

ENGINE (500 CC)

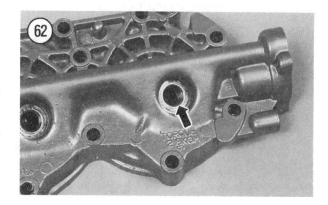

surfaces. Burrs created from improper cleaning will cause preignition and heat erosion.

2. Clean all salt and dirt from the cylinder head cooling passages.

3. With the spark plugs removed, check the spark plug threads in the cylinder head (**Figure 62**) for carbon buildup or cracking. The carbon can be removed with a 14 mm spark plug chaser.

NOTE
Always use an aluminum thread fluid or kerosene on the thread chaser and cylinder head threads when performing Step 3.

4. Once all carbon is removed, clean the cylinder head thoroughly with solvent and a brush.

5. Use a straightedge and feeler gauge and measure the flatness of the cylinder head (**Figure 63**). If the cylinder head warping exceeds the warp limit in **Table 2**, resurface the cylinder head as follows:

   a. Tape a piece of 400-600 grit wet emery sandpaper to a piece of thick plate glass or surface plate.

   b. Slowly resurface the head by moving it in figure-eight patterns on the sandpaper.

   c. Rotate the head several times to avoid removing too much material from one side. Check progress often with the straightedge and feeler gauge.

   d. If the cylinder head warping still exceeds the service limit, it will be necessary to have the head resurfaced by a machine shop. Note that removing material from the cylinder head mating surface will change the compression ratio. Consult the machinist on how much material to remove.

6. Wash the cylinder head in hot soapy water and rinse thoroughly before installation.

7. Check the condition of the anticorrosion paint applied to the head and touch up as required.

Cylinder Block
Cleaning and Inspection

The cylinder block and lower crankcase are matched and align bored assemblies (**Figure 64**). For this reason, do not attempt to assemble an engine with parts salvaged from other blocks. If the following inspection procedure indicates that either the block or crankcase requires replacement, replace both as an assembly.

1. Remove all carbon deposits from the exhaust ports (**Figure 65**) and water jacket passages. Use solvent along with a hardwood scraper or some other rounded scraper. Be careful not to scratch or gouge the areas while cleaning.

2. Clean all gasket residue from the cylinder block and crankcase gasket surfaces with lacquer thinner and a dull putty knife. Clean the aluminum surfaces carefully to avoid nicking them.

3. Clean all salt and dirt from the cylinder block water jackets and cooling passages. Check the condition of the anticorrosion paint applied to the block and touch up as required.

4. Use a straightedge and feeler gauge and check the cylinder block for warping (**Figure 66**). If warping meets or exceeds 0.1 mm (0.004 in.), resurface cylinder block surface on a large sheet of 400-600 grit wet sandpaper as described under *Cylinder Head Cleaning and Inspection*. If the 0.1 mm (0.004 in.) limit cannot be corrected, replace the cylinder block/crankcase cover assembly.

5. Check all of the cylinder block threads for corrosion or damage and clean with a suitable tap. Use an aluminum thread cutting fluid or kerosene when chasing threads.

NOTE
Cylinder measurement requires a precision inside micrometer or bore gauge. If you do not have the right tools, have a dealership or a machine shop take the measurements.

6. Wash each cylinder bore in solvent to remove any oil and carbon particles. The cylinder bore must be cleaned thoroughly before attempting any measurement to prevent incorrect readings.

7. Measure the cylinder bore diameter as described under *Piston/Cylinder Clearance Check* in this chapter.

8. Check each bore for any sign of aluminum transfer from the pistons to the cylinder wall. If

ENGINE (500 CC)

scoring is present but not excessive, the cylinders should be honed. Refer to *Cylinder Honing* in this chapter.

9. After the cylinder has been serviced, wash each bore in hot soapy water. This is the only way to clean the cylinder wall of the fine grit material left from the boring or honing job. After washing the cylinder wall, run a clean white cloth through it. The cylinder wall should show no traces of grit or other debris. If the rag is dirty, the cylinder wall is not clean and must be rewashed. After the cylinder is cleaned, lubricate the cylinder walls with clean engine oil to prevent the cylinder liners from rusting.

CAUTION
A combination of soap and water is the only solution that will completely clean the cylinder wall. Solvent and kerosene cannot wash fine grit out of cylinder crevices. Grit left in the cylinder will act as a grinding compound and cause premature wear to the new rings.

Piston Pin and Needle Bearing Cleaning and Inspection

The piston pins and needle bearings can be reused if they are in good condition.

CAUTION
*Do not intermix the front and rear piston pin and needle bearing assemblies (**Figure 67**) when performing the following procedures.*

1. Clean the piston pin in solvent or kerosene and dry thoroughly.
2. Check the piston pin for excessive wear, scoring or chrome flaking. Also check the piston pin for cracks along the top and side.
3. Clean the needle bearing assembly one set at a time to prevent any possible mix-up. Perform the following:
   a. Place one needle bearing set (including washers) in a suitable container filled with solvent or kerosene.
   b. Agitate the container to loosen all grease, sludge and other contamination.
   c. Remove the bearings from the container, and place them on a clean lint-free cloth or towel and allow to air dry.
   d. Check each needle bearing for flat spots. If one needle bearing is defective, replace the entire set.
4. Check piston pin and bearing play as follows:
   a. Wipe the inside of the connecting rod piston pin bore with a clean rag.
   b. Install the Yamaha connecting rod bearing installer (part No. YB-6107) or equivalent into the connecting rod bore (A, **Figure 68**) and insert needle bearings into rod individually (B, **Figure 68**).
   c. Align the piston pin with the connecting rod. Push pin through connecting rod, and catch the bearing installer as it falls out.

d. Rock the piston pin as shown in **Figure 69**. There should be no radial or axial play. If any play exists, replace the pin and bearing assembly.

e. Remove the piston pin and needle bearing assembly.

CAUTION
If there are signs of piston seizure or overheating, replace the piston pins and bearings. These parts have been weakened from excessive heat and may fail later.

Connecting Rod Inspection

1. Wipe the piston pin bore in the connecting rod with a clean rag, and check it for galling, scratches, or other wear or damage. If any of these conditions exist, replace the connecting rods as described in this chapter.

2. Check the connecting rod big end bearing play. You can make a quick check by simply rocking the connecting rod back and forth (**Figure 70**). If there is more than a very slight rocking motion (some side-to-side sliding is normal), measure the connecting rod side clearance with a feeler gauge. Measure between the side of the crankshaft and the washer (**Figure 71**). If the play exceeds the wear limit specified in **Table 2**, replace the crankshaft as individual replacement parts are not available from Yamaha.

NOTE
While Yamaha does not sell factory parts that allow the rebuilding of WR500 crankshafts, a few watercraft performance dealers are able to rebuild crankshafts.

Piston and Rings
Cleaning and Inspection

1. Carefully check the piston for cracks at the top edge of the transfer cutaways (**Figure 72**). Replace the piston if any are found. Check the piston skirt (A, **Figure 73**) for brown varnish buildup. More than a slight amount is an indication of worn or sticking rings.

2. Check the piston skirt for galling and abrasion which may have resulted from piston seizure. If light galling is present, smooth the affected area

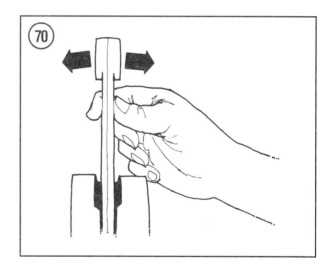

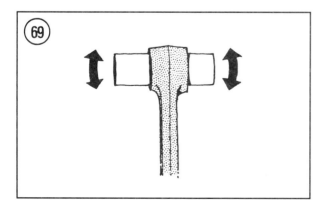

ENGINE (500 CC)

with No. 400 emery paper and oil or a fine oilstone. However, if galling is severe or if the piston is deeply scored, replace it.

3. Check the piston ring locating pins in the piston (**Figure 74**). The pins must be tight and the piston should show no sign of cracking around the pins. If a locating pin is loose, replace the piston. A loose pin will fall out and cause severe engine damage.

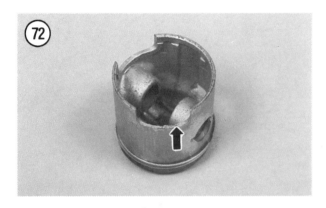

4. Check the piston pin clip grooves (B, **Figure 73**) in the piston for cracks or other damage that could allow a clip to fall out. This would cause severe engine damage. Replace the piston if even one groove is worn or damaged.

5. Lightly lubricate the piston pin with Yamalube Two-Cycle Outboard Oil and insert its end in the piston pin boss. Check for free play by moving the pin in the direction shown in **Figure 75**. The pin should be a hand press-fit with no noticeable vertical play. If play exists or if the pin is excessively loose, replace the pin and/or piston as required.

NOTE
Maintaining proper piston ring end gap helps to ensure peak engine performance. Always check piston ring end gap at the intervals specified in Chapter Three. Excessive ring end gap reduces engine performance and can cause overheating. Insufficient ring end gap will cause the ring ends to butt together and cause the ring to break. This would cause severe engine damage. So that you do not have to wait for parts, always order extra cylinder head and base gaskets to have on hand for routine top end inspection and maintenance.

6. Measure piston ring end gap. Place a ring into the top of the cylinder and push it in about 20

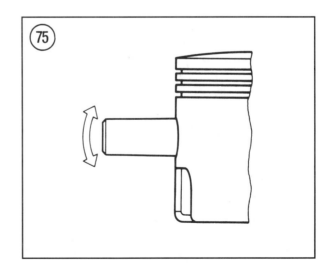

mm (3/4 in.) with the crown of the piston. This ensures that the ring is square in the cylinder bore. Measure the gap with a flat feeler gauge (**Figure 76**) and compare to the wear limit in **Table 2**. If the gap is greater than specified, replace the rings as a set.

NOTE
*When installing new rings, measure the end gap in the same manner as for old ones. If the gap is less than specified, make sure you have the correct piston rings. If the replacement rings are correct but the end gap is too small, carefully file the ends with a fine cut file until the gap is correct (**Figure 77**).*

CAUTION
To remove stubborn carbon deposits in Step 6, carefully scrape the ring groove with the recessed end of a broken ring. Do not use an automotive ring groove cleaning tool, as it will damage the piston ring locating pins.

7. Carefully remove all carbon buildup from the ring grooves with a broken ring (**Figure 78**). Inspect the grooves carefully for burrs, nicks, or broken and cracked lands. Recondition or replace the piston if necessary.

8. Observe the condition of the piston crown (**Figure 79**). Normal carbon buildup can be removed with a wire wheel mounted in a drill. If

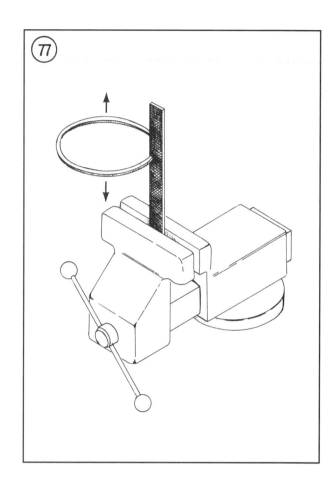

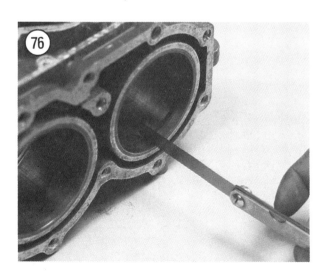

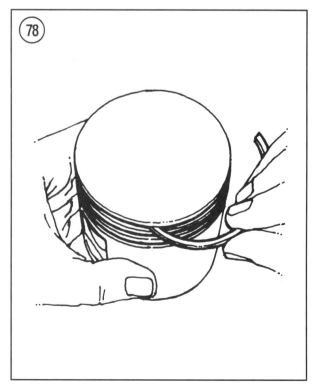

ENGINE (500 CC)

the piston shows signs of overheating, pitting or other abnormal conditions, the engine may be experiencing preignition or detonation; both conditions are discussed in Chapter Two.

9. If the piston checked out okay after performing these inspection procedures, measure the piston outside diameter as described under *Piston/Cylinder Clearance Check* in this chapter.

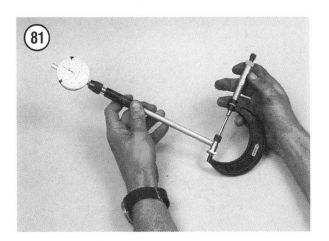

10. If new piston rings are required, hone the cylinders before assembling the engine. Refer to *Cylinder Honing* in this chapter.

Piston/Cylinder Clearance Check

The following procedure requires the use of highly specialized and expensive measuring tools. If such equipment is not readily available, have the measurements performed by at a dealership or machine shop. Always replace both pistons as a set.

1. Measure the outside diameter of the piston with a micrometer approximately 10 mm (0.4 in.) above the bottom of the piston skirt, at a 90° angle to the piston pin (**Figure 80**). If the diameter exceeds the wear limit in **Table 2**, install new pistons.

2. Wash the cylinder block in solvent to remove any oil and carbon particles. The cylinder bore must be cleaned thoroughly before attempting any measurement to prevent incorrect readings.

3. Using an outside micrometer, calibrate the bore gauge to the cylinder bore diameter. See **Figure 81**.

4. Measure the cylinder bore with the bore gauge (**Figure 82**). Measure the cylinder bore at the points shown in **Figure 83**. Measure in 3 axes—aligned with the piston pin and at 90° to the pin. If the bore diameter exceeds the

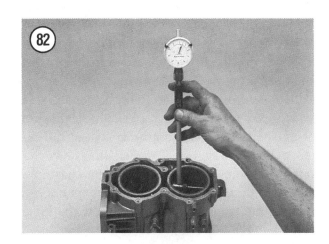

specification (**Table 2**), bore the cylinders to the next oversize and replace the pistons and rings.

NOTE
Always install new rings when installing a new piston.

NOTE
*Purchase the new pistons before the cylinders are bored so the pistons can be measured. The cylinders must be bored to match the pistons. Piston-to-cylinder clearance is specified in **Table 2**.*

4. Piston clearance is the difference between the maximum piston diameter and the minimum cylinder diameter. For a used piston and cylinder, subtract the piston diameter from the cylinder bore diameter. If the clearance exceeds the specification in **Table 2**, rebore the cylinders and install new pistons and rings.

Cylinder Honing

The surface condition of a worn cylinder bore is normally very shiny and smooth. Cylinder honing, often referred to as glaze breaking or deglazing, is required whenever new piston rings are installed. If new piston rings are installed in a cylinder with minimum wear, they will not seat properly if the cylinders are not first deglazed or honed. When a cylinder bore is honed, the surface is slightly roughed up to provide a textured or crosshatched surface. This surface finish controls wear of the new rings and helps them seat and seal properly. *Whenever* new rings are installed, the cylinder surface must be honed. This service can be performed at a Yamaha dealership or independent repair shop. The cost of having the cylinder honed at a dealership is usually minimal compared to the cost of purchasing a hone and doing the job yourself. If you choose to hone the cylinders yourself, follow the hone manufacturer's directions closely.

CAUTION
*After a cylinder is reconditioned by boring or honing, clean the bore to remove all material left from the machining operation. Refer to **Cylinder Inspection** in this chapter. Improper cleaning will not remove all of the machining residue in the cylinder bore and rapid wear of the new piston and rings will result.*

Crankshaft Cleaning and Inspection

Refer to **Figure 29** for this procedure.

NOTE
A set of V-blocks can be made out of hardwood to help hold the crankshaft in place when performing some of the

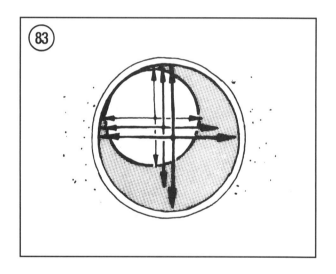

ENGINE (500 CC)

checks in the following steps. However, use only machined V-blocks to check crankshaft runout described later in this procedure.

1. Clean the crankshaft thoroughly with solvent or kerosene and a brush. Blow dry with dry filtered compressed air and lubricate with a light coat of Yamalube Two-Cycle Outboard Oil.

2. Check the crankshaft journals and crankpins for scratches, heat discoloration or other defects.

3. Check crankshaft splines, flywheel taper, threads (A, **Figure 84**) and keyway (B, **Figure 84**) for wear or damage.

4. Check crankshaft oil seal surfaces for grooving, pitting or scratches.

5. Check crankshaft bearing surfaces for rust, water marks, chatter marks and excessive or uneven wear. Minor cases of rust and water or chatter marks can be cleaned up with 320 grit carborundum cloth.

6. If 320 grit cloth is used, clean crankshaft in solvent and recheck surfaces. If they do not clean up properly, replace the crankshaft.

7. Carefully examine the condition of the crankshaft ball bearings (**Figure 85**). Clean the bearings in solvent, and let them dry thoroughly. Oil each bearing with Yamalube Two-Cycle Outboard Oil then turn the outer race. A worn or damaged bearing will sound or feel rough and will not rotate smoothly.

8. Grasp the outer race and try to work it back and forth. Replace the bearing if excessive axial play is noted. In all cases, defective bearings must be replaced.

NOTE
*The bearings installed on the outside of the crank wheels can be replaced as described under **Crankshaft Bearing Replacement** in this chapter. If the bearings installed between the crank wheels are worn or damaged, the crankshaft will either have to be replaced or rebuilt by a qualified dealership or performance shop.*

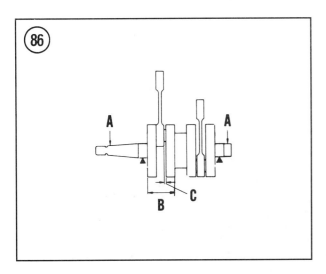

9. Support the crankshaft by placing it onto 2 V-blocks at the points shown in **Figure 86**. Then check runout with a dial indicator at the 2 points, (A, **Figure 86**). Turn the crankshaft slowly and note the gauge reading. The maximum difference recorded is the crankshaft runout at that position. If the runout at any position exceeds the service limit in **Table 2**, replace the crankshaft.

10. Measure the width of the crank wheels with a micrometer or vernier caliper at points marked, (B, **Figure 86**). Compare the readings to the crankshaft width specification listed in **Table 2**.

11. Check connecting rod deflection with a dial indicator. Place the end of a dial indicator so its plunger rests against the connecting rod's small end as shown in **Figure 87**. Then support the

crankshaft and push the connecting rod back and forth, noting the dial indicator gauge movement each time the connecting rod is moved; the difference in the 2 gauge readings is the small end freeplay. Repeat for each rod. If freeplay exceeds the service limit in **Table 2**, replace the crankshaft.

12. Check connecting rod big-end side clearance with a feeler gauge at the point marked, (C, **Figure 86**). Insert the gauge between the crankshaft and the washer as shown in **Figure 71**. If the clearance exceeds the service limit in **Table 2**, replace the crankshaft assembly.

13. Check the sealing ring on the labyrinth seal (A, **Figure 88**) and the circlip (B) on the inner bearing. The circlip can be replaced if it is worn or damaged. A worn or damaged labyrinth seal or ring will require crankshaft replacement.

Crankshaft Bearing Replacement

Replace the front (**Figure 89**) and rear (A, **Figure 90**) outer crankshaft bearings as follows. A bearing puller and splitter are required to remove the bearings. A press is required to install the bearings. Read this procedure completely through before removing the bearings.

NOTE
Before removing the rear bearing, note the position of the pin in the bearing (B, Figure 90). The pin aligns with a groove in the cylinder block when installing the crankshaft. The new bearing must be installed so its pin aligns exactly the same way. If the bearing is turned around, the pin will not align with the cylinder block groove. If so, the bearing must be removed and turned around.

CAUTION
Before removing the front bearing with a bearing puller, thread the flywheel nut onto the crankshaft so that it is flush with the end of the crankshaft. The nut will

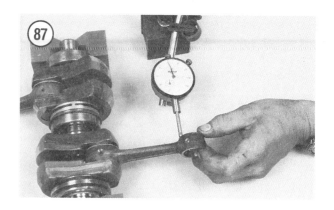

ENGINE (500 CC)

prevent the puller screw from damaging the crankshaft threads.

1. Install a bearing puller and remove the bearing (**Figure 91**). Then reposition the puller and remove the opposite bearing.
2. Check the crankshaft bearing surfaces for rust, water marks, chatter marks or uneven wear. Minor cases of rust and water or chatter marks may be cleaned with 320 grit carborundum cloth.

NOTE
*If 320 grit cloth was used, clean the crankshaft and oil the inner bearing as described under **Crankshaft Cleaning and Inspection** in this chapter.*

3. Clean the crankshaft bearing area with solvent and dry thoroughly.

NOTE
To ease bearing installation, heat the bearings in oil before installation. Completely read Step 4 through before heating and installing bearings. During bearing installation, support the crankshaft securely so the bearings can be installed quickly. If a bearing cools and tightens on the crankshaft before it is completely installed, remove the bearing with the puller and reheat.

4. Install the bearings as follows:
   a. Lay the bearings on a clean, lint-free surface in the order of assembly.
   b. When installing the rear bearing, make sure the bearing pin faces in the direction shown in B, **Figure 90**.
   c. Refer to *Shrink Fit* under *Ball Bearing Replacement* in Chapter One.
   d. After referring to the information in sub-step c, heat and install the bearings. Refer to **Figure 89** (front) or **Figure 90** (rear) during installation.
   e. After the bearings have cooled, check them by rotating the outer bearing race as described in this chapter to make sure the bearings were not damaged during installation.

Piston Assembly

If the pistons were removed from the connecting rods, they must be installed in their original position. The UP mark on the piston crown must face toward the flywheel end of the crankshaft. Install the pistons on their respective connecting rod by following the identification marks made prior to removal. If new pistons are being in-

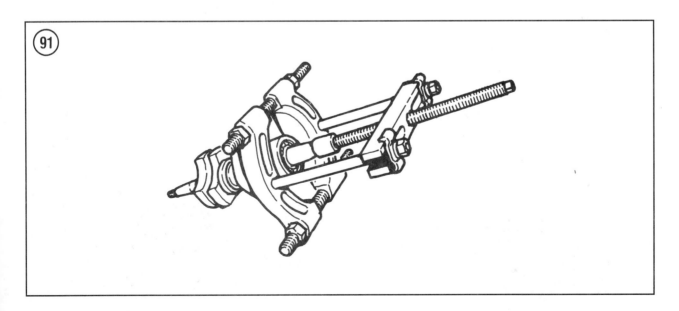

91

stalled and the cylinders were not bored, the pistons can be installed on either rod. If new pistons are being installed and the cylinders were bored out, follow the cylinder identification marks made by the machinist or dealership. Remember, cylinders are bored to fit individual pistons.

Use Yamalube Two-Cycle Outboard Oil whenever the procedure specifies lubrication with oil. Use Yamaha Marine Grease where grease is specified.

1. Check the piston crown number made during disassembly and match each piston with its correct connecting rod. Identify the piston pin and needle bearing assembly with its correct piston (**Figure 92**).

2. Coat the piston pin and piston pin bore with oil.

3. Wipe the connecting rod small end bore with grease.

4. Partially insert the piston pin into the piston (**Figure 93**).

5. Insert the Yamaha connecting rod bearing installer (part No. YB-6107) or equivalent into the connecting rod bore (**Figure 94**) and insert the 34 individual needle bearings into the rod (**Figure 95**). **Figure 96** shows the rod with all of the needles correctly installed.

6. When all bearings are in place, fit the washers on both sides of the bearing assembly (**Figure 97**) so their convex side faces the piston. See **Figure 98**.

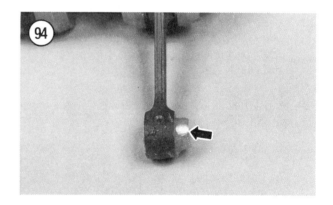

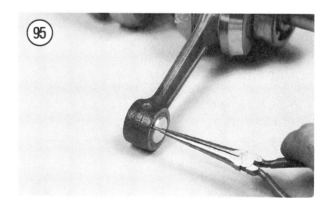

ENGINE (500 CC)

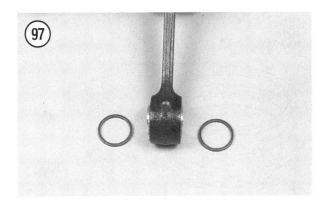

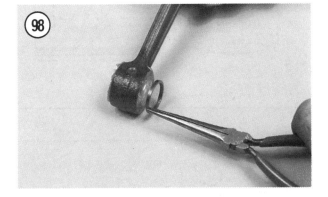

7. Align the piston with its stamped UP mark facing the flywheel end of crankshaft and place the piston carefully over the connecting rod to avoid disturbing the washers or bearings.

8. Align the piston pin bore and the connecting rod bore, then push the piston pin (**Figure 99**) through the connecting rod and catch the bearing installer as it falls out. See **Figure 100**.

WARNING
Wear safety glasses when performing Step 9.

9. Install new piston pin clips with needlenose pliers (**Figure 101**). Make sure they are seated in the grooves and that the end of each clip is *not* opposite the slots in the piston (**Figure 102**).

10. Check the installation by rocking the piston back and forth around the pin axis and from side to side along the axis. It should rotate freely back and forth but not from side to side.

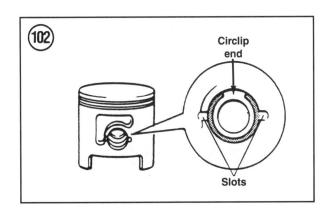

11. Repeat for the opposite piston. **Figure 103** shows both pistons properly installed and facing in the correct position.

Piston Ring Installation

The stock pistons used on the WR500 use a keystone design top ring and a square or flat bottom ring (**Figure 104**). Check new rings carefully before installation and install them in their proper grooves.

1. Check end gap of new rings as described in this chapter. If ring gap is insufficient, the ends of the ring can be filed slightly. Clean the ring thoroughly and recheck its gap. See **Table 2** for specifications.

NOTE
Piston rings must be installed in Step 2 with the mark on the end of the ring facing the piston crown. See Figure 105.

2. If you are installing used rings, install them by referring to the piston and ring identification marks made prior to removal.

3. Install the piston rings-first the bottom one, then the top-with a piston ring expander tool (**Figure 106**). Make sure manufacturer's mark on the piston rings are toward the top of the piston (**Figure 105**).

4. Make sure the rings are seated completely in the grooves, all the way around the circumfer-

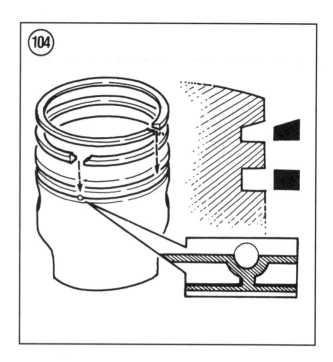

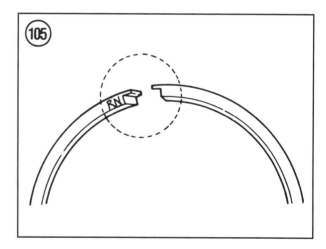

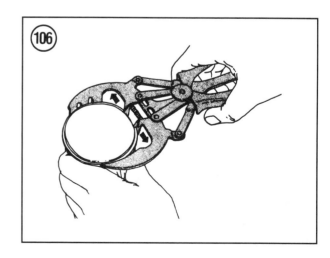

ENGINE (500 CC)

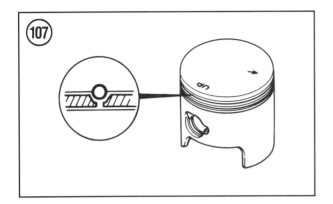

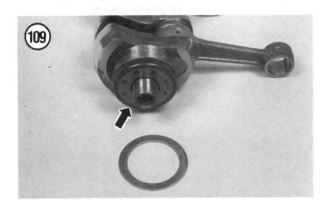

ence, and that the ends are aligned with the locating pins. See **Figure 107**.

Crankshaft Installation

Refer to **Figure 29** for this procedure.

Use Yamalube Two-Cycle Outboard Oil whenever the procedure specifies lubrication with oil. Use Yamaha Marine Grease where grease is specified.

1. Place the crankshaft on a clean workbench surface. Install the crankshaft seal and spacers (**Figure 108**) as follows:
   a. Wipe the spacers with a clean rag and then coat them with oil.
   b. Fill the crankshaft oil seal lip cavity with grease.
   c. Install the first spacer, crankshaft seal and second spacer. Install the seal so its closed end faces out as shown in **Figure 109**.

2. Lubricate the cylinder wall and the connecting rod bearings and the piston assembly with oil.

3. Check each piston ring, and align the ends with its piston locating pin as shown in **Figure 107**. This step is essential so the piston rings' ends will be correctly positioned and will not snag in the ports.

4. Grasp crankshaft assembly with one hand and position it over crankcase. Slowly lower the assembly until the No. 1 piston (front) enters the crankcase, then compress each piston ring by hand and push the piston downward until all rings have entered the cylinder.

5. Rotate the crankshaft in your hand until the No. 2 piston is about to enter its cylinder, then compress each piston ring by hand and push the piston downward until all rings have entered the cylinder.

6. When the piston/ring assemblies have entered the cylinder bores (**Figure 110**), apply sufficient downward pressure to seat the crankshaft in the crankcase.

7. Rotate the center and rear crankshaft ball bearings races to align their locating pin with the pin recess in the crankcase. See A, **Figure 111** and A, **Figure 112**. Then align the bearing washers and the labyrinth seal and center bearing circlips in their respective crankcase grooves. See B, **Figure 111** and B, **Figure 112**.

8. Insert a screwdriver blade or pencil point through the exhaust ports and depress each piston ring slightly (**Figure 113**). The ring should snap back when pressure is released. If it does not, the ring was broken during piston installation. Remove the crankshaft assembly from the crankcase and replace the broken ring.

Engine Assembly

Refer to **Figure 39** for this procedure.

1. Clean all engine fasteners in solvent, and dry them thoroughly, making sure to remove all traces of sealer or Loctite residue from the bolt threads. Examine all of the bolts and screws for head or thread damage. Check the plain washers for cupping, cracks or other damage. Check the lockwashers for weakness or damage. Replace worn or damaged fasteners as required.

2. Install the crankshaft assembly into the cylinder block assembly as described in this chapter.

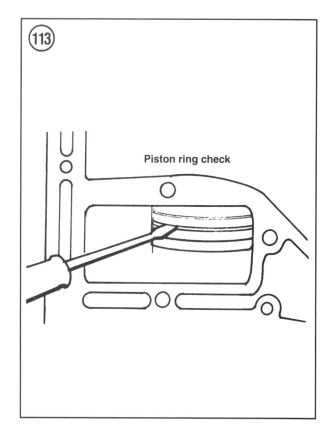

ENGINE (500 CC)

3. Make sure the crankcase mating surfaces are clean and free of all oil residue.

4. Install the 2 crankcase dowel pins (**Figure 114**).

5. Apply a thin coat of Yamabond 4 or equivalent, to the cylinder block and crankcase mating surfaces (**Figure 115**).

6. Align and install the lower crankcase (A, **Figure 116**) onto the cylinder block assembly. Check the mating surfaces all the way around the case halves to make sure they are even.

CAUTION
Do not force the lower crankcase. If the crankcase will not seat all the way around, the shaft is not seated correctly. Remove the lower crankcase and check the crankshaft as described in this chapter. Forcing the case halves together will destroy the expensive aluminum case halves.

7. Install the 4 dowel pins into the lower crankcase as shown in B, **Figure 116**.

8. Place the front and rear engine brackets onto the lower crankcase, making sure to align the brackets with the dowel pins installed in Step 7. See **Figure 117**.

NOTE
*Check once more to make sure the crankcase halves are seated properly (**Figure 118**).*

9. Apply Loctite 242 to the threads of the crankcase bolts and install the bolts with their washers. Run the bolt down finger-tight. Then torque the crankcase bolts in 2 steps in the numerical order shown in **Figure 119**. The bolt tightening sequence is also embossed on the crankcase and engine brackets. Refer to **Table 3** for torque specifications.

CAUTION
*When tightening the crankcase fasteners, make frequent checks to ensure that the crankshaft turns freely. While the pistons and rings will offer a degree of resistance, you should be able to turn the crankshaft by hand. If it appears that the crankshaft is binding, check for one or more broken piston rings. See **Crankshaft Installation**. If the rings are okay, separate the crankcase halves and determine the cause of the problem.*

10. Turn the engine assembly over so the pistons face up. Clean the cylinder block mating surface of all oil and other residue.

11. Install the anode (**Figure 120**) into the cylinder block water passage and secure it with its mounting screw. Make sure the anode is in good condition. Replace if anode is less than 60% of its original size.

12. Install a new cylinder head gasket (**Figure 121**) and place the cylinder head (**Figure 122**) onto the engine. Apply Loctite 242 to the threads of the cylinder head bolts, and install the bolts and their washers finger-tight.

13. Install the muffler assembly as described in this chapter.

14. Wipe the end of the thermosensor with grease, and insert it into the cylinder head. Align the sensor bracket with the cylinder head bolt hole and install the bolt hand-tight.

15. Tighten the cylinder head bolts in the numerical order shown in **Figure 123** in 2 steps. Tighten to the torque specifications listed in **Table 3**.

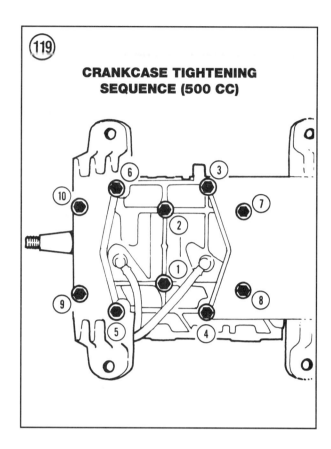

ENGINE (500 CC)

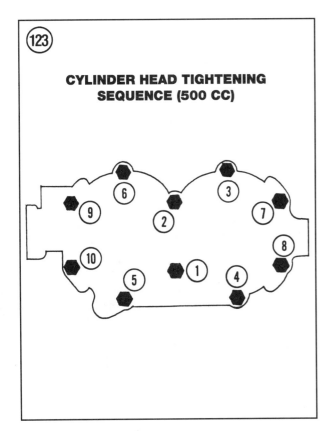

16. Install the intake manifold (**Figure 124**) and reed valve assembly with a new gasket. Wipe the intake manifold bolt threads with Loctite 242 (blue). Install the bolts and tighten securely. Stuff a clean rag into the intake manifold opening until the carburetor can be installed.

17. Before installing the seal housing, inspect the seal for tears or damage. If necessary, replace the seal as follows:

    a. Knock the oil seal out of the housing with a suitable size bearing driver or socket. See **Figure 125**.

    b. Clean the housing bore with solvent and dry thoroughly.

    c. Pack the seal lips with marine grease.

    d. Align the new seal with the housing so the open end faces out as shown in **Figure 126**.

e. Drive the new seal into the housing with a suitable bearing driver or socket (**Figure 127**) until it bottoms out.

18. Install a new gasket onto the oil seal housing (**Figure 128**). Then carefully slide the oil seal housing over the flywheel end of the crankshaft so the oil seal is not damaged. Turn the housing and align the mounting bolt holes and threads. Apply Loctite 242 (blue) to the bolt threads and tighten securely. See **Figure 129**.

19. Install the stator plate and flywheel as described in Chapter Nine.

20. Install the starter motor as described in Chapter Nine.

21. Install the cylinder head end cover (**Figure 130**), if previously removed. Install using a new gasket.

22. Reconnect the 2 engine hoses to the upper (**Figure 131**) and lower (bottom side of crankcase) nozzles.

23. Install the muffler housing and exhaust guide assembly as described in this chapter.

24. Reinstall the engine in the hull as described in this chapter.

25. Refer to *Break-in Procedure* in Chapter Three.

ENGINE (500 CC)

Table 1 ENGINE SPECIFICATIONS (500 CC)

Bore × stroke	72 × 61 mm (2.83 × 2.40 in.)
Displacement	496 cc (30.27 cu. in.)
Compression ratio	7.0:1

Table 2 ENGINE SERVICE SPECIFICATIONS (500 CC)

	New mm (in.)	Wear limit mm (in.)
Cylinder head warp limit	–	0.1 (0.0039)
Cylinder		
Bore	72.00-72.02 (2.834-2.835)	–
Taper limit	–	0.08 (0.003)
Out of round limit	–	0.05 (0.002)
Piston		
Diameter	71.735-71.960 (2.832-2.835)	–
Measuring point	10 (0.39) (from bottom of skirt)	–
Piston clearance	0.06-0.065 (0.0024-0.0026)	0.1 (0.004)
Piston rings		
Type		
Top	Keystone	–
Bottom	Plain (square)	–
End gap		
Top	0.2-0.4 (0.008-0.016)	–
Bottom	0.2-0.35 (0.008-0.014)	–
Side clearance		
Top	0.03-0.05 (0.0012-0.0020)	–
Bottom	0.03-0.07 (0.0012-0.0028)	–
Crankshaft		
Crank width	61.95-62.00 (2.439-2.441)	–
Runout limit	–	0.03 (0.0012)
Big end side clearance	0.3-0.7mm (0.012-0.0028)	–
Small end freeplay limit	–	2.0 (0.08)

Table 3 ENGINE TIGHTENING TORQUE (500 CC)

	N·m	in.-lb.	ft.-lb.
Cylinder head			
1st step	15	–	11
2nd step	30	–	22
Crankcase bolts			
1st step	15	–	11
2nd step	28	–	20
Exhaust cover			
1st step	4	35	–
2nd step	8	71	–
Exhaust guide			
1st step	10	88	–
2nd step	20	–	14
Flywheel	140	–	103
Engine mounting bolts	17	–	12
Coupling half	37	–	27

Chapter Five

Engine (650, 700 and 760 cc)

The 650, 700 and 760 cc engines are 2-stroke twin cylinder marine engines. Each engine is mounted inside the hull and is secured to 2 sets of engine mounts with stainless steel bolts. An aluminum coupler half is mounted onto the end of the crankshaft and to the front of the drive shaft. A rubber coupler is used to engage the crankshaft and drive shaft couplings. The coupler cushions crankshaft and drive shaft engagement and absorbs small amounts of drive train misalignment resulting from engine vibration and drive shaft runout.

This chapter provides information covering routine top-end service as well as crankcase disassembly and crankshaft service.

Work on your water vehicle engine requires considerable mechanical ability. You should carefully consider your own capabilities before attempting any operation involving major disassembly of the engine.

Much of the labor charge for dealer repairs involves the removal and disassembly of other parts to reach the defective component. Even if you decide not to tackle the entire engine overhaul after studying the text and illustrations in this chapter, it can be less expensive to perform the preliminary operations yourself and then take the engine to your dealer. Since dealers have a lengthy waiting list for service (especially during spring and summer), this practice can reduce the time your unit is in the shop. If you have done much of the preliminary work, your repairs can be scheduled and performed much quicker.

General engine specifications are listed in **Table 1**. **Tables 1-4** are found at the end of the chapter.

ENGINE LUBRICATION

On FX700, SJ650, SJ700, SJ700A and WR650, engine lubrication is provided by the fuel/oil mixture used to power the engine. Refer to Chapter Three for oil and ratio recommendations.

On RA700, RA700A, RA700B, RA760, WB700, WB700A, WB760, WRA650, WRA650A, WRA700. WRB650, WRB650A, WRB700 and WVT700, engine lubrication is provided by an oil injection system. Refer to Chapter Ten for details.

ENGINE (650, 700 AND 760 CC)

SERVICE PRECAUTIONS

Whenever you work on the engine, there are several precautions that should be followed to help with disassembly, inspection and reassembly.

1. Before beginning the job, read Chapter One of this manual. You will do a better job with this information fresh in your mind.

2. In the text there is frequent mention of the left-hand and right-hand sides of the engine. This refers to the engine as it is mounted in the hull, not as it sits on your workbench. Left and right refers to the rider's point of view when seated on the watercraft and facing forward. See **Figure 1**.

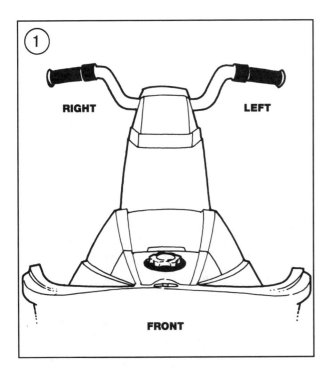

3. Always replace a worn or damaged fastener with one of the same size, type and torque requirements. Stainless steel fasteners are used throughout the watercraft. Make sure to identify each fastener before replacing it with another. Fastener threads should be lubricated with engine oil, unless otherwise specified, before torque is applied. If a tightening torque is not listed in **Table 3** or **Table 4** (end of this chapter), refer to the general torque and fastener information in Chapter One.

4. Use special tools where noted. In some cases, it may be possible to perform the procedure with makeshift tools, but this procedure is not recommended. The use of makeshift tools can damage the components and may cause serious personal injury. Where special tools are required, they may be purchased through any Yamaha watercraft dealership. Other tools can be purchased through your dealer, or from a motorcycle or automotive accessory store. When purchasing tools from an automotive accessory dealer or store, remember that all parts that attach to the engine must have metric threads.

5. Before removing the first bolt, get a number of boxes, plastic bags and containers and store the parts as they are removed (**Figure 2**). Also have on hand a roll of masking tape and a permanent, waterproof marking pen to label each part or assembly as required. If your craft was purchased used and it appears that some of the wiring may have been changed or replaced, label each electrical connection before disconnecting it.

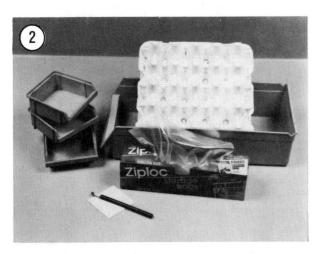

6. Use a vise with protective jaws to hold parts. If protective jaws are not available, insert wooden blocks on each side of the part(s) before clamping them in the vise.

7. Remove and install pressed-on parts with an appropriate mandrel, support and hydraulic

press. Do not try to pry, hammer or otherwise force them on or off.

8. Refer to **Table 3** or **Table 4** at the end of the chapter for torque specifications. Proper torque is essential to ensure long life and satisfactory service from marine components.

9. Discard all O-rings and oil seals during disassembly. Apply a small amount of marine grease to the inner lips of each oil seal to prevent damage when the engine is first started.

10. Keep a record of all shims and where they came from. As soon as the shims are removed, inspect them for damage and write down their thickness and location.

11. Work in an area where there is sufficient lighting and room for component storage.

SERIAL NUMBERS

Yamaha water vehicles are identified by hull and engine identification numbers. The hull number is stamped on a plate mounted on rear of the footrest floor (**Figure 3**) or outside the hull beside the steering nozzle. The engine number is stamped on a plate mounted on the intake manifold box (**Figure 4**) or on the crankcase. The Primary identification number is stamped on a plate mounted inside the engine compartment.

Because Yamaha may make a number of design changes during production or after the craft is sold, always use these numbers when ordering replacement parts.

At the front of this book, write down all of the identification numbers for your craft.

SERVICING ENGINE IN HULL

Some of the components can be serviced while the engine is mounted in the hull:

a. Carburetor.
b. Flywheel.
c. Starter motor.

SPECIAL TOOLS

Where special tools are required or recommended for engine overhaul, the tool part numbers are provided. Yamaha tool part numbers have a YB, YU or YW prefix. These tools can be purchased at a Yamaha water vehicle dealership.

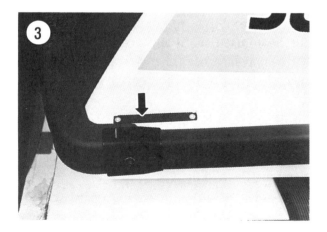

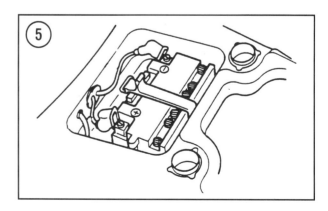

ENGINE (650, 700 AND 760 CC)

PRECAUTIONS

Because of the explosive and flammable conditions that exist around gasoline, always observe the following precautions.

1. Immediately after removing the engine cover, check for the presence of raw gasoline fumes. If you can smell strong fumes, determine their source and correct the problem.
2. Allow the engine compartment to air out before beginning work.

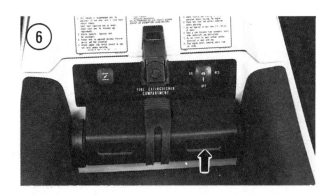

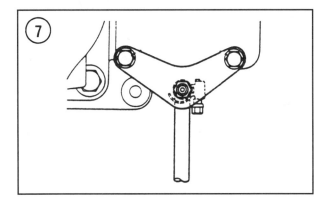

3. Disconnect the negative battery cable. See **Figure 5**.
4. Gasoline dripping onto a hot engine component may cause a fire. Always allow the engine to cool completely before working on any fuel system component.
5. Wipe up spilled gasoline immediately with dry rags. Store the rags in a suitable metal container until they can be cleaned or disposed of. Do not store gasoline or solvent soaked rags in the engine compartment.
6. Do not service any fuel system component while in the vicinity of an open flame, sparks or while anyone is smoking.
7. Always have a Coast Guard approved fire extinguisher close at hand when working on the engine.

ENGINE REMOVAL

Engine removal and crankcase separation is required for repair of the bottom end (crankshaft, connecting rod and bearings) and for removal of the drive shaft.

1. Support the water vehicle on a stand or on wooden boxes so it is secure.
2. Remove the engine cover.
3. *SJ650*—Remove the fire extinguisher box (**Figure 6**).
4. Remove the battery as described in Chapter Nine.
5. Disconnect the spark plug wires from the plugs. Then loosen but do not remove the spark plugs.
6. Remove the carburetor as described in Chapter Eight.
7. Remove the grease nipple bracket from the intake manifold (**Figure 7**).
8A. *FX700, RA700B, SJ650, SJ700, SJ700A WB700, WB700A, WRA650, WRA650A, WRA700, WRB650, WRB650A and WRB700*—If muffler assembly service is required, remove it as described in this chapter. If not, disconnect the water hose from the cylinder head (**Figure 8**), the

inlet water hose (A, **Figure 9**) from the exhaust manifold, the exhaust hose (B, **Figure 9**), from the muffler housing and the bypass hose from the exhaust elbow.

8B. *RA700, RA700A, RA760, WB760 and WVT700*—If muffler assembly service is required, remove it as described in this chapter. If not, disconnect the inlet water hose (A, **Figure 10**) from the exhaust manifold, the exhaust hose (B, **Figure 10**), from the exhaust housing and disconnect the bypass hose (C, **Figure 10**) from the exhaust elbow. Also disconnect the water outlet hose (**Figure 11**) from the exhaust housing.

9. Disconnect the starter switch and stop switch electrical connectors.

10. Disconnect the meter connector on models with display meters.

11. Remove the bolts securing the electrical box to the engine compartment. See **Figure 12**. On FX700, SJ650, SJ700 and SJ700A models, the electrical box mounting bolts are removed from outside of the hull. Set the electrical box aside; it will be removed with the engine.

12. On oil injection models, remove the oil pump and sub-oil tank as described in Chapter Ten. Plug all exposed oil lines to prevent oil leakage and/or contamination.

13. Unbolt and remove the coupling cover from the rear of the engine.

14. If necessary, the engine top end (cylinder head, pistons and cylinder block) can be removed before removing the engine from the hull. Refer to *Cylinder* in this chapter.

NOTE
Shim packs are used underneath the engine mount bolts to achieve proper engine-to-drive shaft alignment. Identify the shim packs during removal so they can be reinstalled in their original position.

15. Loosen the 4 engine mounting bolts (**Figure 13**). Remove the bolts and shims.

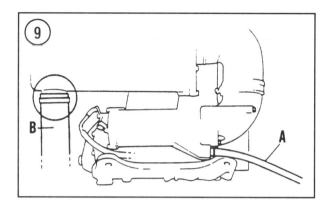

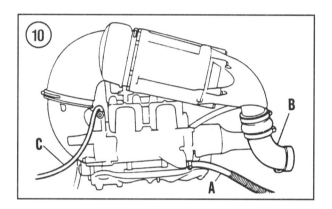

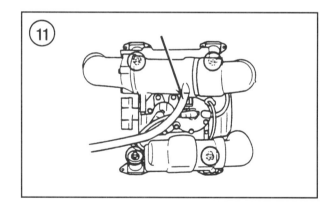

ENGINE (650, 700 AND 760 CC)

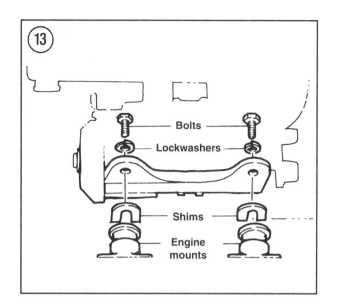

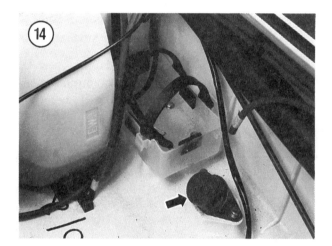

16. Check to make sure all wiring and hoses are disconnected from the engine.

WARNING
A minimum of 2 people are required to remove the engine assembly in Step 17.

17. Slide the engine forward to disengage it from the coupler then lift it up and out of the engine compartment together with the electric box. Take it to a workbench for further disassembly.

18. Remove the rubber coupler.

Installation

1. Wash the engine compartment with clean water.
2. Spray all of the exposed electrical connectors with electrical contact cleaner.
3. Now is a good time to check components which are inaccessible with the engine installed. Check all components for damage and replace as required.
4. Remove and clean the bilge filter as described in Chapter Eleven.
5. Examine the engine mounts and supports for looseness or damage.
6. Check inside the hull for tools or other objects that may interfere with engine installation. Make sure all hoses, cables and wiring connectors are routed and secured properly.
7. *WR650*: Glue the rubber damper to the underside of the engine before installing the engine in the hull.
8. Install the rubber engine mounts (**Figure 14**) into the hull, if previously removed.
9. Check the rubber coupler (**Figure 15**) for wear or damage. Wear grooves or rubber deterioration are signs of wear. Especially check for wear on the round outer knobs; replace if the coupler is worn so much that any rotational play exists between the engine half and the drive shaft coupler half.

CHAPTER FIVE

10. Install the rubber coupler into the forward bearing housing coupling.

NOTE
Before installing the engine into the hull on WR650 models, install a 6 × 25 mm bolt into each of the engine bracket threaded holes as shown in Figure 16. These bolts will be used to jack the engine into position when aligning the couplers and checking coupling alignment.

11. With an assistant, place the engine in the hull and slide it backward, meshing the engine and bearing housing couplings into the rubber coupler.

12. *WR650*—Place a 55 mm (2.17 in.) gauge block underneath the flywheel cover. See **Figure 17**.

NOTE
You can make a gauge block from a piece of aluminum and pieces of shim stock to obtain the correct thickness for your model.

13. Temporarily install the 4 engine mount bolts and washers (**Figure 13**) finger tight. Do not tighten the bolts; they are installed to align the engine brackets with the rubber mounts.

14. Check coupling alignment as follows:

  a. Hold a straightedge against one side of the coupling halves (**Figure 18**). Push the straightedge against the flat sides of the couplings. If you can see a gap between one coupling flat and the straightedge or you can feel the straightedge rock as you push against one coupling flat and then the other, measure the clearance with a feeler gauge. See A, **Figure 18**. If the clearance exceeds 0.6 mm (0.024 in.), the coupling halves are misaligned. Repeat this check with the straightedge against the other side and then on top of the coupler halves.

  b. To correct misalignment, reposition the engine with the jack bolts previously installed in the engine brackets. Repeat substep a to

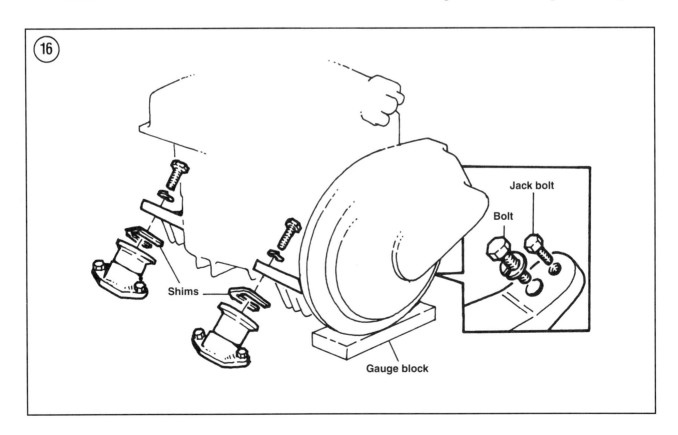

ENGINE (650, 700 AND 760 CC)

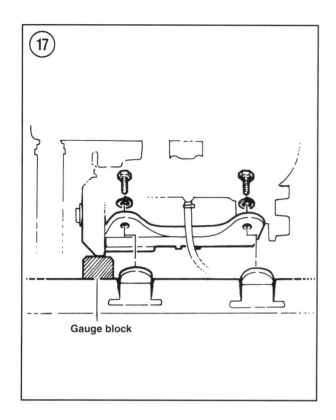

Gauge block

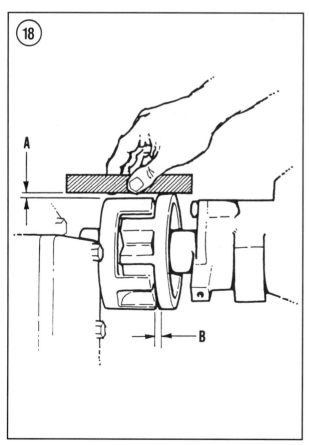

check alignment and continue until alignment is correct.

c. When alignment is correct, install the required shims in the gap between the engine mounts and brackets. Shims can be purchased in the following thicknesses: 0.1, 0.3, 0.5, 1.0 and 2.0 mm.

d. After installing the shims, measure the gap between the front and rear coupling halves as shown in B, **Figure 18**. The correct clearance is 2-4 mm (0.079-0.157 in.). If the clearance is incorrect, reposition the engine by pushing it forward or rearward as required. Recheck clearance.

e. *WR650*—Remove the gauge block from underneath the flywheel cover.

f. Wipe the engine mounting bolt threads with Loctite 271 (red), then install and tighten the bolts to the torque specification listed in **Table 3** or **Table 4**.

g. Recheck engine alignment.

CAUTION
It is important to align the coupler halves as closely as possible. Any significant degree of misalignment will cause excessive vibration and result in damage to the drive train components.

15. Apply Loctite 242 (blue) to each of the coupler cover bolts. Then install the coupler cover with the bolts, flat washers and lockwashers. Turn the drive shaft and check that the coupler assembly does not contact the coupler cover.

16. Reverse Steps 1-12 in *Engine Removal* to complete installation. Note the following:

a. Use new hose or cable guides where required.

b. Adjust the throttle cable and choke cable as described in Chapter Three.

c. On oil injection models, refill the oil tank as described in Chapter Three.

CHAPTER FIVE

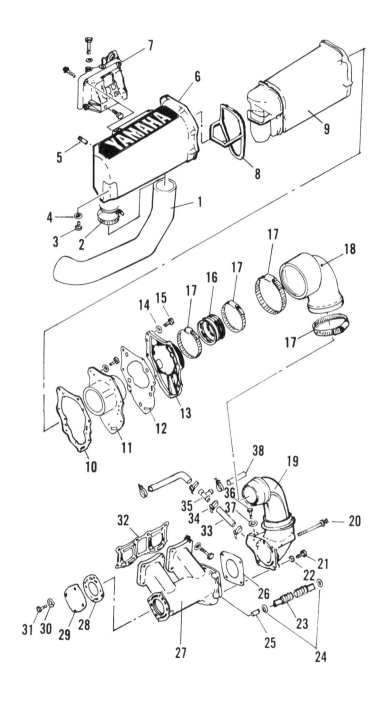

⑲

MUFFLER HOUSING
(RA700B, SJ650, SJ700, SJ700A, FX700, WB700, WB700A, WR650, WRA650, WRA650A, WRA700, WRB650, WRB650A AND WRB700)

1. Exhaust hose
2. Clamp
3. Bolt
4. Washer
5. Nozzle
6. Muffler housing
7. Muffler bracket
8. Seal
9. Silencer
10. Gasket
11. Diffuser
12. Gasket
13. Outer cover
14. Washer
15. Bolt
16. Joint
17. Clamps
18. Exhaust boot
19. Exhaust elbow
20. Bolt
21. Bolt
22. Washer
23. Intake water hose
24. Clamps
25. Nozzle
26. Elbow gasket
27. Exhaust manifold
28. Gasket
29. Exhaust manifold cover
30. Washer
31. Bolt
32. Exhaust manifold gasket
33. Bypass hose
34. Clamps
35. T-connector*
36. Bolt
37. Washer
38. Pilot water hose*

*SJ700, SJ700A, FX700, WB700A only

ENGINE (650, 700 AND 760 CC)

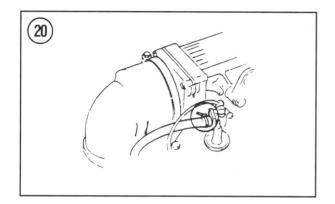

MUFFLER HOUSING

The muffler housing can be removed and installed with the engine installed in the hull.

Removal

FX700, RA700B, SJ650, SJ700, SJ700A, WB700, WB700A, WR650, WRA650, WRA650A, WRA700, WRB650, WRB650A and WRB700

Refer to **Figure 19** for this procedure.
1. Remove the engine cover.
2. Disconnect the water hose from the cylinder head (**Figure 20**).
3. Disconnect the small muffler-to-cylinder head hose (A, **Figure 21**).
4. Disconnect the exhaust hose (B, **Figure 21**) from the muffler housing.
5. Loosen the hose clamp at the opposite end of the muffler housing (**Figure 22**). Then pull back the exhaust boot and loosen the inner hose clamp (**Figure 23**).
6. Remove the muffler housing mounting bolts from the muffler bracket (C, **Figure 21**) and remove the muffler housing (D, **Figure 21**).
7. If necessary, remove the muffler bracket mounting bolts and remove the bracket.
8. *FX700, SJ700, SJ700A and WB700A*—Disconnect the bypass hose (A, **Figure 24**) from the T-connector (B, **Figure 24**).

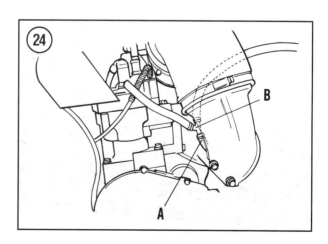

9. Disconnect the inlet water hose from the exhaust manifold then remove the exhaust manifold (**Figure 25**) from the cylinder block.

RA700, RA700A, RA760, WB760 and WVT700

Refer to **Figure 26** and **Figure 27** for this procedure.

1. Remove the engine cover.
2A. *RA700, RA700A and WVT700*—Disconnect the bypass hose (A, **Figure 24**) from the T-connector (B, **Figure 24**).
2B. *RA760 and WB760*—Disconnect the bypass hoses from the exhaust elbow and remove the thermoswitch from the muffler housing.
3A. *RA700, RA700A and WVT700*—Disconnect the water outlet hose (A, **Figure 28**) from the muffler housing.
3B. *RA760 and WB760*—Disconnect the water outlet hose from the exhaust boot (B, **Figure 28**).
4. Loosen the hose clamp and separate the outer cover (B, **Figure 28**) from the exhaust joint. On RA760 and WB760 models, note the position of the water diffuser alignment marks (**Figure 29**). If you remove the water diffusers, you must reinstall them with the alignment marks at 6 and 9 o'clock.
5. Loosen the hose clamp at the opposite end of the muffler housing. Then pull back the exhaust boot and loosen the inner hose clamp (**Figure 30**).

MUFFLER HOUSING (RA700 AND WVT700

1. Exhaust boot
2. Clamps
3. Bolt
4. Washer
5. Outer cover
6. Gasket
7. Muffler bracket
8. Outlet water hose
9. Clamp
10. Bolt
11. Muffler housing
12. Gasket
13. Silencer
14. Gasket
15. Outer cover
16. Washer
17. Bolt
18. Clamps
19. Exhaust joint
20. Exhaust pipe
21. Bolt
22. Washer
23. Outer manifold cover
24. gasket
25. Inner manifold cover
26. Inlet water hose
27. Clamp
28. Nozzle
29. Exhaust manifold
30. Gasket
31. Exhaust manifold gasket
32. Exhaust elbow
33. Nozzle
34. Clamp
35. Bypass hose
36. T-connector
37. Joint
38. Clamps
39. Seal

# ENGINE (650, 700 AND 760 CC) 179

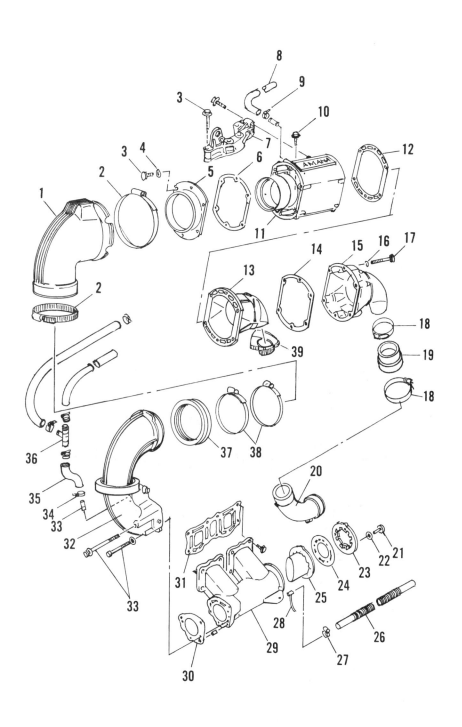

CHAPTER FIVE

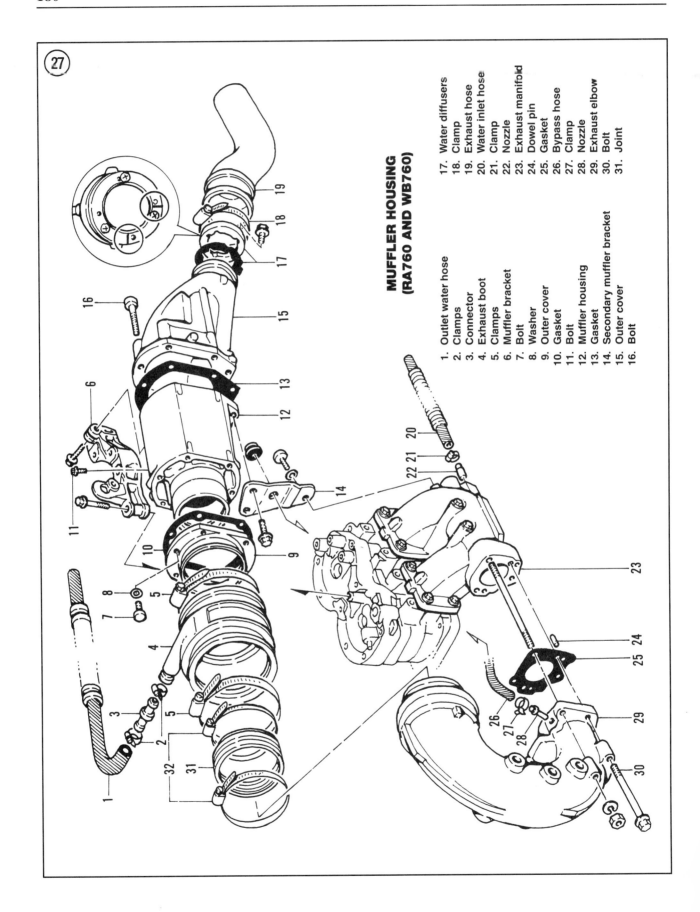

ENGINE (650, 700 AND 760 CC)

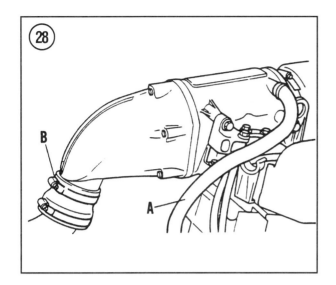

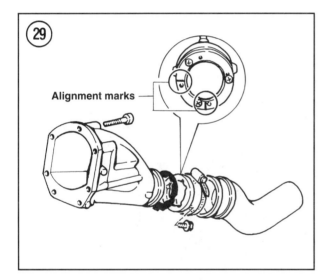

Alignment marks

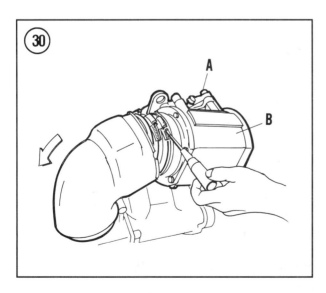

6. Remove the muffler housing mounting bolts from the muffler bracket (A, **Figure 30**) and remove the muffler housing (B, **Figure 30**). On RA760 and WB760 models, remove the secondary muffler bracket from the exhaust manifold (**Figure 27**) then remove the muffler housing.

7. If necessary, remove the muffler bracket mounting bolts and remove the bracket.

8. Disconnect the inlet water hose from the exhaust manifold and remove the exhaust manifold from the cylinder block.

Muffler Housing
Disassembly/Reassembly

Refer to **Figure 19**, **Figure 26** and **Figure 27** for this procedure.

1. Unbolt the outer cover (**Figure 31**) and remove it from the end of the muffler housing.

2A. *FX700, RA700B, SJ650, SJ700, SJ700A, WB700, WB700A, WR650, WRA650, WRA650A, WRA700, WRB650, WRB650A and WRB700*—Remove the diffuser, silencer and seal from the muffler housing.

2B. *RA700, RA700A and WVT700*—Remove the silencer and the seal from one end of the muffler housing and remove outer cover No. 2 from the other end.

2c. *RA760 and WB760*—Remove outer cover No. 2 from the muffler housing.

3. Clean all parts to remove all carbon and oil residue.

4. Carefully remove all gasket residue from all mating surfaces.

5. After all parts are cleaned, visually check for cracks or other damage. Check the seal for cracks or damage. Replace all worn or damaged parts.

6. Assembly is the reverse of these steps. Install new gaskets and tighten the outer cover bolts securely.

Exhaust manifold
Disassembly

FX700, RA700, RA700A, RA700B, SJ650, SJ700, SJ700A, WB700, WB700A, WR650, WRA650, WRA650A, WRA700, WRB650, WRB650A, WRB700, WVT700, RA760 and WB760

1. Loosen the hose clamp (A, **Figure 32**) and pull the exhaust boot (B, **Figure 32**) from the elbow.

2. Remove the outer cover (**Figure 33**) from the exhaust manifold.

3. *RA700 and WVT700*—Remove the inner cover from the manifold.

4. Unbolt and remove the elbow (C, **Figure 32**) from the exhaust manifold.

5. On RA760 and WB760 models, remove the elbow from the exhaust manifold (**Figure 27**).

Exhaust Manifold
Cleaning/Reassembly

1. Clean all parts to remove all carbon and oil residue.

2. Carefully remove all gasket residue from all mating surfaces.

3. Check all parts for cracks or damage and replace as required.

4. Assembly is the reverse of the disassembly procedure. Install new gaskets.

Installation

FX700, RA700B, SJ650, SJ700, SJ700A, WB700, WB700A, WR650, WRA650, WRA650A, WRA700, WRB650, WRB650A and WRB700

Refer to **Figure 19** for this procedure.

1. Install the exhaust manifold (**Figure 25**) onto the cylinder block using a new gasket.

2. Install the muffler bolts and washers. Tighten the bolts in 2 steps to the torque specification listed in **Table 3** or **Table 4**.

3. Reinstall the inlet water hose to the manifold.

4. Install the muffler bracket, if removed. Torque the bolts in 2 steps to the specification listed in **Table 3** or **Table 4**.

5. Align the muffler housing (D, **Figure 21**) with the muffler bracket and install the muffler

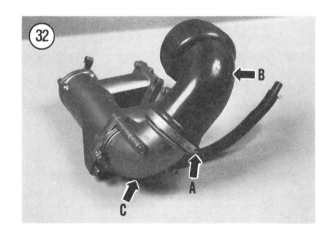

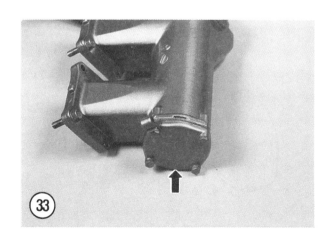

ENGINE (650, 700 AND 760 CC)

housing bolts and washers. Torque the bolts in 2 steps to the specification given in **Table 3** or **Table 4**.

6. Pull back the exhaust boot. Fit the joint onto the muffler housing outer cover (**Figure 23**) and tighten the hose clamp. When the connection is correct, tighten the inner hose clamp. Then slide the exhaust boot over the outer cover and secure it with the hose clamp. See **Figure 22**.

7. *FX700, SJ700, SJ700A and WB700A*—Reconnect the bypass hose (A, **Figure 24**) to the T-connector (B, **Figure 24**).

8. Reconnect the exhaust hose (B, **Figure 21**) at the muffler housing.

9. Reconnect the small muffler-to-cylinder head hose (A, **Figure 21**).

10. Reconnect the water hose to the cylinder head (**Figure 8**).

11. Reinstall the engine cover.

RA700, RA700A, RA760, WB760 and WVT700

Refer to **Figure 26** and **Figure 27** for this procedure.

1. Install the exhaust manifold on the cylinder block. Torque the bolt in 2 steps as specified in **Table 4**.

2. Reconnect the inlet water hose to the manifold.

3. If necessary, reinstall the muffler mounting bracket. Torque the bolt in 2 steps to the specification given in **Table 4**.

4. Align the muffler housing with the muffler bracket and secure the housing in place with the mounting bolts (A, **Figure 34**). Torque the bolts in 2-steps to the specifications given in **Table 4**. On RA760 and WB760 models, also reinstall the secondary muffler bracket to the exhaust manifold.

5. Pull back the exhaust boot and fit the joint over the outer cover No. 2. Tighten the inner hose clamp, slide the boot over the clamp and tighten the outer hose clamp (B, **Figure 34**).

6. Fit the outer cover over the exhaust joint and tighten the hose clamps (B, **Figure 28**). On RA760 and WB760 models, reinstall the water diffusers with the alignment marks at 6 and 9 o'clock if you removed them (**Figure 29**).

7A. *RA700, RA700A and WVT700*—Reconnect the water outlet hose (A, **Figure 28**) to the muffler housing.

7B. *RA760 and WB760*—Reconnect the water outlet hose to the exhaust boot. Apply marine grease to the thermoswitch hole in the muffler housing and reinstall the thermoswitch.

8A. *RA700, RA700B and WVT700*—Reconnect the bypass hose (A, **Figure 24**) to the T-connector (B, **Figure 24**).

8B. *RA760 and WB760*—Reconnect the bypass hoses to the exhaust elbow.

9. Reinstall the engine cover.

ENGINE TOP END

The engine top end consists of the cylinder head, cylinder block, pistons, piston rings, piston pins and the connecting rod small-end bearings. See **Figure 35** and **Figure 36**.

The engine top end can be serviced with the engine installed in the hull.

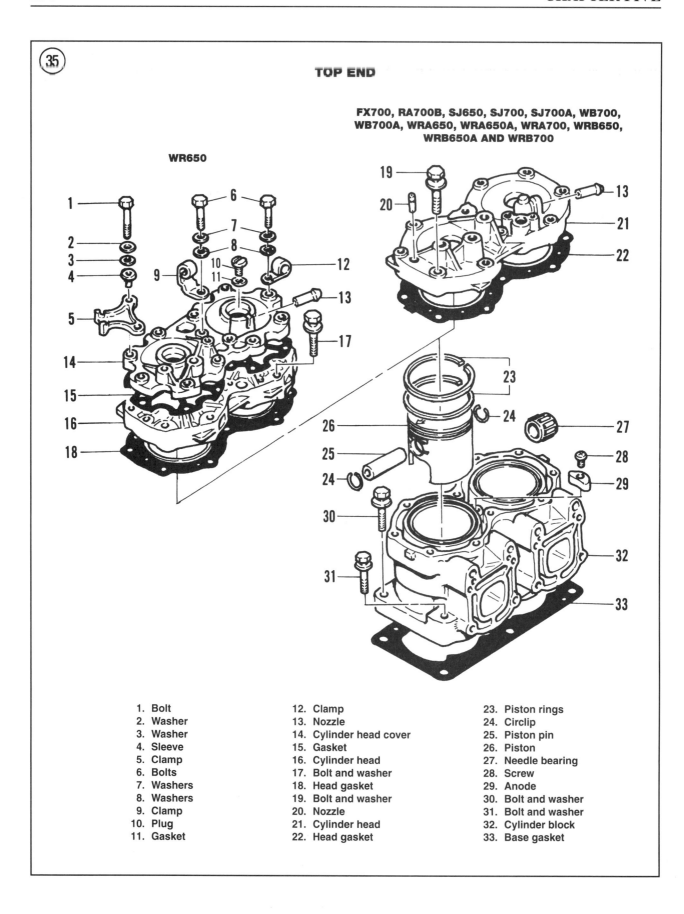

ENGINE (650, 700 AND 760 CC)

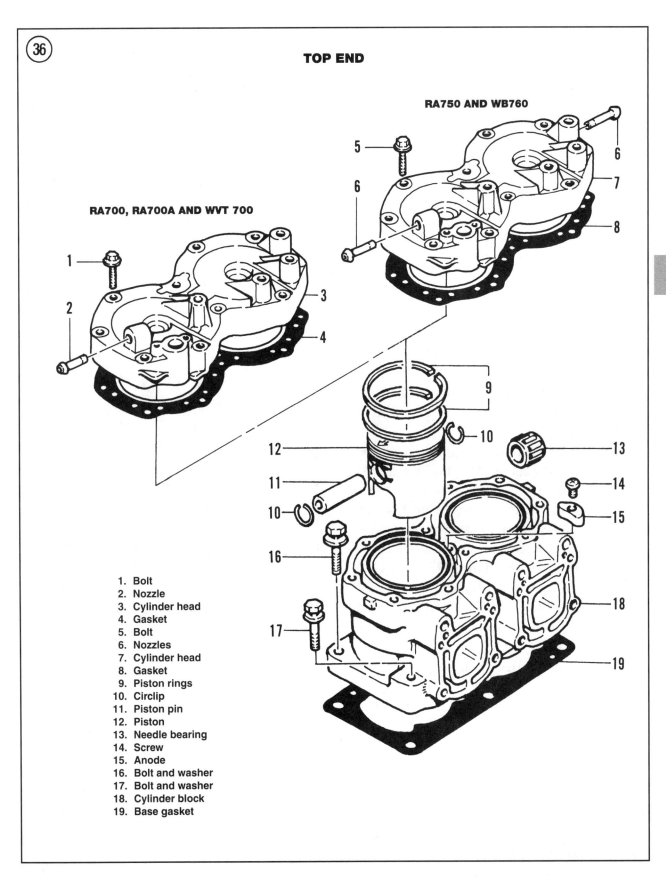

Cylinder Head
Removal/Installation

Refer to **Figure 35** and **Figure 36** for this procedure.

CAUTION
To prevent warping and damage to any component, remove the cylinder head only when the engine is at room temperature.

NOTE
If the engine is being disassembled for inspection procedures, check compression before removing the cylinder head. Refer to Chapter Three.

1. Remove the engine cover.
2. If the engine is mounted in the hull, perform the following:
    a. Disconnect the negative battery cable.
    b. Remove the flame arrestor as described in Chapter Eight.
    c. Remove the thermoswitch bolts and pull the thermoswitch out of the cylinder head (A, **Figure 37**). On RA760 and WB760 models, the thermoswitch is mounted to the muffler housing not to the cylinder head.
    d. Disconnect the coolant hose from the cylinder head nozzle (B, **Figure 37**). On RA760 and WB760 models, remove the coolant hoses from the 2 cylinder head nozzles (one on each side of the head).
    e. Disconnect the 2 spark plug caps.
    f. Remove the muffler housing as described in this chapter.
3. Loosen the spark plugs if they are going to be removed later.

NOTE
Identify and mark all hose guides which are mounted on the cylinder head before removing them in the following steps.

4. Remove the muffler bracket bolts and remove the muffler bracket (**Figure 38**).

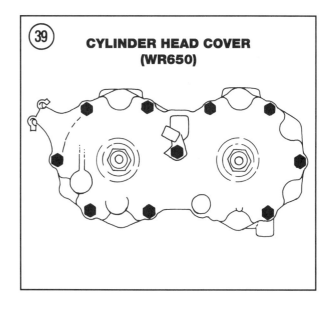

ENGINE (650, 700 AND 760 CC)

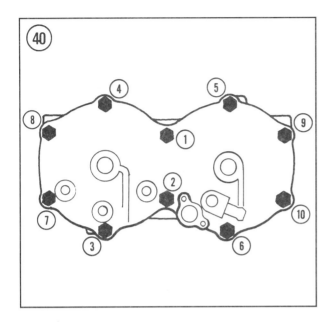

5. *WR650*—Loosen the cylinder head cover mounting bolts and washers (**Figure 39**). Then remove the bolts, washers, cylinder head cover and its gasket.

6. Loosen the cylinder head bolts in a crisscross pattern (**Figure 40**), then remove the bolts and washers.

7. Loosen the cylinder head by tapping around its perimeter with a rubber or plastic mallet, then remove the cylinder head (**Figure 41**).

8. Remove and discard the cylinder head gasket.

9. Lay a rag over the cylinder block to prevent dirt from falling into the cylinders.

10. Clean and inspect the cylinder head as described in this chapter.

11. Before installing the cylinder head gasket in Step 12, make sure the cylinder block and cylinder head mating surfaces are free of all gasket residue.

12. Place a new cylinder head gasket (A, **Figure 42**) onto the cylinder block, without any sealant. Make sure the bolt holes in the gasket align with the bolt holes in the cylinder block. In addition, make sure the tab on the gasket aligns with the boss on the cylinder block. See B, **Figure 42**.

13. Install the cylinder head (**Figure 41**). Apply Loctite 242 to the threads of the cylinder head bolts and install the bolts and washers finger-tight. Tighten the cylinder head bolts, in 2 steps, to the torque specification in **Table 3** or **Table 4**. To prevent cylinder head warping or a blown head gasket, tighten the bolts in the sequence shown in **Figure 40**.

14. *WR650*—Install a new cylinder head cover gasket (without gasket sealer) and install the cylinder head cover (**Figure 39**). Apply Loctite 242 to the threads of the cylinder head cover bolts and install the bolts and washers finger-tight. Tighten the cylinder head cover bolts and washers, in 2 steps, to the torque specification listed in **Table 3** or **Table 4**.

15A. *FX700, SJ650, SJ700, WB700, WB700A, WR650, WRA650, WRA650A, WRA700, WRB650, WRB760A and WRB700*—Apply Loctite 242 (blue) to the threads of the muffler bracket bolts and install the muffler bracket. Finger-tighten the bolts, then torque in 2-steps to the specification given in **Table 3** or **Table 4**. Tighten the bolts in the pattern shown in **Figure 38**.

15B. *RA700, RA700A, RA700B, RA760, WVT70 and WB760*—Apply Loctite 271 (red) to the threads of the muffler bracket bolts and install the muffler bracket. Finger-tighten the bolts, then torque them in 2-steps to the specification given in **Table 3** or **Table 4**.

16. Install the spark plugs and tighten to the torque specification in **Table 3** or **Table 4**. Reinstall the spark plug caps.

17. Reverse Steps 1 and 2 to complete installation. Apply water resistant grease to the thermoswitch hole when installing the thermoswitch.

Inspection

1. Wipe away any soft deposits on the cylinder head (A, **Figure 43**) mating surface. Remove hard deposits with a wire brush mounted in a drill or drill press or with a soft-metal scraper. Be careful not to gouge the aluminum surfaces. Burrs created from improper cleaning will cause preignition and heat erosion.

NOTE
An aluminum thread fluid or kerosene must be applied to a tap if one is used in Steps 2 or 3.

2. With the spark plug removed, check the spark plug threads in the cylinder head (B, **Figure 43**) for carbon buildup or cracking. The carbon can be removed with a 14 mm spark plug thread tap.

3. Check the muffler bracket bolt threads (**Figure 44**) in the top of the cylinder head for corrosion or thread damage. Corrosion can be removed with a tap. If the threads are stripped, a Helicoil or similar thread insert may be required.

4. Check the water hose nozzles on the cylinder head for sludge buildup. Check the water passages in the cylinder head for corrosion or other residue. Clean these passages thoroughly.

5. Use a straightedge and feeler gauge and measure the flatness of the cylinder head (**Figure 45**). If a 0.1 mm (0.004 in.) feeler gauge can be slipped underneath the straightedge, resurface the cylinder head as follows:

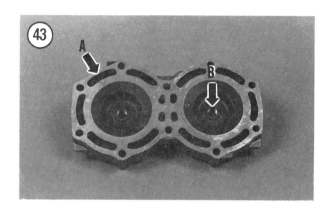

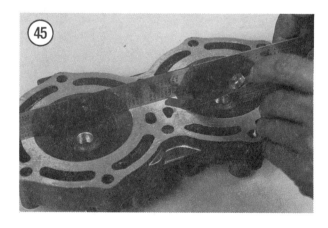

ENGINE (650, 700 AND 760 CC)

a. Tape a piece of 400-600 grit wet emery sandpaper onto a piece of thick plate glass or surface plate.

b. Slowly resurface the head by moving it in figure-eight patterns on the sandpaper. See **Figure 46**.

c. Rotate the head several times to avoid removing too much material from one side. Check progress often with the straightedge and feeler gauge.

d. If the cylinder head warpage still exceeds the service limit, have the head resurfaced at a machine shop. Note that removing material from the cylinder head mating surface will change the compression ratio. Consult with the machinist regarding how much material to remove.

6. Wash the cylinder head in hot soapy water and rinse thoroughly before installation.

CYLINDER

An aluminum cylinder block with cast iron liners is used on all 650, 700 and 760 cc engines. If excessive wear is experienced, the cylinder liner can be bored to 0.25 mm or 0.50 mm oversize and new pistons and rings installed.

Refer to **Figure 35** or **Figure 36** when performing the following procedure.

Removal

1. Remove the cylinder head as described in this chapter.

2. If the engine is installed in the hull, perform the following:

   a. Remove the carburetor and intake manifold (with reed valve) as described in Chapter Eight.

   b. Remove the muffler housing as described in this chapter.

3. The cylinder block is secured to the crankcase with 6 bolts and washers. Loosen the bolts in 1/4 turn increments, in a crisscross pattern, until they are loose. See A, **Figure 47**. Remove the bolts and washers.

4. If the cylinder block is tight to the crankcase, do not pry it off. Instead, tap it around its perimeter with a soft-faced mallet. Then lift the cylinder block (B, **Figure 47**) up, making sure to catch the pistons as they become free of the cylinder bore. This will prevent the pistons from damaging themselves against the crankcase.

5. Remove the base gasket and discard it.

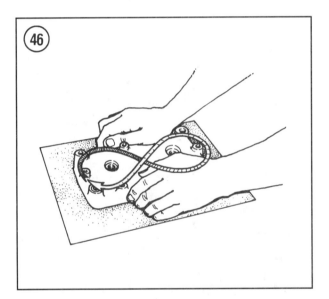

6. Remove the 2 dowel pins (**Figure 48**) from the crankcase.

7. Stuff clean rags around the connecting rods to keep dirt and loose parts from entering the crankcase.

Inspection

Cylinder measurement requires a precision inside micrometer or bore gauge. If you do not have the right tools, have your dealer or a machine shop take the measurements.

1. Using a wooden scraper, remove all carbon residue from the exhaust ports. If you choose to use a wire wheel mounted in some type of drill or hand grinder to clean the ports, make sure you do not damage the cylinder lining.

2. Check all of the tapped holes in the cylinder block for corrosion buildup or damage.

3. Remove all gasket residue from the top (A, **Figure 49**) and bottom (**Figure 50**) gasket surfaces.

4. Visually inspect the anode mounted in the cylinder block water passage (B, **Figure 49**) for severe wear. Replace the anode if it is less than 60% of its original size.

5. Use a straightedge and feeler gauge and measure the flatness of the cylinder block as shown in **Figure 51**. If a 0.1 mm (0.004 in.) feeler gauge can be slipped underneath the straightedge, the cylinder block may require replacement. Refer additional service to your Yamaha dealership.

6. Wash the cylinder block in solvent to remove any oil and carbon particles. The cylinder bores must be cleaned thoroughly before attempting any measurement.

7. Measure the cylinder bore diameter as described under *Piston/Cylinder Clearance Check* in this chapter.

8. If the cylinder is not excessively worn, check the bore carefully for scratches or gouges. The bore may require reconditioning.

ENGINE (650, 700 AND 760 CC)

9. After the cylinder is serviced, wash the bore in hot soapy water. This is the only way to clean the cylinder wall of the fine grit material left from the boring or honing job.

10. After washing the cylinder wall, run a clean white cloth through it. The cylinder wall should show no traces of grit or other debris. If the rag is dirty, the cylinder wall is not clean and must be rewashed. After the cylinder is cleaned, lubricate the cylinder walls with clean engine oil to prevent the cylinder liners from rusting.

CAUTION
A combination of soap and water is the only solution that will completely clean the cylinder wall. Solvent and kerosene cannot wash fine grit out of cylinder crevices. Grit left in the cylinder will cause premature wear to the new rings.

Installation

1. Clean the cylinder bore as described under *Inspection* in this chapter.

2. Make sure the top surface of the crankcase and the bottom cylinder surface are clean prior to installation.

3. Squirt some engine oil into each of the crankcase oil holes shown in **Figure 52**.

4. Install the 2 engine block dowel pins (**Figure 48**).

5. Install a new base gasket, without sealer, making sure all gasket and crankcase bolt holes align.

6. Make sure the piston pin circlips are seated in the piston clip groove correctly. If the pistons were removed from the rods, new piston pin clips must be installed.

7. Make sure the end gaps of the piston rings are aligned with the locating pins in the ring grooves (**Figure 53**). Lightly oil the piston rings and the inside of the cylinder bores with clean engine oil.

8. Rotate the crankshaft so one piston is up and the other is down.

9. Align the cylinder bore with the upper piston. Compress the piston rings with your fingers and slide the cylinder bore over the piston. The cylinder chamfer will compress the rings as necessary.

10. After one piston is installed in its cylinder, rotate the crankshaft until both pistons are at the same height. Lower the cylinder and install the opposite piston. Make sure the cylinder block is seated completely on the crankcase.

11. Apply Loctite 242 to the threads of the cylinder bolts and finger-tighten each bolt and its washer. Then tighten the cylinder bolts in 2 steps to the torque specification listed in **Table 3** or

Table 4. Tighten the bolts in the sequence shown in **Figure 54** to prevent warping.

12. Install the cylinder head as described in this chapter.

PISTON, PISTON PIN, AND PISTON RINGS

The piston is made of aluminum alloy. The piston pin is a precision fit and is held in place by a clip at each end. A caged needle bearing is used on the small end of the connecting rod. See **Figure 35** or **Figure 36** when performing this procedure.

Piston and Piston Ring Removal

1. Remove the cylinder head and cylinder block as described in this chapter.
2. Identify the pistons (**Figure 55**) as to front or rear. In addition, keep each piston together with its own pin, bearing and piston rings to avoid confusion during reassembly.
3. Before removing the piston, hold the rod tight and rock the piston as shown in **Figure 56**. Any rocking motion (do not confuse with the normal sliding motion) indicates wear on the piston pin, needle bearing, piston pin bore or more likely a combination of all three.

NOTE
Wrap a clean shop cloth under the piston so the clip will not fall into the crankcase.

WARNING
Wear safety glasses when performing Step 4.

4. Remove the clips from each side of the piston pin bore (**Figure 57**) with needlenose pliers. Hold your thumb over one edge of the clip when removing it to prevent it from springing out.
5. Use a proper size wooden dowel or socket extension and push out the piston pin (A, **Figure 58**).

CAUTION
If the engine ran hot or seized, the piston pin will probably be difficult to remove.

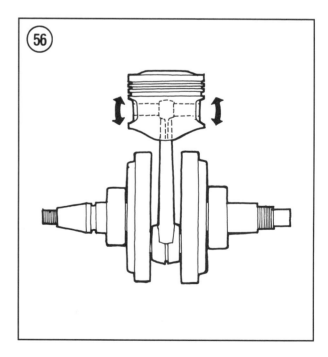

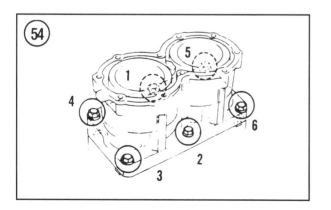

ENGINE (650, 700 AND 760 CC)

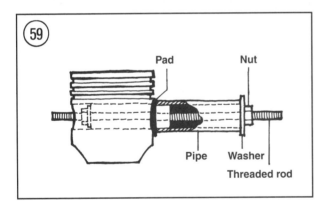

However, do not drive the piston pin out of the piston. This will damage the piston, needle bearing and connecting rod. If the piston pin will not push out by hand, remove it as described in Step 6.

6. If the piston pin is tight, fabricate the tool shown in **Figure 59**. Assemble the tool onto the piston and pull the piston pin out of the piston. Make sure to install a rubber or plastic pad between the piston and piece of pipe to avoid scoring the side of the piston.

7. Lift the piston (B, **Figure 58**) off the connecting rod.

8. Remove the needle bearing from the connecting rod (**Figure 60**).

9. Repeat for the opposite piston.

10. If the pistons are going to be left off for some time, place a piece of foam insulation tube, or shop cloth, over the end of each connecting rod to protect it.

NOTE
Always remove the top piston ring first.

11. Keystone rings (**Figure 61**) are used at the top and bottom positions. If the rings are going to be reused, mark each ring after removal to

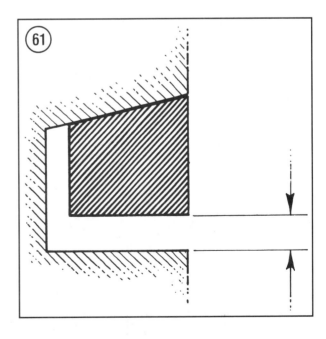

ensure correct installation. Remove the upper ring by spreading the ends with your thumbs just enough to slide it up over the piston (**Figure 62**). Repeat for the lower ring.

Piston Pin and Needle Bearing Inspection

1. Clean the needle bearing (A, **Figure 63**) in solvent and dry it thoroughly. Use a magnifying glass and inspect the bearing cage for cracks at the corners of the needle slots and inspect the needles themselves for cracking. If any cracks are found, the bearing must be replaced.

2. Check the piston pin (B, **Figure 63**) for excessive wear, scoring or chrome flaking. Also check the piston pin for cracks along the top and side. Replace the piston pin if necessary.

3. Oil the needle bearing and piston pin, then install them in the connecting rod. Slowly rotate the pin, and check for radial and axial play (**Figure 64**). If any play exists, replace the pin and bearing, providing the rod bore is in good condition.

CAUTION
If there are signs of piston seizure or overheating, replace the piston pins and bearings. These parts have been weak-

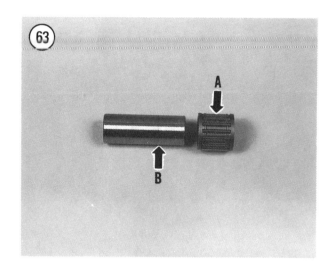

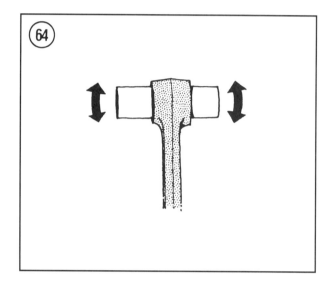

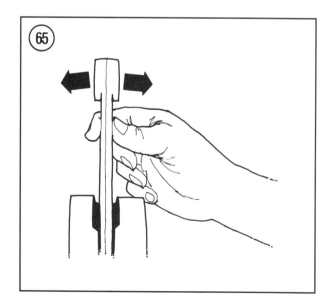

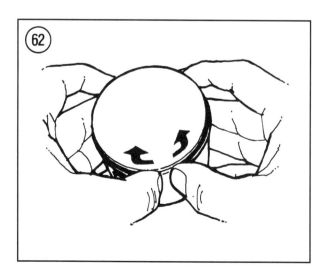

ENGINE (650, 700 AND 760 CC)

ened from excessive heat and may fail later.

Connecting Rod Inspection

1. Wipe the piston pin bore in the connecting rod with a clean rag and check it for galling, scratches, excessive wear or damage. If any of these conditions exist, replace the crankshaft as described in this chapter.

2. Check the connecting rod big end bearing play. You can make a quick check by simply rocking the connecting rod back and forth (**Figure 65**). If there is more than a very slight rocking motion (some side-to-side sliding is normal), measure the connecting rod side clearance with a feeler gauge. Measure between the side of the crankshaft and the washer. **Figure 66** shows the measurement being taken with the crankshaft removed for clarity. If the play exceeds the wear limit specified in **Table 2**, replace the crankshaft; individual replacement parts are not available from Yamaha.

Piston and Ring Inspection

1. Carefully check the piston for cracks at the top edge of the transfer cutaways (**Figure 67**). Replace the piston if any cracks are noted. Check the piston skirt (**Figure 68**) for brown varnish buildup. More than a slight amount is an indication of worn or sticking rings.

2. Check the piston skirt for galling and abrasion which may have resulted from piston seizure. If light galling is present, dress the affected area with 600-800 grit wet sandpaper. However, if galling is severe or if the piston is deeply scored, replace it.

3. Check the piston ring locating pins in the piston (**Figure 69**). The pins must be tight and the piston must not be cracked in the area around the pins. If a locating pin is loose, replace the piston. A loose pin will fall out and cause severe engine damage.

4. Check the piston pin clip grooves in the piston for cracks or other damage that could allow a clip to fall out. This would cause severe engine damage. Replace the piston if any one groove is worn or damaged.

NOTE
Maintaining proper piston ring end gap helps ensure peak engine performance. Always check piston ring end gap at the intervals specified in Chapter Three. Excessive ring end gap reduces engine performance and can cause overheating. Insufficient ring end gap will cause the ring ends to butt together and cause the ring to break. This would cause severe engine damage.

5. Measure piston ring end gap. Place a ring into the bottom of the cylinder and push it in about 20 mm (3/4 in.) with the crown of the piston (**Figure 70**). This ensures that the ring is square in the cylinder bore. Measure the gap with a flat feeler gauge (**Figure 71**) and compare it to the wear limit in **Table 2**. If the gap is greater than specified, replace the rings as a set.

NOTE
*When installing new rings, measure the end gap in the same manner as for old ones. If the gap is less than specified, first make sure you have the correct piston rings. If the replacement rings are correct but the end gap is too small, carefully file the ends with a fine cut file until the gap is correct (**Figure 72**).*

6. Carefully remove all carbon buildup from the ring grooves with a broken ring (**Figure 73**). Inspect the grooves carefully for burrs, nicks, or broken or cracked lands. Recondition or replace the piston if necessary.

7. Measure the side clearance of each ring in its groove with a flat feeler gauge (**Figure 74**) and compare to specifications listed in **Table 2**. If the clearance is greater than specified, replace the rings. If the clearance is still excessive with new rings, replace the piston.

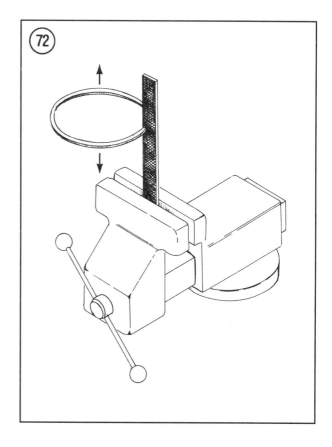

ENGINE (650, 700 AND 760 CC)

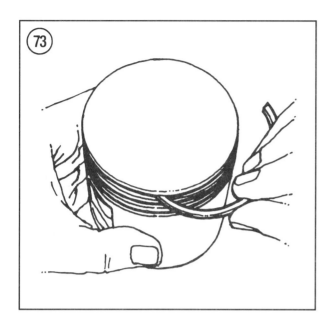

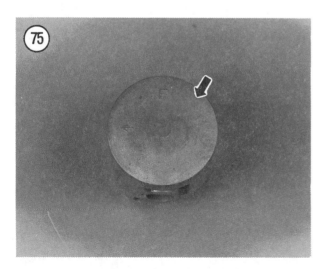

8. Observe the condition of the piston crown (**Figure 75**). Remove normal carbon buildup using a wire wheel mounted in a drill. If the piston shows signs of overheating, pitting or other abnormal conditions, the engine may be experiencing preignition or detonation; both conditions are discussed in Chapter Two.

9. If the piston is in acceptable condition, measure the piston outside diameter as described under *Piston/Cylinder Clearance Check* in this chapter.

10. If new piston rings are required, hone the cylinders before assembling the engine. Refer to *Cylinder Honing* in this chapter.

Piston/Cylinder Clearance Check

The following procedure requires the use of highly specialized and expensive measuring tools. If such equipment is not readily available, have the measurements performed at a dealership or machine shop. Always replace both pistons as a set.

1. Measure the outside diameter of the piston with a micrometer approximately 10 mm (13/32 in.) above the bottom of the piston skirt, at a 90° angle to the piston pin (**Figure 76**). If the diameter exceeds the wear limit in **Table 2**, install new pistons.

NOTE
Always install new rings when installing a new piston.

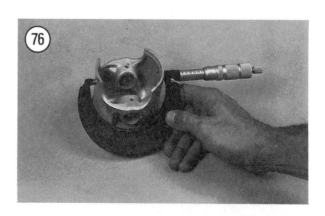

2. Wash the cylinder block in solvent to remove any oil and carbon particles. The cylinder bore must be cleaned thoroughly before attempting any measurement.

3. Measure the cylinder bore with a bore gauge or inside micrometer (**Figure 77**). Measure the cylinder bore at the points shown in **Figure 78**. Measure in 3 axes in line with the piston pin and at 90° to the pin. If the bore is greater than the specification (**Table 2**), bore the cylinders to the next oversize and install oversize pistons and rings.

NOTE
Purchase the new pistons before the cylinders are bored so the pistons can be measured. The cylinders must be bored to match the pistons. Piston-to-cylinder clearance is specified in Table 2.

4. Piston clearance is the difference between the maximum piston diameter and the minimum cylinder diameter. For a used piston and cylinder, subtract the piston diameter from the cylinder diameter. If the clearance exceeds the specification in **Table 2**, bore the cylinders oversize and install new pistons and rings.

Cylinder Honing

The surface condition of a worn cylinder bore is normally very shiny and smooth. Cylinder honing, often referred to as glaze breaking or deglazing, is required whenever new piston rings are installed. If new piston rings are installed in a glazed cylinder, they would not seat properly. When a cylinder bore is honed, the surface is slightly roughed up to provide a textured or crosshatched surface. This surface finish controls wear of the new rings and helps them to seat and seal properly. *Whenever new rings are installed*, the cylinder surface should be honed. This service can be performed at a Yamaha dealership or independent repair shop. The cost of having the cylinder honed is usually minimal compared to the cost of purchasing a hone and doing the job yourself. If you choose to hone the cylinder yourself, follow the hone manufacturer's directions closely.

CAUTION
After a cylinder is reconditioned by boring or honing, the bore should be prop-

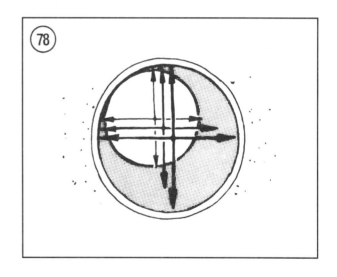

ENGINE (650, 700 AND 760 CC)

erly cleaned to remove all material left from the machining operation. Refer to **Inspection** under **Cylinder** in this chapter. Improper cleaning will not remove all of the machining residue in the cylinder bore and rapid wear of the new piston and rings will result.

Piston Installation

1. Apply assembly oil to the needle bearing and install it in the connecting rod (**Figure 79**).

2. RA700, RA700A, RA700B, RA760, WB760 and WVT700—Install the 2 washers into the connecting rod (**Figure 80**).

3. Oil the piston pin and install it in the piston until the end of the pin extends slightly beyond the inside of the boss (**Figure 81**).

4. Place the piston over the connecting rod with the arrow on the piston crown pointing toward the left-hand side (the exhaust side) of the engine. Align the pin with the bearing and push the pin into the piston until it is even with the piston pin clip grooves.

CAUTION
*If the piston pin will not slide into the piston smoothly, use the home-made tool described under **Piston Removal** to install the piston pin (**Figure 59**). When using the home-made tool, the pipe is not required. Insert the threaded rod through the piston pin and out the other side of the piston. Slide the pad and large washer over the threaded rod so the pad presses against the face of the piston. Run a nut onto the rod so it presses against the large washer. At the other end of the rod, slide the small washer over the threaded rod and against the piston pin. Thread a nut onto the rod and tighten the nut to push the piston pin into the piston. Do not drive the piston pin into the piston or you may damage the needle bearing and connecting rod.*

WARNING
Wear safety glasses when performing Step 5.

5. Install new piston pin clips (**Figure 82**) in the ends of the pin boss. Make sure they are seated

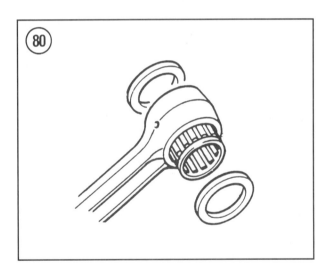

in their grooves and that the end of each clip is *not* opposite the slots in the piston (**Figure 83**).
6. Check the installation by rocking the piston back and forth around the pin axis and from side to side along the axis. It should rotate freely back and forth but not from side to side.

NOTE
The top and bottom rings are the same. If you are installing new rings, they can be installed in either position. If you are installing used rings, install them by referring to the identification marks made prior to removal.

7. Install the piston rings (bottom one first) by carefully spreading the ends of the ring with your thumbs and slipping the ring over the top of the piston. Make sure manufacturer's mark on the piston rings are toward the top of the piston.
8. Make sure the rings are seated completely in the grooves, all the way around the circumference and that the ends are aligned with the locating pins. See **Figure 84**.

CRANKCASE AND CRANKSHAFT

Disassembly of the crankcase—splitting the cases—and removal of the crankshaft assembly requires engine removal from the hull. However, the cylinder head, cylinder and all other attached assemblies should be removed with the engine in the hull.

The crankcase is made in 2 halves of precision diecast aluminum alloy and is of the thin-walled type (**Figure 85**). To avoid damage to them do not hammer or pry on any of the interior or exterior projected walls. These areas are easily damaged if stressed beyond what they are designed for. They are assembled without a gasket; only gasket sealer is used while dowel pins align the crankcase halves when they are bolted together. The crankcase halves are sold as a matched set only (**Figure 86**). If one crankcase half is damaged, both must be replaced.

The procedure which follows is presented as a complete, step-by-step major lower end overhaul.

Special Tools

When splitting the crankcase assembly, a few special tools are required. These tools allow easy disassembly and reassembly of the engine without prying or hammer use. Remember, the crankcase halves can be easily damaged by improper disassembly or reassembly techniques.

a. Yamaha coupling holder (or equivalent) part No. YW-38741 (760 cc models) or part No. YW-06547 (all other models). See **Figure 87**. This tool is used to remove and install the coupling half from the end of the crankshaft or drive shaft. Because the couplings are made of cast aluminum, they can

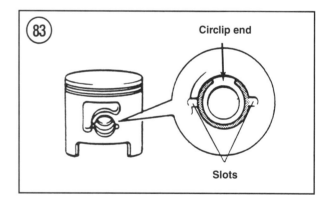

ENGINE (650, 700 AND 760 CC)

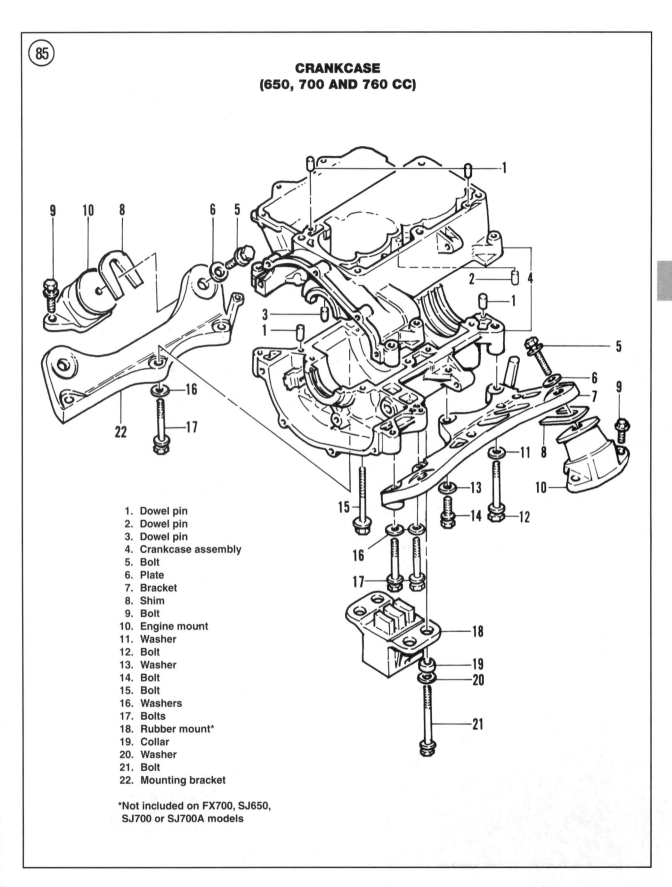

85

**CRANKCASE
(650, 700 AND 760 CC)**

1. Dowel pin
2. Dowel pin
3. Dowel pin
4. Crankcase assembly
5. Bolt
6. Plate
7. Bracket
8. Shim
9. Bolt
10. Engine mount
11. Washer
12. Bolt
13. Washer
14. Bolt
15. Bolt
16. Washers
17. Bolts
18. Rubber mount*
19. Collar
20. Washer
21. Bolt
22. Mounting bracket

*Not included on FX700, SJ650, SJ700 or SJ700A models

be easily damaged by the use of improper tools or removal procedure.

b. Yamaha flywheel holder (or equivalent) part No. YW-06139 (760 cc models [**Figure 88**]) or part No. YW-06547 (all other models [**Figure 89**]). A flywheel holder is required to hold the flywheel during crankshaft couling half removal.

Crankcase Disassembly

This procedure describes disassembly of the crankcase halves and removal of the crankshaft.

1. Remove the engine from the hull as described in this chapter.
2. Remove the starter motor as described in Chapter Nine.
3. On oil injection models, remove the oil pump as described in Chapter Ten.
4. Remove the flywheel cover bolts and the cover (**Figure 90**). Discard the cover gasket.
5. *FX700, SJ650, SJ700, WR650, WB700, WB700A, WRA650, WRA650A, WRB650, WRB650A and WRB700*—Remove the ring nut from the coupler half.
6A. *650 and 700 cc*—Hold the flywheel steady with a holding tool (**Figure 91**) and unscrew the coupler half from the rear end of the crankshaft with the coupling holder. See **Figure 92**.
6B. *760 cc*—Install the flywheel holder (A, **Figure 93**) onto the crankcase and remove the

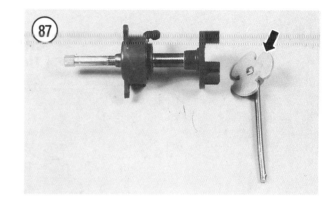

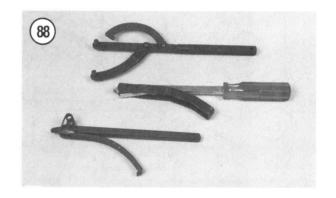

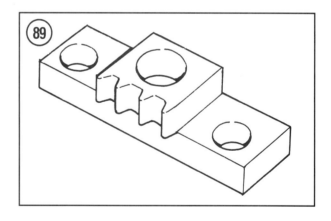

ENGINE (650, 700 AND 760 CC)

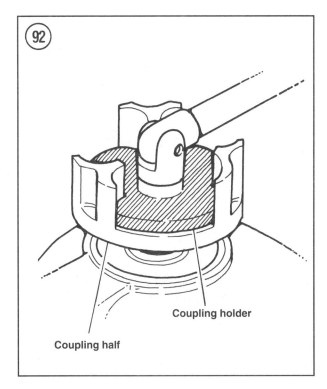

Coupling holder
Coupling half

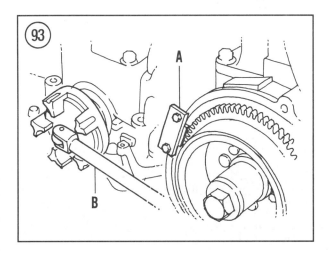

coupler half from the rear end of the crankshaft with the coupling holder (B, **Figure 93**).

7A. *650 and 700 cc*—Remove the flywheel, stator plate and idler gear as described in Chapter Nine.

7B. *760 cc*—Remove the flywheel and idler gear as described in Chapter Nine.

8. Remove the pistons as described in this chapter.

9. Turn the crankcase assembly so it rests up-side-down on wooden blocks.

10. Loosen the bolts securing the engine mounting brackets to the lower crankcase. Remove the mounting brackets (**Figure 94**).

11. Loosen the crankcase bolts (**Figure 95**) in 1/4 turn increments until all bolts are loose. Then remove the bolts and washers.

CAUTION
Make sure you have removed all the fasteners. If the cases are hard to separate,

check for any fasteners you may have missed.

12. Carefully tap the perimeter of the lower crankcase with a soft-faced hammer to break it loose. Then lift the lower crankcase half up and remove it (**Figure 96**).
13. Lift the crankshaft (**Figure 97**) and remove it from the upper crankcase.
14. Remove the 2 case halves dowel pins (**Figure 98**).
15. Remove the 4 short (A, **Figure 99**) and the 1 long (B, **Figure 99**) crankshaft dowel pins.

Crankcase Inspection

Refer to **Figure 85** for this procedure.

1. Clean both crankcase halves with cleaning solvent. Thoroughly dry with compressed air and wipe off with a clean shop cloth. Be sure to remove all traces of old gasket sealer from all mating surfaces.
2. Carefully inspect the case halves (**Figure 86**) for cracks and fractures. Also check for cracks in the areas around the stiffening ribs, around bearing bosses and threaded holes. If any are found, have them repaired by a shop specializing in the repair of precision aluminum castings. If repair is not possible, replace the crankcase assembly.
3. Check all of the upper crankcase threaded bolt holes for thread damage or corrosion. See

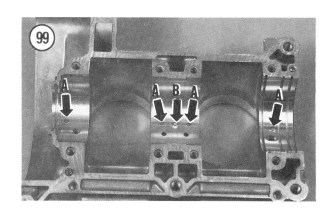

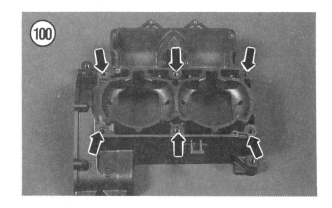

ENGINE (650, 700 AND 760 CC)

Figure 100 and **Figure 101**. If necessary, clean or repair the threads with a suitable size metric tap. Coat the tap threads with kerosene or an aluminum tap fluid before use.

4. Check the idler gear bushing (**Figure 102**) in the lower crankcase for excessive wear, cracks or other damage. If the bushing is damaged, remove it with a blind bearing remover. Then tap the new bushing in using a suitable bushing driver.

Crankshaft Inspection

Refer to **Figure 103** for this procedure. The crankshaft and connecting rod assembly are

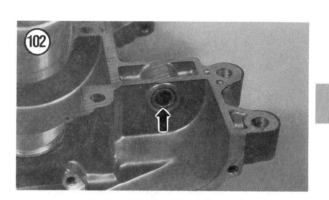

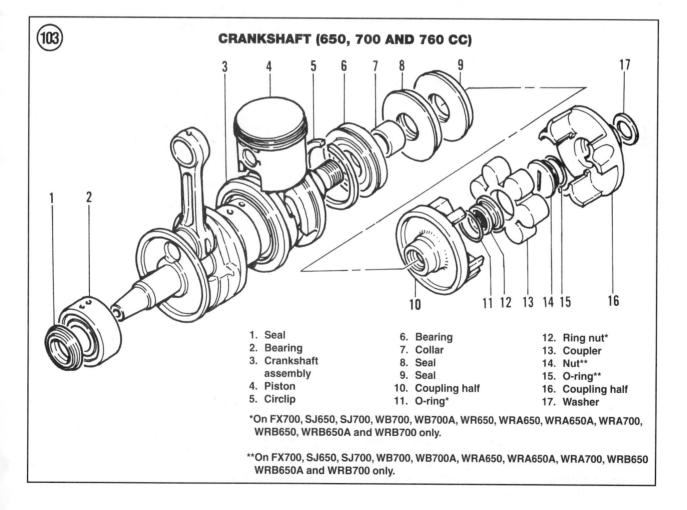

CRANKSHAFT (650, 700 AND 760 CC)

1. Seal
2. Bearing
3. Crankshaft assembly
4. Piston
5. Circlip
6. Bearing
7. Collar
8. Seal
9. Seal
10. Coupling half
11. O-ring*
12. Ring nut*
13. Coupler
14. Nut**
15. O-ring**
16. Coupling half
17. Washer

*On FX700, SJ650, SJ700, WB700, WB700A, WR650, WRA650, WRA650A, WRA700, WRB650, WRB650A and WRB700 only.

**On FX700, SJ650, SJ700, WB700, WB700A, WRA650, WRA650A, WRA700, WRB650 WRB650A and WRB700 only.

pressed together. However, because Yamaha does not sell replacement connecting rods and bearings, the crankshaft must be replaced as a unit if it is worn or damaged.

NOTE
A set of V-blocks can be made out of hardwood to help hold the crankshaft in place when performing some of the checks in the following steps. However, use only precision machined V-blocks to check crankshaft runout described later in this procedure.

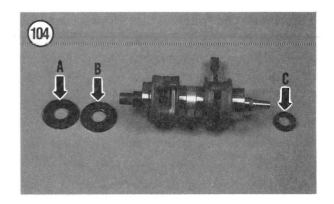

1. See **Figure 104**. The crankshaft is equipped with 3 different seals. Identify the seals then remove them from the end of the crankshaft.
2. Clean the crankshaft thoroughly with solvent or kerosene and a brush. Blow dry with dry filtered compressed air and lubricate with a light coat of Yamalube Two-Cycle Outboard Oil.
3. Check the crankshaft journals and crankpins for scratches, heat discoloration or other defects.
4. Check crankshaft splines, flywheel taper (A, **Figure 105**), threads (A, **Figure 106**) and keyway (B, **Figure 105**) for wear or damage.
5. Check crankshaft seal surfaces for grooving, pitting or scratches.
6. Check crankshaft bearing surfaces for rust, water marks, chatter marks and excessive or uneven wear. Minor cases of rust and water or chatter marks can be cleaned up with 320 grit carborundum cloth.
7. If 320 grit cloth is used, clean the crankshaft in solvent and recheck the surfaces. If they do not clean up properly, replace the crankshaft.

8. Carefully examine the condition of the crankshaft ball bearings (**Figure 107**). Clean the bearings in solvent and allow to dry thoroughly. Oil each bearing with Yamalube Two-Cycle Outboard Oil then turn the outer race. A worn or damaged bearing will sound or feel rough and it will not rotate smoothly.
9. Grasp the outer race and try to work it back and forth. Replace bearing if excessive axial play is noted. In all cases, replace defective bearings.

ENGINE (650, 700 AND 760 CC)

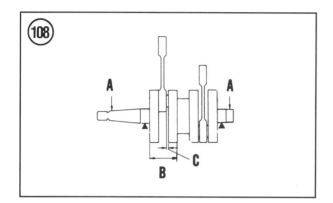

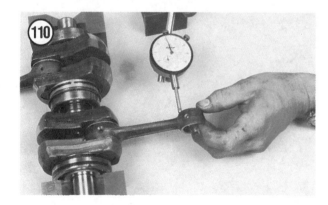

NOTE
*The bearings installed on the outside of the crank wheels can be replaced as described under **Crankshaft Bearing Replacement** in this chapter. If the bearings installed between the crank wheels are worn or damaged, the crankshaft must be replaced.*

10. Support the crankshaft by placing it onto 2 V-blocks at the points shown in **Figure 108**. Then check runout using a dial indicator at the 2 points marked (A, **Figure 108**). Turn the crankshaft slowly and note the indicator reading. The maximum difference recorded is the crankshaft runout at that position. If the runout at any position exceeds the service limit in **Table 2**, replace the crankshaft.

11. Measure the width of the crank wheels with a micrometer or vernier caliper at points marked (B, **Figure 108**) and compare to the specification listed in **Table 2**. See **Figure 109**.

12. Check connecting rod deflection with a dial indicator. Place the end of a dial indicator so its plunger rests against the connecting rod's small end as shown in **Figure 110**. Support the crankshaft and push the connecting rod back and forth, noting the dial indicator gauge movement each time the connecting rod is moved. The difference in the 2 indicator readings is connecting rod small end freeplay. Repeat for each rod. If the freeplay exceeds the service limit in **Table 2**, the crankshaft must be replaced.

13. Check connecting rod big-end side clearance with a feeler gauge at the point marked (C, **Figure 108**). Insert the gauge between the crankshaft and the washer as shown in **Figure 111**. If the clearance exceeds the service limit in **Table 2**, replace the crankshaft assembly.

Crankshaft Bearing Replacement

Replace the front (C, **Figure 105**) and rear (B, **Figure 106**) outer crankshaft bearings as follows. A bearing puller and splitter are required to remove the bearings. A press is required to install the bearings. Read this procedure completely through before removing the bearings.

NOTE
*Before removing the front and rear bearings, note the position of the locating pin hole in the front bearing (**Figure 112**) and the hole and circlip in the rear bearing (**Figure 113**). The pin hole and circlip align with dowel pins and a circlip groove in the upper crankcase during crankshaft installation. The new bearing(s) must be installed so its pin hole and circlip align exactly the same way. If the bearing is turned around during installation, the pin hole and circlip will not align with its mating part or groove, and the bearing will have to be removed and turned around.*

CAUTION
Before removing the front bearing with a bearing puller, thread the flywheel nut onto the crankshaft so it is flush with the end of the crankshaft. The nut will prevent the puller screw from damaging the crankshaft threads.

1. Install a bearing puller and remove the bearing (**Figure 114**). Then reposition the puller and remove the opposite bearing.
2. Check crankshaft bearing surfaces for rust, water marks, chatter marks or uneven wear. Minor cases of rust and water or chatter marks may be cleaned with 320 grit carborundum cloth.

NOTE
*If 320 grit cloth is used dress up the crankshaft bearing surfaces, clean the crankshaft and oil the inner bearing as described under **Crankshaft Cleaning and Inspection** in this chapter.*

3. Clean the crankshaft bearing area with solvent and dry thoroughly.

NOTE
To ease bearing installation, heat the bearings in oil before installation. Completely read Step 4 before heating and installing bearings. During bearing installation, securely support the crankshaft so the bearings can be installed quickly. If a bearing cools and tightens on the crankshaft before it is completely installed, remove the bearing with the puller and reheat.

4. Install the bearings as follows:

   a. Lay the bearings on a clean, lint-free surface in the order of assembly.

   b. When installing the bearings, be sure the pin holes and circlip are positioned as shown in **Figure 112** (front) and **Figure 113** (rear).

   c. Refer to *Shrink Fit* under *Ball Bearing Replacement* in Chapter One.

   d. Heat and install the bearings. Refer to **Figure 112** (front) or **Figure 113** (rear) during installation.

ENGINE (650, 700 AND 760 CC)

Crankshaft Seal Installation

The crankshaft is equipped with 3 different seals (**Figure 104**). While the front seal is not difficult to identify, the 2 rear seals are different in design and can be installed incorrectly. If you purchase the seals in a seal kit, the individual seals will not be identified, thus, if you did not identify the old seals during removal, you cannot use them to identify the new seals. However, if you purchased the seals individually, the seals can be identified by their respective part number.

When installing the rear seals, note the following:

    a. The inner rear seal (B, **Figure 104**) is a single-lip seal.

    b. The outer rear seal (A, **Figure 104**) is a double-lip seal.

    c. The single lip seal has tabs (A, **Figure 115**), whereas the double lip seal (B, **Figure 115**) does not.

1. Lubricate the bearings with Yamaha engine oil (**Figure 116**) before installing the seals.

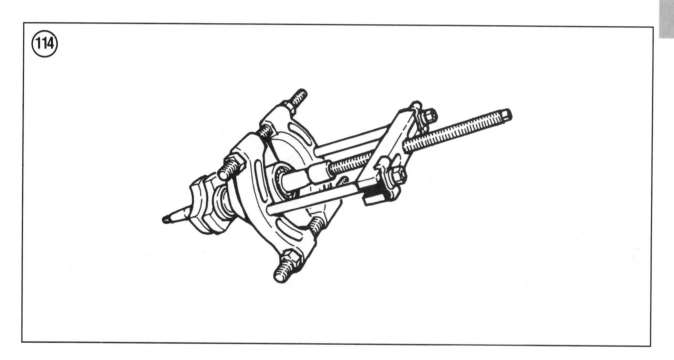

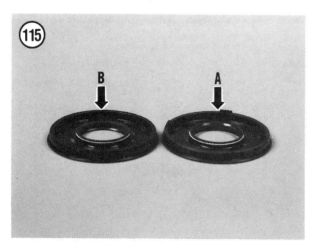

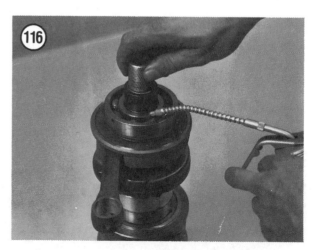

2. Pack a heat durable marine grease between the seal lips as shown in **Figure 117**.

3. Install the front seal so it faces out as shown in **Figure 118**.

4. Install the rear single lip then the rear double lip seal. Be sure both seals face outward as shown in **Figure 119**. See previous introduction for seal identification.

Crankshaft Installation

Refer to **Figure 103** for this procedure.

1. Install the seals onto the crankshaft as described in this chapter.

2. Place the upper crankcase half upside-down on wooden blocks as shown in **Figure 120**.

3. Install the 2 dowel pins (**Figure 121**).

4. Install the 4 short (A, **Figure 122**) and 1 long (B, **Figure 122**) crankshaft bearing dowel pins in the upper crankcase.

NOTE
Step 5 describes crankshaft (Figure 123) installation. However, because of the number of separate steps required during installation, read through Step 5 before actually installing the crankshaft.

5. Align the crankshaft with the upper crankcase half and install the crankshaft (**Figure 123**). Note the following:

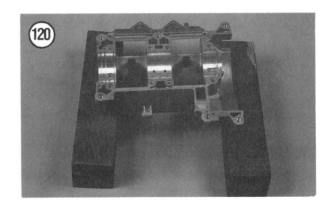

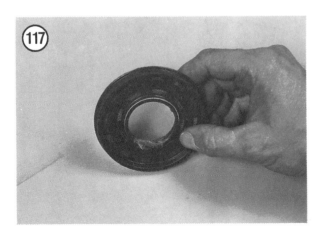

ENGINE (650, 700 AND 760 CC)

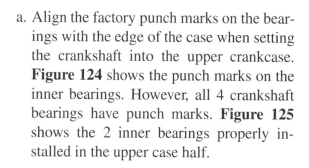

a. Align the factory punch marks on the bearings with the edge of the case when setting the crankshaft into the upper crankcase. **Figure 124** shows the punch marks on the inner bearings. However, all 4 crankshaft bearings have punch marks. **Figure 125** shows the 2 inner bearings properly installed in the upper case half.

NOTE
*Before the crankshaft seats in the case, the marks will actually be slightly above the edge of the case. See **Figure 126**.*

b. Make sure the hole in the labyrinth seal (A, **Figure 127**) and the bearings (B, **Figure 127**) align with the pins in the upper crankcase. See **Figure 122**.

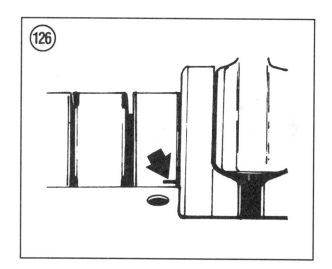

c. Make sure the circlip on the rear bearing and the 2 rear seals seat in the crankcase grooves. See **Figure 128**.
d. Also be sure the circlip opening does *not* align with the case half.
e. Make sure the front seal seats in the crankcase groove as shown in **Figure 129**.

6. Make sure all of the bearings and seals are properly seated.

Crankcase Assembly

1. Install the crankshaft into the upper crankcase half as described in this chapter.
2. Install the 2 dowel pins into the lower crankcase half (**Figure 121**).
3. Make sure the crankcase mating surfaces are completely clean and apply a light coat of Yamabond No. 4 or equivalent to the mating surfaces of one case half.
4. Put the lower case half onto the upper half. Check the mating surfaces all the way around the case halves to make sure they are even and that a locating ring or dowel pin has not worked loose. See **Figure 130**.
5. *RA700, RA700A, RA700B, RA760 WB700, WB700A, WB760, WRA650, WRA650A, WRA700, WRB650, WRB650A, WRB700 and WVT700*—Set the rubber damper in place over the middle 4 crankcase bolt holes. Be sure the F on the mount faces the flywheel end as shown in **Figure 131**.
6. Install a washer onto each crankcase bolt and apply Loctite 242 to the threads of each bolt. Install the crankcase bolts and washers finger-tight, then tighten them in the sequence shown in **Figure 131**. Torque the crankcase bolts, in 2 steps, to the specification given in **Table 3** or **Table 4**.

CAUTION
While tightening the crankcase fasteners, make frequent checks to ensure that the crankshaft turns freely and that the crankshaft locating rings and crankcase dowel pins fit into place in the case halves.

7. Install the 2 engine brackets (**Figure 132**). Apply Loctite 271 (red) to the threads of the engine bracket mounting bolts and install the bolts finger-tight. Tighten the engine bracket mounting bolts, in 2 steps, to the torque specification listed in **Table 3** or **Table 4**. Tighten the bolts in a crisscross pattern.

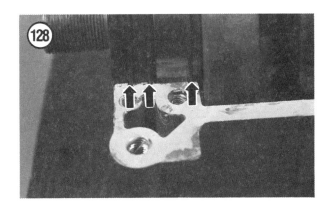

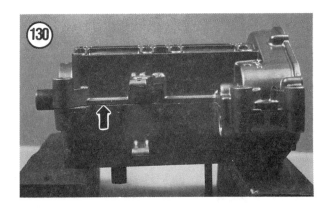

ENGINE (650, 700 AND 760 CC)

8. Check again that the crankshaft turns freely. If it is binding, separate the crankcase halves and determine the cause of the problem.

9. Turn the engine right side up.

10A. *650 and 700 cc models*—Install the flywheel, stator plate and idler gear as described in Chapter Nine.

10B. *760 cc models*—Install the flywheel and idler gear as described in Chapter Nine.

11. Install the engine coupling half as follows:
    a. Install the spacer.
    b. Coat the rear crankshaft threads with Loctite 271.
    c. Screw the coupling half onto the end of the crankshaft.
    d. Hold the flywheel steady with a holding tool and tighten the coupling half with the special tool. Tighten the coupling half to the torque specification in **Table 3** or **Table 4**.
    e. *FX700, SJ650, SJ700, WR650, WB700, WB700A, WRA650, WRA650A, WRB700, WRB650, WRB650A and WRB700*—Apply water resistant grease to a new O-ring and install it in the coupling half. Install the ring nut into the coupling half and torque the nut to the specification in **Table 3** or **Table 4**.

12. Install the flywheel cover as described in Chapter Nine.

13. Install the starter as described in Chapter Nine.

14. Install the engine into the hull as described in this chapter.

15. Reinstall the engine top end as described in this chapter.

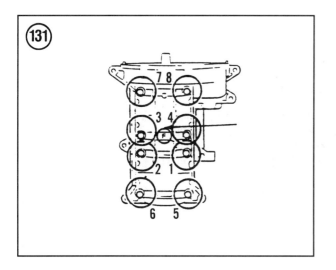

Table 1 ENGINE SPECIFICATIONS (650, 700 AND 760 CC)

Bore × stroke	
650 cc models	77 × 68 mm (3.03 × 2.68 in.)
700 cc models	81 × 68 mm (3.19 × 2.68 in.)
760 cc models	84 × 68 mm (3.30 × 2.68 in.)
Displacement	
650 cc models	633 cc (38.61 cu. in.)
700 cc models	701cc (42.78 cu. in.)
	(continued)

Table 1 ENGINE SPECIFICATIONS (650, 700 AND 760 CC) (continued)

Displacement (continued)	
760 cc models	754 cc (46.0 cu. in.)
Compression ratio	
RA760, WB760	Front-7.2:1 Rear-6.8:1
WR650	7.0:1
All other models	7.2:1

Table 2 ENGINE SERVICE SPECIFICATIONS (650, 700 AND 760 CC)

	New mm (in.)	Wear limit mm (in.)
Cylinder head warp limit	–	0.1 (0.0039)
Cylinder		
Bore		
650 cc models	77.00-77.02 (3.031-3.032)	–
RA700, RA700A, RA700B, WVT700	81.00-81.02 (3.189-3.190)	81.10 (3.193)
Other 700 cc models	81.00-81.02 (3.189-3.190)	–
RA760, WB760	84.00-84.02 (3.307-3.308)	84.10 (3.311)
Taper limit	–	0.08 (0.003)
Out of round limit	–	0.05 (0.002)
Piston		
Diameter		
650 cc models	76.915-76.940 (3.028-3.029)	–
700 cc models	80.925-80.950 (3.186-3.187)	–
RA760, WB760	83.897-83.916 (3.3030-3.3083)	
Measuring point	10 (0.39) (from bottom skirt)	–
Piston clearance		
650 cc models	0.080-0.085 (0.0031-0.0033)	0.1 (0.004)
RA700, RA700A, WVT700	0.080-0.085 (0.0031-0.0033)	0.13 (0.005)
Other 700 cc models	0.070-0.075 (0.0028-0.0030)	0.1 (0.004)
RA760, WB760	0.100-0.105 (0.0039-0.0041)	0.115 (0.0061)
Piston rings		
Type	Keystone	–
End Gap	0.2-0.4 (0.008-0.016)	–
Side clearance		
SJ650	0.03-0.07 (0.0012-0.0028)	–
SJ700A	0.01-0.03 (0.0004-0.0012)	–
WR650, WRA650, WRA650A	0.03-0.05 (0.0012-0.0020)	–
All other models	0.02-0.06 (0.0008-0.0024)	–
Crankshaft		
Crank width	61.95-62.00 (2.429-2.441)	–
Runout limit		
SJ760, WB700, WB700A, WRA650, WRA650A, WRA700, WRB650, WRB650A, WRB700	0.03 (0.0012)	–
FX700, RA700, RA700A, RA700B, RA760, SJ700, SJ700A, WB760, WR650, WVT700	0.05 (0.0020)	–
Big end side clearance		
RA700, RA700A, RA700B, RA760, SJ700A, WB760, WVT700	0.25-0.75 (0.010-0.030)	–
All other models	0.03-0.07 (0.012-0.028)	–
Small end freeplay limit	–	2.0 (0.08)

Table 3 ENGINE TIGHTENING TORQUE (650 CC)

	N·m	in.-lb	ft.-lb.
Cylinder head			
1st step	15	–	11
2nd step	28	–	20
Cylinder block bolts			
SJ650, WR650			
1st step	20	–	14
2nd step	40	–	29
WRA650, WRA650A			
1st step	15	–	11
2nd step	28	–	20
WRB650, WRB650A			
1st step	24	–	17
2nd step	30	–	22
Crankcase bolts			
SJ650, WR650			
1st step	15	–	11
2nd step	28	–	20
WRA650, WRA650A,			
WRB650, WRB650A			
1st step	14	–	10
2nd step	28	–	20
Exhaust cover			
SJ650, WR650, WRA650,			
WRA650A, WRB650, WRB650A			
1st step	20	–	4
2nd step	40	–	29
Exhaust guide			
WR650			
1st step	10	88	–
2nd step	20	–	14
Flywheel	70	–	51
Engine mounting bolts	17	–	12
Engine mounting bracket			
SJ650			
1st step	15	–	11
2nd step	30	–	22
WRA650, WRA650A,			
WRB650, WRB650A			
1st step	23	–	17
2nd step	47	–	34
Coupling half	37	–	27
Muffler-to-stay bolts			
SJ650, WR650, WRA650,			
WRA650A, WRB650,			
WRB650A			
1st step	25	–	18
2nd step	45	–	33
Muffler stay			
SJ650, WR650, WRA650,			
WRA650A, WRB650,			
WRB650A			
1st step	4	35	–
2nd step	40	–	29
(continued)			

Table 3 ENGINE TIGHTENING TORQUE (650 CC) (continued)

	N·m	in.-lb.	ft.-lb.
Muffler-to-cylinder-block bolts			
SJ650, WR650, WRA650,			
WRA650A, WRB650,			
WRB650A			
1st	15	–	11
2nd	28	–	20

Table 4 ENGINE TIGHTENING TORQUE (700 AND 760 CC)

	N·m	in.-lb.	ft.-lb.
FX700, SJ700, WB700, WB700A,			
WRA700, WRB700			
1st step	15	–	11
2nd step	28	–	20
RA700, RA700A, RA700B, RA760,			
SJ700A, WB760, WVT700			
1st step	15	–	11
2nd step	36	–	26
Cylinder block bolts			
FX700, SJ700, SJ700A, WB700,			
WB700A, WRA700, WRB700			
1st step	23	–	17
2nd step	30	–	22
RA700, RA700A, RA700B,			
RA760, WB760, WVT700			
1st step	23	–	17
2nd step	40	–	29
Crankcase bolts			
WRB700			
1st step	14	–	10
2nd step	28	–	20
All other models			
1st step	15	–	11
2nd step	28	–	20
Exhaust cover			
RA700, RA700A, RA760, WB760			
1st step	15	–	11
2nd step	30	–	22
FX700, SJ760, SJ700, WRA700,			
WRB700			
1st step	20	–	14
2nd step	40	–	29
Flame arrestor cover			
WB700, WB700A, WRA700,			
WRB700	1	8.8	–
RA700, RA700A, RA700B, RA760,			
SJ700A, WB760, WVT700	2	18	–
Flame arrestor			
WB700, WB700A, WRB700			
1st step	4	35	13
2nd step	40	–	29
Flywheel	70	–	51
(continued)			

Table 4 ENGINE TIGHTENING TORQUE (700 AND 760 CC) (continued)

	N·m	in.-lb.	ft.-lb.
Engine mounting bolts			
RA700, RA700A, RA700B, RA760, WB760, WRA700, WRB700, WVT700	17	–	12
SJ700, SJ700A, WB700, WB700A			
1st step	17	–	12
2nd step	34	–	25
FX700			
1st step	8	71	–
2nd step	16	–	12
Engine mounting bracket			
FX700, RA700B, SJ700, SJ700A, WB700, WB700A, WRA700, WRB700			
1st step	23	–	17
2nd step	47	–	34
RA700, RA700A, RA760, WB760, WVT700			
1st step	23	–	17
2nd step	53	–	39
Coupling half	37	–	27
Exhaust chamber bolt			
RA700, RA700A, RA760, WB760, WVT700	40	–	29
Exhaust chamber-to-stay bolt			
RA700, RA700A, RA760, WB760, WVT700			
1st step	3	26	–
2nd step	47	–	34
Muffler stay			
RA700, RA700A, RA760, WB760, WVT700	40	–	29
All other models			
1st step	4	35	–
2nd step	40	–	29
Muffler-to-stay bolts			
FX700, RA700B, SJ700, WB700, WB700A, WRA700, WRB700			
1st step	28	–	20
2nd step	53	–	39
Muffler-to-cylinder-block bolts			
RA700, RA700A, RA760, WB760, WVT700			
1st step	22	–	16
2nd step	40	–	29
RA700B			
1st step	15	–	11
2nd step	30	–	22
FX700, SJ700, SJ700A, WB700, WB700A, WRA700, WRB700			
1st	15	–	11
2nd	28	–	20

(continued)

Table 4 ENGINE TIGHTENING TORQUE (700 AND 760 CC) (continued)

	N·m	in.-lb.	ft.-lb.
Crankcase bolts			
WRB700			
1st step	14	–	10
2nd step	28	–	20
All other models			
1st step	15	–	11
2nd step	28	–	20

Chapter Six

Engine (1100 cc)

The 1100 cc engine is a 2-stroke triple cylinder marine engine. It has 3 ignition coils, 3 pulse coils and features a 7 amp charging system. This engine also has 3 carburetors with a single fuel pump. The electrical box has been relocated on 1100 cc models. It is integrated into the flywheel cover.

The engine is mounted inside the hull and is secured to 2 sets of engine mounts with stainless steel bolts. An aluminum coupler half is mounted onto the end of the crankshaft and to the front of the drive shaft. A rubber coupler is used to engage the crankshaft and drive shaft couplings. The coupler cushions crankshaft and drive shaft engagement and absorbs small amounts of drive train misalignment, resulting from engine vibration and drive shaft runout.

This chapter covers information to provide routine top-end service as well as crankcase disassembly and crankshaft service.

Work on your water vehicle engine requires considerable mechanical ability. You should carefully consider your own capabilities before attempting any operation involving major disassembly of the engine.

Much of the labor charge for dealer repairs involves the removal and disassembly of other parts to reach the defective component. Even if you decide not to tackle the entire engine overhaul after studying the text and illustrations in this chapter, it can be less expensive to perform the preliminary operations yourself, then take the engine to your dealer. Since dealerships have a lengthy waiting list for service (especially during spring and summer), this practice can reduce the time your unit is in the shop. If you have done much of the preliminary work, your repairs can be scheduled and performed much quicker.

General engine specifications are listed in **Table 1**. **Tables 1-3** are found at the end of the chapter.

ENGINE LUBRICATION

Engine lubrication is provided by an oil injection system. Refer to Chapter Ten for details.

SERVICE PRECAUTIONS

Whenever you work on the engine, there are several precautions that should be followed to help with disassembly, inspection, and reassembly.

1. Before beginning the job, read Chapter One of this manual. You will do a better job with this information fresh in your mind.

2. In the text there is frequent mention of the left-hand and right-hand side of the engine. This refers to the engine as it is mounted in the hull, not as it sits on your workbench. Left and right refers to the rider's point of view when seated on the watercraft facing forward. See **Figure 1**.

3. Always replace a worn or damaged fastener with one of the same size, type and torque requirements. Stainless steel fasteners are used throughout the watercraft. Make sure to identify each fastener before replacing it with another. Lubricate fastener threads with engine oil, unless otherwise specified, before torque is applied. If a tightening torque is not listed in **Table 3**, refer to the torque and fastener information in Chapter One.

4. Use special tools where noted. In some cases, it may be possible to perform the procedure with makeshift tools, but this practice is not recommended. The use of makeshift tools can damage the components and may cause serious personal injury. Where special tools are required, they may be purchased through any Yamaha watercraft dealership. Other tools can be from a motorcycle or automotive accessory store. When purchasing tools from an automotive accessory dealer or store, remember that all parts that attach to the engine must have metric threads.

5. Before removing the first bolt, get a number of boxes, plastic bags and containers and store the parts as they are removed (**Figure 2**). Also have on hand a roll of masking tape and a permanent, waterproof marking pen to label each part or assembly as required. If your craft was purchased second hand and it appears that some of the wiring may have been changed or replaced, it will be to your advantage to label each electrical connection before disconnecting it.

6. Use a vise with protective jaws to hold parts. If protective jaws are not available, insert wooden blocks on each side of the part(s) before clamping them in the vise.

7. Remove and install pressed-on parts with an appropriate mandrel, support and hydraulic press. *Do not* try to pry, hammer or otherwise force them on or off.

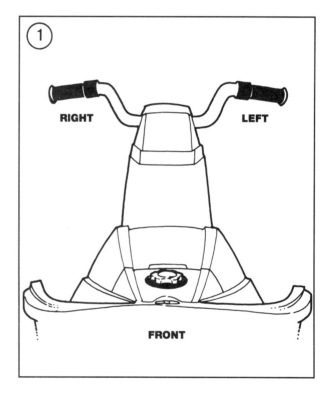

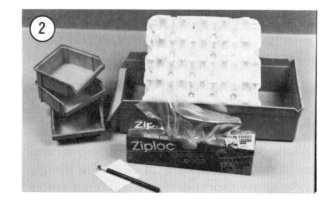

ENGINE (1100 CC)

8. Refer to **Table 3** for torque specifications. Proper torque is essential to ensure long life and satisfactory service from marine components.

9. Discard all O-rings and oil seals during disassembly. Apply a small amount of marine grease to the inner lips of each oil seal to prevent damage when the engine is first started.

10. Keep a record of all shims and where they came from. As soon as the shims are removed, inspect them for damage and write down their thickness and location.

11. Work in an area where there is sufficient lighting and room for component storage.

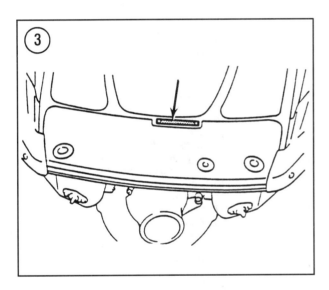

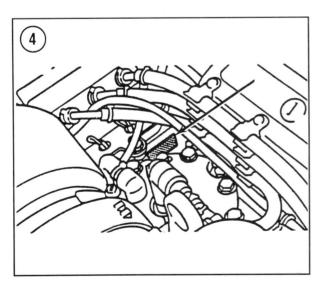

SERIAL NUMBERS

Yamaha water vehicles are identified by hull and engine identification numbers. The hull number is stamped on a plate mounted on the rear of the footrest floor (**Figure 3**). The engine number is stamped on a plate mounted on the rear of the electrical box (**Figure 4**). The Primary identification number is stamped on a plate mounted to the hull at the front of the engine compartment.

Because Yamaha may make a number of design changes during production or after the craft is sold, always use these numbers when ordering replacement parts.

SERVICING ENGINE IN HULL

Some of the components can be serviced while the engine is mounted in the hull:

a. Carburetor.
b. Flywheel.
c. Starter motor.

SPECIAL TOOLS

Where special tools are required or recommended for engine overhaul, the tool part numbers are provided. Yamaha tool part numbers have a YB, YU or YW prefix. These tools can be purchased from a Yamaha water vehicle dealership.

PRECAUTIONS

Because of the explosive and flammable conditions that exist around gasoline, always observe the following precautions.

1. Immediately after removing the engine cover, check for the presence of raw gasoline fumes. If strong fumes can be smelled, determine their source and correct the problem.

2. Allow the engine compartment to air out before beginning work.

3. Disconnect the negative battery cable. See **Figure 5**, typical.

4. Gasoline dripping onto a hot engine component may cause a fire. Always allow the engine to cool completely before working on any fuel system component.

5. Wipe up spilled gasoline immediately with dry rags. Then store the rags in a suitable metal container until they can be cleaned or disposed of. Do not store gasoline or solvent soaked rags in the engine compartment.

6. Do not service any fuel system component while in the vicinity of an open flame, sparks or while anyone is smoking.

7. Always have a Coast Guard-approved fire extinguisher close at hand when working on the engine.

ENGINE REMOVAL

Engine removal and crankcase separation is required for repair of the bottom end (crankshaft, connecting rod and bearings) and for removal of the drive shaft.

Refer to **Figure 6** when performing this procedure.

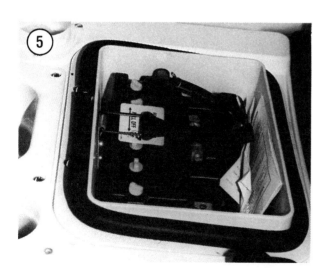

ENGINE REMOVAL (1100 CC)

1. Positive battery lead
2. Fuel hose
3. Water outlet hose
4. Bolt
5. Grease nipple
6. Coupling cover
7. Bolt
8. Collar
9. Exhaust hose
10. Clamp
11. Clamp
12. Water inlet hose
13. Negative battery lead
14. Rear engine bracket
15. Bolt
16. Washer
17. Shim
18. Engine mount
19. Bolt
20. Engine mount spacer
21. Pilot water hose
22. Oil hoses
23. Fuel return hose
24. Front engine bracket
25. Throttle cable
26. Choke cable
27. Starter/stop switch connectors
28. Meter connector

ENGINE (1100 CC)

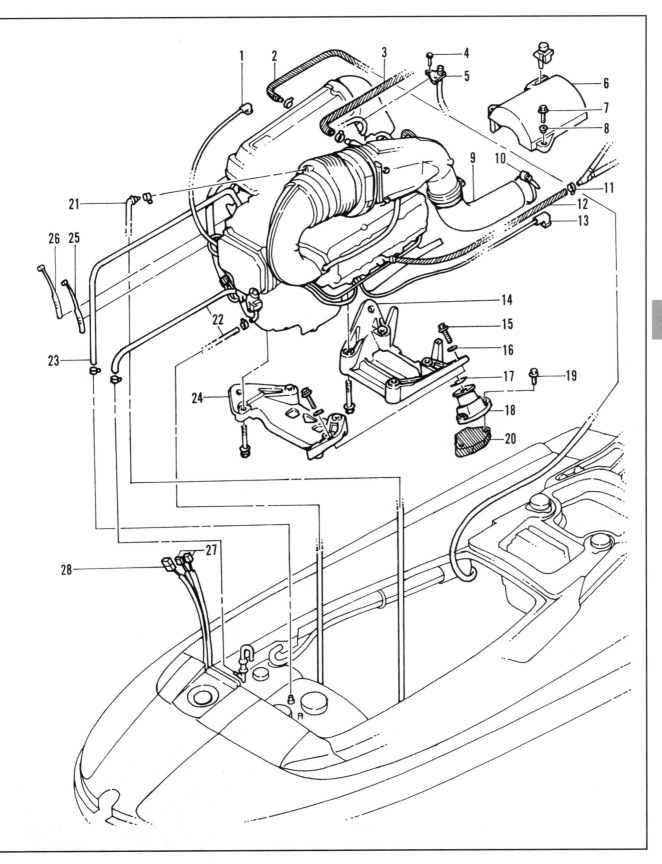

1. Support the water vehicle on a stand or on wooden boxes so it is secure.
2. Remove the engine cover.
3. Remove the negative battery cable from the battery, then remove the positive cable (**Figure 5**).
4. Disconnect the starter switch and stop switch connectors. Then disconnect the meter connector (A, **Figure 7**). Clip the 2 cable ties that secure the wiring harness to the hull so the harness is free (B. **Figure 7**).
5. Disconnect the fuel hose from the fuel pump and disconnect the fuel return hose from the No. 1 carburetor (**Figure 8**).
6. Remove the oil hose and the oil return hose from the oil pump.
7. Disconnect the throttle cable from the throttle lever and the choke cable from the choke lever (**Figure 9**).
8. Remove the grease nipple bracket from the carburetor cover (**Figure 10**).
9. Disconnect the water inlet hose from the exhaust manifold cover (**Figure 6**).
10. Disconnect the pilot water hose from the exhaust boot (A, **Figure 11**) and disconnect the water outlet hose from the muffler housing (B, **Figure 11**).
11. Disconnect the exhaust hose from the muffler housing.
12. Remove the bolts holding the coupling cover to its mounting brackets and remove the cover.

NOTE
Shim packs are used underneath the engine mounting bolts to achieve proper engine-to-drive shaft alignment. Use different colored markers or other appropriate means to mark each shim pack so it can be identified and reinstalled in its original position.

13. Loosen the 4 engine mounting bolts. Remove the bolts and shims.
14. Check to make sure all of the wiring and hoses are disconnected from the engine.

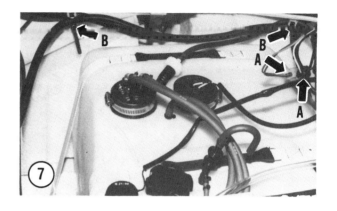

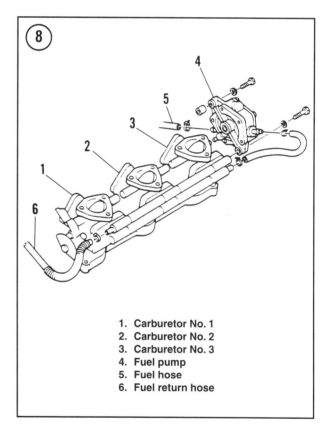

1. Carburetor No. 1
2. Carburetor No. 2
3. Carburetor No. 3
4. Fuel pump
5. Fuel hose
6. Fuel return hose

ENGINE (1100 CC)

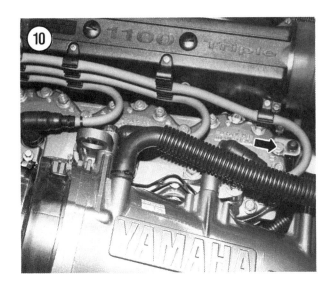

WARNING
A minimum of 2 people are required to remove the engine assembly in Step 16.

15. Slide the engine forward to disengage it from the coupler then lift it up and out of the engine compartment. Take it to a workbench for further disassembly.

16. Remove the rubber coupler from the coupling half (**Figure 12**).

Installation

1. Wash the engine compartment with clean water.

2. Spray all of the exposed electrical connectors with electrical contact cleaner.

3. Before installing the engine, check components which are normally inaccessible for excessive wear or damage.

4. Remove and clean the bilge filter as described in Chapter Eleven.

5. Examine the engine mounts and supports for looseness or damage.

6. Check inside the hull for tools or other objects that may interfere with engine installation. Make sure all hoses, cables and wiring connectors are routed and secured properly.

7. Check the rubber damper on the front engine bracket before installing the engine in the hull.

8. Install the engine mounts (**Figure 6**) into the hull, if they were previously removed.

9. Check the rubber coupler (**Figure 12**) for wear grooves or rubber deterioration. Especially check for wear on the round outer knobs; replace if the coupler is worn so much that any rotational play exists between the engine half and the drive shaft coupler half.

10. Install the rubber coupler into the forward bearing housing coupling.

11. With an assistant, place the engine in the hull and slide it rearward. Be sure the engine and bearing housing couplings engage the rubber coupler.

12. Temporarily install the 4 engine mounting bolts and washers finger-tight (**Figure 13**). Do not tighten the bolts now. They are installed to align the engine brackets with the rubber mounts.

13. Check coupling alignment as follows:

   a. Hold a straightedge against one side of the coupling halves (**Figure 14**). Push the straightedge against the flat sides of the couplings. If you can see a gap between one coupling flat and the straightedge or you can feel the straightedge rock as you push against one coupling flat then the other, measure the clearance with a feeler gauge; see A, **Figure 14**. If the clearance exceeds 0.6 mm (0.024 in.), the coupling halves are misaligned. Repeat this check with the straightedge against the other side, then on top of the coupler halves.

   b. To correct misalignment, reposition the engine. Repeat substep a to check alignment and continue until alignment is correct.

   c. When alignment is correct, install the required shims in the gap between the engine mounts and engine brackets. Shims can be purchased through a Yamaha water vehicle dealership in the following thicknesses: 0.1, 0.3, 0.5, 1.0 and 2.0 mm.

   d. After installing the shims, measure the gap between the front and rear coupling halves as shown in B, **Figure 14**. The correct clearance is 2-4 mm (0.079-0.157 in.). If the clearance is incorrect, reposition the engine by pushing it forward or rearward as required. Recheck clearance.

   e. Wipe the engine mounting bolt threads with Loctite 271 (red), then install and tighten the bolts to the torque specification listed in **Table 3**.

CAUTION
It is important to align the coupler halves as closely as possible. Any significant degree of misalignment will cause excessive vibration and result in damage to the drive train components.

14. Apply Loctite 242 (blue) to each of the coupling cover bolts. Then install the coupling cover with the bolts, flat washers and lockwashers. Turn the drive shaft and check that the coupler assembly does not contact the coupler guard.

15. Reverse Steps 1-12 in *Engine Removal* to complete installation. Note the following:

   a. Use new hose or cable guides where required.

   b. Adjust the throttle cable and choke cable as described in Chapter Three.

   c. Refill the oil tank as described in Chapter Three.

MUFFLER HOUSING

The muffler housing can be removed and installed with the engine installed in the hull.

NOTE
This engine has a second muffler that is downline from the water lock box. Refer to Chapter Eight for information about this muffler.

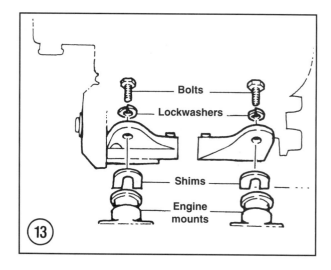

ENGINE (1100 CC)

Removal

Refer to **Figure 15** for this procedure.

1. Disconnect the pilot water hose from the exhaust boot (A **Figure 11**) and disconnect the water outlet hose from the muffler housing (B **Figure 11**).

2. Disconnect the 2 water hoses from the cylinder head. See **Figure 16**.

3. Remove the thermoswitch from the rear boss of the muffler housing (**Figure 17**).

4. Disconnect the exhaust hose from the muffler housing.

5. Loosen the hose clamp (C, **Figure 11**) securing the exhaust boot to the muffler housing outer cover. Pull back the exhaust boot, loosen the inner hose clamp and disconnect the joint from the outer cover.

6. At the other end of the boot, loosen the hose clamp (D, **Figure 11**) securing the boot to the exhaust elbow. Pull back the exhaust boot, loosen the inner hose clamp and disconnect the exhaust joint from the exhaust elbow.

7. Remove the 2 muffler housing-to-bracket bolts and the 2 muffler housing-to-cylinder block bolts.

8. Remove the muffler housing from the engine.

Muffler Housing
Disassembly/Reassembly

Refer to **Figure 18** when performing this procedure.

1. Remove the 6 muffler bolts from the muffler housing.

2. Remove the outer muffler cover, the inner muffler cover, the silencer and the seal.

3. Clean all parts to remove carbon and oil residue.

4. Carefully remove all gasket residue from all mating surfaces.

5. After all parts have been cleaned, visually check for cracks or other damage. Check the seal for cracks or damage. Replace all worn or damaged parts.

6. Assembly is the reverse of these steps. Install new gaskets. Apply Loctite 271 (red) to the threads of the muffler bolts and torque them in the 2 steps to the specification given in **Table 3**.

Exhaust Manifold
Removal

Refer to **Figure 19** when performing this procedure.

1. Remove the muffler housing as described in this chapter.

2. Remove the water inlet hose from the exhaust manifold cover.

3. Remove the 12 exhaust manifold bolts and remove the manifold along with the exhaust elbow.

4. Remove the 2 dowel pins from the cylinder block.

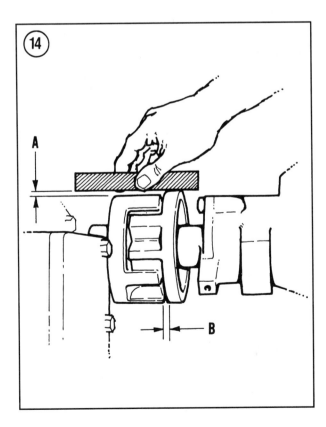

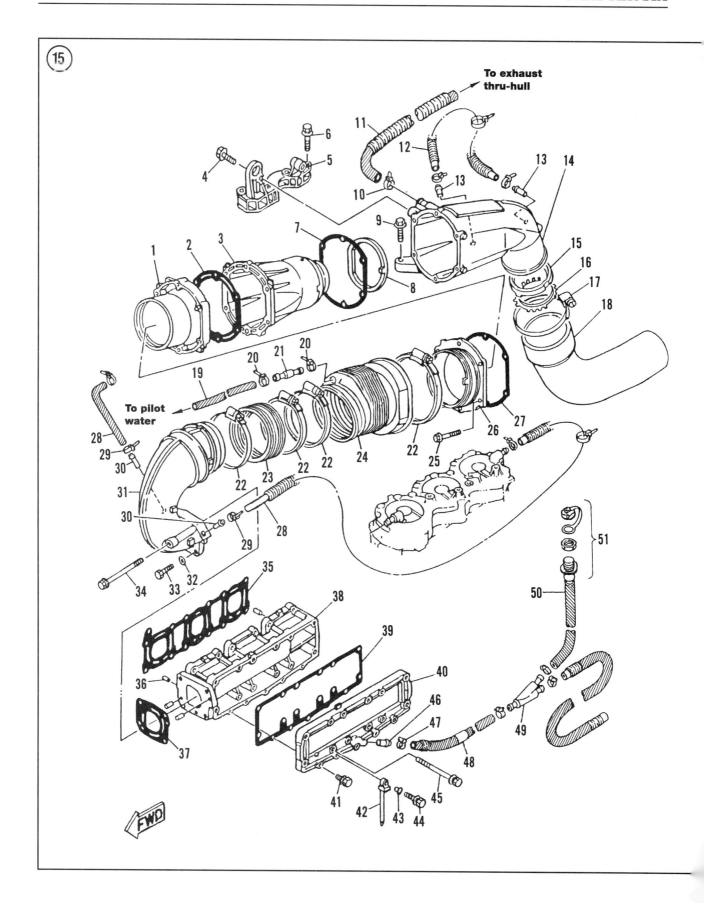

ENGINE (1100 CC)

EXHAUST SYSTEM (1100 CC)

1. Inner cover
2. Gasket
3. Silencer
4. Bolt
5. Muffler bracket
6. Bolt
7. Gasket
8. Seal
9. Bolt
10. Clamp
11. Water outlet hose
12. Water hose
13. Nozzle
14. Muffler housing
15. Seal
16. Stopper
17. Clamp
18. Exhaust hose
19. Pilot water hose
20. Clamp
21. Connector
22. Clamp
23. Exhaust joint
24. Exhaust boot
25. Bolt
26. Outer cover
27. Gasket
28. Water hose
29. Clamp
30. Nozzle
31. Exhaust elbow
32. Washer
33. Bolt
34. Bolt
35. Gasket
36. Dowel pin
37. Gasket
38. Exhaust manifold
39. Gasket
40. Exhaust manifold cover
41. Bolt
42. Plastic tie
43. Collar
44. Bolt
45. Bolt
46. Nozzle
47. Clamp
48. Water inlet hose
49. Y-connector
50. Flush hose
51. Flush connector

Exhaust manifold
Disassembly/Reassembly/Installation

Refer to **Figure 19** when performing this procedure.

1. Remove the exhaust elbow from the exhaust manifold and remove the 2 dowel pins.
2. Remove the exhaust manifold cover from the manifold.
3. Clean all parts to remove carbon and oil residue.
4. Carefully remove all gasket residue from all mating surfaces.
5. After all parts have been cleaned, visually check for cracks or other damage. Check the seal for cracks or damage. Replace all worn or damaged parts.
6. Install the manifold cover along with a new gasket to the exhaust manifold. Apply Loctite 242 to the threads of the manifold cover bolts.
7. Install the exhaust elbow and a new gasket onto the exhaust manifold. Apply Loctite 271 to the threads of the exhaust elbow bolts, and torque the bolts to the specification given in **Table 3**. Torque the bolts in the sequence embossed on the cover.
8. Install the exhaust manifold along with a new gasket to the cylinder block. Apply Loctite 271 (red) to the threads of the exhaust manifold bolts and torque them in the sequence shown in **Figure 20**. Refer to **Table 3** for the correct torque specification.
9. Insert the 2 dowels into the exhaust manifold and install the exhaust elbow along with a new gasket. Apply Loctite 271 (red) to the exhaust elbow bolts and torque them to the specification given in **Table 3**.

Muffler Housing
Installation

Refer to **Figure 15** when performing this procedure.

1. Install the muffler bracket onto the cylinder head, if it was removed. Apply Loctite 271 (red) to the bolt threads and torque the bolts to the specification given in **Table 3**.
2. Set the muffler housing in place, and reconnect the exhaust hose to the muffler housing.
3. Align the muffler housing with the muffler bracket. Apply Loctite 271 (red) to the threads of the muffler housing-to-bracket bolts. Install the bolts and torque them in 2 steps to the specification given in **Table 3**.
4. Apply Loctite 271 (red) to the threads of the 2 muffler housing-to-cylinder block bolts. Install the bolts and torque them to the specification given in **Table 3**.
5. Pull the exhaust boot back and fit the exhaust joint over the exhaust elbow. When the connec-

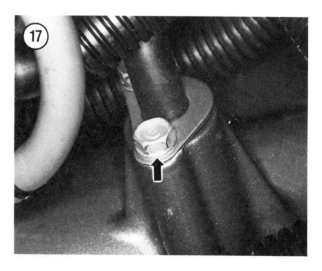

ENGINE (1100 CC)

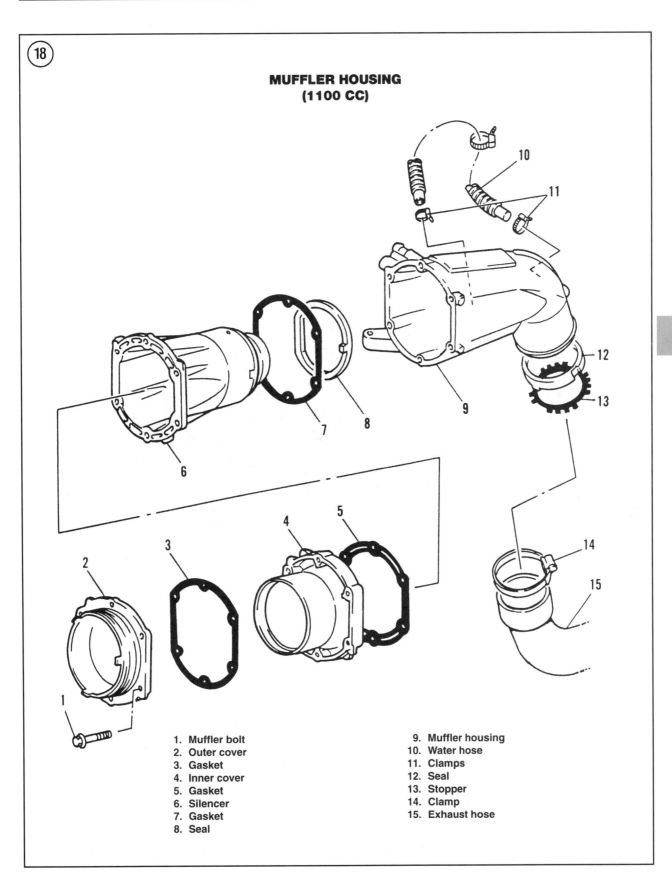

MUFFLER HOUSING (1100 CC)

1. Muffler bolt
2. Outer cover
3. Gasket
4. Inner cover
5. Gasket
6. Silencer
7. Gasket
8. Seal
9. Muffler housing
10. Water hose
11. Clamps
12. Seal
13. Stopper
14. Clamp
15. Exhaust hose

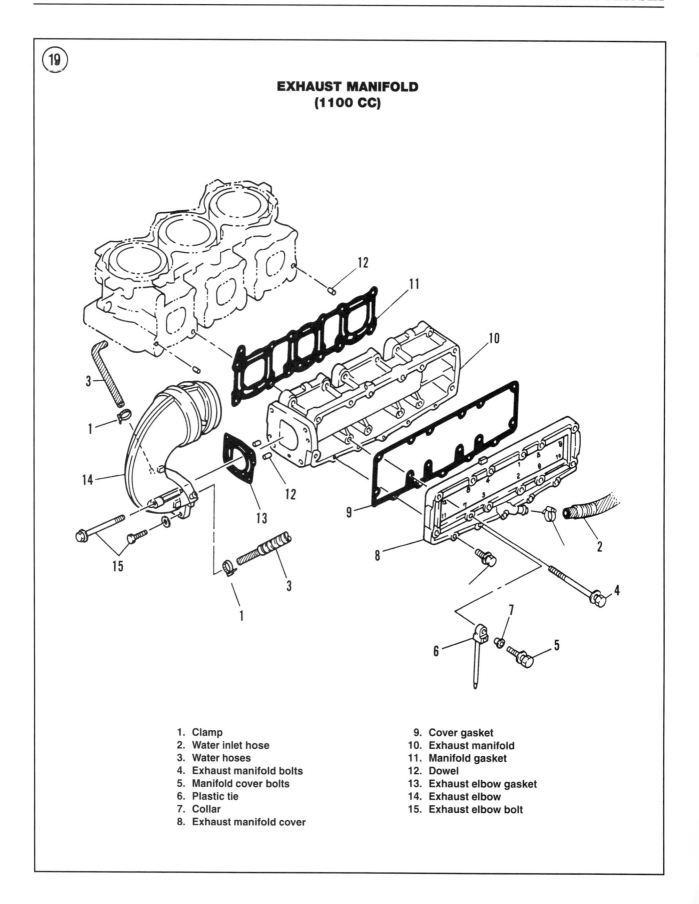

ENGINE (1100 CC)

tion is correct, tighten the inner hose clamp. Then slide the exhaust boot over the elbow and secure it with the hose clamp. See D, **Figure 11**.

6. Pull back the exhaust boot and fit the joint over the outer cover on the muffler housing. When the connection is correct, tighten the inner hose clamp. Then slide the boot over the muffler housing and secure it with the hose clamp. See C, **Figure 11**.

7. Reconnect the pilot water hose to the exhaust boot (A, **Figure 11**). Reconnect the water outlet hose to the muffler housing (B, **Figure 11**).

8. Reconnect the water hoses to the nozzles on the cylinder head cover. See **Figure 16**.

9. Apply marine grease to the thermoswitch hole in the muffler housing, and reinstall the thermoswitch (**Figure 17**).

ENGINE TOP END

The engine top end consists of the cylinder head, cylinder block, pistons, piston rings, piston pins and the connecting rod small-end bearings. See **Figure 21**.

The engine top end can be serviced with the engine installed in the hull.

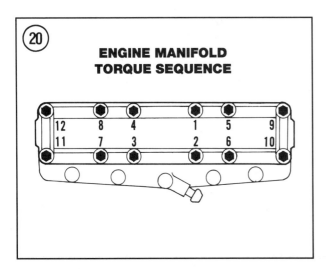

Cylinder Head Removal/Installation

Refer to **Figure 21** for this procedure.

CAUTION
To prevent warping and damage to any component, remove the cylinder head only when the engine is at room temperature.

NOTE
If the engine is being disassembled for inspection procedures, check compression before removing the cylinder head. Refer to Chapter Three.

NOTE
Identify and mark all hose guides which are mounted on the cylinder head before removing them in the following steps.

1. Remove the engine cover.
2. If the engine is mounted in the hull, perform the following:
   a. Disconnect the negative battery cable.
   b. Remove the flame arrestor as described in Chapter Eight.
   c. Disconnect the 3 spark plug caps.
   d. Remove the muffler housing as described in this chapter.
3. Loosen the spark plugs if they are going to be removed later.
4. Remove the muffler bracket bolts and remove the muffler bracket.
5. Loosen the cylinder head bolts (**Figure 22**) in a crisscross pattern then remove the bolts and washers.
6. Loosen the cylinder head by tapping around its perimeter with a rubber or plastic mallet, then remove the cylinder head.
7. Remove and discard the cylinder head gasket.
8. Loosen the cylinder head cover bolts and washers. Remove the bolts, washers, cylinder head cover and its gasket.
9. Lay a rag over the cylinder block to prevent dirt from falling into the cylinders.

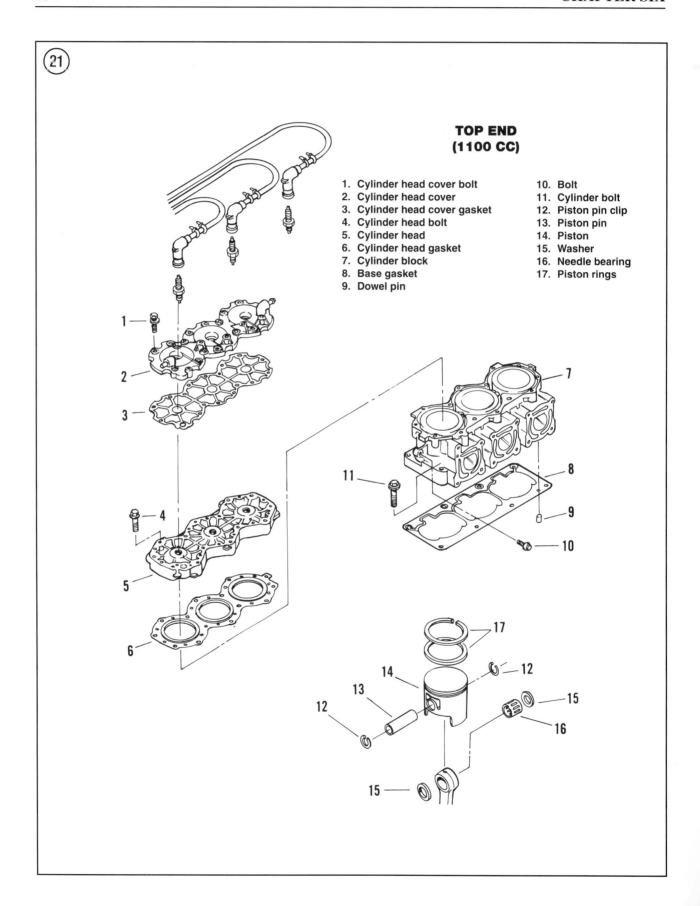

ENGINE (1100 CC)

10. Clean and inspect the cylinder head as described in this chapter.

11. Before installing the cylinder head cover gasket in Step 12, make sure the cylinder block and cylinder head mating surfaces are free of all gasket residue.

12. Install a new cylinder head cover gasket (without gasket sealer) and set the cylinder head cover onto the cylinder head. Apply Loctite 242 to the cylinder head cover bolts, and install the bolts and washers finger-tight.

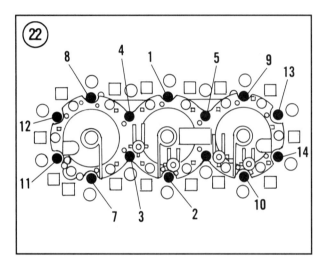

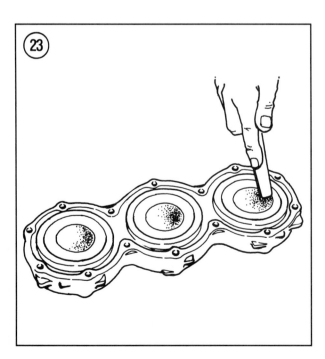

13. Tighten the cylinder head cover bolts and washers, in 2 steps, in a crossing pattern, to the torque specification listed in **Table 3**.

14. Place a new cylinder head gasket onto the cylinder block, without sealant. Be sure the bolt holes in the gasket align with the bolt holes in the cylinder block. In addition, make sure the tab on the gasket aligns with the boss on the cylinder block.

15. Set the cylinder head onto the cylinder block. Apply Loctite 271 to the threads of the cylinder head bolts then install the cylinder head bolts and washers finger-tight.

16. Tighten the cylinder head bolts in 2 steps to the torque specification in **Table 3**. To prevent cylinder head warping or a blown head gasket, tighten the cylinder bolts in the sequence shown in **Figure 22**.

17. Install the muffler bracket onto the cylinder head cover. Apply Loctite 271 (red) to the bolt threads and torque the bracket bolts to the specification shown in **Table 3**.

18. Install the spark plugs and tighten to the torque specification in **Table 3**. Reinstall the spark plug caps.

19. Reverse Steps 1 and 2 to complete installation.

Inspection

1. Wipe away any soft deposits on the cylinder head (**Figure 23**) mating surface. Remove hard deposits in the combustion chambers using a soft-metal scraper. Be careful not to gouge the aluminum surfaces. Burrs created from improper cleaning will cause preignition and heat erosion.

NOTE
An aluminum thread fluid or kerosene must be applied to a tap if one is used in Steps 2 or 3.

2. Check the spark plug threads for any signs of carbon buildup or cracking. The carbon can be removed with a 14 mm spark plug tap.

3. Check the muffler bracket bolt threads (A, **Figure 24**) in the top of the cylinder head cover for corrosion or thread damage. Corrosion buildup can be removed with a tap. If the threads are stripped, a Helicoil or similar thread insert may be required.

4. Check the water hose nozzles (B, **Figure 24**) on the cylinder head cover for sludge buildup. Check the water passages (**Figure 25**) in the cylinder head for corrosion or other residue. Clean these passages thoroughly.

5. Use a straightedge and feeler gauge and measure the flatness of the cylinder head (**Figure 26**). If a 0.1 mm (0.004 in.) feeler gauge can be slipped underneath the straightedge, resurface the cylinder head as follows:

   a. Tape a piece of 400-600 grit wet emery sandpaper onto a piece of thick plate glass or surface plate.

   b. Slowly resurface the head by moving it in figure-eight patterns on the sandpaper. See **Figure 27**.

   c. Rotate the head several times to avoid removing too much material from one side. Check progress often with the straightedge and feeler gauge.

   d. If the cylinder head warpage still exceeds the service limit, it will be necessary to have the head resurfaced by a machine shop. Note that removing material from the cylinder head mating surface will change the compression ratio. Consult with the machinist on how much material to remove.

6. Wash the cylinder head in hot soapy water and rinse thoroughly before installation.

CYLINDER

An aluminum cylinder block is used on all 1100 cc engines with cast iron liners pressed into

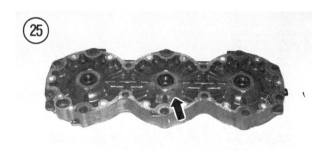

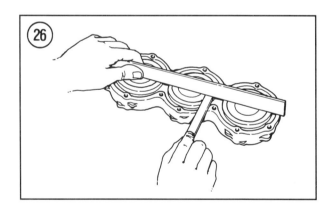

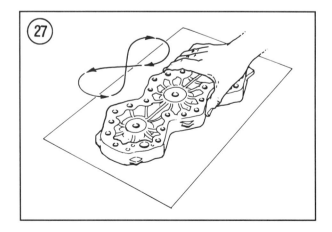

ENGINE (1100 CC)

the block. If excessive wear is present, the cylinder liner can be bored to 0.25 mm and 0.5 mm oversize and new pistons and rings installed. See **Figure 20**.

Removal

1. Remove the cylinder head as described in this chapter.

2. If the engine is installed in the hull, perform the following:

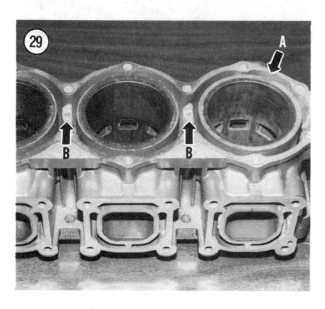

a. Remove the carburetor and intake manifold (with reed valve) as described in Chapter Eight.
b. Remove the muffler as described in this chapter.

3. The cylinder block is secured to the crankcase with 8 bolts and washers. Loosen the bolts in 1/4 turn increments, in a crisscross pattern, until they are loose. Remove the bolts and washers.

4. If the cylinder block is still tight to the crankcase, do not pry it off. Instead, tap around its perimeter with a soft-faced mallet. Then lift the cylinder block up, making sure to catch the pistons as they become free of the cylinder bore. This will prevent the pistons from damaging themselves against the crankcase.

5. Remove the base gasket and discard it.

6. Remove the 2 dowel pins (**Figure 28**).

7. Stuff clean rags around the connecting rods to keep dirt and loose parts from entering the crankcase.

Inspection

Cylinder measurement requires a precision inside micrometer or bore gauge. If you do not have the right tools, have your dealer or a machine shop take the measurements.

1. Using a wooden scraper, remove all carbon residue from the exhaust ports. If you use a wire wheel mounted in some type of drill or hand grinder to clean the ports, make sure you do not damage the cylinder lining.

2. Check all of the tapped holes in the cylinder block for corrosion buildup or damage.

3. Remove all gasket residue from the top (A, **Figure 29**) and bottom gasket surfaces of the cylinder block.

4. Visually inspect the anode mounted in the cylinder block water passage (B, **Figure 29**) for severe wear. Replace the anode if it is less than 60% of its original size.

5. Use a straightedge and feeler gauge and measure the flatness of the cylinder block as

shown in **Figure 30**. If a 0.1 mm (0.004 in.) feeler gauge can be slipped underneath the straightedge, the cylinder block may require replacement. Refer additional service to a Yamaha dealership.

6. Wash the cylinder block in solvent to remove any oil and carbon particles. The cylinder bores must be cleaned thoroughly before attempting any measurement.

7. Measure the cylinder bore diameter as described under *Piston/Cylinder Clearance Check* in this chapter.

8. Check the bore carefully for scratches or gouges.

9. After the cylinder has been serviced, wash the bore in hot soapy water. This is the only way to clean the cylinder wall of the fine grit material left from the boring or honing job. After washing the cylinder wall, run a clean white cloth through it. The cylinder wall should show no traces of grit or other debris. If the rag is dirty, the cylinder wall is not clean and must be rewashed. After the cylinders are cleaned, lubricate the cylinder walls with clean engine oil to prevent the cylinder liners from rusting.

CAUTION
A combination of soap and water is the only solution that will completely clean the cylinder wall. Solvent and kerosene cannot wash fine grit out of cylinder crevices. Grit left in the cylinder will cause premature wear to the new rings.

Installation

1. Clean the cylinder bore as described under *Inspection* in this chapter.

2. Make sure the top surface of the crankcase and the bottom cylinder surface are clean prior to installation.

3. Squirt some engine oil into each of the crankcase oil holes (**Figure 31**).

4. Install the 2 engine block dowel pins (**Figure 28**).

5. Install a new base gasket, without sealer, making sure all gasket and crankcase bolt holes align.

6. Make sure the piston pin circlips are seated in the piston clip groove correctly. If the pistons were removed from the rods, new circlips must be installed.

7. Make sure the end gaps of the piston rings are aligned with the locating pins in the ring grooves

ENGINE (1100 CC)

(**Figure 32**). Lightly oil the piston rings and the inside of the cylinder bores with clean engine oil.

8. Rotate the crankshaft so one piston is at the top of its stroke.

9. Align the cylinder bore with this upper piston. Compress the piston rings with your fingers and slide the cylinder bore over the piston. The cylinder chamfer will compress the rings as necessary.

10. After the first piston is installed in its cylinder, rotate the crankshaft until the first and second pistons are at the same height. Lower the cylinder and install the second piston.

11. Repeat this procedure for the remaining piston, then lower the cylinder block into place on the crankcase. Make sure the cylinder block is seated completely on the crankcase.

12. Wipe the cylinder bolt threads with Loctite 242, then install each bolt and its washer. Install the bolts finger-tight. Tighten the bolts, in 2 steps, to the torque specification listed in **Table 3**. Tighten the bolts in a crisscross pattern to prevent warping.

13. Install the cylinder head as described in this chapter.

PISTON, PISTON PIN, AND PISTON RINGS

The piston is made of aluminum alloy. The piston pin is a precision fit and is held in place by a clip at each end. A caged needle bearing is used on the small end of the connecting rod. See **Figure 21**.

Piston and Piston Ring Removal

1. Remove the cylinder head and cylinder block as described in this chapter.

2. Identify the pistons (**Figure 33**) as to front, center or rear. In addition, keep each piston together with its own pin, bearing and piston rings to avoid confusion during reassembly.

3. Before removing the piston, hold the rod tight and rock the piston as shown in **Figure 34**. Any rocking motion (do not confuse with the normal sliding motion) indicates wear on the piston pin, needle bearing, piston pin bore, or more likely a combination of all three.

NOTE
Wrap a clean shop cloth under the piston so the clip will not fall into the crankcase.

WARNING
Wear safety glasses when performing Step 4.

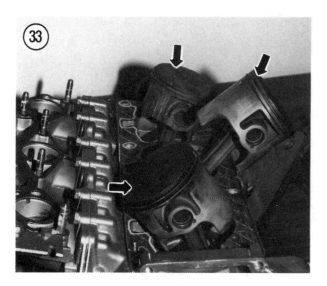

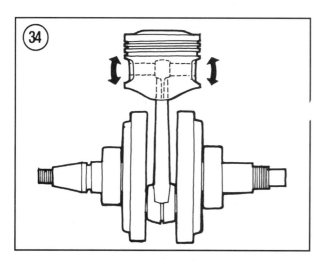

4. Remove the clips from each side of the piston pin bore (**Figure 35**) using needlenose pliers. Hold your thumb over one edge of the clip when removing it to prevent it from springing out.

5. Use a proper size wooden dowel or socket extension and push out the piston pin (A, **Figure 36**).

CAUTION
If the engine ran hot or seized, the piston pin will probably be difficult to remove. However, do not drive the piston pin out of the piston. This will damage the piston, needle bearing and connecting rod. If the piston pin will not push out by hand, remove it as described in Step 6.

6. If the piston pin is tight, fabricate the tool shown in **Figure 37**. Assemble the tool onto the piston and pull the piston pin out of the piston. Make sure to install a rubber or plastic pad between the piston and piece of pipe to avoid scoring the side of the piston.

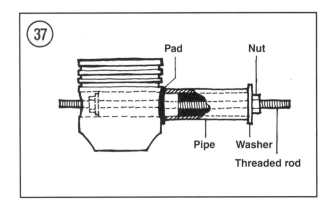

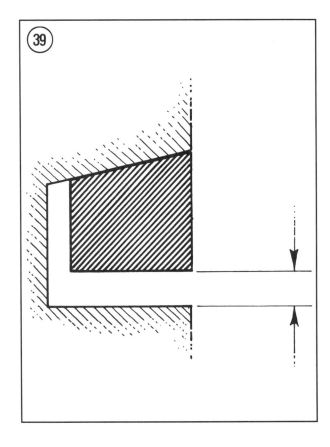

ENGINE (1100 CC)

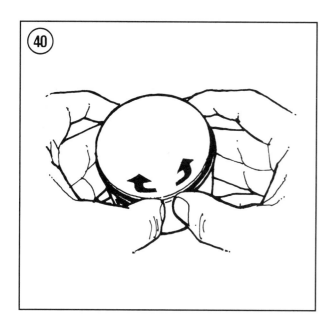

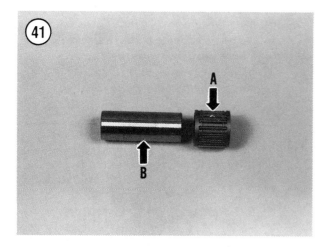

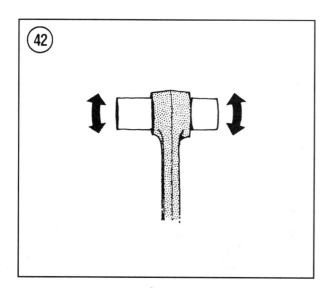

7. Lift the piston (B, **Figure 36**) off the connecting rod.

8. Remove the needle bearing from the connecting rod (**Figure 38**).

9. Repeat for the remaining pistons.

10. If the pistons are going to be left off for some time, place a piece of foam insulation tube, or shop cloth over the end of each rod to protect it.

NOTE
Always remove the top piston ring first.

11. Keystone rings (**Figure 39**) are used at the top and bottom positions. If the rings are going to be reused, mark each ring after removal to ensure correct installation. Remove the upper ring by spreading the ends with your thumbs just enough to slide it up over the piston (**Figure 40**). Repeat for the lower ring.

Piston Pin and Needle Bearing Inspection

1. Clean the needle bearing (A, **Figure 41**) in solvent and dry it thoroughly. Use a magnifying glass and inspect the bearing cage for cracks at the corners of the needle slots and inspect the needles themselves for cracking. If any cracks are found, replace the bearing.

2. Check the piston pin (B, **Figure 41**) for excessive wear, scoring or chrome flaking. Also check the piston pin for cracks along the top and side. Replace the piston pin if necessary.

3. Oil the needle bearing and pin and install them in the connecting rod. Slowly rotate the pin and check for radial and axial play (**Figure 42**). If any play exists, replace the pin and bearing, providing the rod bore is in good condition.

CAUTION
If there is evidence of piston seizure or overheating, replace the piston pins and bearings. These parts have been weakened from excessive heat and may fail later on.

Connecting Rod Inspection

1. Wipe the piston pin bore in the connecting rod with a clean rag and check it for galling, scratches, or any other signs of wear or damage. If any of these conditions exist, replace the crankshaft as described in this chapter.

2. Check the connecting rod big end bearing play. You can make a quick check by simply rocking the connecting rod back and forth (**Figure 43**). If there is more than a very slight rocking motion (some side-to-side sliding is normal), measure the connecting rod side clearance with a feeler gauge. Measure between the side of the crankshaft and the washer. **Figure 44** shows the crankshaft removed for clarity. If play exceeds the wear limit specified in Table 2, replace the crankshaft.

Piston and Ring Inspection

1. Carefully check the piston for cracks at the top edge of the transfer cutaways (**Figure 45**). Replace the piston if any cracks are noted. Check the piston skirt (**Figure 46**) for brown varnish buildup. More than a slight amount is an indication of worn or sticking rings.

2. Check the piston skirt for galling and abrasion which may have resulted from piston seizure. If light galling is present, smooth the affected area with 600-800 grit wet sandpaper. However, if galling is severe or if piston is deeply scored, replace it.

3. Check the piston ring locating pins in the piston (**Figure 47**). The pins must be tight and the piston should show no signs of cracking around the pins. If a locating pin is loose, replace the piston. A loose pin will fall out and cause severe engine damage.

4. Check the piston pin clip grooves in the piston for cracks or other damage that could allow a clip to fall out. This would cause severe engine damage. Replace the piston if any one groove is worn or damaged.

NOTE
Maintaining proper piston ring end gap helps ensure peak engine performance. Always check piston ring end gap at the intervals specified in Chapter Three. Excessive ring end gap reduces engine performance and can cause overheating. Insufficient ring end gap will cause the ring ends to butt together and cause the

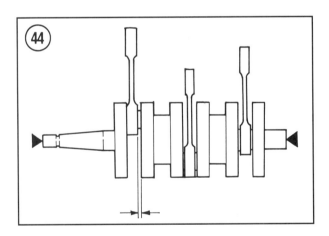

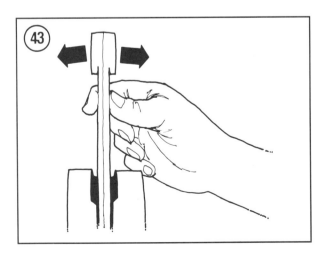

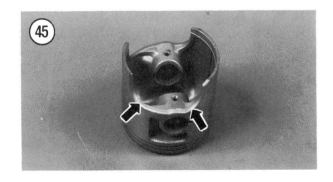

ENGINE (1100 CC)

ring to break. This will cause severe engine damage.

5. Measure piston ring end gap. Place a ring into the bottom of the cylinder and push it in about 20 mm (3/4 in.) with the crown of the piston. This ensures that the ring is square in the cylinder bore. Measure the gap with a flat feeler gauge (**Figure 48**) and compare to the wear limit in **Table 2**. If the gap is greater than specified, replace the rings as a set.

NOTE
*When installing new rings, measure the end gap in the same manner as for old ones. If the gap is less than specified, first make sure you have the correct piston rings. If the replacement rings are correct but the end gap is too small, carefully file the ends with a fine cut file until the gap is correct (**Figure 49**).*

6. Carefully remove all carbon buildup from the ring grooves with a broken ring (**Figure 50**).

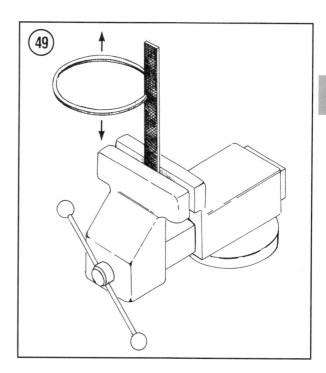

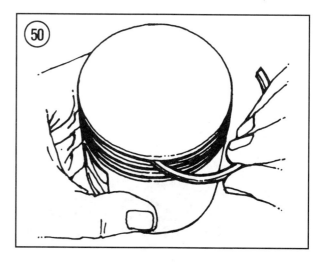

Inspect the grooves carefully for burrs, nicks, or broken and cracked lands. Recondition or replace the piston if necessary.

7. Measure the side clearance of each ring in its groove with a flat feeler gauge (**Figure 51**) and compare to specifications listed in **Table 2**. If the clearance is greater than specified, replace the rings. If the clearance is still excessive with new rings, replace the piston.

8. Observe the condition of the piston crown (**Figure 52**). Normal carbon buildup can be removed with a wire wheel mounted in a drill. If the piston shows signs of overheating, pitting or other abnormal conditions, the engine may be experiencing preignition or detonation; both conditions are discussed in Chapter Two.

9. Measure the piston outside diameter as described under *Piston/Cylinder Clearance Check* in this chapter.

10. If new piston rings are required, hone the cylinders before assembling the engine. Refer to *Cylinder Honing* in this chapter.

Piston/Cylinder Clearance Check

The following procedure requires the use of highly specialized and expensive measuring tools. If such equipment is not readily available, have the measurements performed at a dealership or machine shop. Always replace all pistons as a set.

1. Measure the outside diameter of the piston with a micrometer approximately 10 mm (13/32 in.) above the bottom of the piston skirt, at a 90° angle to the piston pin (**Figure 53**). If the diameter is less than the wear limit in **Table 2**, install new pistons.

NOTE
Always install new rings when installing a new piston.

2. Wash the cylinder block in solvent to remove any oil and carbon particles. The cylinder bore must be cleaned thoroughly before attempting any measurement.

3. Measure the cylinder bore with a bore gauge or inside micrometer (**Figure 54**). Measure the cylinder bore at the points shown in **Figure 55**. Measure in 3 axes in line with the piston pin and at 90° to the pin. If the bore is greater than specification (**Table 2**), bore the cylinders to the next oversize and install new pistons and rings.

NOTE
*Purchase the new pistons before the cylinders are bored so the pistons can be measured. The cylinders must be bored to match the pistons. Piston-to-cylinder clearance is specified in **Table 2**.*

ENGINE (1100 CC)

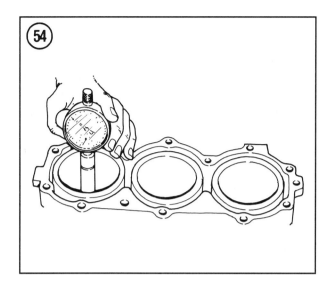

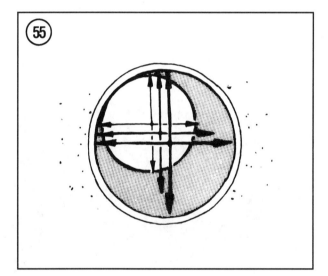

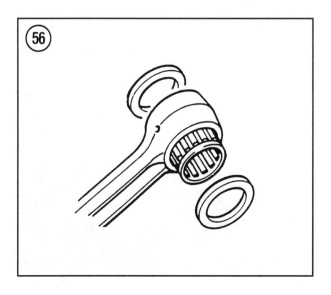

4. Piston clearance is the difference between the maximum piston diameter and the minimum cylinder diameter. Subtract the diameter of the piston from the cylinder bore diameter. If the clearance exceeds the specification in **Table 2**, bore the cylinders to the next oversize and install new pistons and rings.

Cylinder Honing

The surface condition of a worn cylinder bore is normally very shiny and smooth. Cylinder honing, often referred to as glaze breaking or deglazing, is required whenever new piston rings are installed. If new piston rings were installed in a glazed cylinder, they would not seat properly. When a cylinder bore is honed, the surface is slightly roughed up to provide a textured or crosshatched surface. This surface finish controls wear of the new rings and helps them to seat and seal properly. *Whenever new rings are installed*, the cylinder surface must be honed. This service can be performed at a Yamaha dealership or independent repair shop. The cost of having the cylinder honed is usually minimal compared to the cost of purchasing a hone and doing the job yourself. If you choose to hone the cylinder yourself, follow the hone manufacturer's directions closely.

CAUTION
*After a cylinder is reconditioned by boring or honing, the bore must be properly cleaned to remove all material left from the machining operation. Refer to Inspection under **Cylinder** in this chapter. Improper cleaning will not remove all of the machining residue in the cylinder bore and rapid wear of the new piston and rings will result.*

Piston Installation

1. Apply assembly oil to the needle bearing. Install the bearing and 2 washers into the connecting rod (**Figure 56**).

2. Oil the piston pin and install it in the piston until the end of it extends slightly beyond the inside of the boss (**Figure 57**).

3. Place the piston over the connecting rod with the arrow on the piston crown pointing toward the left-hand (exhaust) side of the engine. Align the pin with the bearing and push the pin into the piston until it is even with the piston pin clip grooves.

CAUTION
*If the piston pin will not slide in the piston smoothly, use the tool described during **Piston Removal** to install the piston pin (**Figure 37**). When using the home-made tool, the pipe is not required. Insert the threaded rod through the piston pin and out the other side of the piston. Slide the pad and large washer over the threaded rod so the pad presses against the face of the piston. Run a nut onto the rod so it presses against the large washer. At the other end of the rod, slide the small washer over the threaded rod and against the piston pin. Thread a nut onto the rod, and tighten the nut to push the piston pin into the piston. Do not drive the piston pin into the piston or you may damage the needle bearing and connecting rod.*

WARNING
Wear safety glasses when performing Step 4.

4. Install new piston pin clips (**Figure 58**) in the ends of the pin boss. Make sure they are seated in the grooves and that the end of each clip is *not* opposite the slots in the piston (**Figure 59**).

5. Check the installation by rocking the piston back and forth around the pin axis and from side to side along the axis. It should rotate freely back and forth but not from side to side.

NOTE
The top and bottom rings are the same. If you are installing new rings, they can be installed in either position. If you are installing used rings, install them by referring to the identification marks made prior to removal.

6. Install the piston rings—first the bottom one, then the top—by carefully spreading the ends of the ring with your thumbs and slipping the ring over the top of the piston. Make sure manufacturer's mark on the piston rings are toward the top of the piston.

7. Make sure the rings are seated completely in the grooves, all the way around the circumference, and that the ends are aligned with the locating pins. See **Figure 60**.

CRANKCASE AND CRANKSHAFT

Disassembly of the crankcase (splitting the case) and removal of the crankshaft assembly

ENGINE (1100 CC)

requires engine removal from the hull. However, the cylinder head, cylinder and all other attached assemblies should be removed with the engine in the hull.

The crankcase is made in 2 halves of precision diecast aluminum alloy and is of the thin-walled type (**Figure 61**). To avoid damage to them do not hammer or pry on any of the interior or exterior projected walls. These areas are easily damaged if stressed beyond what they are designed for. They are assembled without a gasket; only gasket sealer is used while dowel pins align the crankcase halves when they are bolted together. The crankcase halves are sold as a matched set only. If one crankcase half is severely damaged, both must be replaced.

The procedure which follows is presented as a complete, step-by-step major lower end overhaul. Use this procedure if the engine is to be completely reconditioned.

Special Tools

When splitting the crankcase assembly, a few special tools are required. These tools allow easy disassembly and reassembly of the engine without prying or hammer use. Remember, the crankcase halves can be easily damaged by improper disassembly or reassembly techniques.

    a. Yamaha coupling holder (part No. YW-06546 or equivalent) (**Figure 62**). This tool is used to remove and install the coupling half installed on the end of the crankshaft or drive shaft. Because the couplings are made of cast aluminum, they can be easily damaged by the use of improper tools or removal procedures.

    b. Yamaha flywheel holder (part No. YW-41528 or equivalent) (**Figure 63**). A flywheel holder is required to hold the flywheel while the crankshaft coupling half is removed.

Crankcase Disassembly

This procedure describes disassembly of the crankcase halves and removal of the crankshaft.

1. Remove the engine from the hull as described in this chapter.

2. Remove the starter motor as described in Chapter Nine.

3. Remove the oil pump as described in Chapter Ten.

4. Remove the flywheel cover bolts and the flywheel cover (**Figure 64**). Discard the cover gasket.

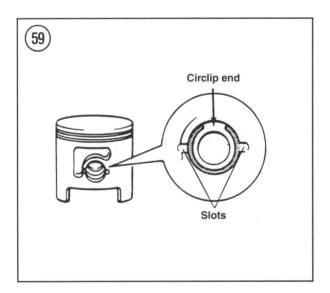

248

CHAPTER SIX

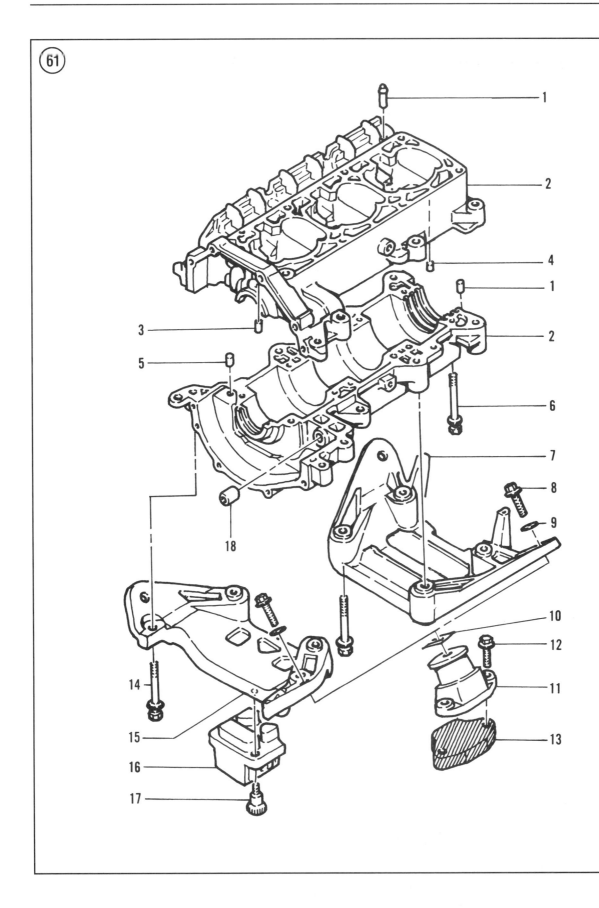

ENGINE (1100 CC)

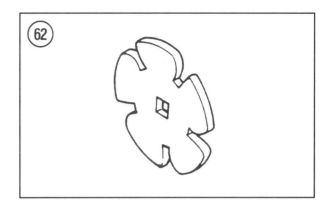

CRANKCASE (1100 CC)

1. Nipple
2. Crankcase assembly
3. Dowel pin
4. Dowel pin
5. Dowel pin
6. Bolt
7. Rear engine bracket
8. Bolt
9. Washer
10. Shim
11. Engine mount
12. Bolt
13. Engine mount spacer
14. Bolt
15. Front engine bracket
16. Damper
17. Bolt
18. Bushing

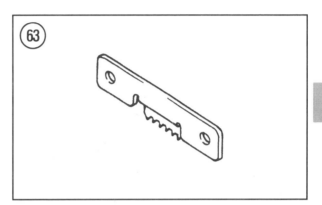

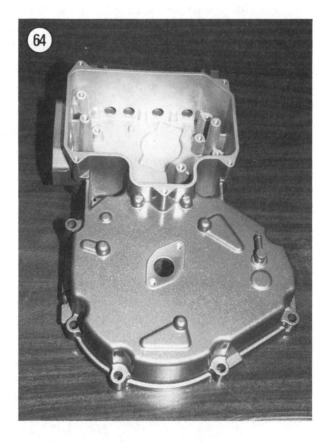

5. Install the flywheel holder onto the engine case (A, **Figure 65**) and remove the coupler half from the rear end of the crankshaft with the coupling holder. (B, **Figure 65**).

6. Remove the flywheel and idler gear as described in Chapter Nine.

7. Remove the pistons as described in this chapter.

8. Turn the crankcase assembly so it rests upside-down on wooden blocks.

9. Loosen the bolts securing the front and rear engine brackets to the lower crankcase (**Figure 66**). Remove the brackets.

10. Loosen the 12 crankcase bolts (**Figure 67**) in 1/4 turn increments until all bolts are loose. Then remove the bolts and washers.

CAUTION
Make sure that you have removed all the fasteners. If the cases are hard to separate, check for any fasteners you may have missed.

11. Carefully tap the perimeter of the lower crankcase with a soft-faced hammer to break it loose. Then lift the lower crankcase half up and remove it.

12. Lift the crankshaft and remove it from the upper crankcase.

13. Remove the 2 case halves dowel pins (**Figure 68**).

14. Remove the 7 short dowels (A, **Figure 69**) and the 1 long dowel pin (B, **Figure 69**) from the upper crankcase.

Crankcase Inspection

Refer to **Figure 61** for this procedure.

1. Clean both crankcase halves with cleaning solvent. Thoroughly dry with compressed air and wipe off with a clean shop cloth. Be sure to remove all traces of old gasket sealer from all mating surfaces.

2. Carefully inspect the case halves for cracks and fractures. Also check the areas around the stiffening ribs, around bearing bosses and threaded holes. If any are found, have them repaired by a shop specializing in the repair of precision aluminum castings. If repair is not possible, replace the crankcase assembly.

3. Check all of the upper crankcase threaded bolt holes for thread damage or corrosion. If necessary, clean or repair the threads with a suitable size metric tap. Coat the tap threads with kerosene or an aluminum tap fluid before use.

4. Check the idler gear bushing (**Figure 70**) in the lower case half for excessive wear, cracks or other types of damage. If the bushing is damaged, remove it with a blind bearing remover. Then tap the new bushing in with a suitable bushing driver.

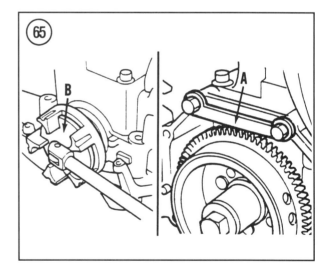

ENGINE (1100 CC)

Crankshaft Inspection

Refer to **Figure 71** for this procedure. The crankshaft and connecting rods are pressed together. However, because Yamaha does not sell replacement connecting rods and bearings, the crankshaft must be replaced as a unit if worn or damaged.

NOTE
A set of V-blocks can be made out of hardwood to help hold the crankshaft in place when performing some of the checks in the following steps. However, use only precision machined V-blocks to check crankshaft runout described later in this procedure.

1. The crankshaft is equipped with 3 different seals. Identify the seals then remove them from the end of the crankshaft.

2. Clean the crankshaft thoroughly with solvent or kerosene and a brush. Blow dry with dry filtered compressed air and lubricate with a light coat of Yamalube Two-Cycle Outboard Oil.

3. Check the crankshaft journals and crankpins for scratches, heat discoloration or other defects.

4. Check crankshaft splines, flywheel taper (A, **Figure 72**), keyway (B, **Figure 72**) and threads (A, **Figure 73**) for wear or damage.

5. Check crankshaft seal surfaces for grooving, pitting or scratches.

6. Check crankshaft bearing surfaces for rust, water marks, chatter marks and excessive or uneven wear. Clean up minor cases of rust and water or chatter marks with 320 grit carborundum cloth.

7. If 320 grit cloth is used, clean crankshaft in solvent and recheck surfaces. If they did not clean up properly, replace the crankshaft.

8. Carefully examine the condition of the crankshaft ball bearings (B, **Figure 73**). Clean the bearings in solvent and allow to dry thoroughly. Oil each bearing with Yamalube Two-Cycle Outboard Oil and then turn the outer

252 CHAPTER SIX

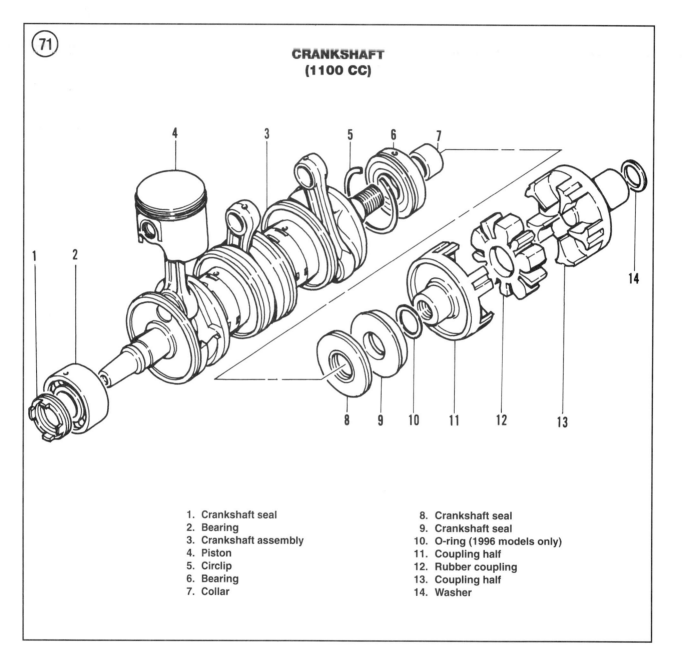

CRANKSHAFT (1100 CC)

1. Crankshaft seal
2. Bearing
3. Crankshaft assembly
4. Piston
5. Circlip
6. Bearing
7. Collar
8. Crankshaft seal
9. Crankshaft seal
10. O-ring (1996 models only)
11. Coupling half
12. Rubber coupling
13. Coupling half
14. Washer

ENGINE (1100 CC)

race. A worn or damaged bearing will sound or feel rough and will not rotate smoothly.

9. Grasp the outer race and try to work it back and forth. Replace the bearing if excessive axial play is noted. In all cases, defective bearings must be replaced.

NOTE
*The bearings installed on the outside of the crank wheels can be replaced as described under **Crankshaft Bearing Replacement** in this chapter. If the bearings installed between the crank wheels are worn or damaged, the crankshaft must be replaced.*

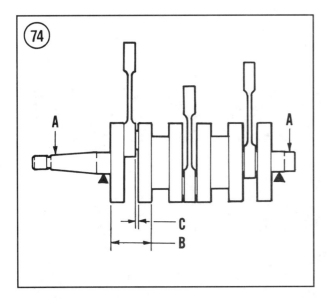

10. Support the crankshaft by placing it onto 2 V-blocks at the points shown in **Figure 74**. Then check runout with a dial indicator at the 2 points (A, **Figure 74**). Turn the crankshaft slowly and note the gauge reading. The maximum difference recorded is the crankshaft runout at that position. If the runout at any position exceeds the service limit in **Table 2**, repair or replace the crankshaft.

11. Measure the width of the crank wheels with a micrometer or vernier caliper at points (B, **Figure 74**) and compare to specifications listed in **Table 2**. See **Figure 75**.

12. Check connecting rod deflection with a dial indicator. Place the end of a dial indicator so its plunger rests against the connecting rod's small end as shown in **Figure 76**. Then support the crankshaft and push the connecting rod back and forth, noting the dial indicator gauge movement each time the connecting rod is moved. The difference in the 2 gauge readings is the small end freeplay. Repeat for each rod. If freeplay exceeds the service limit in **Table 2**, replace the crankshaft.

13. Check the connecting rod big-end side clearance with a feeler gauge at the point marked (C, **Figure 74**). Insert the gauge between the crankshaft and the washer as shown

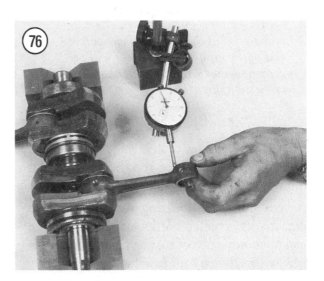

in **Figure 77**. If the clearance exceeds the service limit in **Table 2**, replace the crankshaft assembly.

Crankshaft Bearing Replacement

Replace the front (C, **Figure 72**) and rear (B, **Figure 73**) outer crankshaft bearings as follows. A bearing puller and splitter are required to remove the bearings. A press is required to install the bearings. Read this procedure completely through before removing the bearings.

NOTE
*Before removing the front and rear bearings, note the position of the locating pin hole in the front bearing (**Figure 78**) and the hole and circlip in the rear bearing (**Figure 79**). The locating pin hole and circlip align with dowel pins and a circlip groove in the upper crankcase half during crankshaft installation. Install the new bearing(s) so its pin hole and circlip align exactly the same way. If the bearing is turned around during installation, the pin hole and circlip will not align with its mating part or groove. If so, remove the bearing and turn it around.*

CAUTION
Before removing the front bearing with a bearing puller, thread the flywheel nut onto the crankshaft so it is flush with the end of the crankshaft. The nut will prevent the puller screw from damaging the crankshaft threads.

1. Install a bearing puller and remove the bearing (**Figure 80**). Then reposition the puller and remove the opposite bearing.

2. Check crankshaft bearing surfaces for rust, water marks, chatter marks or uneven wear. Clean up minor cases of rust and water or chatter marks with 320 grit carborundum cloth.

3. Clean the crankshaft bearing area with solvent and dry thoroughly.

NOTE
To ease bearing installation, heat the bearings in oil before installation. Completely read Step 4 before heating and installing bearings. During bearing installation, securely support the crankshaft so the bearings can be installed quickly. If a bearing cools and tightens on the crankshaft before it is completely

ENGINE (1100 CC)

installed, remove the bearing with the puller and reheat.

4. Install the bearings as follows:
   a. Lay the bearings on a clean, lint-free surface in the order of assembly.
   b. When installing the bearings, make sure they face in the direction shown in **Figure 78** (front) and **Figure 79** (rear).
   c. Refer to *Bearing Replacement* in Chapter One.
   d. After referring to the information in substep c, heat and install the bearings. Refer to **Figure 78** (front) or **Figure 79** (rear) during installation.

Crankshaft Seal Installation

The crankshaft is equipped with 3 different seals (**Figure 71**). While the front seal is not difficult to identify, the 2 rear seals are different in design and can be installed incorrectly. If you purchase the seals in a seal kit, the individual seals are not be identified, thus, if you did not identify the old seals during removal, you cannot use them to identify the new seals. However, if you purchase the seals individually, the seals are identified by their respective part number. When installing the rear seals, note the following:

   a. The inner rear seal is a single lip seal.
   b. The outer rear seal is a double lip seal.
   c. The single lip seal has arms (A, **Figure 81**), whereas the double lip seal (B, **Figure 81**) does not.

1. Oil the bearings with Yamaha engine oil (**Figure 82**) before installing the seals.

2. Pack a heat durable marine grease between the seal lips as shown in **Figure 83**.

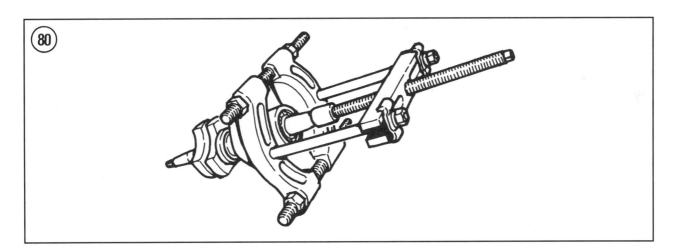

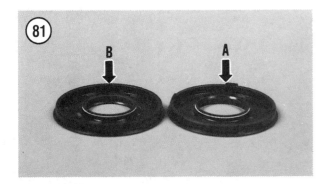

3. Install the front seal so it faces out as shown in **Figure 84**.
4. Install the rear single lip seal and then the rear double lip seal. See **Figure 85**. See previous introduction for seal identification.

Crankshaft Installation

Refer to **Figure 71** for this procedure.
1. Install the seals onto the crankshaft as described in this chapter.
2. Place the upper crankcase half upside-down on wooden blocks.
3. Install the 2 case halves dowel pins (**Figure 86**).
4. Install the 7 short (A, **Figure 69**) and 1 long (B, **Figure 69**) crankshaft bearing dowel pins in the upper crankcase.

NOTE
Step 5 describes crankshaft installation. However, because of the number of separate procedures required during installation, read through Step 5 before actually installing the crankshaft.

5. Align the crankshaft with the upper crankcase half and install the crankshaft. Note the following:
   a. Align the factory punch marks on the bearings with the edge of the case when setting the crankshaft into the upper crankcase. **Figure 87** shows the punch marks on 2 inner bearings. However, all 6 crankshaft

ENGINE (1100 CC)

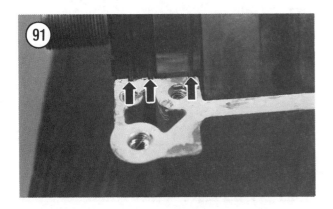

bearings have punch marks. **Figure 88** shows 2 inner bearings properly installed in the upper case half.

NOTE
*Before the crankshaft seats in the case, the marks will actually be slightly above the edge of the case. See **Figure 89**.*

b. Make sure the holes in the labyrinth seal (A, **Figure 90**) and the bearings (B, **Figure 90**) align with the pins in the upper crankcase half. See **Figure 69**. **Figure 90** shows the inner bearings.

c. Make sure the circlip on the rear bearing and the 2 rear seals seat in the crankcase grooves. See **Figure 91**.

d. Also make sure that the circlip opening does *not* align with the case half.

e. Check that the front seal seats in the crankcase groove as shown in **Figure 92**.

6. Make sure that all of the bearings and seals are seated.

Crankcase Assembly

Refer to **Figure 61** for this procedure.

1. Install the crankshaft into the upper crankcase half as described in this chapter.

2. Install the 2 dowel pins into the lower crankcase half (**Figure 86**).

3. Make sure the crankcase mating surfaces are completely clean and apply a light coat of

Yamabond 4 or equivalent to the mating surface of one case half.

4. Put the lower case half onto the upper half. Check the mating surfaces all the way around the case halves to make sure they are even and that a locating ring or dowel pin has not worked loose.

5. Install a washer onto each crankcase bolt and apply Loctite 242 to the threads of each bolt.

6. Install the crankcase bolts and washers finger-tight (**Figure 67**). Then install the 2 engine brackets (**Figure 66**) and their mounting bolts. Apply Loctite 271 (red) to the engine bracket mounting bolts and tighten bolts finger-tight.

CAUTION
While tightening the crankcase fasteners, make frequent checks to ensure that the crankshaft turns freely and the crankshaft locating rings and crankcase dowel pins fit into place in the case halves.

7. Tighten the engine bracket bolts, in 2 steps, to the torque specification listed in **Table 3**. Tighten bolts in a crisscross pattern.

8. Tighten the crankcase bolts, in 2 steps, to the torque specification listed in **Table 3**. Tighten bolts in the sequence embossed on the engine case.

9. Check again that the crankshaft turns freely. If it is binding, separate the crankcase halves and determine the cause of the problem.

10. Reinstall the rubber damper if it was removed. Be sure the side marked F faces the flywheel. Apply Loctite 242 (blue) to the threads of the mounting bolts, and torque the bolts to the specification given in **Table 3**.

11. Turn the engine right side up.

12. Install the stator plate, flywheel and starter clutch as described in Chapter Nine.

13. Install the engine coupling half as follows:
   a. Install the seal.
   b. Install the flywheel holder (A, **Figure 65**) onto the crankcase.
   c. Coat the rear crankshaft threads with Loctite 271. Screw the coupling half onto the end of the crankshaft and tighten the coupling half with the special tool (B, **Figure 65**). Tighten the coupling half to the torque specification in **Table 3**.

14. Install the flywheel cover as described in Chapter Nine.

15. Install the starter as described in Chapter Nine.

16. Install the engine into the hull as described in this chapter.

17. Reinstall the engine top end as described in this chapter.

Table 1 ENGINE SPECIFICATIONS (1100 CC)

Bore × stroke	81 × 68 mm (3.19 x 2.68 in.)
Displacement	1051cc (64.14 cu. in.)
Compression ratio	5.8:1

Table 2 ENGINE SERVICE SPECIFICATIONS (1100 CC)

	New mm (in.)	Wear limit mm (in.)
Cylinder head warp limit	–	0.1 (0.0039)
Cylinder		
Bore	81.00-81.02 (3.189-3.190)	–
Taper limit	–	0.08 (0.003)
Out of round limit	–	0.05 (0.002)
Piston		
Diameter	80.885-80.890 (3.184-3.185)	–
Measuring point	10 (0.39) (from bottom of skirt)	–
Piston clearance	0.110-0.115 (0.0043-0.0045)	0.16 (0.006)
Piston rings		
Type	Keystone	–
End Gap	0.2-0.4 (0.008-0.016)	–
Side clearance	0.02-0.06 (0.0008-0.0024)	
Crankshaft		
Crank width	61.95-62.00 (2.429-2.441)	–
Runout limit		0.05 (0.0020)
Big end side clearance	0.25-0.75 (0.010-0.030)	–
Small end freeplay limit	–	2.0 (0.08)

Table 3 ENGINE TIGHTENING TORQUE (1100 CC)

	N·m	in.-lb.	ft.-lb.
Cylinder head			
1st step	15	–	11
2nd step	36	–	26
Cylinder head cover			
M6			
1st step	4	35	–
2nd step	8	71	–
M8			
1st step	15	–	11
2nd step	30	–	22
Cylinder block bolts			
1st step	23	–	17
2nd step	40	–	29
Crankcase bolts			
1st step	15	–	11
2nd step	28	–	20
Exhaust manifold bolts			
1st step	15	–	11
2nd step	30	–	22
Flame arrestor cover	2	18	–
Flywheel	70	–	51
Engine mounting bolts			
1st step	23	–	17
2nd step	53	–	39
Engine mounting bracket			
1st step	23	–	17
2nd step	53	–	39
Coupling half	37	–	27
(continued)			

Table 3 ENGINE TIGHTENING TORQUE (1100 CC) (continued)

	N·m	in.-lb.	ft.-lb.
Muffler bracket bolts	40	–	29
Muffler housing-to-bracket bolts			
1st step	2	18	–
2nd step	47	–	34
Muffler housing-to-cylinder bolt	40	–	29
Crankcase bolts			
1st step	15	–	11
2nd step	28	–	20
Rubber damper mounts	8	71	–
Exhaust elbow bolt	30	–	22

Chapter Seven

Drive Train

The drive train assembly consists of the engine/drive shaft coupling, intermediate housing and jet pump.

Jet pump specifications are listed in **Table 1** and **Table 2**. **Tables 1-5** are at the end of the chapter.

INTERMEDIATE SHAFT AND HOUSING (WR500 AND WR650)

The intermediate housing is bolted to the rear bulkhead and is used as a coupling device to connect the engine to the stainless steel drive shaft. The housing is equipped with 2 sealed ball bearings and an intermediate shaft. The intermediate shaft is threaded to accept a coupling half at the front and machined with a female spline at the rear to accept engagement with the drive shaft. A rubber coupler engages the crankshaft and intermediate housing couplings. The coupler cushions crankshaft and drive shaft engagement, and it absorbs small amounts of drive train misalignment, resulting from engine vibration and drive shaft runout.

Intermediate Housing Removal

Refer to **Figure 1** for this procedure.

1. Remove the engine from the hull as described in Chapter Four or Chapter Five.
2. Remove the rubber coupler (A, **Figure 2**) from the intermediate housing coupling half.
3. Disconnect the hose from the grease nipple.

NOTE
After removing the intermediate housing in Step 3, you may find shims installed between the housing and the hull bulkhead. These shims ensure engine-to-drive shaft alignment. To maintain alignment, each shim must be reinstalled

in its original position during reassembly. Shims are available in 3 thicknesses—0.3, 0.5 and 1.0 mm. To avoid intermixing the shims at the 3 mounting positions, identify and label each shim after its removal.

4. Remove the intermediate housing mounting bolts and remove the housing (B, **Figure 2**) by pulling it straight back and away from the hull bulkhead. Label and remove the shims positioned under the intermediate housing mounting lugs, if so equipped.

5. Remove and discard the bulkhead seal (**Figure 3**).

6. Remove the O-ring (**Figure 4**) from the intermediate housing.

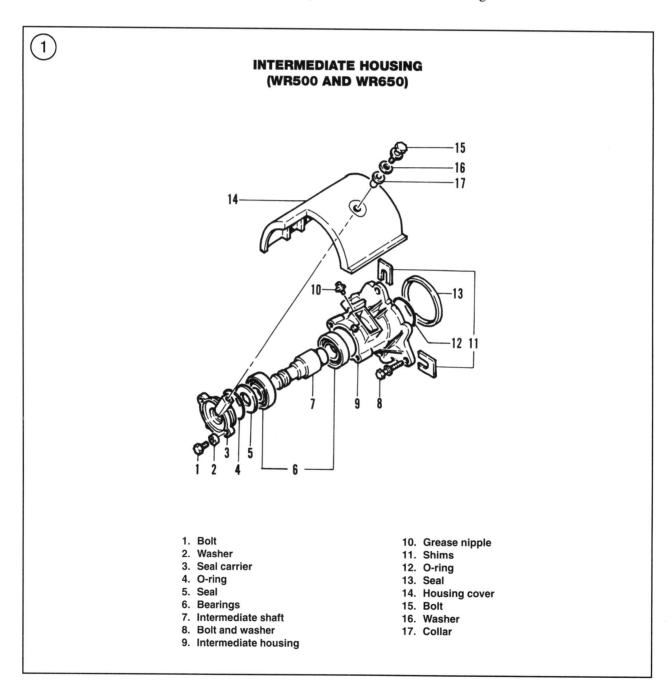

INTERMEDIATE HOUSING (WR500 AND WR650)

1. Bolt
2. Washer
3. Seal carrier
4. O-ring
5. Seal
6. Bearings
7. Intermediate shaft
8. Bolt and washer
9. Intermediate housing
10. Grease nipple
11. Shims
12. O-ring
13. Seal
14. Housing cover
15. Bolt
16. Washer
17. Collar

DRIVE TRAIN

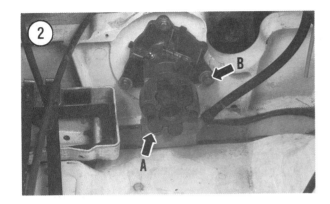

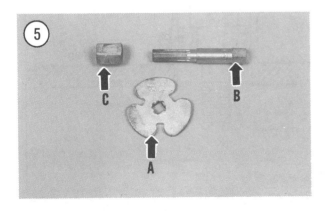

Intermediate Housing Disassembly

Overhaul of the intermediate housing requires the use of the following special tools:

a. Yamaha coupling holder part No. YW-06365 (WR500) or part No. YW-38741 (WR650). See A, **Figure 5**.

b. Yamaha shaft holder (part No. YW-06355). See B, **Figure 5**.

c. Hydraulic press.

d. A universal bearing driver/installer set.

Refer to **Figure 1** for this procedure.

1. Clean the intermediate housing in solvent and dry thoroughly.

2. Remove the coupling half (**Figure 6**) as follows:

 a. Clamp the shaft holder vertically in a vise and fit the intermediate shaft into the holder (**Figure 7**).

 b. Attach the coupling holder onto a breaker bar and then fit the holder into the coupling half. Turn the breaker bar counterclockwise to loosen and remove the coupling half and washer.

 c. Remove the tools from the intermediate housing.

3. Remove the 2 seal carrier bolts (A, **Figure 8**) and remove the carrier (B, **Figure 8**) from the housing.

4. Install the Yamaha shaft holder into the intermediate shaft and support the intermediate housing in the press so the Yamaha tool faces up. Position the press blocks so you have adequate room to allow for removal of the intermediate shaft and the front bearing (**Figure 9**). Press the shaft and front bearing out of the housing. The shaft and bearing will fall out when free from housing. Remove the Yamaha shaft holder from the intermediate shaft.

5. Support the intermediate shaft on wooden blocks and drive the rear bearing out of the housing with a suitable bearing driver or socket placed on the outer bearing race.

6. Remove the O-ring (A, **Figure 10**) and the seal (B, **Figure 10**) from the seal carrier. Support the seal carrier and pry the seal out of the carrier with a large wide-blade screwdriver. Place a rag underneath the screwdriver to avoid damaging the bore. A hardened seal can be difficult to remove; after applying pressure at one point, move the screwdriver to another point on the seal to avoid damaging the housing bore.

7. Remove the outer O-ring from the intermediate housing (**Figure 11**).

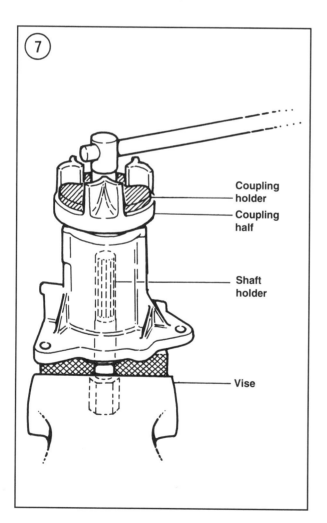

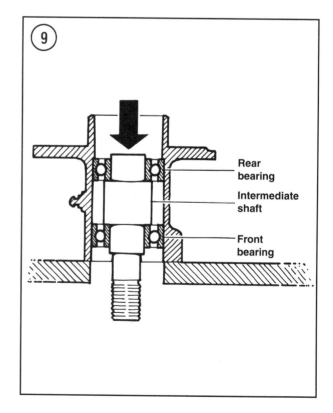

DRIVE TRAIN

Inspection

1. Clean all components (**Figure 12**) in solvent and dry them thoroughly. Place the bearings and shaft on a clean lint-free cloth after drying.

2. Clean the seal cover in solvent and dry thoroughly. Check the seal bore for damage.

3. Check each bearing by rotating the outer or inner bearing race. Bearings should turn smoothly and quietly and have no rough spots or other damage. If any problem is found, replace the damaged bearing(s). If the front bearing is damaged, remove it from the intermediate shaft with a press or bearing splitter.

4. Check the intermediate shaft surfaces for cracks, deep scoring or heat discoloration. Check the threads and splines for cracks or other damage.

5. Check the rubber coupler (**Figure 13**) for wear grooves or rubber deterioration. Especially check for wear on the round outer knobs.

6. Replace worn or damaged parts.

7. Install a new seal and new O-rings.

Intermediate Housing Assembly

Refer to **Figure 1** for this procedure.

NOTE
Refer to the information listed under ***Ball Bearing Replacement*** *in Chapter One when pressing the bearings onto the shaft and when pressing the shaft into the housing in the following steps.*

NOTE
Wipe the inner bearing races and the intermediate shaft bearing surface with

oil before installing the bearings in Step 1.

1. The front and rear intermediate housing bearings have the same part number. Press both bearings onto the shaft separately until they bottom. See **Figure 14**.

2. Support the intermediate housing in the press, and press the intermediate shaft/bearing assembly into the housing until the bearings bottom. See **Figure 15**. Turn the intermediate shaft by hand; it should turn freely. If the shaft is tight or turns roughly, a bearing may be damaged. Remove the shaft and bearings and inspect all parts.

3. Install a new outer O-ring (**Figure 11**) onto the intermediate housing.

4. Install a new seal into the seal carrier.
   a. Fill the seal cavity with marine grease.
   b. Align the seal with the carrier so the closed end faces up (B, **Figure 10**). Drive the seal squarely into the carrier with a bearing driver or a suitable size socket. Drive the seal into the bore until it bottoms.

5. Wipe the seal carrier O-ring with marine grease and install the O-ring into the carrier groove. See A, **Figure 10**.

6. Wipe the carrier seal lip with marine grease and place the carrier (B, **Figure 8**) over the intermediate shaft and align it with the housing bolt bosses. Apply Loctite 242 to the threads of the mounting bolts (A, **Figure 8**) and tighten them securely.

7. Install the coupling half (**Figure 6**) as follows:
   a. Clamp the shaft holder (part No. YW-06355) vertically in a vise and fit the intermediate shaft into the holder (**Figure 7**).
   b. Apply Loctite 242 to the threads of the intermediate shaft and thread the coupling half onto the shaft by hand as far as it will go.
   c. Attach the coupling holder (part No. YW-06365) to a torque wrench. Fit the coupling holder into the coupling half and tighten the coupling half to 37 N•m (27 ft.-lb.).
   d. Remove the tools from the intermediate housing.

8. Fill the intermediate housing with marine grease applied from a grease gun through the housing nipple.

Intermediate Housing Installation

1. Wipe marine grease onto a new hull bulkhead seal and insert the seal into the bulkhead as shown in **Figure 3**. The seal should seat squarely in the bulkhead groove.

2. Wipe marine grease onto a new O-ring and install the O-ring onto the intermediate housing. See **Figure 4**.

3. Coat the drive shaft splines with marine grease. Then align the drive shaft with the intermediate shaft and join the 2 components.

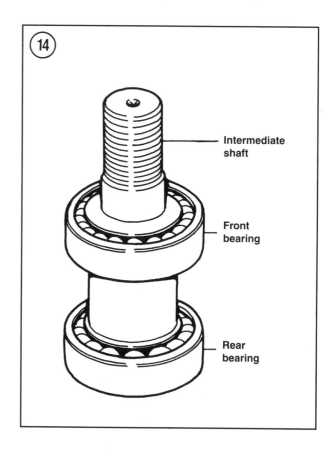

DRIVE TRAIN

4. Loosely install the 3 intermediate housing mounting bolts (B, **Figure 2**) and washers. Reinstall the shims in their original position. Tighten the bolts securely.

NOTE
Always keep the original factory installed shims between the intermediate housing and bulkhead. These shims affect engine and drive shaft alignment. If it is necessary to replace a shim, measure the old shim with a vernier caliper or micrometer and purchase one with the same thickness.

5. Reinstall the hose onto the grease nipple.

6. Insert the rubber coupler into the coupling half (A, **Figure 2**).

7. Install the engine and check the engine alignment as described in Chapter Four or Chapter Five.

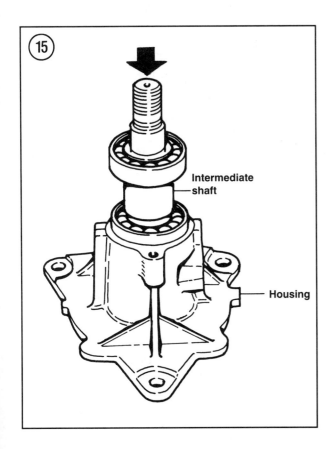

INTERMEDIATE SHAFT AND HOUSING (WRA650, WRA650A AND WRA700)

The intermediate housing is bolted to the rear bulkhead and is used as a coupling device to connect the engine to the drive shaft. The intermediate housing assembly consists of a 2-piece housing (a bearing housing and a support housing), a single ball bearing, intermediate shaft and 2 seals (**Figure 16**). The intermediate shaft is press-fit into the intermediate ball bearing so that a specified distance can be maintained between the end of the shaft and a fixed point on the housing to ensure proper drive shaft engagement. The intermediate shaft is threaded to accept a coupling half at the front and machined with a female spline at the rear to accept engagement with the drive shaft. A rubber coupler is used to engage the crankshaft and intermediate housing couplings. The coupler cushions crankshaft and drive shaft engagement and absorbs small amounts of drive train misalignment, engine vibration and drive shaft runout.

Intermediate Housing Removal

Refer to **Figure 16** for this procedure.

1. Remove the engine from the hull as described in Chapter Five.

2. Remove the rubber coupler (A, **Figure 2**) from the intermediate housing coupling half.

3. Disconnect the hose from the grease nipple.

NOTE
After removing the intermediate housing in Step 3, you may find shims installed between the housing and the hull bulkhead. These shims ensure engine to drive shaft alignment; to maintain alignment, each shim must be reinstalled in its original position. Shims are available in 2 thicknesses—0.5 and 1.0 mm. To avoid intermixing the shims at the 3 mounting

positions, identify and label each shim after its removal.

4. Remove the intermediate housing mounting bolts and remove the housing (B, **Figure 2**). Pull the housing off of the bulkhead while at the same time disconnecting the drive shaft and intermediate shaft.

5. Measure the thickness of each shim with a micrometer or vernier caliper. If all of the shims have the same thickness, labeling them is not necessary. If one or more shims have a different thickness, label and identify each shim indicating their mounting position, then set them aside.

6. Remove and discard the bulkhead seal (**Figure 3**).

7. Remove the O-ring from the intermediate housing (See **Figure 4**).

8. Turn the intermediate shaft by hand. A rough turning shaft indicates bearing wear or damage. Replace the bearing as described in this chapter.

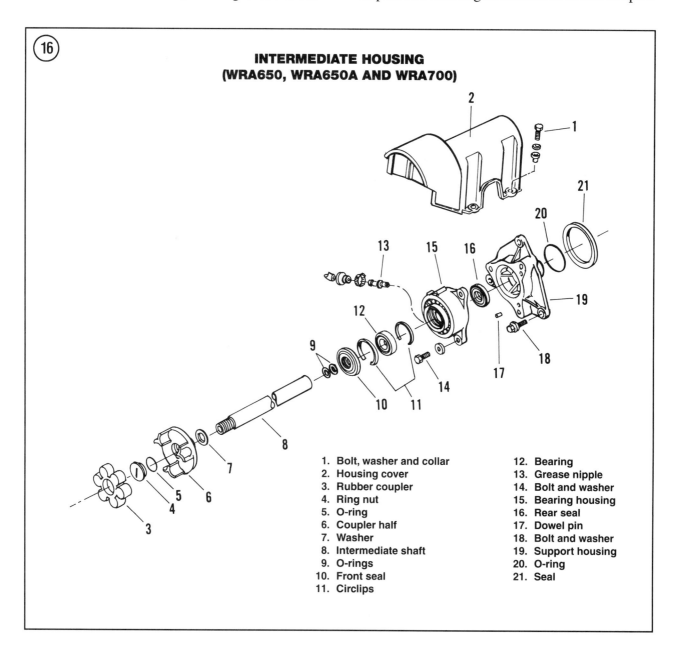

INTERMEDIATE HOUSING (WRA650, WRA650A AND WRA700)

1. Bolt, washer and collar
2. Housing cover
3. Rubber coupler
4. Ring nut
5. O-ring
6. Coupler half
7. Washer
8. Intermediate shaft
9. O-rings
10. Front seal
11. Circlips
12. Bearing
13. Grease nipple
14. Bolt and washer
15. Bearing housing
16. Rear seal
17. Dowel pin
18. Bolt and washer
19. Support housing
20. O-ring
21. Seal

DRIVE TRAIN

Intermediate Housing Disassembly

Complete disassembly of the intermediate housing involves pressing out the intermediate shaft and removing the seals and bearing. Removal of the seals will destroy them, so purchase new seals to have on hand for reassembly. Because the bearing is sealed, its removal for periodic lubrication is unnecessary. Keep in mind too, that there is always the risk of damaging a good bearing when removing it. Bearing removal should only be performed when it is necessary to replace the bearing or if the intermediate housing is damaged.

Complete disassembly of the intermediate housing requires the following special tools:

a. Yamaha coupling holder (part No. YW-38741). See A, **Figure 17**.
b. Yamaha shaft holder (part No. YW-38742). See B, **Figure 17**.
c. Hydraulic press.
d. A universal bearing driver/installer set.

CAUTION
If you do not have access to the special tools, overhaul of the intermediate housing should be left to a dealer or machine shop. Under no condition should you attempt to drive the shaft out of the housing with a hammer.

Refer to **Figure 16** for this procedure.

1. Clean the intermediate housing in solvent and dry thoroughly.
2. Remove the bearing housing from the support housing.
3. Remove the coupling half (**Figure 6**) as follows:
   a. Remove the ring nut and O-ring from the end of the coupling half.
   b. Clamp the shaft holder (part No. YW-38742) vertically in a vise and fit the intermediate shaft into the holder (**Figure 7**).
   c. Attach the coupling holder (part No. YW-38741) to a breaker bar then fit the holder into the coupling half. Turn the breaker bar counterclockwise to loosen and remove the coupling half. See **Figure 7**.
4. The intermediate shaft is press-fit in the bearing and its removal is not required unless the bearing, seals or shaft must be replaced. Furthermore, removal of the shaft requires replacement of both seals. To remove the shaft, support the housing in a press so the shaft faces down (**Figure 18**). Then place a driver between the press ram and shaft and press the shaft out of the bearing. Catch the shaft to prevent it from falling to the floor when it is free of the bearing.
5. Referring to **Figure 18**, note the 2 seals installed in the bearing housing. When removing the seals, first support the housing on a soft flat surface, such as a wooden block. Then carefully pry the first seal out of the housing with a large wide-blade screwdriver. Place a rag underneath the screwdriver to avoid damaging the bore.

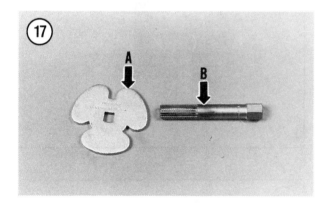

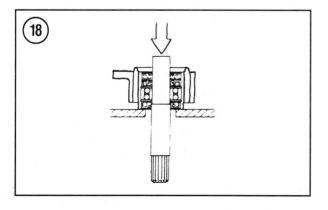

Normally, the seal(s) will easily pop from the bore when pried. Occasionally, however, you will encounter a seal which has hardened. A seal in this condition can be very difficult to remove, and if you're not careful, you can damage the housing. If a seal will not pop out, work the screwdriver around the seal when applying pressure, making sure to keep the rag underneath the screwdriver. Remove the 2 seals installed in the rear of the housing separately. Do not attempt to remove both seals at the same time.

6. Remove the 2 circlips positioned next to the bearing.

7. Support the housing securely. Then place a bearing driver or socket on the bearing's outer race and drive it out of the housing (**Figure 19**).

Inspection

1. Clean all sealer residue from the housing and bulkhead mating surfaces.

2. Clean all of the components in solvent and dry thoroughly. The intermediate shaft bore is used to store grease for lubrication of the drive shaft during engagement. Remove all of the grease in the shaft as it will become contaminated with solvent when cleaning the outside surface.

3. After cleaning the bearing and shaft, place them on a clean lint-free cloth and allow to air dry.

4. Inspect the housing for cracks or damage. Check the circlip grooves in the bearing bore for cracks or other damage that may allow a circlip to unseat. Remove any burrs from the grooves with emery cloth.

5. Turn the inner bearing race by hand. The bearing must turn smoothly without sticking, excessive noise or other damage. Replace the bearing if it is worn or damaged.

6. Check the shaft surface for cracks, deep scoring and excessive wear. Especially check the shaft's bearing support area. Check the threads and splines for cracks or damage.

7. Check the rubber coupler (**Figure 13**) for wear grooves or rubber deterioration. Especially check for wear on the round outer knobs.

8. Replace worn or damaged parts as required.

Intermediate Housing Assembly

Refer to **Figure 16** for this procedure.

NOTE
*Refer to the information listed under **Ball Bearing Replacement** in Chapter One when pressing the shaft into the bearing.*

1. Make sure the intermediate housing is clean.

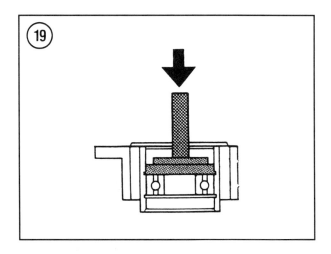

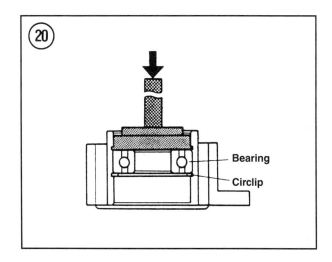

DRIVE TRAIN

2. Install the bearing as follows:
   a. Place the intermediate housing on a flat surface so that the crankshaft end (front) faces up.
   b. Install the rear circlip into the housing groove. Make sure the circlip seats in the groove completely.
   c. Wipe a light coat of marine grease on the outer bearing race.
   d. Align the bearing with the housing bore and drive it squarely into the bore with a bearing driver or socket placed on the outer bearing race. Drive the bearing into the housing until it seats against the circlip (**Figure 20**).
   e. Install the front circlip into the housing groove, making sure it seats in the groove completely.
   f. Rotate the inner bearing race and check it for roughness. If the bearing race does not turn smoothly, the bearing was damaged during installation. Remove and replace bearing.

NOTE
On 1993 WRA650 and WRA650A models, the front (crankshaft side) and rear intermediate shaft seals are identical.

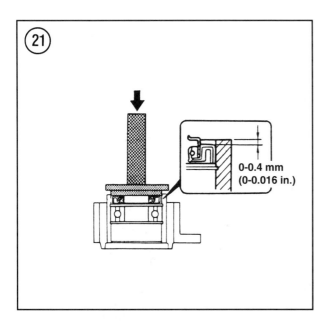

On later WRA650, WRA650A and WRA700 models, they are different sizes. The diameter of the hole in the front seal is 10 mm (0.38 in.). In the rear seal, it is 8 mm (0.31 in.). Do not confuse these seals during installation.

NOTE
When installing the seals in the following steps, pack each seal lip cavity with marine grease.

3. Install the front seal into the bearing housing with a suitable bearing driver or socket. Thoroughly pack the bearing grease, then install the seal so the manufacturer's marks face out. Be sure the gap between the seal and the housing is within the specification shown in **Figure 21**.

CAUTION
Do not install the rear seal until the intermediate shaft is installed. If the rear seal is installed first, the shaft may tear the seal lips when it is pressed into place.

4. Install the intermediate shaft as follows:
   a. Wipe the seal lip (installed in Step 3) with marine grease.
   b. Make sure the shaft is clean, then wipe the bearing end of the shaft with marine grease.
   c. Support the bearing housing in a press and adjust the press table height to accept the bearing and shaft.
   d. Align the nonthreaded shaft end with the bearing and place a piece of aluminum between the press ram and shaft threads. Hold the shaft as near vertical as possible and slowly operate the press, making sure the shaft enters the bearing squarely. If the shaft is at an angle as it enters the bearing, stop and release pressure; realign shaft before pressing into the bearing. When the shaft enters the bearing squarely, continue to press the shaft into the bearing until it ex-

tends 14.5-15.5 mm (0.57-0.61 in.) beyond the housing as shown in **Figure 22**.

   e. After installing the shaft, turn it to check bearing operation. The bearing should rotate smoothly. If the bearing turns roughly, remove the shaft and check the bearing for damage.

5. Secure the housing so the rear side faces up and install the rear seal into the housing with a suitable driver. Apply marine grease to the seal and add enough marine grease on top of the bearing to fill the area between the circlip with grease, then install the seal so the manufacturer's marks face out. Be sure the gap between the seal and the housing is within the specification shown in **Figure 23**.

6. Apply marine grease to 2 new O-rings and install them onto the intermediate shaft. (**Figure 23**).

7. Install the coupling half as follows:
   a. Clamp the shaft holder (part No. YW-38742) vertically in a vise and fit the intermediate shaft into the holder (**Figure 7**).
   b. Install the washer onto the shaft.
   c. Apply Loctite 242 to the threads of the intermediate shaft threads and thread the coupling half onto the shaft by hand as far as it will go.
   d. Attach the coupling holder (part No. YW-38741) to a torque wrench. Fit the coupling holder into the coupling half and tighten to 37 N•m (27 ft.-lb.). See **Figure 7**.
   e. Apply marine grease to the coupling half O-ring, and install the O-ring.
   f. Install the ring nut into the coupling half and torque the nut to the specification given in **Table 3**.
   f. Remove the tools from the bearing housing.

8. Place the dowel pin into the support housing and mate the bearing and support housings. Install the mounting bolts and washers. Apply Loctite 242 to the threads of the bolts and torque the bolts to the specification given in **Table 3**.

9. Fill the intermediate shaft with approximately 8.2 cc (0.27 oz.) of marine grease.

Intermediate Housing Installation

1. Wipe marine grease on a new hull bulkhead seal and insert the seal into the bulkhead as shown in **Figure 3**. The seal should seat squarely in the bulkhead groove.

2. Coat the drive shaft splines with marine grease. Then align the drive shaft with the intermediate shaft and join the 2 components.

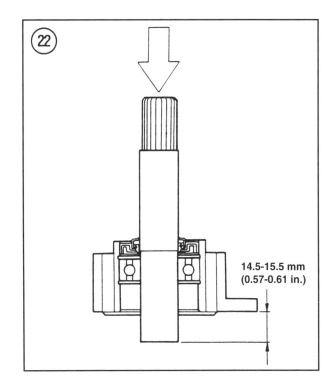

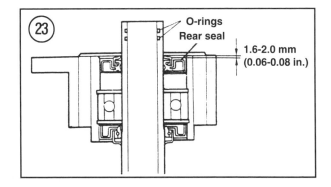

DRIVE TRAIN

3. Wipe marine grease on a new O-ring and install the O-ring onto the intermediate housing. See **Figure 4**.

4. Apply Loctite 242 to the threads of the intermediate housing mounting bolts and loosely install the 3 intermediate housing mounting bolts (B, **Figure 2**) and washers. Reinstall any shims in their original position. Torque the bolts to the specification given in **Table 3**.

NOTE
Always keep the original factory installed shims between the intermediate housing and bulkhead. These shims affect engine and drive shaft alignment. If it is necessary to replace a shim, measure the old shim with a vernier caliper or micrometer and purchase one with the same thickness.

5. Install the rubber hose onto the grease nipple.
6. Insert the rubber coupler into the coupling half (A, **Figure 2**).
7. Install the engine and check engine alignment as described in Chapter Five.

INTERMEDIATE SHAFT AND HOUSING
(FX700, SJ650, SJ700, SJ700A, WB700, WB700A, WB760, WRB650, WRB650A, AND WRB700)

The intermediate housing is bolted to the rear bulkhead and is used as a coupling device to connect the engine to the drive shaft. The intermediate housing assembly consists of a bearing housing, a single ball bearing, intermediate shaft and 3 seals (**Figure 24**). The ball bearing is

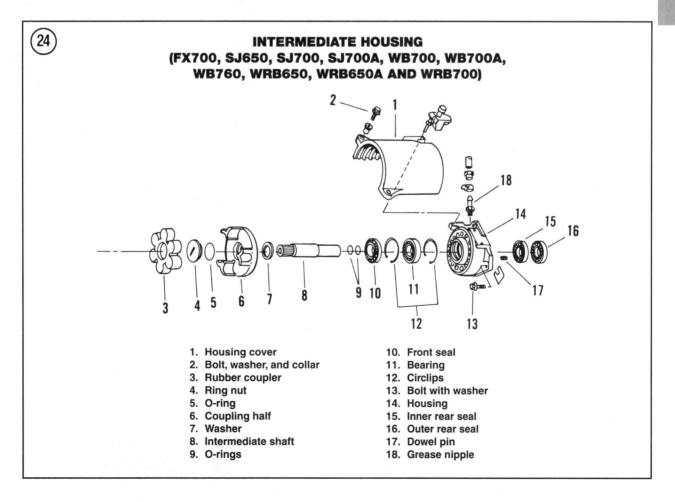

INTERMEDIATE HOUSING (FX700, SJ650, SJ700, SJ700A, WB700, WB700A, WB760, WRB650, WRB650A AND WRB700)

1. Housing cover
2. Bolt, washer, and collar
3. Rubber coupler
4. Ring nut
5. O-ring
6. Coupling half
7. Washer
8. Intermediate shaft
9. O-rings
10. Front seal
11. Bearing
12. Circlips
13. Bolt with washer
14. Housing
15. Inner rear seal
16. Outer rear seal
17. Dowel pin
18. Grease nipple

press-fit in the housing with a specified distance maintained from the end of the shaft to a point on the housing. This ensures proper intermediate shaft-to-drive shaft engagement. The intermediate shaft is threaded to accept a coupling half at the front and machined with a female spline at the rear for drive shaft engagement. A rubber coupler is used to engage the crankshaft and intermediate housing couplings. The coupler cushions crankshaft and drive shaft engagement and absorbs small amounts of drive train misalignment from engine vibration and drive shaft runout.

Intermediate Housing Removal

Refer to **Figure 24** for this procedure.

1. Remove the engine from the hull as described in Chapter Five.

2. Remove the rubber coupler (A, **Figure 2**) from the intermediate housing coupling half.

3. Disconnect the hose from the grease nipple.

> *NOTE*
> *Shims placed between the housing and the hull bulkhead ensure engine-to-drive shaft alignment. In order to maintain alignment during reassembly, each shim must be reinstalled in its original position. Identify and label each shim after its removal. Shims are available in 3 thicknesses: 0.3, 0.5 and 1.0 mm.*

4. Remove the intermediate housing mounting bolts and remove the housing (**Figure 25**); pull the housing off of the bulkhead while at the same time disconnecting the drive shaft and intermediate shaft. Remove the 2 dowel pins (**Figure 26**) from the bulkhead.

5. Measure the thickness of each shim with a micrometer or vernier caliper. If all of the shims are the same thickness, labeling them is not necessary. If one or more shims have a different

DRIVE TRAIN

thickness, label and identify each shim indicating their mounting position, then set them aside.

6. Turn the intermediate shaft by hand. A rough turning shaft indicates bearing wear or damage. Replace the bearing as described in this chapter.

Intermediate Housing Disassembly

Complete disassembly of the intermediate housing involves pressing out the intermediate shaft and the removal of the shaft seals and bearing. Removal of the seals will destroy them, so purchase new oil seals to have on hand for reassembly. Because the bearing is sealed, its removal for periodic lubrication is unnecessary. Keep in mind too, that there is always the risk of damaging a good bearing when removing it. Bearing removal should only be performed if it is necessary to replace the bearing or if the intermediate housing is damaged.

Complete disassembly of the intermediate housing requires the following special tools:

a. Yamaha coupling holder part No. YW-06546 (WB760) or part No. YW-38741 (all other models). See A, **Figure 17**.

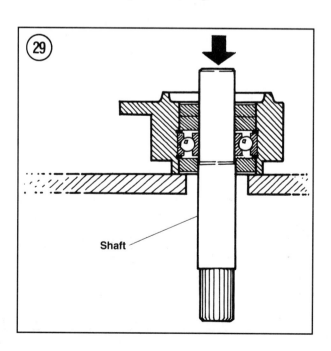
Shaft

b. Yamaha shaft holder part No. YW-38742. See B, **Figure 17**.
c. Hydraulic press.
d. A universal bearing driver/installer set.

CAUTION
If you do not have access to the special tools, overhaul of the intermediate housing should be left to a dealer or machine shop. Under no condition should you attempt to drive the shaft out of the housing with a hammer.

Refer to **Figure 24** for this procedure.

1. Clean the intermediate housing in solvent and dry thoroughly.
2. Remove the coupling half (**Figure 27**) as follows:

   a. Remove the ring nut and O-ring from the end of the coupling half.

   NOTE
   The ring nut and O-ring is not used on 1995 WRB650, 1995 FX700 and 1996 SJ700A.

   b. Clamp the shaft holder (part No. YW-38742) vertically in a vise and fit the intermediate shaft into the holder (**Figure 7**).

   c. Attach the coupling holder (part No. YW-06546 [WB760] or part No. YW-38741 [all other models]) onto a breaker bar, then fit the holder into the coupling half. turn the breaker bar counterclockwise to loosen and remove the coupling half and washer.

   d. Remove the tools from the intermediate housing.

3. The intermediate shaft (**Figure 28**) is press-fit in the bearing and its removal is not required unless the bearing, seals or shaft must be replaced. Furthermore, removal of the shaft will require replacement of all 3 seals. To remove the shaft, support the housing in a press so the shaft faces down (**Figure 29**). Then place a driver between the press ram and shaft and press the shaft out of the bearing. Catch the shaft to

prevent it from falling to the floor when it is free of the bearing.

4. Referring to **Figure 29**, note the 3 seals installed in the housing. When removing the seals, first support the housing on a soft flat surface, such as a wooden block. Then carefully pry the first seal out of the housing with a large wide-blade screwdriver. Place a rag underneath the screwdriver to avoid damaging the bore. Normally, the seal(s) will easily pop out of the bore when pried. Occasionally, however, you will encounter a seal which has hardened. A seal in this condition can be very difficult to remove, and if you are not careful, you can damage the housing. If a seal will not pop out, work the screwdriver around the seal when applying pressure, making sure to keep the rag underneath the screwdriver. Then remove the 2 seals installed in the rear of the housing separately. Do not attempt to remove both seals at the same time.

5. Remove the 2 circlips positioned next to the bearing.

6. Support the housing securely. Then place a bearing driver or socket on the bearing's outer race and drive it out of the housing. If you plan on reusing the bearing, catch it as it is released from the housing.

Inspection

1. Clean all sealer residue from the housing and bulkhead mating surfaces.

2. Clean all of the components in solvent and dry thoroughly. The intermediate shaft bore is used to store grease for lubrication of the drive shaft during engagement. Remove all of the grease in the shaft as it will become contaminated with solvent when cleaning the outside surface.

3. After cleaning the bearing and shaft, place them on a clean lint-free cloth and allow to air dry.

4. Inspect the housing for cracks or damage. Check the circlip grooves in the bearing bore for cracks or other damage that may allow a circlip to unseat. Remove any burrs from the grooves with emery cloth.

5. Turn the inner bearing race by hand. The bearing should turn smoothly without sticking, excessive noise or other damage. Replace the bearing if it is worn or damaged.

6. Check the shaft surface for cracks, deep scoring and excessive wear. Especially check the shaft's bearing support area. Check the threads and splines for cracks or damage.

7. Check the rubber coupler (**Figure 13**) for wear grooves or rubber deterioration. Especially check for wear on the round outer knobs.

8. Replace worn or damaged parts as required.

9. Always install new oil seals and new O-rings during reassembly.

Intermediate Housing Assembly

Refer to **Figure 24** for this procedure.

NOTE
*Refer to the information listed under **Bearing Replacement** in Chapter One when installing the bearing in the housing in the following steps.*

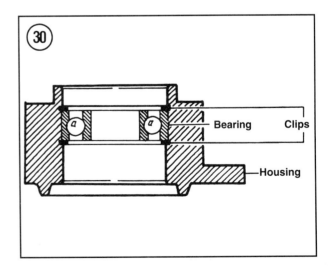

DRIVE TRAIN

1. Make sure the intermediate housing is clean.
2. Install the bearing as follows:
   a. Place the intermediate housing on a flat surface so the crankshaft end (front) faces up.
   b. Install the rear circlip into the housing groove (**Figure 30**). Make sure the circlip seats in the groove completely.
   c. Wipe a light coat of marine grease on the outer bearing race.
   d. Align the bearing with the housing bore and drive it squarely into the bore with a bearing driver or socket placed on the outer bearing face. Drive the bearing into the housing until it seats against the circlip.
   e. Install the front circlip into the housing groove, making sure it seats in the groove completely. See **Figure 30**.
   f. Rotate the inner bearing race and check it for roughness. If the bearing race does not turn smoothly, the bearing was damaged during installation. Remove and replace bearing.

NOTE
Two different sized seals are used in the intermediate housing. The diameter of the hole in the front seal and in the inner rear seal is 10 mm (0.38 in.). In the outer rear seal, the hole is 8 mm (0.31 in.). Do not confuse these seals during installation.

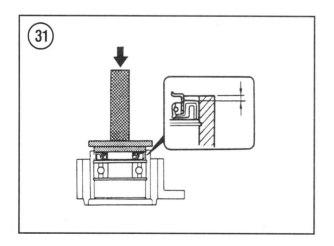

NOTE
When installing the seals in the following steps, pack each seal lip cavity with marine grease.

3. Thoroughly pack the front bearing with water-resistant marine grease. Position the front seal so its manufacturer's marks face outward, then install the seal using a suitable driver. Drive the seal into the housing until the outer face of the seal is positioned as follows (Figure 31):
   a. WRB650, WRB650A and WRB700—flush with the outer edge of the housing.
   b. FX700, SJ650, SJ700 and SJ700A—4.8-5.2 mm (0.19-0.20 in.) below the outer edge of the housing.
   c. WB700, WB700A and WB760—0.5-0.9 mm (0.020-0.035 in.) below the outer edge of the housing.

CAUTION
Do not install the 2 rear seals until the intermediate shaft is installed. If the seals are installed first, the shaft may tear the seal lips when it is pressed into place.

4. Install the intermediate shaft as follows:
   a. Wipe the front seal lip (installed in Step 3) with marine grease.
   b. Make sure the shaft is clean, then wipe the bearing end of the shaft with marine grease.
   c. Support the bearing housing in a press and adjust the press table height to accept the bearing and shaft.
   d. Align the nonthreaded shaft end with the bearing and place a piece of aluminum between the press ram and shaft threads. Hold the shaft as near vertical as possible and slowly operate the press, making sure the shaft enters the bearing squarely. If the shaft is at an angle as it enters the bearing, stop and release pressure and realign the shaft before pressing it into the bearing. When the shaft enters the bearing squarely, con-

tinue to press the shaft into the bearing until it extends 19.5-20.5 mm (0.77-0.81 in.) beyond the housing as shown in **Figure 32**.

e. After installing the shaft, turn it to check bearing operation. The bearing should rotate smoothly. If the bearing turns roughly, remove the shaft and check the bearing for damage.

5. Secure the housing so the rear side faces up. Install the 2 rear seals separately into the housing with a suitable driver. Add enough marine grease on top of the bearing to fill the area between the circlips with grease, and then install the inner rear seal until it bottoms against the circlip. Install the outer rear seal until it bottoms against the first seal. Be sure the gap between the outer rear seal and the housing is within the specification shown in **Figure 33**. Install both seals so their manufacturer's marks face out.

6. Install new O-rings onto the intermediate shaft.

7. Install the coupling half (**Figure 27**) as follows:

a. Clamp the shaft holder (part No. YW-38742) vertically in a vise and fit the intermediate shaft into the holder (**Figure 7**).

b. Install the washer onto the intermediate shaft.

c. Apply Loctite 242 to the threads of the intermediate shaft and thread the coupling half onto the shaft by hand as far as it will go.

d. Attach the coupling holder (part No. YW-06546 or YW-38741) to a torque wrench. Fit the coupling holder into the coupling half and tighten to 37 N•m (27 ft.-lb.). See **Figure 7**.

e. Apply marine grease to the O-ring and install the O-ring into the coupling half. Install the ring nut and torque it to the specification given in **Table 3**.

f. Remove the tools from the intermediate housing.

8. Fill the intermediate shaft with approximately 8.2 cc (0.27 oz.) of marine grease.

Intermediate Housing Installation

1. Install the 2 intermediate housing dowel pins (**Figure 26**).

NOTE
When you install the intermediate housing in Step 2, you will also be engaging the drive shaft splines with the intermediate shaft. Coat the drive shaft splines with marine grease before installation.

2. Coat the intermediate housing mating surface with silicone sealant and place the housing against the bulkhead, aligning the housing with the bulkhead dowel pins. Coat the threads of the bolts with Loctite 242 (blue) and install the bolts by hand until the housing is secure. While there

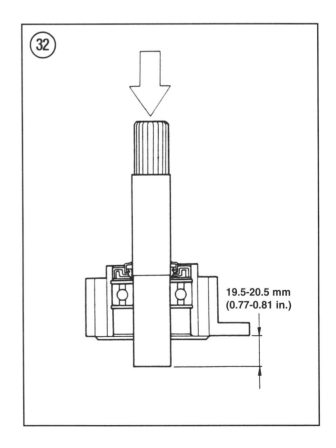

DRIVE TRAIN

is still a gap between the housing and bulkhead, install the housing shims in their correct position; follow the marks made during removal. With shims positioned correctly, tighten the housing bolts to the torque specification listed in **Table 3**.

3. Insert the rubber coupler into the coupling half (A, **Figure 2**).

4. Install the engine and check engine alignment as described in Chapter Five.

INTERMEDIATE SHAFT AND HOUSING (RA700, RA700A, RA700B, RA760, RA1100, WVT700, AND WVT1100)

The intermediate housing is bolted to the rear bulkhead and is used as a coupling device to connect the engine to the drive shaft. The intermediate housing assembly consists of a 2-piece housing (a bearing housing and a support housing), a single ball bearing, intermediate shaft and 2 oil seals (**Figure 34**). The intermediate shaft is press-fit in the intermediate ball bearing so that a specified distance can be maintained between the end of the shaft and a fixed point on the housing to ensure proper drive shaft engagement. The intermediate shaft is threaded to accept a coupling half at the front and machined with a female spline at the rear to accept engagement with the drive shaft. A rubber coupler is used to engage the crankshaft and intermediate housing couplings. The coupler cushions crankshaft and drive shaft engagement and absorbs small amounts of drive train misalignment engine vibration and drive shaft runout.

Intermediate Housing Removal

Refer to **Figure 34** for this procedure.

1. Remove the engine from the hull as described in Chapter Five or Chapter Six.

2. Remove the rubber coupler (A, **Figure 2**) from the intermediate housing coupling half.

3. Disconnect the hose from the grease nipple. Cable ties are used to secure the hose to the grease nipple on some Wave Raiders and Wave Ventures. If this is the case with your model, clip and discard the tie.

NOTE
After removing the intermediate housing in Step 4, you may find shims installed between the housing and the hull bulkhead. These shims ensure engine to drive shaft alignment; to maintain alignment, each shim must be reinstalled in its original position. Factory installed shims are available in 2 thicknesses: 0.5 and 1.0 mm. To avoid intermixing the shims at the 4 mounting positions, identify and label each shim after its removal.

4. Remove the intermediate housing mounting bolts and remove the housing (B, **Figure 2**). Pull the housing off of the bulkhead while at the same time disconnecting the drive shaft and intermediate shaft.

5. Measure the thickness of each shim with a micrometer or vernier caliper. If all of the shims have the same thickness, labeling them is not necessary. If one or more shims have a different thickness, label and identify each shim indicating their mounting position, then set them aside.

6. Remove and discard the bulkhead seal (**Figure 3**).

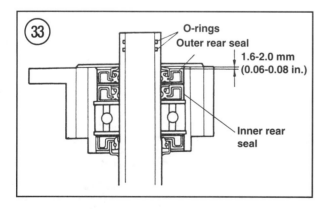

CHAPTER SEVEN

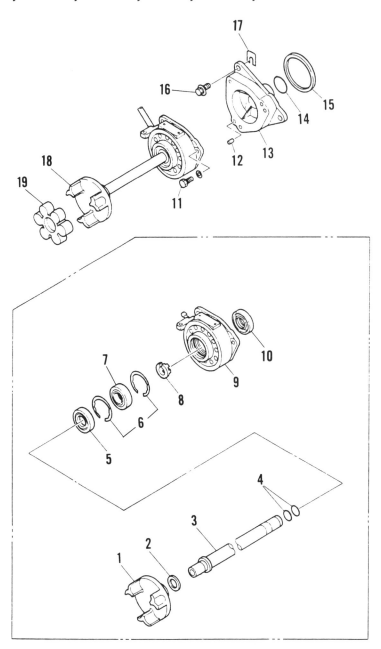

**INTERMEDIATE HOUSING
(RA700, RA700A, RA700B, RA760, RA1100, WVT700 AND WVT1100)**

1. Coupling half
2. Washer
3. Intermediate shaft
4. O-rings
5. Front seal
6. Circlips
7. Bearing
8. Spacer (except 1994 RA700 models)
9. Bearing housing
10. Rear seal
11. Bolt and washer
12. Dowel pin
13. Support housing
14. O-ring
15. Seal
16. Bolt with washer
17. Shim
18. Coupler half
19. Rubber coupler

DRIVE TRAIN

7. Remove the O-ring from the support housing (C, **Figure 35**).

8. Turn the intermediate shaft by hand. A rough turning shaft indicates bearing wear or damage. Replace the bearing as described in this chapter.

Intermediate Housing Disassembly

Complete disassembly of the intermediate housing involves pressing out the intermediate shaft and the removal of the seals and bearing. Removal of the seals will destroy them, so purchase new seals to have on hand for reassembly. Because the bearing is sealed, its removal for periodic lubrication is unnecessary. Keep in mind too, that there is always the risk of damaging a good bearing when removing it. Bearing removal should only be performed if it is necessary to replace the bearing or if the intermediate housing is damaged.

Complete disassembly of the intermediate housing requires the following special tools:

a. Yamaha coupling holder part No. YW-38741 (RA700, RA700A, RA700B and WVT700) or part No. YW-06546 (RA760, RA1100, WVT1100). See A, **Figure 17**.
b. Yamaha shaft holder part No. YW-38742. See B, **Figure 17**.
c. Hydraulic press.
d. A universal bearing driver/installer set.

CAUTION
If you do not have access to the special tools, overhaul of the intermediate housing should be left to a dealer or machine shop. Under no condition should you attempt to drive the shaft out of the housing with a hammer.

Refer to **Figure 34** for this procedure.

1. Clean the intermediate housing in solvent and dry thoroughly.

2. Remove the bearing housing (A, **Figure 35**) from the support housing (B, **Figure 35**).

3. Remove the coupling half (**Figure 6**) as follows:

   a. Clamp the shaft holder (part No. YW-38742) vertically in a vise and fit the intermediate shaft into the holder (**Figure 7**).
   b. Attach the appropriate coupling holder to a breaker bar and fit the holder into the coupling half. Turn the breaker bar counterclockwise to loosen and remove the coupling half. See **Figure 7**.

4. The intermediate shaft is press-fit into the bearing. Its removal is not required unless the bearing, seals or shaft must be replaced. Furthermore, removal of the shaft will require replacement of both seals. See **Figure 36**. To remove the

shaft, support the housing in a press so the shaft faces down (**Figure 18**). Then place a driver between the press ram and shaft and press the shaft out of the bearing. Catch the shaft to prevent it from falling to the floor when it is free of the bearing.

5. Referring to **Figure 18**, note the 2 seals installed in the bearing housing. When removing the seals, first support the housing on a soft flat surface, such as a wooden block. Then carefully pry the first seal out of the housing with a large wide-blade screwdriver. Place a rag underneath the screwdriver to avoid damaging the bore. Normally, the seal(s) will easily pop out of the bore when pried. Occasionally, however, you will encounter a seal which has hardened. A seal in this condition can be very difficult to remove, and if you are not careful, you can damage the housing. If a seal will not pop out, work the screwdriver around the seal when applying pressure, making sure to keep the rag underneath the screwdriver. Then remove the 2 seals installed in the rear of the housing separately. Do not attempt to remove both seals at the same time.

6. Remove the spacer that was behind the rear oil seal.

7. Remove the 2 circlips positioned next to the bearing.

8. Support the housing securely. Place a bearing driver or socket on the bearing's outer race and drive it out of the housing. If you plan on reusing the bearing, catch it as it is released from the housing.

Inspection

1. Clean all sealer residue from the housing and bulkhead mating surfaces.

2. Clean all components in solvent and dry thoroughly. The intermediate shaft bore is used to store grease for lubrication of the drive shaft during engagement. Remove all of the grease as it will become contaminated with solvent when cleaning the outside surface.

3. After cleaning the bearing and shaft, place them on a clean lint-free cloth and allow to air dry.

4. Inspect the housing for cracks or damage. Check the circlip grooves in the bearing bore for cracks or other damage that may allow a circlip to unseat. Remove any burrs from the grooves with emery cloth.

5. Turn the inner bearing race by hand. The bearing should turn smoothly without sticking, excessive noise or other damage. Replace the bearing if it is worn or damaged.

6. Check the shaft surface for cracks, deep scoring and excessive wear. Especially check the shaft's bearing support area. Check the threads and splines for cracks or damage.

7. Check the rubber coupler (**Figure 13**) for wear grooves or rubber deterioration. Especially check for wear on the round outer knobs.

8. Replace worn or damaged parts as required.

Intermediate Housing Assembly

Refer to **Figure 34** for this procedure.

NOTE
*Refer to the information listed under **Bearing Replacement** in Chapter One when pressing the shaft into the bearing.*

1. Make sure the intermediate housing is clean.
2. Install the bearing as follows:

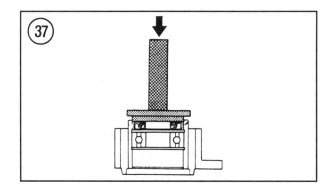

DRIVE TRAIN

a. Place the intermediate housing on a flat surface so the crankshaft end (front) faces up.
b. Install the rear circlip into the housing groove. Make sure the circlip seats in the groove completely.
c. Wipe a light coat of marine grease on the outer bearing race.
d. Align the bearing with the housing bore and drive it squarely into the bore with a bearing driver or socket placed on the outer bearing face. Drive the bearing into the housing until it seats against the circlip (**Figure 20**).
e. Install the front circlip into the housing groove, making sure it seats in the groove completely.
f. Rotate the inner bearing race and check it for roughness. If the bearing race does not turn smoothly, the bearing was damaged during installation. Remove and replace bearing.

NOTE
The front (crankshaft side) and rear intermediate shaft seals are different sizes.

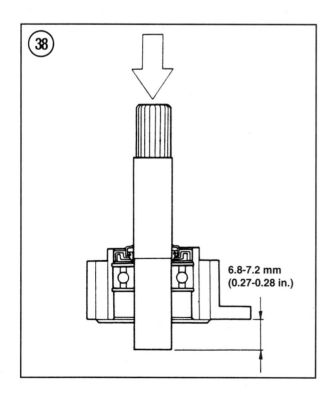

The diameter of the hole in the front seal is 10 mm (0.38 in.). In the rear seal, the diameter is 8 mm (0.31 in.). Do not confuse these seals during installation.

NOTE
When installing the seals in the following steps, pack each seal lip cavity with marine grease.

3. Install the front seal into the bearing housing with a suitable bearing driver or socket. Add enough marine grease on top of the bearing to fill the area between the circlips with grease, then install the seal its manufacturer's marks face out. See **Figure 37**.

CAUTION
Do not install the rear seal until the intermediate shaft is installed. If the rear seal is installed first, the shaft may tear the seal lips when it is pressed into place.

4. Install the intermediate shaft as follows:
   a. Wipe the seal lip (installed in Step 3) with marine grease.
   b. Make sure the shaft is clean, then wipe the bearing end of the shaft with marine grease.
   c. Support the bearing housing in a press and adjust the press table height to accept the bearing and shaft.
   d. Align the nonthreaded shaft end with the bearing and place a piece of aluminum between the press ram and shaft threads. Hold the shaft as near vertical as possible and slowly operate the press, making sure the shaft enters the bearing squarely. If the shaft is at an angle as it enters the bearing, stop and release pressure; realign the shaft before pressing it into the bearing. When the shaft enters the bearing squarely, continue to press the shaft into the bearing until it extends the specified distance (**Figure 38**) beyond the housing.
   e. After installing the shaft, turn it to check bearing operation. The bearing should rotate smoothly. If the bearing turns roughly,

remove the shaft and check the bearing for damage.

5. Secure the housing so the rear side faces up and install the spacer.

6. Apply marine grease to the rear seal and install the seal into the housing with a suitable driver. Add enough marine grease on top of the bearing to fill the area between the circlips with grease, then install the seal so that its manufacturer's marks face out. Be sure the gap between the seal and the housing is within the specification shown in **Figure 39**.

7. Apply marine grease to 2 new O-rings and install them onto the intermediate shaft. (**Figure 39**).

8. Install the coupling half as follows:
   a. Clamp the shaft holder (part No. YW-38742) vertically in a vise and fit the intermediate shaft into the holder (**Figure 7**).
   b. Install the washer onto the shaft.
   c. Apply Loctite 242 to the intermediate shaft threads and thread the coupling half onto the shaft by hand as far as it will go.
   d. Attach the appropriate coupling holder to a torque wrench. Fit the coupling holder into the coupling half and tighten to 37 N•m (27 ft.-lb.). See **Figure 7**.
   e. Remove the tools from the bearing housing.

9. Fill the intermediate shaft with approximately 30-40 cc (1.0-1.4 oz.) of marine grease. See **Figure 40**.

10. Place the dowel pin into the support housing and mate the bearing and support housings. Install the mounting bolts and washers. Apply Loctite 242 to the threads of the bolts and torque the bolts to the specification given in **Table 3**.

Intermediate Housing Installation

1. Wipe marine grease on a new bulkhead seal and insert the seal into the bulkhead as shown in **Figure 3**. The seal should seat squarely in the bulkhead groove.

2. Coat the O-ring with marine grease and install it onto the support housing (C, **Figure 35**).

3. Coat the drive shaft splines with marine grease. Then align the drive shaft with the intermediate shaft and join the 2 components.

4. Apply Loctite 242 (blue) to the threads of the intermediate housing mounting bolts and loosely install the 3 intermediate housing mounting bolts (B, **Figure 2**) and washers. Reinstall any shims in their original position. Torque the bolts to the specifications given in **Table 3**.

NOTE
Always keep the original factory installed shims between the intermediate housing and bulkhead. These shims affect engine and drive shaft alignment. If it is necessary to replace a shim, measure the old shim with a vernier caliper or micrometer and purchase one with the same thickness.

5. Install the hose onto the grease nipple if the hose was removed. Secure it in place with a hose clamp. If the grease hose on your model was secured with a cable tie, install a new cable tie.

6. Insert the rubber coupler into the coupling half (A, **Figure 2**).

7. Install the engine and check engine alignment as described in Chapter Five or Chapter Six.

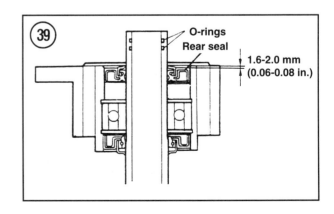

DRIVE TRAIN

JET PUMP
(WR500 AND WR650)

The jet pump is a mixed-flow, single-stage design. The jet pump assembly consists of the impeller housing, impeller, drive shaft grille, nozzle and steering nozzle. The drive axle is supported by 5 sealed ball bearings—2 bearings are installed in the intermediate housing and 3 bearings are used in the impeller housing assembly. The sealed bearings do not require periodic lubrication.

The drive train consists of the crankshaft-to-drive shaft connection, drive shaft, impeller and jet pump. As the impeller rotates, water is drawn in through the intake area at the bottom of the jet pump housing. Water is then forced through the impeller housing where it exits through the steering nozzle assembly. As engine speed increases, a higher volume of water is forced through the jet pump and the vehicle's speed increases. Stationary vanes used in the impeller housing control water flow for improved thrust and efficiency. Steering is accomplished by a steering cable which is attached to the steering nozzle assembly.

The pump is carefully manufactured so there is very little clearance between the impeller and the impeller housing. As the pump wears from normal use, the clearance between the impeller and housing will increase, thus decreasing pump thrust. Inspect the impeller and housing according to the maintenance schedule in Chapter Three. Replace the housing if it is excessively worn.

Service Note

The following sections describe removal and complete overhaul of the WR500 and WR650 jet pump assemblies. If you intend to just measure the impeller housing clearance at this time, refer to *Impeller Clearance Check and Adjustment* in this chapter. It is possible to check impeller housing clearance without removing the jet pump from the hull.

Jet Pump/Impeller Removal

Refer to **Figure 41** for this procedure.

1. Remove the battery to prevent acid spillage. See Chapter Nine.

CAUTION
If the exhaust system is installed on the engine, do not turn the craft onto its right side when performing Step 2. Water in the exhaust system may drain into the engine's exhaust ports and cause serious engine damage.

2. Turn the fuel petcock OFF and turn the watercraft onto its *left* side. Support the craft securely on a suitable stand. To protect the hull from damage, support the craft with 2 discarded motorcycle tires, heavy blankets or similar object that can accept the craft's weight.

3. Remove the bolts holding the water valve body to the ride plate. See **Figure 42**.

4. Remove the bolts holding the ride plate (**Figure 43**) to the bottom of the hull and remove the ride plate. Remove the 2 washers installed between the ride plate and hull at the front mounting hole positions, if used.

5. Disconnect the cooling system feed hose from the jet pump (**Figure 44**) and disconnect the bilge hoses.

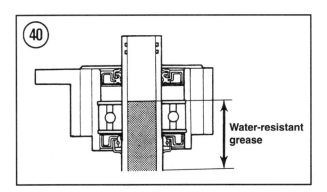

CHAPTER SEVEN

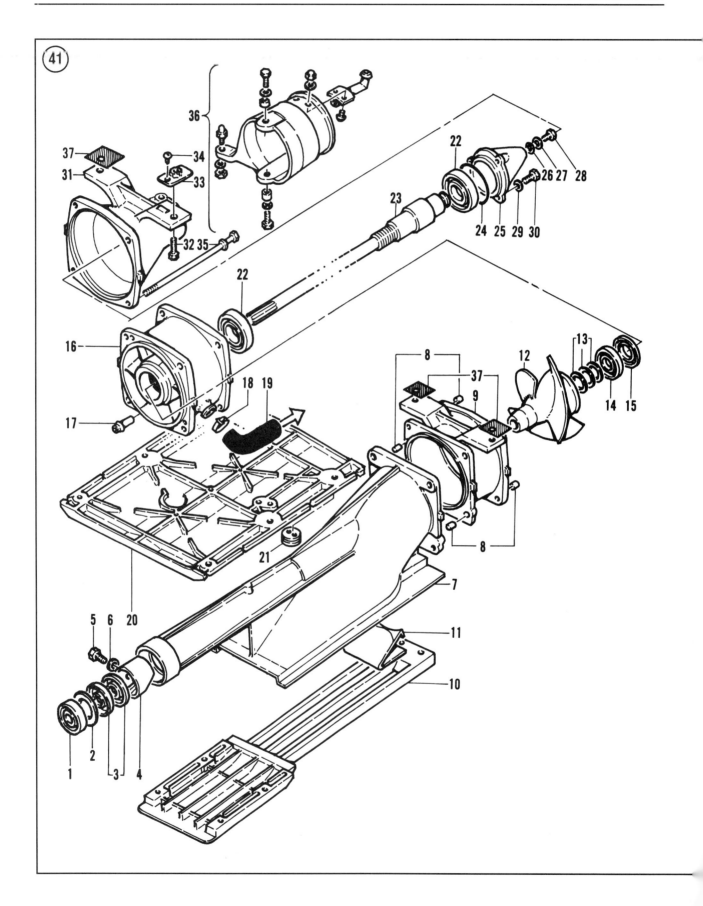

JET PUMP
TYPE II—WR500 AND WR650

1. Bearing
2. Washer
3. Seals
4. Spacer
5. Bolt
6. Washer
7. Intake duct
8. Dowel pins
9. Impeller housing
10. Intake grate
11. Impeller duct
12. Impeller
13. Shim
14. Seal
15. Seal
16. Impeller duct
17. Pipe
18. Hose clamp
19. Hose
20. Ride plate
21. Adjuster
22. Bearing
23. Drive shaft
24. O-ring
25. Cap
26. Gasket
27. Washer
28. Bolt
29. Washer
30. Bolt
31. Nozzle
32. Bolt
33. Nut
34. Screw
35. Bolt and washer
36. Steering nozzle assembly
37. Shims

288

CHAPTER SEVEN

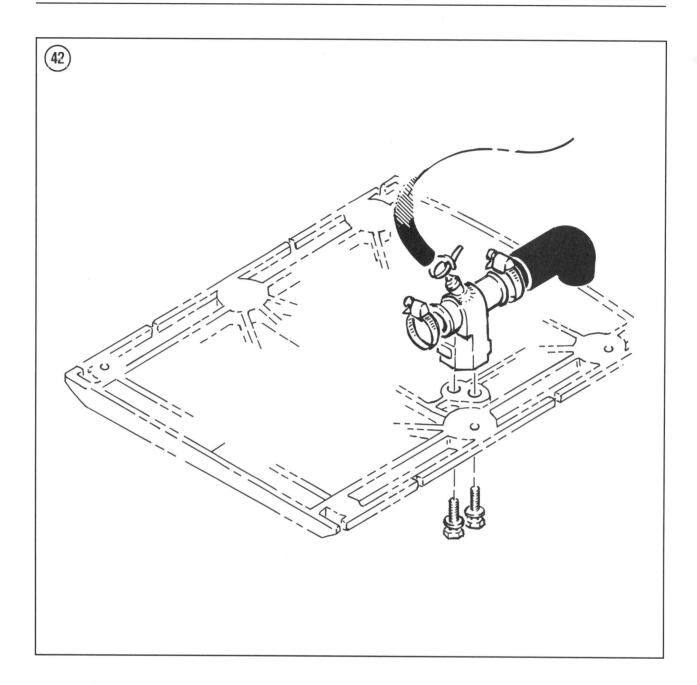

DRIVE TRAIN

6. Detach the steering cable/nozzle ball joint from the steering nozzle. Slide its spring-loaded sleeve forward and remove the connector from the ball (**Figure 45**).

CAUTION
When removing the water intake grate in Step 6, do not pry the grate up from one end. The grate is made of aluminum and will snap in half. Pull up on the grate evenly. Because silicone sealant is used during grate installation, this step will not be easy to do. You may have to use a razor blade to cut as much of the silicone away from the grate as possible. Remember though, prying the grate from one end will break it.

7. Remove the front and rear bolts holding the water intake grate (**Figure 46**) to the hull. Then grasp the grate on *both* ends and pull it away from the hull.

8. Remove the impeller duct (**Figure 47**) positioned underneath the grate.

CAUTION
*The threads used on the jet pump mounting bolts have a tendency to corrode. Because these bolts thread into an insert installed in the hull, a corroded bolt will usually break the insert loose from its mounting position, allowing the bolt and insert to turn together. If this should happen, remove the fire extinguisher compartment to gain access to the insert, then hold the insert with a pair of locking pliers so you can remove the bolt (very time consuming). Before attempting to loosen the jet pump bolts, spray **each** bolt with WD-40 or a thread penetrating lubricant, and allow sufficient time for the lubricant to work its way down the bolt threads. When attempting to loosen the bolts, do not use impact wrench as the power of the tool can rip the insert away from its mounting position. After allowing the penetrating lubricant to soak in, loosen the bolts carefully with a socket, ratchet and suitable extension.*

9. Apply a penetrating lubricant to each jet pump mounting bolt (see previous *CAUTION*). Allow plenty of time for the lubricant to work its way down the bolt threads. Then carefully loosen the bolts with a ratchet.

CAUTION
When removing the jet pump, label any shims positioned underneath the pump mounting lugs. These shims will have to be reinstalled in exactly the same posi-

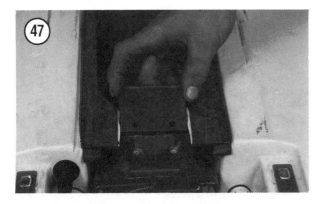

tion for the pump and impeller to align properly.

10. Cut any sealant at the pump mounting lugs.

11. Remove the jet pump assembly by pulling it rearward. See **Figure 48**.

Pump Case Disassembly

The jet pump assembly is made up of a number of separate subassemblies. See **Figure 41**. To prevent confusion during the following service procedure, refer to **Figure 49** and the following identification list:

a. Steering nozzle.
b. Nozzle.
c. Impeller duct.
d. Impeller housing.
e. Intake duct.

1. Before disassembling the jet pump, check the impeller tip-to-impeller housing clearance as described in this section. Performing this check before you disassemble the pump will allow you to purchase new impeller shims, if required, for use during pump reassembly.

2. The jet pump assembly is held together with 4 long bolts and washers (**Figure 50**). Loosen and remove each of the bolts and washers.

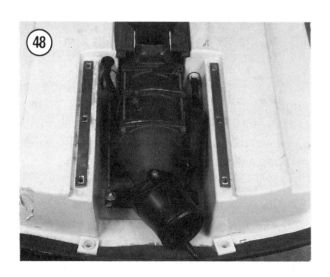

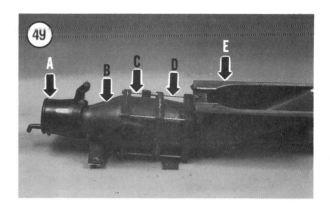

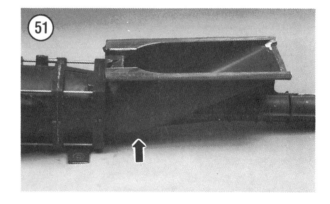

DRIVE TRAIN

3. Slide the intake duct (**Figure 51**) off of the impeller housing and remove the 2 dowel pins.

4. Separate and remove the nozzle (A, **Figure 52**), impeller duct (B) and impeller housing (C). Locate and remove the dowel pins from the housing mating surfaces (where used). See A, **Figure 53**, typical. Store the dowel pins in a plastic bag, tagged for reassembly reference.

5. Remove the impeller (B, **Figure 53**) from the drive shaft as follows:

a. Slide the Yamaha drive shaft holder (part No. YB-6079 [C, **Figure 5**]) onto the end of the drive shaft, then secure the holder horizontally in a vise.

CAUTION
*The impeller uses left-hand threads. Turn the impeller **clockwise** to loosen it.*

b. Turn the impeller *clockwise* with a wrench (**Figure 54**) and remove it from the drive shaft.

c. Remove the shims (A, **Figure 55**) and store them in a plastic bag, tagged for reassembly.

6. Unbolt and remove the cap (**Figure 56**) from the impeller duct.

Inspection

The jet pump is a vital link in the overall performance of your water vehicle. Normal wear of the impeller housing and impeller can reduce jet pump thrust and top speed, even when the engine is running perfectly. Secondary thrust problems can be caused by obstructions in the pump, a damaged impeller housing or impeller blades. The jet pump assembly should be inspected carefully and all worn or damaged parts repaired or replaced as required.

1. Clean sealant residue from all affected surfaces.

2. Remove and discard all grease from inside the impeller duct cap (**Figure 57**).

3. Clean all parts in solvent, and blow them dry with compressed air.

4. Check the impeller blades (**Figure 58**) for nicks or gouges, and if minor, smooth them out with abrasive paper or a file. It is especially important for the blade edges to be smooth. If there is major damage to the impeller, replace it.

5. Check the inside of the impeller housing (**Figure 59**) for deep scratches. If the pump case is damaged, either replace it or have the pump remachined by a qualified marine dealer.

6. Check the intake duct (A, **Figure 60**) for cracks or damage. If the intake duct bearing assembly was not removed, check that the spacer mounting bolt (B, **Figure 60**) is tight.

7. Check the impeller and intake duct bearings as follows:

   a. *Impeller duct*—The drive shaft is a press fit into the impeller duct bearings. Support the impeller duct and turn the drive shaft by hand. The drive shaft should turn smoothly with no sign of bearing wear or damage. If the shaft turns roughly, suspect worn or damaged bearings. Service the impeller duct as described in this chapter.

   b. *Intake duct*—The intake duct is equipped with a single ball bearing and 2 seals.

   c. Replace the seals whenever the impeller duct or intake duct is disassembled.

8. Inspect the drive shaft as follows:

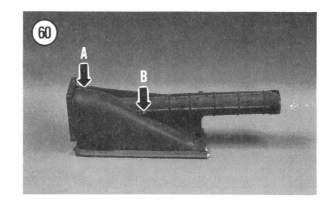

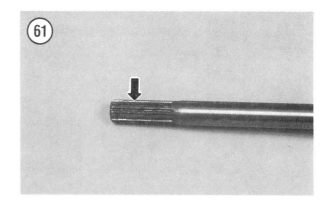

DRIVE TRAIN

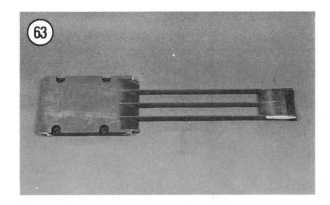

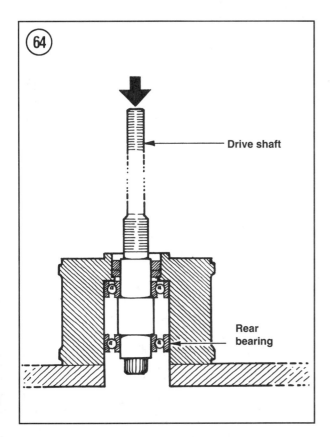

a. Inspect the drive shaft splines (**Figure 61**) for cracks, excessive wear or other damage.

b. Yamaha lists a runout specification for the drive shaft. To check runout, however, the drive shaft must first be pressed out of the impeller duct assembly. Refer to *Drive Shaft Removal/Installation and Impeller Duct Bearing Replacement* in this chapter to service the drive shaft.

9. Check the nozzle (A, **Figure 62**) and steering nozzle (B) for cracks or other damage. Pivot the steering nozzle from side to side to make sure it moves smoothly. If the steering nozzle does not pivot smoothly, remove the pivot bolts and check all parts for wear or damage. Replace worn or damaged parts as required. Wipe the pivot bolts with marine grease before installing them.

10. Inspect the intake grate (**Figure 63**) for cracks, breaks or other damage. Pay close attention to the small arms, particularly where they attach to the mounting plates. Replace the intake grate if necessary.

Drive Shaft Removal/Installation and Impeller Duct Bearing Replacement

The impeller duct (B, **Figure 55**) is equipped with 2 bearings and 2 seals. The drive shaft (C, **Figure 55**) is a press fit in the impeller duct bearings. A hydraulic press is required to remove and install the drive shaft. When the drive shaft is pressed out, the rear impeller duct bearing should come out with the drive shaft. A bearing remover is required to remove the front impeller duct bearing. If you do not have the special tools, refer all impeller duct and drive shaft service to a Yamaha dealership or machine shop.

1. Support the impeller duct in a press so the drive shaft faces toward the press ram (**Figure 64**). Press the drive shaft and the rear bearing out of the impeller duct. Make sure to catch the drive shaft once it is free of the impeller duct. It could be damaged if it drops to the floor.

2. Remove the impeller duct from the press and place it on a piece of wood laying flat across a workbench. Using a bearing driver or suitable socket, drive the front bearing out of the impeller duct.

NOTE
The 2 seals installed in the impeller duct have different part numbers. Identify the seals after removing them so you can use them to identify the new seals for reassembly.

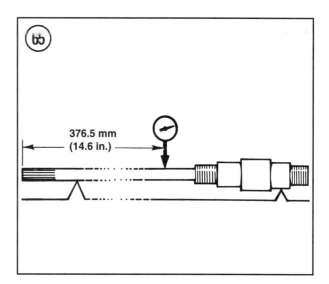

3. Turn the impeller duct over and remove the 2 seals by driving them out with a bearing driver or suitable size socket.

4. The rear bearing is pressed onto the drive shaft; remove the rear bearing with a press.

5. Mount the drive shaft on V-blocks at the points indicated in **Figure 65**. Turn the drive shaft slowly and check runout with a dial indicator at the test point indicated in **Figure 65**. Runout is the difference between the highest and lowest dial indicator reading. Replace the drive shaft if the runout is greater than the specification given in **Table 2**.

6. Check the impeller duct bearings. Rotate the inner bearing race by hand; the bearing should turn smoothly. Replace the bearing if it is damaged.

7. Clean the impeller duct in solvent and dry with compressed air, if available.

NOTE
After all parts are cleaned and inspected, assemble the impeller duct and install the drive shaft as described in the following steps.

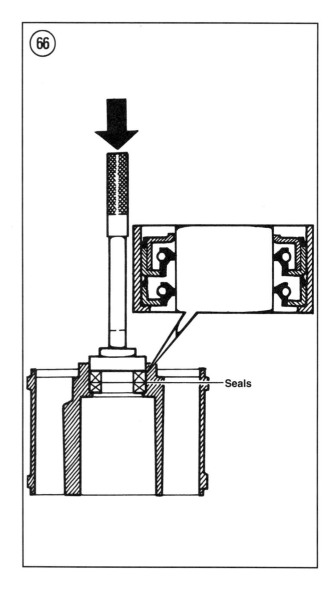

8. Install the impeller duct seals as follows:
   a. Pack the seal lip cavity of each seal with marine grease.

DRIVE TRAIN

b. Install both seals so their *closed* sides (the side with the manufacturer's marks) face outward, to the front of the impeller duct. See **Figure 66**.

c. Install the lower seal by driving it into the impeller duct using a driver with the same outside diameter as the seal. Drive the seal in squarely until it bottoms.

d. Repeat to install the upper seal. Drive the seal into the impeller duct until it bottoms against the lower seal.

9. Apply marine grease to the bearings and press both impeller duct bearings onto the drive shaft, then press the drive shaft into the impeller duct.

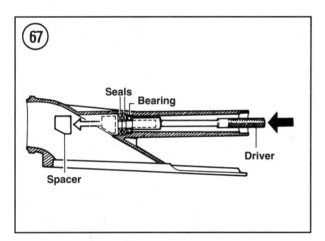

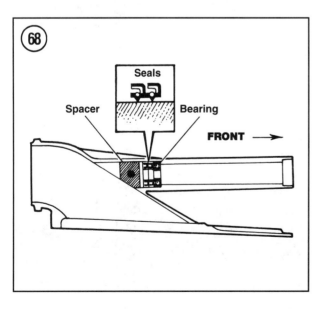

10. Turn the drive shaft by hand. It should turn smoothly without sticking. A rough turning or sticking bearing indicates bearing damage.

Intake Duct
Bearing and Oil Seal Replacement

The intake duct is equipped with a bearing, washer, 2 seals (same part number) and spacer. The spacer is secured with a bolt and washer.

1. Loosen and remove the spacer bolt and washer (B, **Figure 60**) from the intake duct.

2. Using a suitable bearing driver and extension, drive the spacer, seals, washer and bearing out of the intake duct. See **Figure 67**.

3. Check the intake duct bearing. Rotate the inner bearing race by hand; the bearing should turn smoothly. Replace the bearing if damaged.

4. Remove all sealer residue from the intake duct assembly. Clean the impeller duct, spacer and washer in solvent.

5. Inspect the impeller duct, spacer and washer for damage. Replace worn or damaged parts.

NOTE
After all parts are cleaned and inspected, assemble the intake duct bearing assembly as follows.

6. Place the intake duct on a workbench. Block the front end to prevent it from sliding when installing the bearing and seals.

7. Apply marine grease to the bearing and install the bearing into the intake duct housing using a suitable bearing driver and extension. Drive the bearing into the housing until it bottoms.

8. Install the washer against the bearing on models so equipped.

9. Install the intake duct seals as follows:

   a. Pack the seal lip cavity of each seal with marine grease.

   b. Install both seals so their *closed* side faces forward (**Figure 68**). Both seals have the same part number.

c. Install the first seal by driving it into the intake duct using a driver with the same outside diameter as the seal. Drive the seal in squarely until it bottoms against bearing (or the washer).

d. Repeat to install the second seal. Drive the seal into the intake duct until it bottoms against the first seal.

e. Install the spacer into the intake duct, aligning the intake duct bolt hole with the spacer threads. Secure the spacer with the intake spacer bolt and washer. Apply Loctite 242 to the threads of the bolt and tighten it securely.

Pump Case Assembly

Refer to **Figure 41** for this procedure.

1. Apply marine grease to a new O-ring and fit the O-ring onto the impeller duct cap (**Figure 57**). Fill the cap with marine grease, and install the cap onto the impeller duct. Apply Loctite 242 to the threads of the mounting bolts, install the bolts with washers and torque the bolts to the specification given in **Table 3**.

2. Install the impeller as follows:

a. If you checked impeller clearance before disassembly, install the new shims (if required) onto the drive shaft (A, **Figure 55**). If you did not check impeller clearance, install the impeller and impeller housing as described in this section, then check clearance as described in this chapter.

CAUTION
*The impeller uses left-hand threads. Turn the impeller **counterclockwise** to tighten it.*

b. Apply Loctite 242 to the impeller threads, then slide the impeller onto the drive shaft. Thread the impeller (B, **Figure 53**) onto the drive shaft by turning it *counterclockwise*.

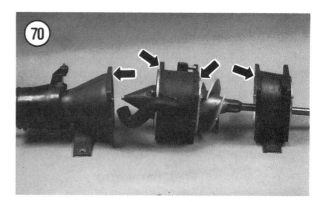

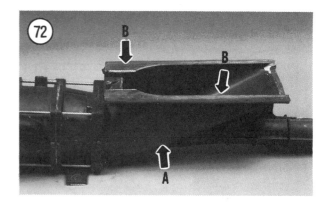

DRIVE TRAIN

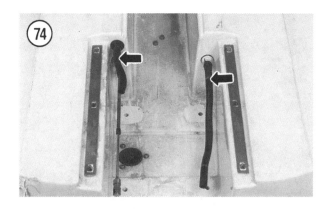

c. Slide the Yamaha drive shaft holder (part No. YB-6079 [C, **Figure 5**]) onto the end of the drive shaft, then secure the holder horizontally in a vise.

d. Tighten the impeller with a wrench (**Figure 54**) to the torque specification listed in **Table 3**.

e. Turn the drive shaft to make sure it turns freely.

f. Temporarily install the impeller housing (**Figure 69**) and check impeller clearance as described in this chapter. Remove impeller and readjust shim(s) thickness as required.

g. When impeller clearance is correct, proceed to Step 3.

3. Install the dowel pins in the mating surface pin holes. Then apply Yamabond No. 4 to the impeller housing, impeller duct and nozzle mating surfaces (**Figure 70**).

4. Assemble the nozzle, impeller duct and impeller housing (**Figure 71**).

5. Apply Yamabond No. 4 to the impeller housing and intake duct mating surfaces. Press the intake duct (A, **Figure 72**) against the mating surface of the impeller housing, making sure to engage the dowel pins previously installed.

6. When all of the subassemblies are assembled, install the 4 through-bolts and washers (**Figure 73**). Apply Loctite 242 (blue) to the threads of the bolts, hand-tighten each bolt, and then tighten the bolts evenly in a crisscross pattern.

Jet Pump Installation

1. Remove all of the old sealant from the pump intake area.

2. Install the bilge hoses (**Figure 74**) to their outlet nozzles. Secure the hoses with new plastic ties (**Figure 75**).

3. Make sure the large bilge hose (**Figure 76**) is secured to the impeller duct with a hose clamp.

4. Apply silicone sealant to the jet pump intake duct sealing surfaces (B, **Figure 72**).

5. Wipe the end of the drive shaft with marine grease, then slide the jet pump forward into place in the hull. Be sure the drive shaft splines engage the shaft in the intermediate housing.

NOTE
Before installing the jet pump mounting bolts, check the threads on each bolt for corrosion or damage. Clean the threads thoroughly, or if necessary, replace bolts before installing them. Likewise, the threaded inserts installed in the hull should be cleaned of all corrosion. It is also important to note that some models use an uncommon M9 × 1.25 mm thread on the jet pump mounting bolts. Do not attempt to use an 8 or 10 mm bolt if you are working on such a model. When purchasing replacement bolts, confirm bolt diameter and pitch as described under **Fasteners** *in Chapter One. Apply Loctite 242 (blue) to the threads of the 4 pump mounting bolts, then install the bolts together with their flat washers and lockwashers into the hull. Insert the alignment shims through the pump mounting bolts at their original mounting position.*

6. Tighten the jet pump mounting bolts with a torque wrench to the specification listed in **Table 3**. Do not overtighten the bolts.

7. Check the foam sealing strips placed on the bottom side of the hull (A, **Figure 77**) for cracks or damage. Replace damaged or missing strips.

8. Seal the 4 corners indicated in B, **Figure 77** with marine-grade RTV sealant.

9. Install the impeller duct (**Figure 78**), making sure to align the 2 holes in the duct with the mounting holes in the intake duct.

10. Apply RTV sealant to the bottom side of the intake grate (**Figure 79**) and install the grate onto the hull; tighten its mounting bolts securely.

11. Reconnect the cooling system feed hose (A, **Figure 80**) at the impeller duct.

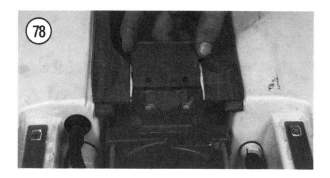

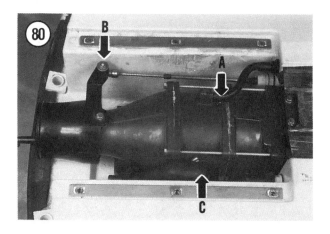

DRIVE TRAIN

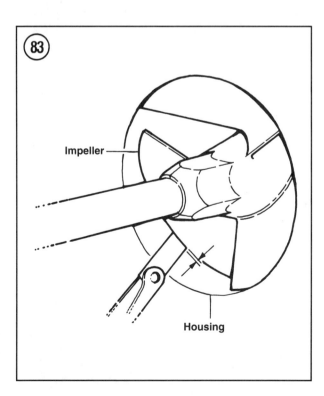

12. Wipe the steering cable ball-joint ball with marine grease and reconnect the steering cable to the steering nozzle. Slide the cable's spring-loaded sleeve forward and insert the connector onto the ball (B, **Figure 80**). Release the sleeve and check to make sure the cable is attached properly. Operate the handlebar to make sure the steering nozzle pivots correctly.

13. If you removed the water valve body from the ride plate, reattach it now (**Figure 42**). Apply Loctite 242 (blue) to the bolt threads and tighten the bolts securely.

14. Turn the ride plate over and reconnect the 2 bilge hoses (**Figure 81**) to the water valve body. Secure each hose with new a plastic tie or with a hose clamp as necessary.

CAUTION
If the small bilge hose should become kinked, the bilge system will not operate. Check bilge hose routing and reposition as required.

15. Position the ride plate on the hull and install the 2 washers (if used), at the front of the ride plate as shown in **Figure 82**. Then position the ride plate into its mounting position and install its bolts and washers. Make sure the 2 front mounting bolts engage the 2 washers properly. Apply Loctite 242 (blue) to the threads of the ride plate bolts and torque the bolts to the specification given in **Table 3**.

16. Turn the water vehicle upright and install the battery. See Chapter Nine.

Impeller Clearance
Check and Adjustment

Impeller clearance can be checked without removing the jet pump assembly.

1. Remove the intake grill as described under *Jet Pump Removal* in this chapter.

2. Measure the impeller-to-impeller housing clearance with a feeler gauge (**Figure 83**). The correct clearance for all WR500 and WR650

models is 0.3 mm (0.012 in.). If the clearance is incorrect, refer to the shim selection chart in **Table 4** and cross reference your recorded clearance with the new shim pack thickness listed underneath it. For example, suppose your recorded clearance is 0.5 mm. Therefore, to obtain the correct impeller clearance, you must increase the existing shim thickness by 1.5 mm as shown in **Table 4**. Replacement shims can be purchased from a Yamaha marine dealership in 0.1, 0.3, 0.5 and 1.0 mm thicknesses..

3. If the clearance is incorrect, remove the jet pump as described in this chapter and remove the impeller to change the shim pack thickness. Reassemble and recheck the clearance as described in the appropriate procedure.

4. If the clearance is correct, reinstall the intake grill as described under *Jet Pump Installation* in this chapter.

JET PUMP
(RA700, RA700A, RA700B, RA760, RA1100, WRA650, WRA650A, WRA700, WVT700 AND WVT1100)

The jet pump is an axial-flow, single-stage design. The jet pump assembly consists of the impeller housing, impeller, drive shaft, nozzle, steering nozzle and reverse assembly. Three sealed ball bearings support the drive shaft: a single bearing in the intermediate housing and 2 bearings in the impeller housing assembly. Sealed bearings do not require periodic lubrication.

The drive train consists of the crankshaft-to-drive shaft connection, drive shaft, impeller and jet pump. As the impeller rotates, water is drawn into the pump through the intake area at the bottom of the jet pump housing. Water is then forced through the impeller housing where it exits through the steering nozzle assembly. As engine speed increases, a higher volume of water is forced through the jet pump and the vehicle's speed increases. Stationary vanes in the impeller housing better control water flow for improved thrust and efficiency. A steering cable, attached from the steering column to the steering nozzle, controls steering movement.

WRA650, WRA650A, WRA700, WVT700 and WVT1100 models are equipped with a reverse shift assembly that allows the craft to operate in reverse for slow speed maneuvering. The reverse shift assembly consists of a shift lever mounted on the starboard deck, a nozzle deflector mounted on the jet pump and a cable connecting the shift lever to the nozzle deflector (**Figure 84**). The shift lever has 2 operating positions—down (forward) and up (reverse). When the vehicle is running forward, the jet thrust exits the steering nozzle in a backward direction, forcing the craft forward. When the shift lever is pulled UP, the nozzle deflector drops over the end of the steering nozzle. Jet thrust exiting the steering nozzle then changes direction (rearward to forward) as it follows the nozzle deflector, forcing the vehicle to move backward. To shift into reverse, the engine must be turned OFF or it must be operating at idle speed; do not operate the shift lever when the engine is running above idle speed. Remember too, there is no NEUTRAL engine speed, there will always be some forward or reverse thrust when the engine is running at idle.

The pump is manufactured with very little clearance between the impeller and the impeller housing. As the pump wears from normal use, the clearance between the impeller and housing will increase, thus decreasing pump thrust. Therefore, inspect the impeller and housing according to the maintenance schedule in Chapter Three.

Jet Pump Identification

The jet pump used on the models described in this section are basically the same with minor differences It is important to pay particular atten-

DRIVE TRAIN

tion to the location and order of parts during service. The jet pumps are identified as follows:

a. *Type III jet pump* (**Figure 85**)—WRA650, WRA650A and WRA700.
b. *Type IV jet pump* (**Figure 86**)—RA700, RA700A, RA700B, RA760, RA1100, WVT700 and WVT1100.

Service Note

The following sections describe removal and complete overhaul of Type III and IV jet pump assemblies. If measuring the impeller housing clearance is the only service required at this time, refer to *Impeller Housing Clearance Check and Adjustment* in this chapter. You can check impeller housing clearance without removing the jet pump assembly.

Jet Pump/Impeller Removal

Refer to **Figure 85** or **Figure 86** for this procedure.

1. Remove the battery to prevent acid spillage. See Chapter Nine.
2. Remove the oil tank as described in Chapter Ten.
3. Drain the fuel tank and store the fuel in a container approved for gasoline storage.

> *CAUTION*
> *If the exhaust system is installed on the engine, do not turn the craft onto its right side when performing Step 4. Water in the exhaust system may drain into the engine's exhaust ports and cause serious engine damage.*

4. Turn the fuel petcock OFF and turn the watercraft onto its *left* side. Support the craft securely on a suitable stand. To protect the hull from damage, support the craft with 2 discarded motorcycle tires, heavy blankets or similar object that can accept the craft's weight.
5. On models with a water valve body, remove the bolts holding the valve body to the ride plate. See **Figure 87**.
6. Remove the screws holding the speed sensor to the ride plate on models so equipped.
7. Unbolt and remove the ride plate from the bottom of the hull.
8. Disconnect the engine cooling and bilge hoses (**Figure 88**).
9. Detach the steering cable/nozzle ball joint from the steering nozzle. Slide its spring-loaded

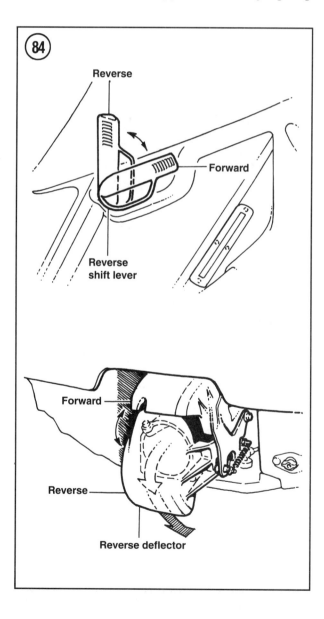

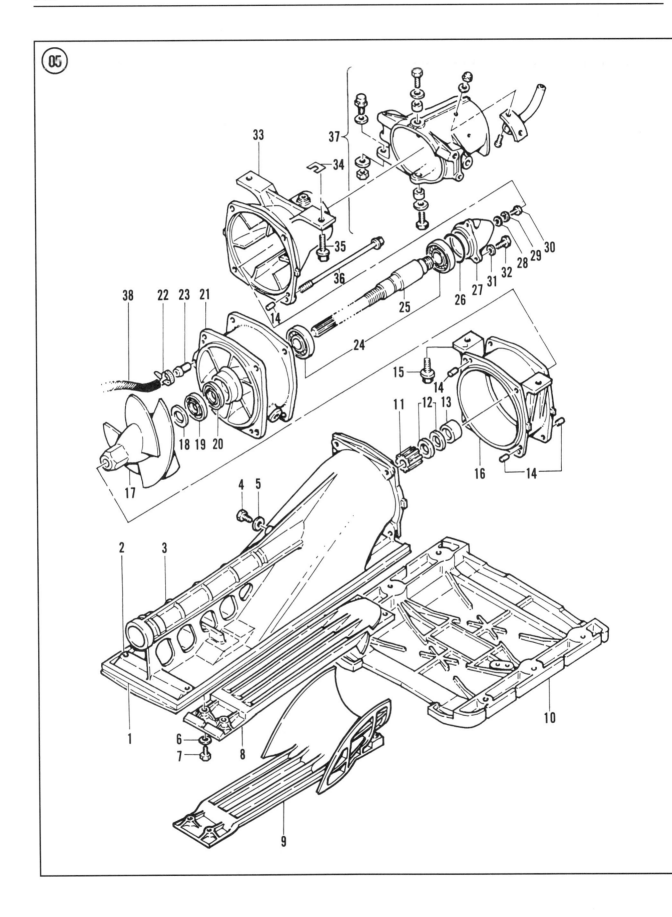

DRIVE TRAIN

JET PUMP
(TYPE III—WRA650, WRA650A AND WRA700)

1. Felt packing
2. Felt packing
3. Intake duct
4. Bolt
5. Washer
6. Washer
7. Bolt
8. Intake grate (WRA650)
9. Intake grate (WRA700)
10. Ride plate
11. Bushing
12. Seals
13. Spacer
14. Dowel pins
15. Bolt and washer
16. Impeller housing
17. Impeller
18. Washer
19. Seal
20. Seal
21. Impeller duct
22. Clamp
23. Nozzle
24. Bearings
25. Drive shaft
26. O-ring
27. Cap
28. Seal
29. Washer
30. Bolt
31. Washer
32. Bolt
33. Nozzle
34. Shim
35. Bolt
36. Bolt and washer
37. Steering nozzle assembly
38. Hose

CHAPTER SEVEN

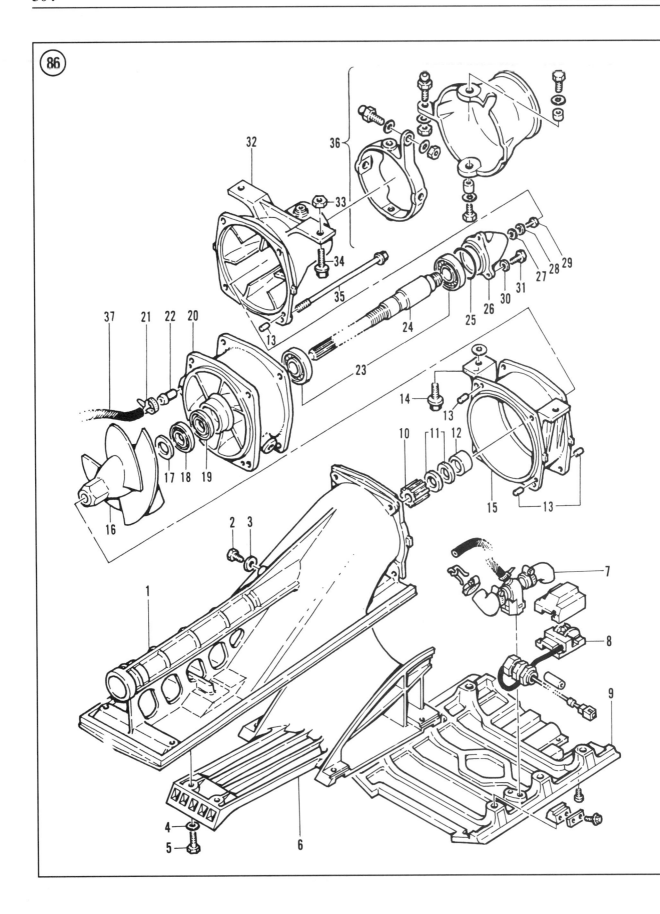

DRIVE TRAIN

**JET PUMP
(TYPE IV—RA700, RA700A, RA700B,
RA760, RA1100, WVT700 AND WVT1100)**

1. Intake duct
2. Bolt
3. Washer
4. Washer
5. Bolt
6. Intake grate
7. Water valve
8. Speed sensor
9. Ride plate
10. Bushing
11. Seals
12. Spacer
13. Dowel pins
14. Bolt and washer
15. Impeller housing
16. Impeller
17. Washer
18. Seal
19. Seal
20. Impeller duct
21. Clamp
22. Nozzle
23. Bearings
24. Drive shaft
25. O-ring
26. Cap
27. Seal
28. Washer
29. Bolt
30. Washer
31. Bolt
32. Nozzle
33. Shim
34. Bolt
35. Bolt and washer
36. Steering nozzle assembly
37. Hose

sleeve forward and pull the connector off of the ball (**Figure 88**).

> *CAUTION*
> *Do not pry the intake grate up from one end when removing it in Step 10. The grate is made of aluminum and will break in half. Pull up on the grate evenly from both ends.*

10. Remove the front and rear bolts holding the water intake grate (**Figure 88**) to the hull. Grasp the grate on *both ends* and pull it away from the hull.

> *CAUTION*
> *The threads used on the jet pump mounting bolts corrode easily. Because these bolts thread into an insert installed in the hull, a corroded bolt may force the insert to twist and turn in its mounting position, allowing the bolt and insert to turn together. If the insert breaks away, you will have to access the insert and then hold it with a pair of locking pliers or similar tool to remove the bolt. Before loosening the jet pump mounting bolts, spray **each bolt** with WD-40 or a thread penetrating lubricant. Allow sufficient time for the lubricant to work its way down the bolt threads. When attempting to loosen the bolt, do not use an air impact wrench as the power of the tool can rip the insert away from its mounting position. After applying a penetrating lubricant, loosen the bolt carefully with a socket, ratchet and suitable extension.*

11. Apply a penetrating lubricant to each jet pump mounting bolt (see previous *CAUTION*). Allow plenty of time for the lubricant to work its way down the bolt threads. Then carefully loosen the bolts with a ratchet.

> *CAUTION*
> *When removing the jet pump, label any shims positioned underneath the pump mounting lugs. These shims must be reinstalled in exactly the same position for the pump to align properly.*

12. Cut any sealant at the pump mounting lugs.
13. Remove the jet pump assembly by pulling it rearward. See **Figure 88**.

Pump Case Disassembly

The jet pump assembly consists of a number of separate subassemblies. To prevent confusion during the following service procedure, refer to **Figure 89** and the following identification list:

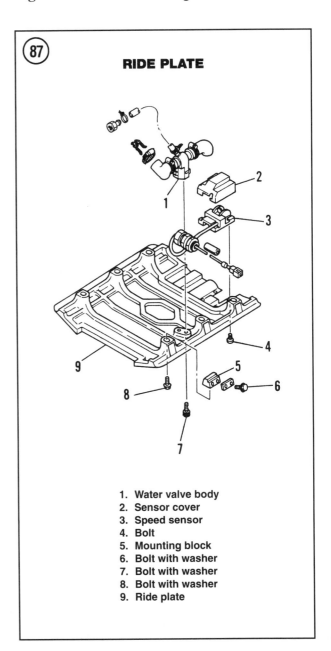

87
RIDE PLATE

1. Water valve body
2. Sensor cover
3. Speed sensor
4. Bolt
5. Mounting block
6. Bolt with washer
7. Bolt with washer
8. Bolt with washer
9. Ride plate

DRIVE TRAIN

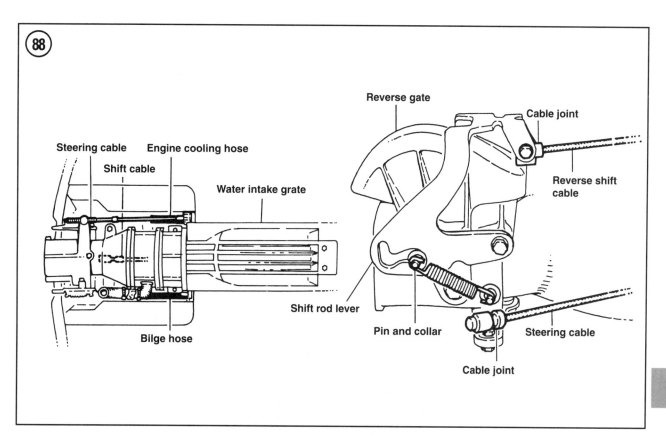

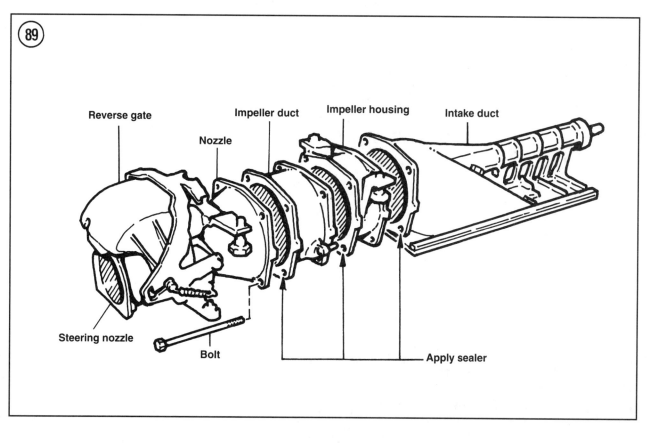

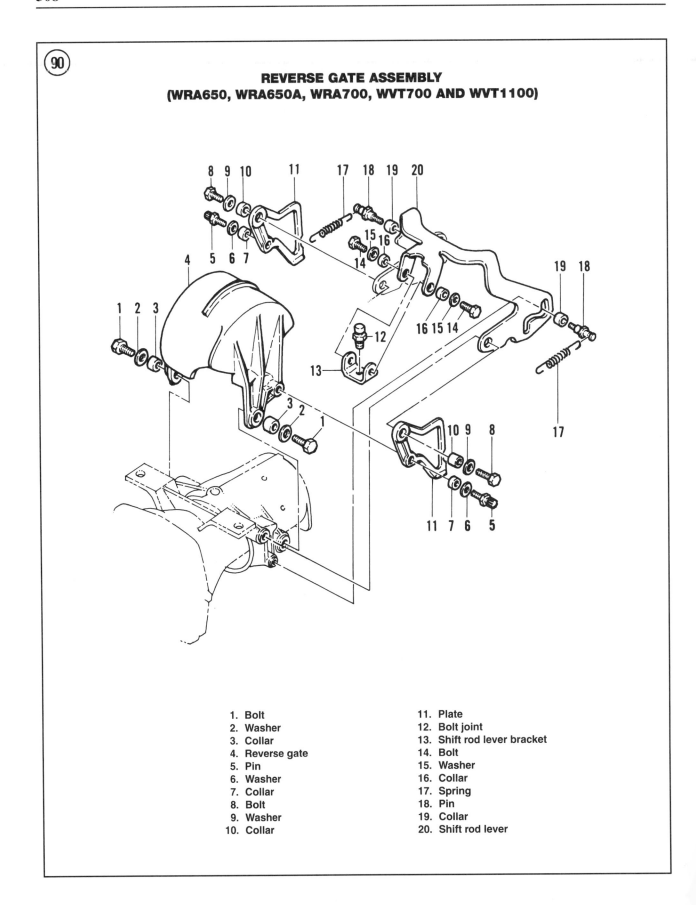

DRIVE TRAIN

a. Steering nozzle.

b. Nozzle.

c. Impeller duct.

d. Impeller housing.

e. Intake duct.

1. Before disassembling the jet pump, check the impeller tip-to-impeller housing clearance as described in this section. Performing this check before you disassemble the pump will allow you to purchase new parts, if required, and install them during pump reassembly.

2. *WRA650, WRA650A, WRA700, WVT700 and WVT1100*—Remove the reverse gate assembly as described below. Refer to **Figure 88** and **Figure 90** when performing the following procedure:

    a. Disconnect the reverse cable from the shift rod lever.

    b. Disconnect the shift rod lever springs from the side of the reverse gate.

    c. Unbolt and remove the plates from the shift rod lever. Account for the collars installed in the plates.

    d. Unbolt and remove the shift rod lever.

    e. Remove the pivot bolts securing the reverse gate to the steering nozzle. Remove the reverse gate.

3. Loosen the hose clamp and disconnect the hose from the impeller duct.

4. The jet pump assembly is held together with 4 long bolts and washers (**Figure 89**). Loosen and remove each of the bolts and washers.

5. Slide the intake duct (**Figure 89**) off of the impeller housing and remove the 2 dowel pins.

6. Separate and remove the nozzle, impeller duct and impeller housing. See **Figure 88**. Locate and remove the dowel pins from the housing mating surfaces (where used). See A, **Figure 91**. Store the dowel pins in a plastic bag, tagged for reassembly.

7. Remove the impeller (B, **Figure 91**) from the drive shaft as follows:

    a. Slide the Yamaha drive shaft holder (part No. YB-6049) onto the end of the drive shaft, then secure the holder horizontally in a vise.

CAUTION
The impeller uses left-hand threads. Turn the impeller clockwise to loosen it.

    b. Turn the impeller *clockwise* with a wrench (**Figure 92**) and remove it from the drive shaft.

    c. Remove the impeller washer (**Figure 93**).

8. Unbolt and remove the cap (**Figure 94**) from the impeller duct.

Inspection

The jet pump is a vital link in the overall performance of your water vehicle. Normal wear of the impeller housing and impeller can reduce jet pump thrust and top speed, even when the engine is running perfectly. Secondary thrust problems can be caused by obstructions in the pump, a damaged impeller housing or damaged impeller blades. Therefore, inspect the jet pump assembly carefully, and replace all worn or damaged parts.

1. Clean sealant residue from all affected surfaces.
2. Remove and discard all grease from inside the impeller duct cap.
3. Clean all parts in solvent and blow dry with compressed air, if available.
4. Check the impeller blades (**Figure 95**) for nicks or gouges. Smooth minor damage with abrasive paper or a file. It is especially important for the blade edges to be smooth. If there is major damage to the impeller, replace it.
5. Check the impeller housing bore (**Figure 96**) for deep scratches or excessive wear. If the pump case is damaged, either replace it or have the pump remachined by a qualified marine dealer.
6. Check the intake duct for cracks or damage. If the intake duct bearing assembly is not removed, check the tightness of the spacer mounting bolt.
7. Visually inspect the rubber seal on the intake duct (**Figure 97**). Replace the seal if it is worn or cracked.
8. *WRA650, WRA650A, WRA700, WVT700 and WVT1100*—Check the reverse gate and shift rod lever for cracks or damage. Check all of the collars used in the reverse gate assembly for corrosion, cracks or damage. Replace worn or damaged parts.

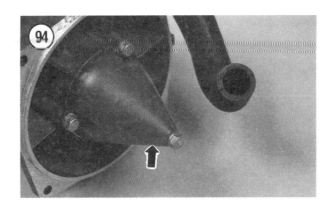

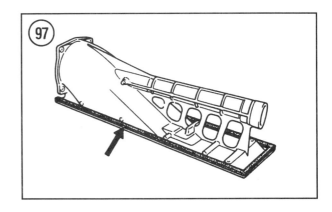

DRIVE TRAIN

9. *WRA650, WRA650A, WRA700, WVT700 and WVT1100*—Check the shift rod lever springs for fatigue or damage. Replace both springs as a set, as required.

10. The drive shaft is a press fit into the impeller duct bearings. Support the impeller duct and spin the drive shaft. The drive shaft should turn smoothly with no sign of bearing wear or damage. If the shaft turns roughly, suspect worn or damaged bearings. Service the impeller duct as described in this chapter.

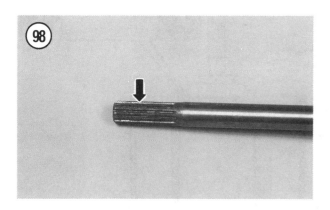

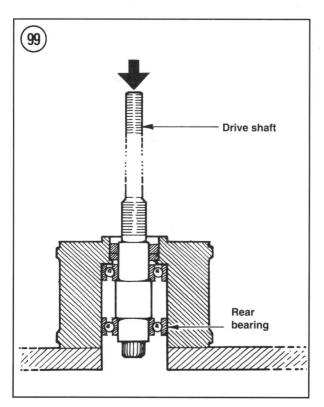

11. The intake duct is equipped with a spacer, 2 seals and a bushing. Because these components are inaccessible to normal observation, perform periodic removal to detect worn or damaged parts.

12. Inspect the drive shaft as follows:
    a. Inspect the drive shaft splines (**Figure 98**) for cracks, excessive wear or other damage.
    b. To check drive shaft runout, the shaft must be pressed out of the impeller duct assembly. Refer to *Drive Shaft Removal/Installation and Impeller Duct Bearing Replacement* in this chapter to service the drive shaft.

13. Check the nozzle and steering nozzle for cracks or other damage. Pivot the steering nozzle from side to side. It should move smoothly. If the steering nozzle does not pivot smoothly, remove the pivot bolts and check all parts for wear or damage. Replace worn or damaged parts as required. Wipe the pivot bolts with marine grease before installing them.

14. Inspect the intake grate, especially at the small arm attachment points, for cracks, breaks or other damage. Replace if necessary.

Drive Shaft Removal/Installation and Impeller Duct Bearing Replacement

The impeller duct is equipped with 2 bearings and 2 seals. The drive is a press fit in the impeller duct bearings. A hydraulic press is required to remove and install the drive shaft. When the drive shaft is pressed out, the rear impeller duct bearing should come out with the drive shaft. A bearing remover is required to remove the front impeller duct bearing. If you do not have the special tools, refer all impeller duct and drive shaft service to a Yamaha dealership or machine shop.

1. Support the impeller duct in a press so the drive shaft faces toward the press ram (**Figure 99**). Press the drive shaft and the rear impeller duct bearing out of the impeller duct. Support the

drive shaft so it cannot fall to the floor when it is pressed free.

2. Remove the impeller duct from the press and place it on a piece of wood laying flat across a workbench. Drive the front bearing out of the impeller duct with a bearing driver.

NOTE
The 2 impeller duct seals have different part numbers. Identify the seals immediately after removing them so they can be used to identify the new seals.

3. Turn the impeller duct over, then drive the 2 seals out with a bearing driver.

4. If necessary, remove the rear bearing from the drive shaft with a press or bearing splitter.

5. Mount the drive shaft on V-blocks at the points indicated in **Figure 100**. Turn the drive shaft slowly and check runout with a dial indicator at the test point indicated in **Figure 100**. Runout is the difference between the highest and lowest dial indicator reading. Replace the drive shaft if runout is greater than the specification given in **Table 2**.

6. Check the impeller duct bearings by rotating the inner bearing race by hand. The bearing should turn smoothly without excessive play. Replace bearings if necessary.

7. Clean the impeller duct in solvent, and dry it with compressed air, if available.

NOTE
After all parts are cleaned and inspected, assemble the impeller duct and install the drive shaft as described in the following steps.

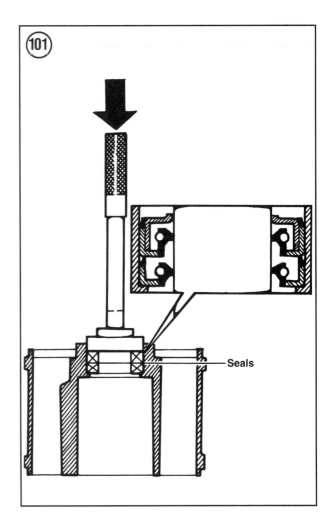

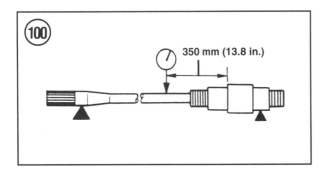

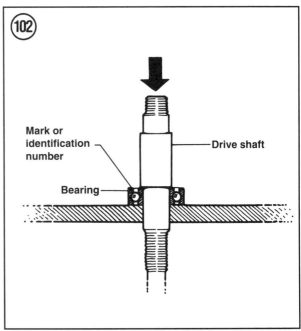

DRIVE TRAIN

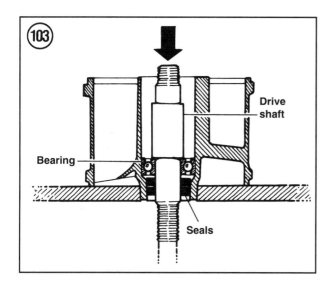

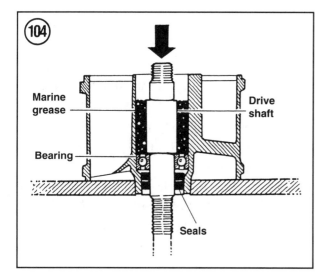

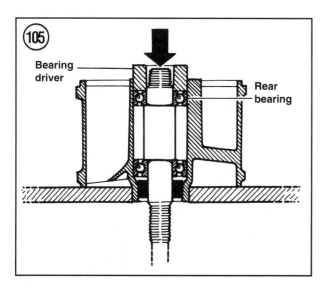

8. Install the impeller duct seals as follows:
   a. The 2 impeller duct seals have different part numbers. If you labeled the old seals during removal, use them to identify the new seals. If not, ask your parts vendor to identify the new seals by their part number. After identifying the seals, remove them from their wrapping and pack the seal lip cavity of each seal with marine grease.
   b. Install both seals with their *closed* side facing to the front of the impeller duct. See **Figure 101**.
   c. Install the lower seal by driving it into the impeller duct with a bearing driver. Drive the seal in squarely until it bottoms.
   d. Repeat substep c to install the upper seal. Drive the seal into the impeller duct until it bottoms against the lower seal.
9. Press the front impeller duct bearing onto the drive shaft as shown in **Figure 102**. Make sure the mark or identification number on the bearing faces outward.
10. Press the drive shaft/front bearing assembly into the impeller duct as shown in **Figure 103**. Press the shaft into the duct until the bearing bottoms.
11. Install the rear impeller duct bearing as follows:
    a. Support the impeller duct in a press bed so the drive shaft faces down as shown in **Figure 104**.
    b. Add marine grease to the impeller duct until the grease level is even with the rear bearing seat on the drive shaft.
    c. Align the rear bearing with the impeller duct and place the Yamaha bearing driver (part No. YB-34473) onto the bearing (**Figure 105**). Be sure the manufacturer's mark on the bearing faces outward.

CAUTION
The rear bearing is installed over the drive shaft and against the impeller duct housing at the same time. To prevent

*bearing damage during installation, the Yamaha tool (substep b) or equivalent must be used. This tool is designed to apply pressure against both races as the bearing is being installed. If the bearing is installed improperly, the balls will be forced against the inner or outer bearing race and damage it. Refer to **Ball Bearing Replacement** in Chapter One for additional information.*

d. Bring the press ram into contact with the bearing driver (**Figure 105**) and press the bearing into the housing until it bottoms.

e. Release the press ram and remove the bearing driver.

12. Turn the drive shaft by hand. It should turn smoothly. A rough turning or tight bearing indicates damage.

Intake Duct
Bushing and Oil Seal Replacement

The impeller housing is equipped with a bushing, 2 seals (same part number) and a spacer. The spacer is secured with a bolt and washer (**Figure 106**).

1. Loosen and remove the spacer bolt and washer from the intake duct.

2. Drive the spacer, seals and bearing out of the intake duct at the same time with a bearing driver and extension.

3. Remove all sealer residue from the intake duct assembly. Then clean the impeller duct in solvent and allow to dry thoroughly.

4. Visually inspect the bushing for excessive wear, cracks or other damage.

5. Inspect the impeller duct for cracks or damage.

NOTE
After all parts are cleaned and inspected, assemble the intake duct bushing and seal assembly as follows.

6. Place the intake duct on a workbench and block the front end to prevent it from sliding forward when installing the bushing and seals.

7. Wipe the bushing with marine grease. Then drive the bushing into the intake duct using a bearing driver and extension (**Figure 107**). Drive the bushing into the housing until it bottoms.

8. Install the intake duct seals as follows:

a. Pack the seal lip cavity of each seal with marine grease.

b. Install the front seal so its *closed* side faces the front of the housing as shown in **Figure 108**. Install the seal by driving it into the intake duct with a bearing driver. Drive the seal in squarely until it bottoms against the bushing.

c. Install the rear seal so its *open* side faces the front of the housing as shown in **Figure 108**. Again, use a bearing driver to drive the seal into the intake duct until it bottoms.

d. Install the spacer into the intake duct so that it faces the direction shown in **Figure 106**. Then align the hole in the spacer with the hole in the side of the intake duct. Wipe the spacer threads with Loctite 242 and thread the bolt (with its washer) into the spacer. Tighten the bolt securely.

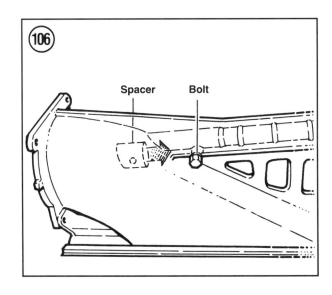

DRIVE TRAIN

Pump Case Assembly

Refer to **Figure 85** or **Figure 86** for this procedure.

1. Apply marine grease to a new O-ring and install the O-ring onto the impeller duct cap. Fill the cap with marine grease and install the cap to the impeller duct. Secure it with its mounting bolts and washers. Apply Loctite 242 to the threads of the bolts and torque the bolts to the specification given in **Table 3**. See **Figure 94**.
2. Install the impeller as follows:

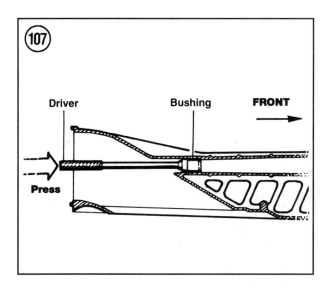

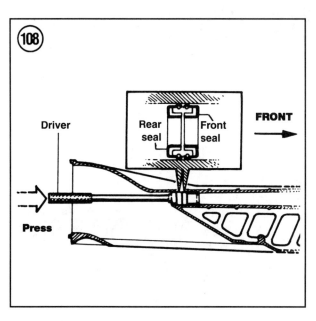

NOTE
If you checked impeller clearance before disassembly and it was incorrect, you will have to install a new impeller, impeller housing or both. If you did not check impeller clearance, install the impeller and impeller housing as described in this section, then check clearance as described in this chapter.

a. Install the impeller washer (**Figure 93**).

CAUTION
*The impeller uses left-hand threads. Turn the impeller **counterclockwise** to tighten it.*

b. Wipe the impeller threads with Loctite 242, then slide the impeller onto the drive shaft. Thread the impeller *counterclockwise* onto the drive shaft to install it.
c. Slide the Yamaha drive shaft holder (part No. YB-6049) onto the end of the drive shaft, then secure the holder horizontally in a vise.
d. Tighten the impeller with a wrench (**Figure 92**) to the torque specification listed in **Table 3**.
e. Turn the drive shaft to make sure it turns freely.
f. Temporarily install the impeller housing and check impeller clearance as described in this chapter.
g. When impeller clearance is correct, proceed to Step 3.

3. Install the dowel pins in the impeller duct pin holes. Then apply Yamabond No. 4 to the mating surfaces of the impeller housing, impeller duct and nozzle (**Figure 89**).
4. Assemble the nozzle, impeller duct and impeller housing (**Figure 89**). Be sure all dowel pins correctly engage their mating holes.
5. When all of the subassemblies are assembled, install the 4 through bolts and washers (**Figure 89**) hand-tight. Then tighten the bolts evenly in a crisscross pattern.

6. *WRA650, WRA650A, WRA700, WVT700 and WVT1100*—Install the reverse gate assembly (**Figure 88** and **Figure 90**) as follows:

NOTE
If the steering and shift cable ball joints were removed from their mounting positions, apply Loctite 242 (blue) to their threads before reinstalling and tightening them.

a. Apply marine grease to the reverse gate collars and pivot bolts.
b. Fit the reverse gate into position and install its collars, washers and pivot bolts. Tighten the pivot bolts securely.
c. Install the shift rod lever. Apply Loctite 242 to the shift rod lever bolts and tighten securely.
d. Apply marine grease to the shift rod lever plate collars. Install the collars into the plates. Then install the bolts and washers and tighten securely.
e. Reconnect the shift rod lever springs (**Figure 88**).
f. Reconnect the reverse cable to the shift rod lever (**Figure 88**).

7. Reconnect the hose to the impeller duct and secure it with its hose clamp.

Jet Pump Installation

1. Remove all of the old sealant from the pump intake area.

2. Install the bilge hoses to their outlet nozzles. Secure the hoses with new plastic ties (**Figure 109**).

3. Make sure the large bilge hose is secured to the impeller duct with a hose clamp.

4. Wipe the end of the drive shaft with marine grease, then slide the jet pump forward into position in the hull. Be sure the splines of the drive shaft engage the shaft of the intermediate housing.

NOTE
Before installing the jet pump mounting bolts, check the threads on each bolt for corrosion or damage. Clean the threads thoroughly, or if necessary, replace bolts before installing them. Likewise, the threaded inserts installed in the hull should be cleaned of all corrosion. It is also important to note that some models use an uncommon M9 × 1.25 mm thread on the jet pump mounting bolts. Do not attempt to use an 8 or 10 mm bolt if you

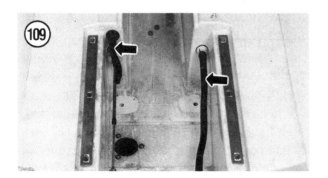

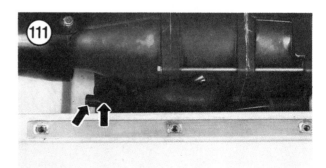

DRIVE TRAIN

are working on such a model. When purchasing replacement bolts, confirm bolt diameter and pitch as described under **Fasteners** *in Chapter One.*

5. Loosely install the 4 pump mounting bolts with flat washers and lockwashers. Insert the alignment shims through the pump mounting bolts at their original mounting position.

6. Tighten the jet pump mounting bolts with a torque wrench to the torque specification listed in **Table 3**. Do not overtighten the bolts.

7. Check the foam sealing strips placed on the bottom side of the hull for cracks or damage (A, **Figure 110**); replace strips if damaged or missing.

8. Apply marine-grade RTV sealant to the sealing surface of the intake grate and install the intake grate. Apply Loctite 242 to the intake grate bolts and torque the bolts to the specification given in **Table 3**.

9. Reconnect the engine cooling and bilge hoses (**Figure 85** or **Figure 86**).

10. Apply marine grease to the steering cable ball joint on the steering nozzle, and reconnect the steering cable. Slide its spring-loaded sleeve forward and place the connector onto the ball. Release the sleeve and check to make sure the cable is attached properly. Operate the handlebar to make sure the steering nozzle pivots correctly.

11. *WRA650, WRA700, WVT700 and WVT1100*—Reconnect the reverse shift cable to the shift rod lever. Slide the spring-loaded sleeve forward and insert the connector onto the ball. Release the sleeve and check that the cable is secured properly. Operate the shift lever to make sure the reverse nozzle operates properly. Refer to Chapter Three for cable adjustment.

12. On models with a water valve body:
    a. Align the valve body with the ride plate bolt holes. Install and tighten the bolts securely. See **Figure 87**.
    b. Turn the ride plate over and reconnect the 2 bilge hoses (**Figure 111**) to the water valve body. Secure each hose with new a plastic tie or with a hose clamp as necessary. Be sure the bilge hoses are not kinked and that they are properly routed through the hull.

13. On models with a speed sensor, reinstall the speed sensor to the ride plate.

14. Position the ride plate on the hull. Apply Loctite 242 (blue) to the threads of the ride plate bolts and torque the bolts to the specification given in **Table 3**.

15. Turn the vehicle upright and perform the following:
    a. Reinstall the oil tank as described in Chapter Ten.
    b. Refill the fuel tank.
    c. Reinstall and reconnect the battery as described in Chapter Nine.

Impeller Clearance Check and Adjustment

The impeller clearance can be checked without having to remove the jet pump assembly.

1. Remove the intake grill as described under *Jet Pump Removal* in this chapter.

2. Measure the impeller-to-impeller housing clearance with a feeler gauge (**Figure 112**).

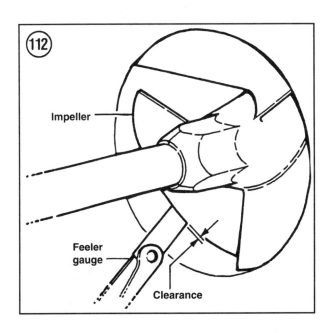

CHAPTER SEVEN

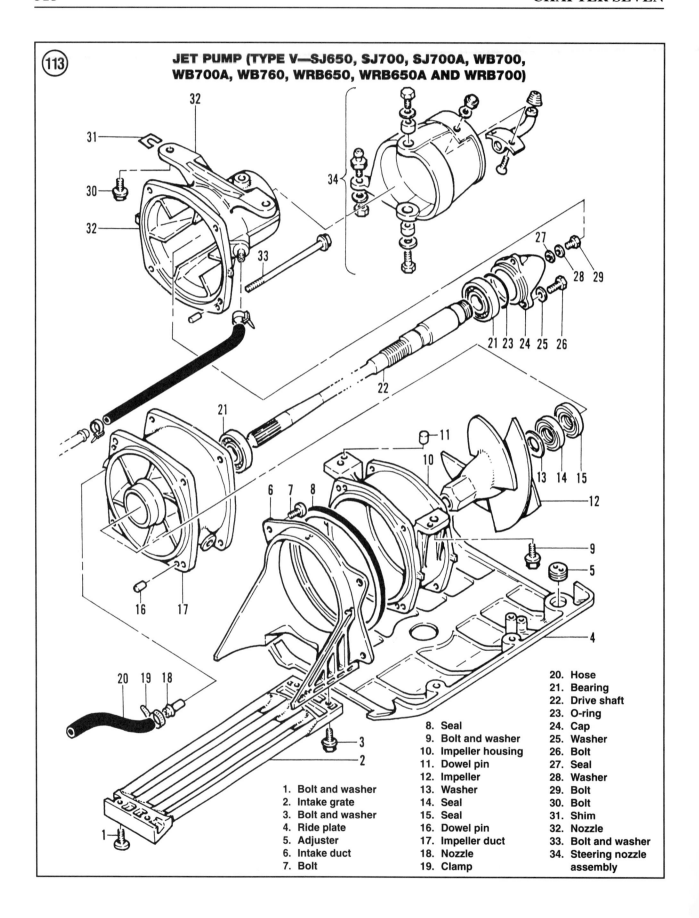

DRIVE TRAIN

Measure the clearance at each of the 4 impeller blade points. The correct clearance is listed in **Table 2**. If the clearance at any one blade point is incorrect, impeller housing or both parts.

3. Reinstall the intake grate as described in this chapter.

JET PUMP
(FX700, SJ650, SJ700, SJ700A, WB700, WB700A, WB760, WRB650, WRB650A AND WRB700)

The jet pump is an axial-flow, single-stage design. The jet pump assembly consists of the impeller housing, impeller, drive shaft, nozzle and steering nozzle. The drive shaft is supported by 3 sealed ball bearings. A single bearing is used in the intermediate housing and 2 bearings are used in the impeller duct housing. Sealed bearings do not require periodic lubrication.

The drive train consists of the crankshaft-to-drive shaft connection, drive shaft, impeller and jet pump. As the impeller rotates, water is drawn into the pump through the intake area at the bottom of the jet pump housing. Water is then forced through the impeller housing where it exits through the steering nozzle assembly. As engine speed increases, a higher volume of water is forced through the jet pump and the vehicle's speed increases. Stationary vanes used in the impeller housing help control water flow for improved thrust and efficiency. A steering cable, attached from the steering column to the steering nozzle, controls steering movement.

The pump is manufactured with very little clearance between the impeller and the impeller housing. As the pump wears from normal use, the clearance between the impeller and housing will increase, thus decreasing pump thrust. Therefore, inspect the impeller and housing according to the maintenance schedule in Chapter Three. Replace the housing if it is excessively worn.

Jet Pump Identification

The jet pumps used on the models described in this section are basically the same with minor differences. It is important to pay particular attention to the location and order of parts during service. The jet pumps are identified as follows:

The Type V jet pump (**Figure 113**) is used on the following models:

a. SJ650, SJ700 and SJ700A.
b. WRB650 and WRB700.
c. WB700, WB700A and WB760.

The Type VI jet pump (**Figure 114**) is used on the FX700 models.

Service Note

The following sections describe removal and complete overhaul of Type V and VI jet pump assemblies. If measuring impeller housing clearance is the only service required at this time, refer to *Impeller Housing Clearance Check and Adjustment* in this chapter. Impeller housing clearance can be checked with the jet pump assembly installed in the hull; jet pump removal is not required.

Jet Pump/Impeller Removal

Refer to **Figure 113** or **Figure 114** for this procedure.

1. Remove the battery to prevent acid spillage. See Chapter Nine.
2. Drain the fuel tank and store the fuel in a container approved for gasoline storage.

CAUTION
If the exhaust system is installed on the engine, do not turn the craft onto its right side when performing Step 3. Water in the exhaust system may drain into the engine's exhaust ports and cause serious engine damage.

320 CHAPTER SEVEN

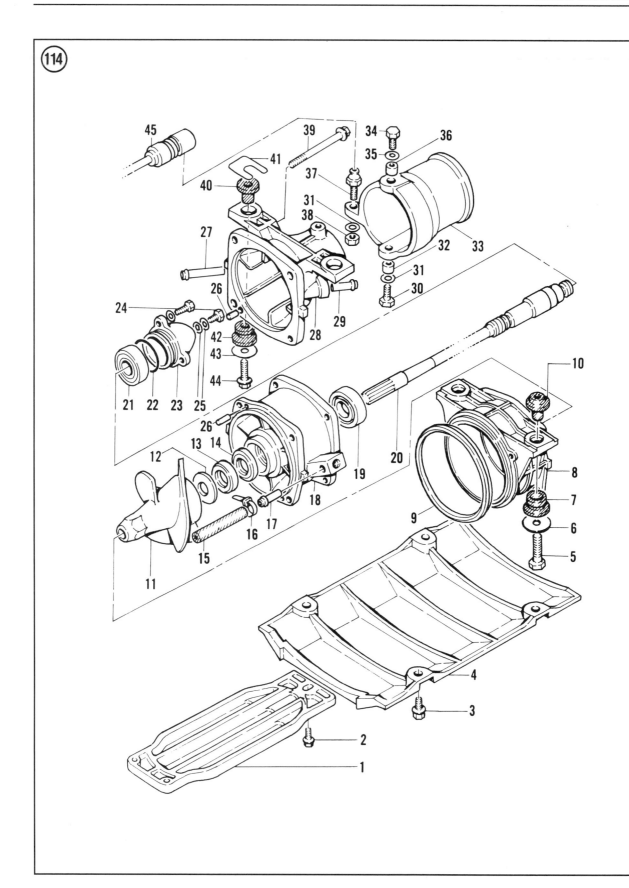

DRIVE TRAIN

JET PUMP (TYPE VI—FX700)

1. Intake grate
2. Bolt and washer
3. Bolt and washer
4. Ride plate
5. Bolt
6. Washer
7. Grommet
8. Impeller housing
9. Seal
10. Collar
11. Impeller
12. Washer
13. Seal
14. Seal
15. Hose
16. Clamp
17. Nozzle
18. Impeller duct
19. Bearing
20. O-ring
21. Bearing
22. O-ring
23. Cap
24. Bolt
25. Washers
26. Dowel pin
27. Nozzle
28. Output nozzle
29. Nozzle
30. Bolt
31. Washer
32. Bushing
33. Steering nozzle
34. Bolt
35. Washer
36. Bushing
37. Cable connector stud
38. Nut
39. Bolt
40. Collar
41. Shim
42. Grommet
43. Washer
44. Bolt
45. Steering cable

3. Turn the fuel petcock OFF and turn the watercraft onto its *left side*. Support the craft securely on a suitable stand. To protect the hull from damage, support the craft with 2 discarded motorcycle tires, a heavy blanket or similar object that can accept the craft's weight.

NOTE
*Washers placed between the hull and ride plate are used for ride plate adjustment. See **Ride Plate Adjustment** in this chapter.*

4. Unbolt and remove the ride plate (B, **Figure 115**) from the bottom of the hull.

5. Unbolt and remove the intake grate (C, **Figure 115**).

6. Detach the steering cable from the ball joint on the steering nozzle. Slide the cable's spring-loaded sleeve forward and pull the connector off of the ball (**Figure 116**).

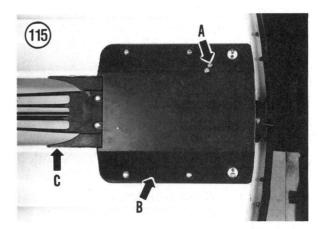

7. Disconnect the water hose from the jet pump (**Figure 117**).

8. Disconnect the bilge hoses (**Figure 118**) from the jet pump.

> *CAUTION*
> *The threads used on the jet pump mounting bolts corrode easily. Because these bolts thread into an insert installed in the hull, a corroded bolt may force the insert to twist and turn in its mounting position, allowing the bolt and insert to turn together. If the insert breaks away, you must access the insert and hold it with a pair of locking pliers or similar tool to remove the bolt. Before loosening the jet pump mounting bolts, spray **each bolt** with WD-40 or a thread penetrating lubricant. Allow sufficient time for the lubricant to work its way down the bolt threads. When attempting to loosen the bolt, do not use an air impact wrench as the power of the tool can rip the insert away from its mounting position. After applying a penetrating lubricant, loosen the bolt carefully with a socket, ratchet and suitable extension.*

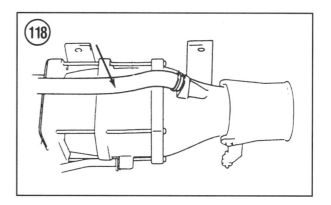

9. Apply a penetrating lubricant to each jet pump mounting bolt. Allow plenty of time for the lubricant to work its way down the bolt threads. Then carefully loosen the bolts with a ratchet. *On FX700 models*, remove the collars along with the jet pump mounting bolts.

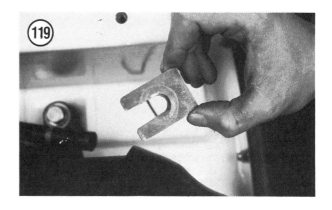

> *CAUTION*
> *When removing the jet pump, label the shims (**Figure 119**) positioned underneath the pump mounting lugs. These shims must be reinstalled in exactly the same position for the pump and impeller to align properly.*

10. Lift the jet pump from the hull until it clears the dowel pins (collars on FX700 models) in the impeller housing, then remove the jet pump assembly by pulling it rearward (**Figure 120**).

DRIVE TRAIN   323

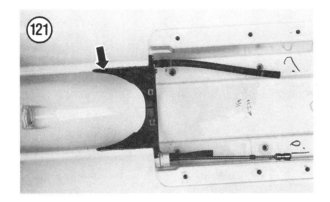

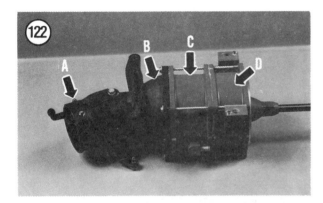

NOTE
*Do not remove the intake duct (**Figure 121**) unless it is damaged or if its sealing surface has loosened against the hull.*

11. Remove the Phillips screws holding the intake duct (**Figure 121**) to the hull. Then carefully pull the intake duct away from the hull, breaking the silicone seal as you remove it.

Pump Case Disassembly

The jet pump assembly consists of a number of separate subassemblies. To prevent confusion during the following service procedure, refer to **Figure 122** and the following identification list:
   a. Steering nozzle.
   b. Nozzle.
   c. Impeller duct.
   d. Impeller housing.

1. Before disassembling the jet pump, check the impeller tip-to-impeller housing clearance as described in this section. Performing this check before you disassemble the pump will allow you to purchase new parts, if required, and install them during pump reassembly.

2. Remove the 2 dowel pins (A, **Figure 123**) from the impeller housing. *On FX700 models*, remove the 2 collars.

3. Remove and discard the intake duct seal (B, **Figure 123**).

4. Remove the 4 jet pump bolts and washers (**Figure 124**) and remove the nozzle.

5. Remove the impeller (**Figure 125**) from the drive shaft as follows:

a. Slide the Yamaha drive shaft holder (part No. YB-6049) onto the end of the drive shaft, then secure the holder horizontally in a vise (**Figure 126**).

CAUTION
*The impeller uses left-hand threads. Turn the impeller **clockwise** to loosen it.*

b. Turn the impeller *clockwise* with a wrench (**Figure 126**) and remove it from the drive shaft.

c. Remove the washer from the drive shaft (**Figure 127**).

6. Unbolt and remove the cap (**Figure 128**) from the impeller duct.

7. Separate the impeller duct and impeller housing using a wide blade flat-tipped screwdriver at the pry point shown in **Figure 129**. Do not pry between the mating surfaces.

8. Remove the dowel pins from between the impeller duct and impeller housing.

Cleaning and Inspection

The jet pump is a vital link in the overall performance of your water vehicle. Normal wear of the impeller housing and impeller can reduce jet pump thrust and top speed, even if the engine is running perfectly. Secondary thrust problems can be caused by obstructions in the pump, a damaged impeller housing or damaged impeller blades. Inspect the jet pump assembly carefully and replace all worn or damaged parts.

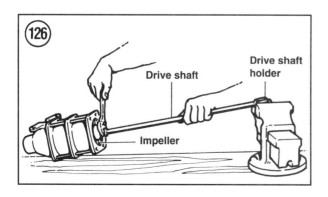

DRIVE TRAIN

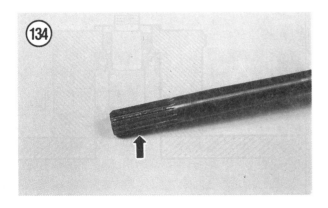

1. Clean sealant residue from all affected surfaces.
2. Remove and discard all grease from inside the impeller duct cap (**Figure 130**).
3. Clean all parts in solvent and blow them dry with compressed air, if available.
4. Check the impeller blades (**Figure 131**) for nicks or gouges; smooth minor damage with abrasive paper or a file. It is important for the blade edges to be smooth. If there is major damage to the impeller, replace it.
5. Check the impeller housing bore (**Figure 132**) for excessive wear or scratches. If the housing bore is worn or damaged, either replace it or have it remachined by a qualified marine dealer.
6. The drive shaft (A, **Figure 133**) is a press fit into the impeller duct bearings (B, **Figure 133**). Support the impeller duct and turn the drive shaft by hand. The drive shaft should turn smoothly with no sign of bearing wear or damage. If the shaft turns roughly, suspect worn or damaged bearings; service the impeller duct as described in this chapter.
7. Inspect the drive shaft as follows:
    a. Inspect the drive shaft splines (**Figure 134**) for cracks, excessive wear or other damage.
    b. The drive shaft must be pressed out of the impeller duct to check runout; refer to *Drive Shaft Removal/Installation and Impeller Duct Bearing Replacement* in this chapter to service the drive shaft.
8. Check the output nozzle and steering nozzle (**Figure 135**) for cracks or other damage. Pivot

the steering nozzle from side to side. It should move smoothly without roughness or tightness. If the steering nozzle does not pivot smoothly, remove the pivot bolts and check all parts for wear or damage. Replace worn or damaged parts as required. Wipe the pivot bolts with marine grease before installing them.

9. Inspect the intake grill, especially where the small arms attach at opposite sides for cracks, breaks or other damage. Replace if necessary.

Drive Shaft Removal/Installation and Impeller Duct Bearing Replacement

The impeller duct is equipped with 2 bearings and 2 seals. The drive shaft (A, **Figure 133**) is a press fit into the impeller duct bearings. A hydraulic press is required to remove and install the drive shaft. When the drive shaft is pressed-out, the rear impeller duct bearing should come out with the drive shaft. A bearing remover is required to remove the front impeller duct bearing. If you do not have the special tools, refer all impeller duct and drive shaft service to a dealer or machine shop.

1. Support the impeller duct in a press so the drive shaft faces toward the press ram (**Figure 136**). Press the drive shaft and the rear bearing out of the impeller duct. Support the drive shaft so it cannot fall to the floor when it is pressed out of the impeller duct bearing.

2. Remove the impeller duct from the press and remove the front bearing with a blind bearing remover.

NOTE
The 2 impeller duct seals have different part numbers. Identify the seals immediately after removing them so they can be used to identify the new seals.

3. Turn the impeller duct over and drive the 2 seals out of the impeller duct with a bearing driver.

4. Remove the rear bearing from the drive shaft with a press or bearing splitter.

5. Mount the drive shaft on V-blocks at the points indicated in **Figure 137**. Turn the drive shaft slowly and check runout with a dial indicator at the test point indicated. Runout is the difference between the highest and lowest dial indicator reading. Replace the drive shaft if the

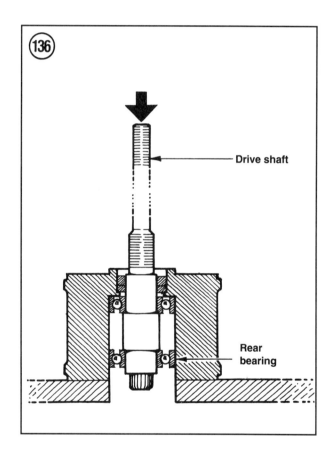

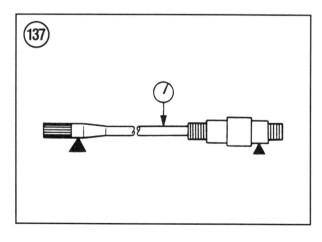

DRIVE TRAIN 327

runout is greater than the specification given in **Table 2**.

6. Check the impeller duct bearings. Rotate the inner bearing race by hand; the bearing should turn smoothly. Replace bearings as a set if any one bearing is damaged.

7. Clean the impeller duct in solvent and dry with compressed air, if available.

NOTE
After all parts are cleaned and inspected, assemble the impeller duct and install the drive shaft as described in the following steps.

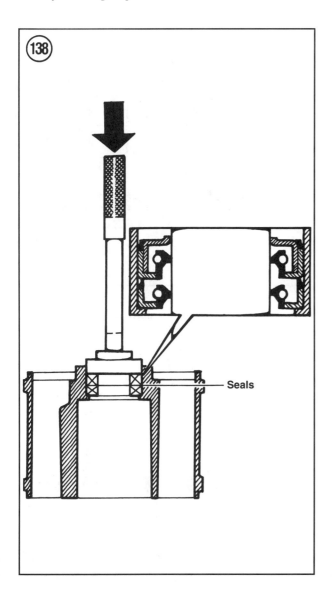

8. Install the impeller duct seals as follows:
   a. Pack the seal lip cavity of each seal with marine grease.
   b. Install both seals with their *closed* side facing to the front of the impeller duct. See **Figure 138**.
   c. Install the lower seal by driving it into the impeller duct using a driver with the same outside diameter as the seal. Drive the seal in squarely until it bottoms.
   d. Repeat substep c to install the upper seal. Drive the seal into the impeller duct until it bottoms against the lower seal.

9. Press the front impeller duct bearing onto the drive shaft. Be sure the manufacturer's marks on the bearing face outward, toward the threaded end of the drive shaft as shown in **Figure 139**.

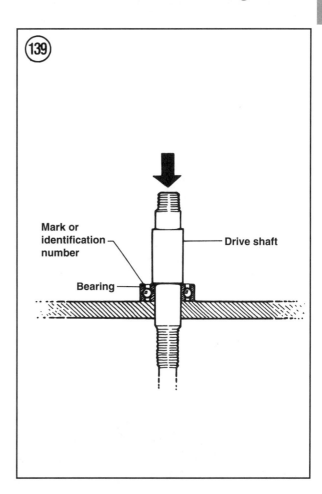

10. Press the drive shaft/front bearing assembly into the impeller duct as shown in **Figure 140**. Press the shaft into the duct until the bearing bottoms.

11. Install the rear impeller duct bearing as follows:

   a. Support the impeller duct in a press bed so the drive shaft faces down as shown in **Figure 141**.

   b. Add marine grease to the impeller duct until the grease level is even with the rear bearing seat on the drive shaft.

   c. Align the rear bearing with the impeller duct. Be sure the manufacturer's marks on the bearing face outward, toward the threaded end of the drive shaft. Then place the Yamaha bearing driver (part No. YB-34474) onto the bearing (**Figure 142**).

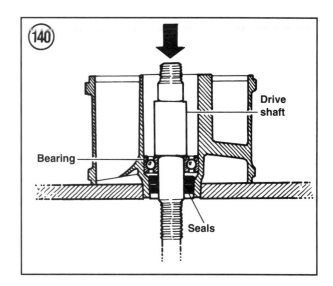

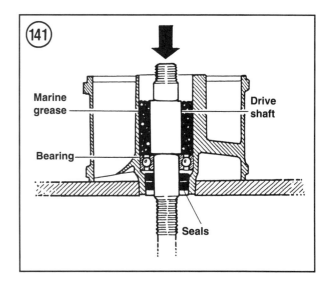

CAUTION
*The rear bearing is installed over the drive shaft and against the impeller duct housing at the same time. To prevent bearing damage during installation, the Yamaha tool (part No. YB-34474) or equivalent must be used. This tool is designed to apply pressure against both races as the bearing is being installed. If the bearing is installed improperly, the balls will be forced against the inner or outer bearing race and damage it. Refer to **Bearing Replacement** in Chapter One for additional information.*

   d. Bring the press ram into contact with the bearing driver (**Figure 142**) and press the bearing into the housing until it bottoms.

   e. Release the press ram and remove the bearing driver.

12. Turn the drive shaft by hand. It must turn smoothly. A rough turning or sticking bearing indicates bearing damage.

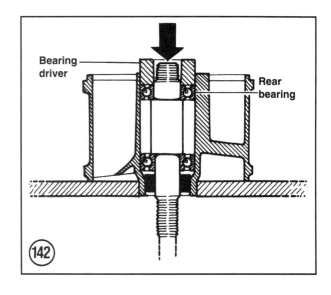

DRIVE TRAIN

Pump Case Assembly

Refer to **Figure 113** or **Figure 114** for this procedure.

1. Apply marine grease to a new O-ring, and fit a new O-ring onto the impeller duct cap. Fill the cap with marine grease (**Figure 130**) and install the cap onto the impeller with the mounting bolts and washers. Apply Loctite 242 to all bolt threads and tighten the bolts securely. See **Figure 128**.

2. Install the impeller as follows:

NOTE
If you checked impeller clearance before disassembly and it was incorrect, install a new impeller, impeller housing or both. If you did not check impeller clearance, install the impeller and impeller housing as described in this section, then check clearance as described in this chapter.

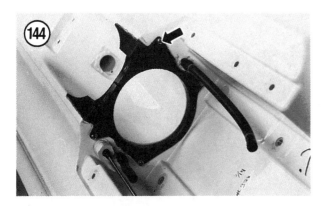

 a. Install the impeller washer over the drive shaft making sure the flat washer side faces toward the impeller. See **Figure 143**.

CAUTION
The impeller uses left-hand threads. Turn the impeller counterclockwise to tighten it.

 b. Wipe the impeller threads with Loctite 242, then slide the impeller onto the drive shaft. Turn the impeller *counterclockwise* to install it.
 c. Slide the Yamaha drive shaft holder (part No. YB-6049) onto the end of the drive shaft, then secure the holder horizontally in a vise (**Figure 126**).
 d. Tighten the impeller to the torque specification listed in **Table 3**.
 e. Turn the drive shaft to make sure it turns freely.
 f. Temporarily install the impeller housing and check impeller clearance as described in this chapter.
 g. When impeller clearance is correct, proceed to Step 3.

3. Install the mating surface dowel pins.
4. Apply Yamabond No. 4 between the impeller housing and intake duct mating surfaces and then join the 2 parts together. Make sure the dowel pins align with their respective mating holes.
5. Apply Yamabond No. 4 between the impeller duct and nozzle mating surfaces and join them together. Wipe the threads of the 4 housing bolts with Loctite 242, then install the bolts and washers into the housing finger-tight. Final tightening of bolts will take place during jet pump installation.

Jet Pump Installation

1. If the intake duct (**Figure 144**) was removed, apply silicone sealant to its hull side mating

surface and press it into position against the hull. Install the intake duct Phillips screws and tighten them securely.

2. Check the bilge hoses for damage or looseness. Replace damaged hoses as required. Secure hoses in place with new cable ties (**Figure 145**).

3. Install the 2 impeller housing mounting arm dowel pins (A, **Figure 146**).

4. Install a new intake duct seal onto the impeller housing (B, **Figure 146**).

5. Install the jet pump:
   a. Wipe the drive shaft splines with marine grease.
   b. FX700: Install the collars in the mount dogs on the impeller housing and nozzle. See **Figure 114.**
   c. Place the jet pump into the hull, aligning the drive shaft (A, **Figure 147**) with the intermediate shaft. Rotate the drive shaft as required to mate parts after alignment.
   d. Position the jet pump in the hull, making sure to engage the dowel pins (collars in FX700 models) in the impeller housing with the pin holes in the hull. Also make sure the pump seal (B, **Figure 147**) mates against the intake duct (C, **Figure 147**) evenly.

NOTE
*Before installing the jet pump mounting bolts, check the threads on each bolt for corrosion or damage. Clean the threads thoroughly, or if necessary, replace bolts before installing them. Likewise, clean the threaded inserts installed in the hull of all corrosion. It is also important to note that some models use an uncommon M9 × 1.25 mm thread on the jet pump mounting bolts. Do not attempt to use an 8 or 10 mm bolt if you are working on such a model. When purchasing replacement bolts, confirm bolt diameter and pitch as described under **Fasteners** in Chapter One.*

   e. Wipe the threads of each pump mounting bolt with Loctite 242.
   f. Install the pump mounting bolts by hand to make sure they do not cross thread. Then insert the alignment shims (**Figure 119**) through the bolts at their original mounting position. On *FX700 models*, install the collar along with each mounting bolt.

6. The proper jet pump housing and pump mounting tightening sequence and torque are critical to prevent warpage and leaks and to ensure the proper alignment of pump components. In addition, the tightening sequence is not the same if the jet pump assembly was completely disassembled rather than simply removed and reinstalled in the hull. Therefore, tighten the pump housing and mounting fasteners in the following sequence to the torque listed.

7A. *SJ650, WRB650, WRB650A and WRB700*:
   a. If the jet pump assembly was disassembled, tighten the 2 front pump mounting bolts to 17 N•m (12 ft.-lb.), then tighten the 4 pump housing bolts to 11 N•m (97 in.-lb.). Next,

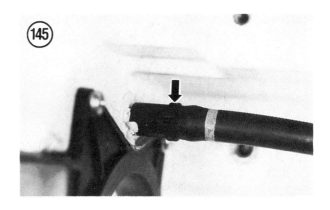

DRIVE TRAIN

tighten the 2 rear pump mounting bolts to 17 N•m (12 ft.-lb.). Last, tighten the 2 front pump mounting bolts, then the 2 rear mounting bolts to 34 N•m (25 ft.-lb.).

b. If the pump was not disassembled, tighten the front, then the rear pump mounting bolts to 17 N•m (12 ft.-lb.). Then, tighten all 4 pump mounting bolts to 34 N•m (25 ft.-lb.).

7B. *WB700 and WB700A*:

a. If the jet pump was disassembled, first tighten the 2 front pump mounting bolts to 17 N•m (12 ft.-lb.), then tighten all 4 pump housing bolts to 10 N•m (88 in.-lb.). Next, tighten the 2 rear pump mounting bolts to 17 N•m (12 ft.-lb.). Last, tighten the 2 front, then the 2 rear mounting bolts to 34 N•m (25 ft.-lb.).

b. If the jet pump was not disassembled, tighten the 2 front, then the 2 rear pump mounting bolts to 17 N•m (12 ft.-lb.). Then, tighten all 4 pump mounting bolts to 34 N•m (25 ft.-lb.).

7C. *SJ700 and SJ700A*:

a. If the jet pump was disassembled, first tighten the 2 front, then the 2 rear pump mounting bolts to 17 N•m (12 ft.-lb.). Next, tighten all 4 pump housing bolts to 10 N•m (88 in.-lb.). Last, tighten the 4 pump mounting bolts to 34 N•m (25 ft.-lb.).

b. If the jet pump was not disassembled, tighten the 2 front, then the 2 rear pump mounting bolts to 17 N•m (12 ft.-lb.). Then, tighten all 4 pump mounting bolts to 34 N•m (25 ft.-lb.).

7D. *FX700*:

a. If the jet pump was disassembled, first tighten the 2 front, then the 2 rear pump mounting bolts to 8 N•m (71 in.-lb.). Then, tighten all 4 pump housing bolts to 16 N•m (11 ft.-lb.). Next, tighten all 4 pump mounting bolts to 16 N•m (11 ft.-lb.).

b. If the jet pump was not disassembled, tighten the 2 front, then the 2 rear pump mounting bolts to 8 N•m (71 in.-lb.). Then, tighten all 4 pump mounting bolts to 16 N•m (11 ft.-lb.).

7E. *WB760*:

a. If the jet pump was disassembled, first, tighten the 2 front pump mounting bolts to 17 N•m (12 ft.-lb.). Next, tighten all 4 pump housing bolts to 20 N•m (15 ft.-lb.), then tighten the 2 rear pump mounting bolts to 17 N•m (12 ft.-lb.). Last, tighten all 4 pump mounting bolts to 34 N•m (25 ft.-lb.).

b. If the jet pump was not disassembled, tighten the 2 front, then the 2 rear pump mounting bolts to 17 N•m (12 ft.-lb.). Then, tighten all 4 pump mounting bolts to 34 N•m (25 ft.-lb.).

8. Reconnect the water inlet hose to the impeller duct. (**Figure 117**).

9. Reconnect the bilge hose to the impeller duct (**Figure 118**).

10. Apply marine grease to the ball on the steering nozzle and reconnect the steering cable to the steering nozzle (**Figure 116**). Slide the cable's spring-loaded sleeve forward and insert the connector onto the ball. Release the sleeve and check to make sure the cable is attached properly. Operate the handlebar to make sure the steering nozzle pivots correctly.

11. On models equipped with a water valve body, align the valve body with the ride plate bolt holes. Install and tighten the bolts securely. See A, **Figure 115**.

12. Turn the ride plate over and reconnect the 2 bilge hoses to the water valve body, on models so equipped. Secure each hose with new a plastic tie or with a hose clamp as necessary.

CAUTION
If a bilge hose should become kinked, the bilge system will not operate. Check bilge hose routing and reposition as required.

NOTE
*Washers installed between the hull and ride plate are for ride plate adjustment. Install these washers, if used, as described under **Ride Plate Adjustment** in this chapter.*

13. Position the ride plate (B, **Figure 115**) onto the hull. Coat the ride plate bolts with Loctite 242, then install and tighten the bolts securely.

14. Install the water intake grate (C, **Figure 115**). When the intake grate is fully seated, install and tighten its mounting bolts.

15. Turn the vehicle upright and perform the following:
   a. Refill the fuel tank.
   b. Reinstall and reconnect the battery. See Chapter Nine.

Impeller Clearance Check and Adjustment

The impeller clearance can be checked without removing the jet pump assembly.

1. Remove the intake grate as described under *Jet Pump Removal* in this chapter.

2. Measure the impeller-to-impeller housing clearance with a feeler gauge (**Figure 148**). Measure the clearance at each of the 4 impeller blade points. The correct clearance is listed in **Table 2**. If the clearance at any one blade point is incorrect, replace the impeller, impeller housing or both parts.

3. Reinstall the intake grate as described in this chapter.

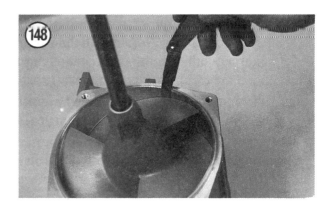

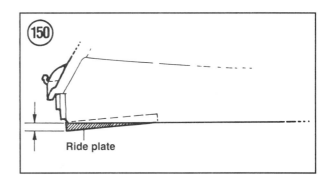

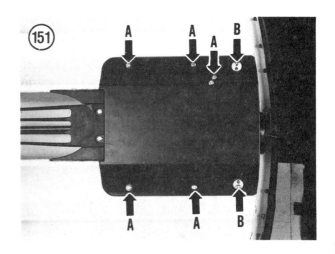

DRIVE TRAIN

RIDE PLATE ADJUSTMENT

On SJ650, SJ700, WR650, WRB650, WBR650A and WRB700 models, adjuster bolts (**Figure 149**) are provided in the ride plate to control how far the rear of the vehicle will sink or squat during operation (**Figure 150**). When adjusting the ride plate, note the following:

a. Loosening the adjuster bolt decreases ride plate height. The craft will have a tendency to squat on water.

b. Tightening the adjuster bolt increases ride plate height. The craft will have a tendency to ride higher on water.

1. Remove the battery to prevent acid spillage. See Chapter Nine.
2. Drain the fuel tank and store the fuel in a container approved for gasoline storage.

CAUTION
If the exhaust system is installed on the engine, do not turn the craft onto its right side when performing Step 3. Water in the exhaust system may drain into the engine's exhaust ports and cause serious engine damage.

3. Turn the fuel petcock OFF and turn the watercraft onto its *left side*. Support the craft securely on a suitable stand. To protect the hull from damage, support the craft with 2 discarded motorcycle tires, a heavy blanket or similar object that can accept the craft's weight.
4. Measure ride plate height at the rear of the ride plate. Record measurement for later comparison.
5. Loosen and remove the ride plate mounting bolts (A, **Figure 151**). On models equipped with a water valve body, do not loosen the 2 water valve bolts (B, **Figure 151**) unless you wish to remove the ride plate.
6A. *SJ650, SJ700, WRB650, WB650A and WRB700*—Ride plate adjustment is made by turning the 2 ride plate adjusters in equal amounts. Ride plate height will change 1.0 mm (0.04 in.) for each turn of the adjuster. When adjusting the ride plate, note that 1.0 mm (0.04 in.) thick washers must be placed between the ride plate and hull at the 2 middle bolt positions (**Figure 152**). These washers are selected according to the desired amount of adjustment required. See **Table 5** for adjustment guidelines.
6B. *WR650*—Ride plate adjustment is made by turning the 2 front and 2 rear adjusters (**Figure 153**) in a 2:1 ratio. For example, to move the ride plate 1.0 mm (0.04 in.), turn the 2 front adjuster bolts 1/2 turn and turn the 2 rear adjuster bolts 1

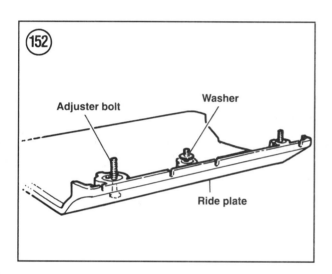

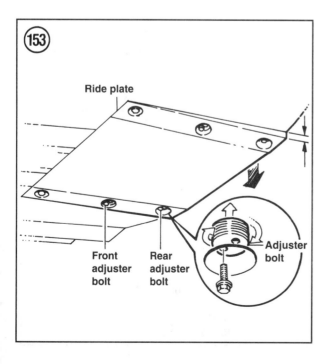

turn. Follow this ratio when making all adjustments. When adjusting the ride plate, never turn the 2 rear adjuster bolts more than 7.5 turns from the initial factory setting.

7. After adjusting the ride plate, wipe the ride plate mounting bolts with Loctite 242 (blue), reinstall the bolts and tighten securely.
8. Reverse Steps 1-3 to complete assembly.

Table 1 JET PUMP GENERAL SPECIFICATIONS

Transmission	Direct drive from engine
Type	
WR500, WR650	Mixed flow, single stage
All other models	Axial flow, single stage
Impeller rotation (rear view)	Counterclockwise
Steering nozzle angle	
WR500, WR650	$29° \pm 1°$
FX700, SJ650, SJ700, SJ700A	$18.5° \pm 24.5°$
RA700, RA700A, RA700B, RA760,	
RA1100, WRA650, WRA650A,	
WRA700	$23° \pm 1°$
WB700, WB700A	$24.5° \pm 1°$
WRB650, WRB650A, WRB700	$21° \pm 1°$
WB760, WVT700, WVT1100	$28° \pm 1°$

Table 2 JET PUMP SERVICE SPECIFICATIONS

	New mm (in.)	Wear limit mm (in.)
Impeller clearance		
WR500, WR650	0.3 (0.012)	0.6 (0.024)
SJ650	0.2 (0.008)	0.6 (0.024)
RA700, RA700A, RA700B, RA760,		
RA1100, SJ700, SJ700A, WB760,		
WVT700, WVT1100	0.3-0.4 (0.012-0.016)	0.6 (0.024)
FX700	0.25-0.35 (0.010-0.014)	0.6 (0.024)
WB700, WB700A, WRA650,		
WRA650A, WRA700, WRB650,		
WRB650A, WRB700	0.2-0.6 (0.008-0.006)	0.6 (0.024)
Drive shaft runout		
SJ650, WR500, WR650	—	0.5 (0.020 in.)
All other models	—	0.3 (0.011 in.)

DRIVE TRAIN

Table 3 DRIVE SYSTEM TIGHTENING TORQUE

	N·m	in.-lb.	ft.-lb.
Impeller	18	–	13
Jet pump mounting bolts	34	–	25
Intake grate			
RA700, RA760, RA1100, WVT700, WVT1100	11	97	–
All other models	7	62	–
Ride plate			
RA700, RA760, RA1100, WVT700, WVT1100	17	–	12
WR500, WR650, WRA650, WRB650, WRB700	7	62	–
WB700, WB760	11	97	–
SJ650, SJ700, FX700	8	71	–
Bearing housing-to-support housing	16	–	11
Intermediate housing mounting bolts			
FX700, SJ650, SJ700	36	–	26
All other models	16	–	11

Table 4 IMPELLER SHIM ADJUSTMENT—WR500 AND WR650

Clearance mm (in.)	0.0 (0.000)	0.1 (0.004)	0.2 (0.008)	0.3 (0.012)	0.4 (0.016)	0.5 (0.020)	0.6 (0.024)	0.7 (0.027)	0.8 (0.031)	0.9 (0.035)	1.0 (0.039)
Shim Adjustment mm (in.)	-2.2 (0.086)	-1.5 (0.059)	-0.7 (0.027)	0.0 (0.000)	+0.7 (0.027)	+1.5 (0.059)	+2.2 (0.086)	+2.9 (0.113)	+3.7 (0.144)	+4.4 (0.172)	+5.1 (0.199)

Table 5 RIDE PLATE WASHER SELECTION—FX700, SJ650, SJ700, SJ700A, WRB650, WRB650A AND WRB700

Ride plate height	Number of washers used (each side)
0-1 mm (0-0.04 in.)	0
1.5-2.0 mm (0.06-0.08 in.)	1
2.5-3.0 mm (0.10-0.12 in.)	2
See text for adjustment procedures	

Chapter Eight

Fuel and Water Box Systems

This chapter includes removal and repair procedures for the carburetor, fuel pump, fuel tank and water box system. See Chapter Three for idle speed and mixture adjustment. **Tables 1-3** are at the end of the chapter.

PRECAUTIONS

Because of the explosive and flammable conditions that exist around gasoline, always observe the following precautions.

1. Immediately after removing the engine hatch, check for the presence of raw gasoline fumes. If you smell strong fumes, determine their source and correct the problem.
2. Allow the engine compartment to air out before beginning work.
3. Disconnect the negative battery cable.
4. Gasoline dripping onto a hot engine component may cause a fire. Always allow the engine to cool completely before working on any fuel system component.
5. Wipe up spilled gasoline immediately with dry rags. Store the rags in a suitable metal container until they can be cleaned or disposed of. Do not store gasoline or solvent soaked rags in the engine compartment.
6. Do not service any fuel system component while in the vicinity of open flames, sparks or while anyone is smoking.
7. Always have a Coast Guard-approved fire extinguisher close at hand when working on the engine.

FLAME ARRESTOR

The flame arrestor assembly consists of the flame arrestor, cover, holder, gaskets and mounting fasteners. Routine service requires removal of the flame arrestor only. When carburetor removal is required, the holder must also be removed. Before removing the holder, make sure to purchase new gaskets for reassembly. Never run the engine without the flame arrestor and cover properly installed.

Removal/Cleaning/Installation (Type I—WR500 Models)

Refer to **Figure 1** for this procedure.

FUEL AND WATER BOX SYSTEMS

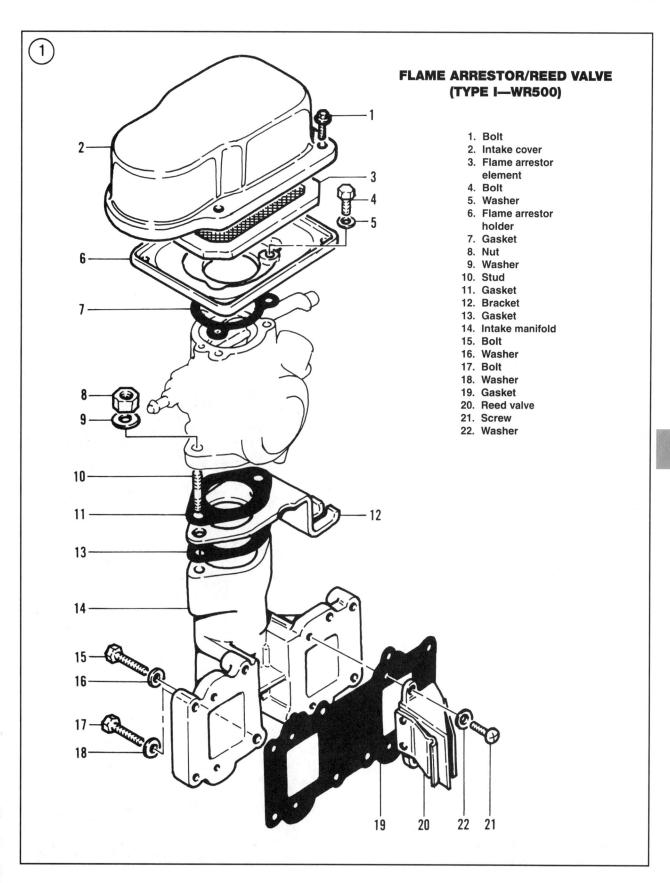

1. Remove the engine hood.
2. Disconnect the battery negative cable.
3. Remove the bolts holding the intake cover to the flame arrestor holder. Remove the intake cover (**Figure 2**).
4. Lift the flame arrestor element (**Figure 3**) out of the holder. Handle the element carefully to prevent damage.
5. If necessary, remove the holder as follows:
   a. Pull the choke knob out to close the choke (prevents objects from dropping undetected into the carburetor).
   b. Remove the bolts (A, **Figure 4**) securing the holder to the carburetor. Remove the holder (B, **Figure 4**).
   c. Remove and discard the holder-to-carburetor gasket.
   d. Place a rag over the carburetor.
6. Clean the flame arrestor with compressed air. Check the flame arrestor for tearing or other damage. Replace if necessary.
7. If the holder was removed, install it as follows:
   a. Remove gasket residue from all gasket mating surfaces.
   b. Install a new gasket and place the holder (B, **Figure 4**) onto the carburetor.
   c. Install the holder mounting bolts and their washers (A, **Figure 4**). Tighten bolts securely.
8. Place the flame arrestor into its holder (**Figure 3**).
9. Install the intake cover (**Figure 2**) and torque the bolts to 1.0 N•m (9 in.-lb.).

Removal/Cleaning/Installation (Type II—FX700, RA700B, SJ650, SJ700, WB700, WB700A, WR650, WRA650, WRA650A, WRA700, WRB650, WRB650A AND WRB700)

Refer to **Figure 5** for this procedure.
1. Remove the engine hood.
2. Disconnect the battery negative cable.

3. Remove the bolts holding the intake cover to the flame arrestor holder. Remove the intake cover (**Figure 6**).
4. Lift the flame arrestor element (**Figure 7**) out of the holder. Handle the element carefully to prevent damage.
5. If necessary, remove the holder as follows:
   a. Pull the choke knob out to close the choke (prevents objects from dropping undetected into the carburetor).
   b. *On oil injection models,* remove the 2 oil hoses from the flame arrestor holder.
   c. Remove the bolts (A, **Figure 8**) securing the holder to the carburetor. Remove the holder (B, **Figure 8**).

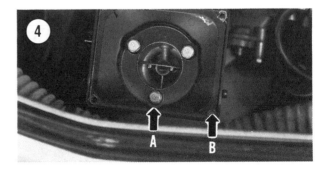

FUEL AND WATER BOX SYSTEMS

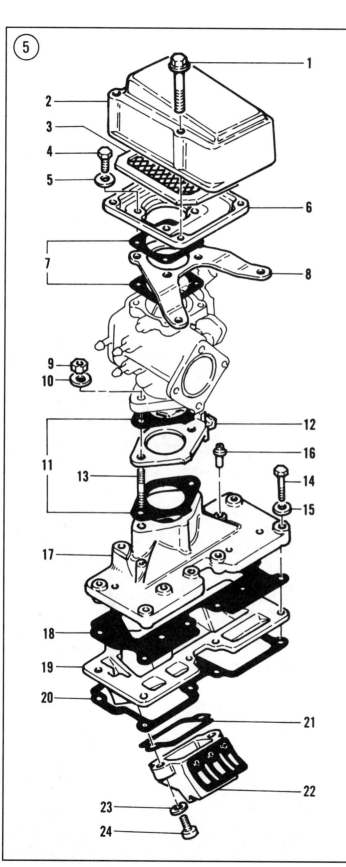

**FLAME ARRESTOR/REED VALVE
(TYPE II—FX700, RA700B, SJ650, SJ700,
WB700, WB700A, WR650, WRA650,
WRA650A, WRA700, WRB650, WRB650A
AND WRB700)**

1. Bolt
2. Intake cover
3. Flame arrestor element
4. Bolt
5. Washer
6. Flame arrestor holder
7. Gaskets
8. Bracket
9. Nut
10. Washer
11. Gaskets
12. Plate
13. Stud
14. Bolt
15. Washer
16. Nozzle
17. Intake manifold
18. Gasket
19. Manifold
20. Gasket
21. Gasket
22. Reed valve
23. Washer
24. Screw

d. Remove and discard the holder-to-carburetor gasket.

6. If necessary, remove the carburetor bracket as follows:
   a. Loosen and remove the 2 cylinder head-to-bracket bolts (A, **Figure 9**).
   b. Lift the bracket (B, **Figure 9**) off of the carburetor.
   c. Discard the bracket gasket.

7. Place a rag over the carburetor.

8. Clean the flame arrestor with compressed air. Check the flame arrestor for tearing or other damage. Replace if necessary.

9. Remove gasket residue from all gasket mating surfaces.

10. If the carburetor bracket was removed, install as follows:
    a. Install a new bracket gasket.
    b. Place the bracket onto the carburetor and align the bracket arms with the 2 cylinder head bolt holes.
    c. Install the cylinder head bolts and washers and tighten to the torque specification listed in Chapter Five.

11. If the holder was removed, install it as follows:
    a. Install a new gasket and place the holder (B, **Figure 8**) onto the carburetor bracket.
    b. Install the holder mounting bolts and their washers (A, **Figure 8**). Tighten each bolt securely.
    c. *On oil injection models*, reinstall the 2 oil hoses to the flame arrestor holder.

12. Place the flame arrestor into its holder (**Figure 7**).

13. Install the intake cover (**Figure 6**), and torque the bolts to 1 N•m (9 in.-lb.).

Flame Arrestor Identification

The flame arrestors used on the models described in this section are basically the same with minor differences. It is important to pay particular attention to the location and order of parts during service. The flame arrestors are identified as follows:

The Type III flame arrestor (**Figure 10**) is used on the following models:

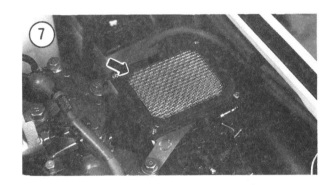

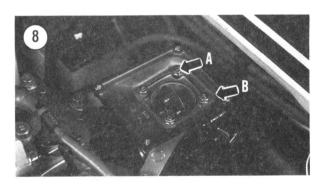

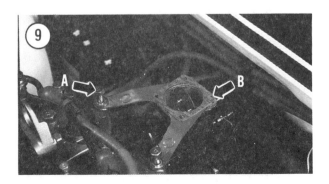

FUEL AND WATER BOX SYSTEMS

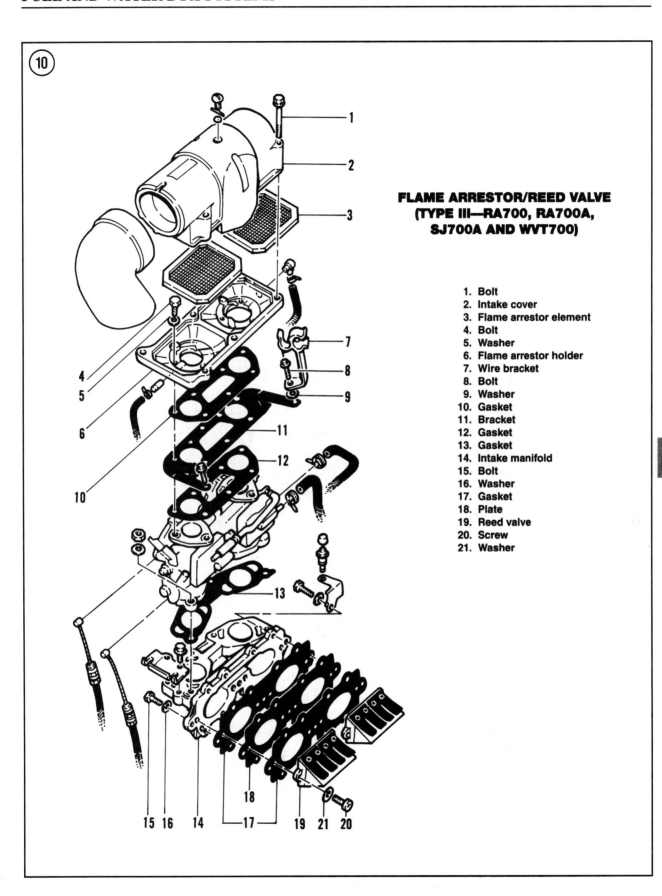

FLAME ARRESTOR/REED VALVE (TYPE III—RA700, RA700A, SJ700A AND WVT700)

1. Bolt
2. Intake cover
3. Flame arrestor element
4. Bolt
5. Washer
6. Flame arrestor holder
7. Wire bracket
8. Bolt
9. Washer
10. Gasket
11. Bracket
12. Gasket
13. Gasket
14. Intake manifold
15. Bolt
16. Washer
17. Gasket
18. Plate
19. Reed valve
20. Screw
21. Washer

342 CHAPTER EIGHT

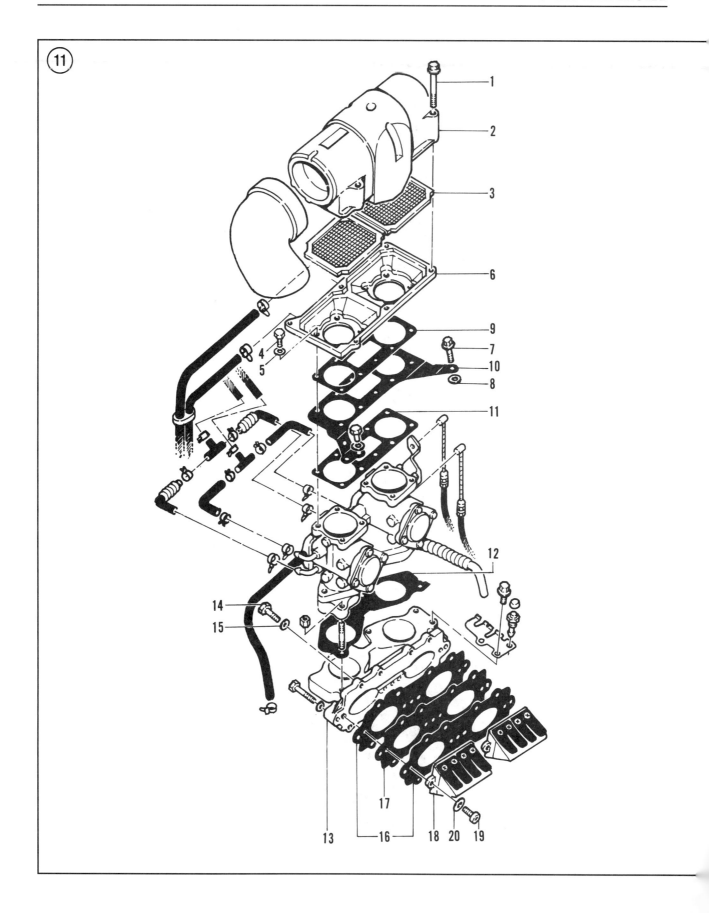

FUEL AND WATER BOX SYSTEMS

**FLAME ARRESTOR/REED VALVE
(TYPE IV—RA760 AND WB760)**

1. Bolt
2. Intake cover
3. Flame arrestor
4. Bolt
5. Washer
6. Flame arrestor holder
7. Bolt
8. Washer
9. Gasket
10. Bracket
11. Gasket
12. Gasket
13. Intake manifold
14. Bolt
15. Washer
16. Gasket
17. Plate
18. Reed valve
19. Screw
20. Washer

a. RA700 and RA700A.
b. SJ700A.
c. WVT700.

The Type IV flame arrestor (**Figure 11**) is used on the following models:

a. RA760.
b. WB760.

The Type V flame arrestor (**Figure 12**) is used on the following models:

a. RA1100.
b. WVT1100.

Removal/Cleaning/Installation
(Type III, Type IV and Type V)

Refer to **Figure 10**, **Figure 11** or **Figure 12** for this procedure.

1. Remove the engine hood.

2. Disconnect the battery negative cable.

3. *On RA1100 and WVT1100 models*, remove the grease nipple plate from the intake cover (**Figure 13**).

4. emove the bolts holding the intake cover to the flame arrestor holder. Remove the intake cover.

5. Lift the flame arrestor elements out of the holder. Handle the elements carefully to prevent damage.

6. If necessary, remove the holder as follows:

   a. Pull the choke knob out to close the choke (prevents objects from dropping undetected into the carburetor).

   b. Remove the 2 oil hoses (3 hoses on RA1100 and WVT1100) from the flame arrestor holder. See **Figure 14**. (The SJ700A is a premix model. There are no oil hoses on these models.)

   c. On RA1100 and WVT1100 models, remove the bolt securing the remote fuel pump to the holder (B, **Figure 15**), and remove the bolt collar.

CHAPTER EIGHT

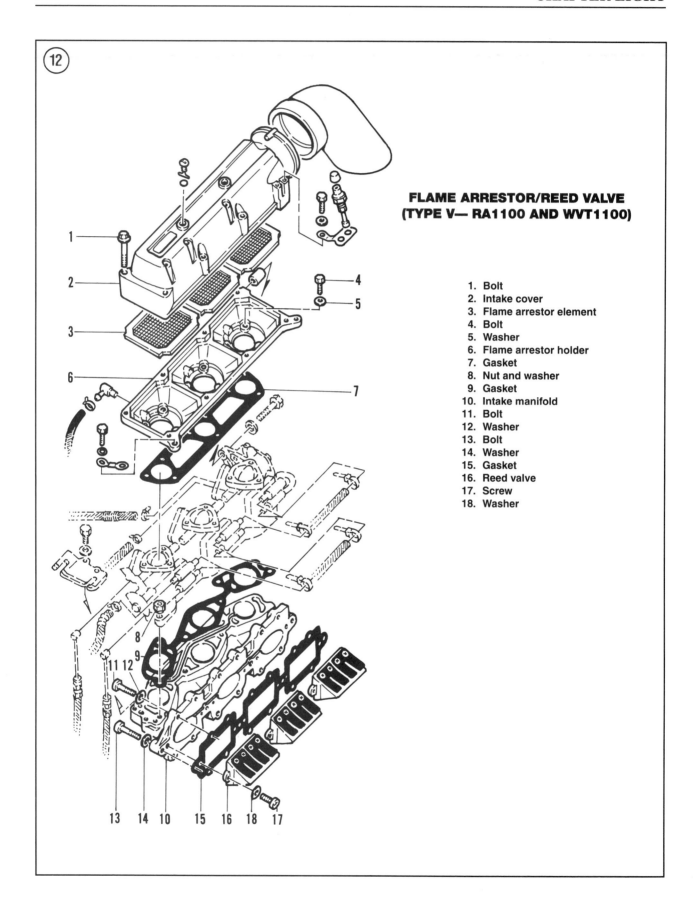

FLAME ARRESTOR/REED VALVE (TYPE V— RA1100 AND WVT1100)

1. Bolt
2. Intake cover
3. Flame arrestor element
4. Bolt
5. Washer
6. Flame arrestor holder
7. Gasket
8. Nut and washer
9. Gasket
10. Intake manifold
11. Bolt
12. Washer
13. Bolt
14. Washer
15. Gasket
16. Reed valve
17. Screw
18. Washer

FUEL AND WATER BOX SYSTEMS

d. Remove the bolts (A, **Figure 15**) securing the holder to the carburetors. Remove the holder (B, **Figure 15**).

e. Remove and discard the holder-to-carburetor gasket.

7. *RA700, RA700A, RA760, SJ700A, WB760 and WVT700*—If necessary, remove the carburetor bracket as follows:

   a. Loosen and remove the 2 bracket mounting bolts (A, **Figure 16**).

b. On RA700, RA700A, SJ700A and WVT700 models, remove the wire bracket.

c. Lift the bracket (B, **Figure 16**) off of the carburetor.

d. Discard the bracket gasket.

8. Place a rag over the carburetor.

9. Clean the flame arrestor with compressed air. Check the flame arrestor for tearing or other damage. Replace if necessary.

10. Remove gasket residue from all gasket mating surfaces.

11. *RA700, RA700A, RA760, SJ700A, WB760 and WVT700*—If the carburetor bracket was removed, install it as follows:

   a. Install a new bracket gasket.

   b. Place the bracket onto the carburetor and align the bracket arms with the mounts on the cylinder.

   c. On RA700, RA700A and WVT700 models, install the wire bracket along with the bracket.

   d. Apply Loctite 242 to the bracket mounting bolts. Install the bolts along with their washers and tighten the bolts securely.

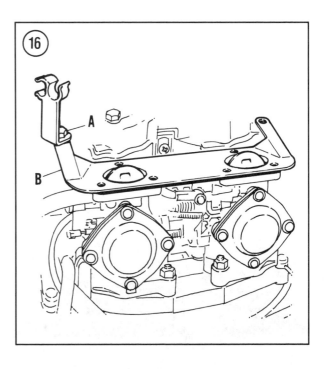

12. If the holder was removed, install it as follows:
    a. Install a new gasket and place the holder (B, **Figure 15**) onto the carburetor bracket.
    b. Install the holder mounting bolts and their washers (A, **Figure 15**). Tighten each bolt securely.
    c. On RA1100 and WVT1100 models, secure the remote fuel pump to the holder with the fuel pump mounting bolt along with its collar. Apply Loctite 242 to the bolt threads, and tighten the bolt securely.
    d. Reinstall the 2 oil hoses (3 oil hoses on RA1100 and WVT1100) to the flame arrestor holder. (There are no oil hoses on SJ700A models).
13. Place the flame arrestor into its holder.
14. Install the intake cover. Apply Loctite 242 to the intake cover bolts, and torque the bolts to 2 N•m (18 in.-lb.).
15. *On RA1100 and WVT1100 models*, reattach the grease nipple plate to the intake cover (**Figure 13**).

INTAKE MANIFOLD AND REED VALVE

The reed valve assembly is mounted between the intake manifold and crankcase on all models. See **Figure 1** (Type I), **Figure 5** (Type II), **Figure 10** (Type III), **Figure 11** (Type IV) or **Figure 12** (Type V).

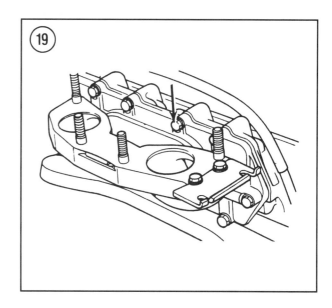

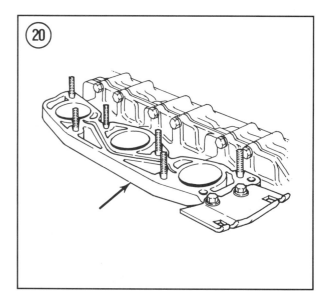

FUEL AND WATER BOX SYSTEMS

Reed valves control the passage of the air/fuel mixture into the crankcase by opening and closing as crankcase pressure changes. When crankcase pressure is high, the reeds maintain contact with the reed plate to which they are attached. As crankcase pressure drops on the compression stroke, the reeds move away from the plate and allow air/fuel mixture to pass. Reed travel is limited by the reed stop. As crankcase pressure increases, the reeds return to the reed plate.

The intake manifold/reed valve assembly can be removed with the engine mounted in the hull.

Removal/Installation

Refer to **Figure 1**, **Figure 5** or **Figures 10-12** for this procedure.

1. Remove the flame arrestor as described in this chapter.
2. Remove the carburetor as described in this chapter.
3. Remove the bolts holding the intake manifold to the crankcase. Remove the intake manifold/reed valve assembly and gasket from the crankcase. See **Figure 17** (Type I models), **Figure 18** (Type II models), **Figure 19** (Type III and Type IV models) or **Figure 20** (Type V models). Discard the gaskets.
4. Clean and inspect the reed block assembly as described in this chapter.
5. Installation is the reverse of these steps. Note the following:
6A. *WR500*—On these models, it is necessary to remove the reed blocks (**Figure 21**) from the intake manifold to replace and install a new gasket. Install the new gasket so the sealer strip faces out as indicated in **Figure 22**. Apply Loctite 242 (blue) to the reed block Phillips screws and tighten the screws securely.
6B. *RA700, RA700A, SJ700A, RA1100, WB760, WVT700 and WVT1100*—On these models, it is necessary to remove the reed valves from the intake manifold to replace and install new gaskets. See **Figure 10**, **Figure 11** or **Figure 12**. Apply Loctite 242 (blue) to the reed valve Phillips screws and tighten the screws securely.
6C. *650 cc*—Install a new manifold gasket (**Figure 23**) before installing the reed valve onto the manifold.
7. Apply Loctite 242 to the threads of the intake manifold mounting bolts. Torque the manifold

bolts in the sequence shown in **Figure 24** (Type II) or **Figure 25** (Type V) where applicable.

Cleaning and Inspection

1. *RA700, RA700A, RA760, SJ700A, RA1100, WB760, WR500, WVT700 and WVT1100*—Remove the reed valves from the manifold.

NOTE
Removal of the reed valves from the manifolds on other models is unnecessary unless reed block replacement or other service is required.

2. Remove all gasket and sealant residue from the manifold mating surface. See A, **Figure 26**, typical.
3. Clean the reed block assembly with solvent and blow dry with low-pressure compressed air, taking care not to direct the air stream directly on or through the reed valves.
4. Check the carburetor studs (B, **Figure 26**) for corrosion or thread damage. Clean corrosion with a brush or die. Replace damaged studs as described in Chapter One.
5. The screws securing the reed blocks to the manifold are normally secured with Loctite; if these screws are removed, all Loctite residue must be removed from the manifold screw holes with a tap (**Figure 27**).

6. Check the intake side of the assembly to make sure the reeds are not sticking tightly to the valve face.
7. Check the crankcase side of the assembly to make sure that the reeds are lying flat on the valve face with no preload. To check flatness, gently push each reed petal out. Constant resistance should be felt with no audible noise.

CAUTION
Always replace reeds in sets. Never turn a reed over for reuse.

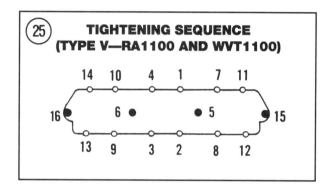

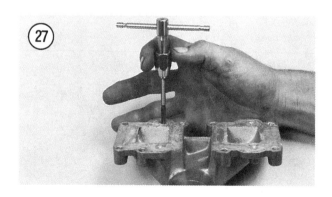

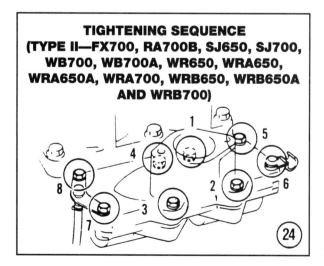

FUEL AND WATER BOX SYSTEMS

8. Check for cracked or broken reeds. Replace if any defects are noted. See **Figure 28** (Type I models), **Figure 29** (Type II models) or **Figure 30** (Type V models).

9. Measure the gap between the reeds and reed block with a flat feeler gauge (**Figure 31**). Replace reeds if the gap is excessive. Refer to **Table 2**.

10. Check each reed stop opening by measuring from inside the reed stop to the top of the reed block (**Figure 32**). If the reed stop is not within specification (**Table 2**), remove the reed stop and carefully bend the stop to obtain the specified opening.

Reed and Reed Stop Replacement

1. Remove the screws holding the reed stop and reeds to the valve seat.
2. Remove the reed stop and reeds.
3. Clean the reed block screw threads of all Loctite residue with a tap.
4. Place a new reed on the valve seat and check for flatness.
5. Center the reed over the valve seat opening.
6. Wipe the reed stop screw threads with Loctite 242 (blue). Install the reed stop and torque the screws to 1.0 N•m (9 in.-lb.).
7. Check reed tension and opening. See *Cleaning and Inspection* in this chapter.

CHAPTER EIGHT

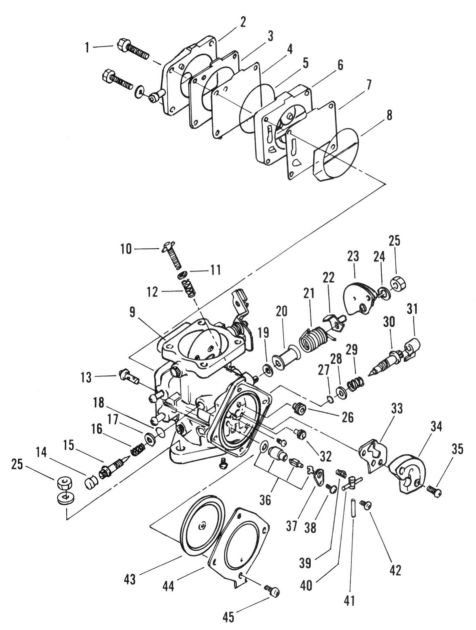

**CARBURETOR
(TYPE IV—FX700, RA700B, RA760, SJ650, SJ700, WB700, WB760, WR500, WR650, WRA650, WRA650A, WRA700, WRB650, WRB650A AND WRB700)**

1. Screw
2. Pump Cover
3. Gasket
4. Diaphragm
5. O-ring
6. Body
7. Diaphragm
8. O-ring
9. Carburetor body
10. Idle adjust screw
11. Washer
12. Spring
13. Filter
14. Cap
15. High-speed mixture screw
16. Spring
17. Washer
18. O-ring
19. Washer
20. Collar
21. Spring
22. Bracket
23. Throttle assembly
24. Washer
25. Nut
26. Main jet
27. O-ring
28. Washer
29. Spring
30. Low-speed mixture screw
31. Limiter cap
32. Pilot jet
33. Gasket
34. Body
35. Screw
36. Needle valve assembly
37. Holder
38. Screw
39. Spring
40. Control arm
41. Pivot pin
42. Screw
43. Diaphragm
44. Carburetor cover
45. Screw

FUEL AND WATER BOX SYSTEMS

CARBURETOR

Carburetor Identification

The carburetors used on the models described in this section are basically the same with minor differences. It is important to pay particular attention to the location and order of parts during service. The carburetors are identified as follows:

The Type IV carburetor (**Figure 33**) is used on the following models:
a. SJ650, SJ700 and FX700.
  b. WR500 and WR650.
  c. WRA650, WRA650A, WRA700, WRB650, WRB650A and WRB700.
  d. WB700 and WB760.
  e. RA700B and RA760.

The Type V carburetor (**Figure 34**) is used on the following models:
a. SJ700A.
b. RA700, RA700A and RA1100.
c. WB700A.
d. WVT700 and WVT1100.

Carburetor Removal

WARNING
Some fuel may spill during this procedure. Work in a well-ventilated area at least 50 ft. (15 m) from any sparks or flames, including gas appliance pilot lights. Do not smoke in the area. Keep a B:C rated fire extinguisher handy.

NOTE
On multicarburetor models (RA700, RA700A, RA760, RA1100, WB760, WVT700 and RA1100), the carburetors in an assembly are identical except for the fuel pump cover. Each pump cover in an assembly has a unique set of fuel hose fittings. A particular carburetor can be identified by its fuel pump cover. Note the location of each carburetor before disassembling the carburetor assembly. Each carburetor must have to be reinstalled at its original location in the assembly.

1. Remove the engine cover.
2. Remove the flame arrestor as described in this chapter.

WARNING
*Before disconnecting any fuel lines, loosen the fuel filler cap (**Figure 35**) to relieve any pressure in the fuel tank.*

3. Label and disconnect the fuel lines from the carburetor.
4. Label and disconnect all vent lines from the carburetor.
5. Loosen the control cable set screws and pull the inner cables out of their fittings on the carburetor.
6A. *Single-carburetor models*—Unscrew the nuts securing the carburetor to the intake manifold. Remove the carburetor (**Figure 36**).
6B. *Multicarburetor models*—Unscrew the nuts securing the carburetors to the intake manifold. Remove the carburetor assembly (**Figure 37**, twin-carb model).
7. Remove and discard the carburetor gasket.
8. Cover the intake ports to keep foreign objects out of the engine.
9. Installation is the reverse of these steps. Note the following.
10. *On multicarburetor models*, if the carburetor assembly was disassembled, synchronize the carburetors as described in *Carburetor Synchronization* in Chapter Three.
11. Install a new carburetor gasket. Tighten mounting nuts securely to prevent carburetor warpage or an air leak.
12. Insert the control cables in their fittings, tighten the set screws and adjust the throttle and choke cables as described in Chapter Three.
13. Reconnect all hoses, following the marks recorded during disassembly.

CAUTION
Check the routing of all hoses. They must be routed properly and without any sharp bends or kinks and placed away from engine components that could cause damage.

352

CHAPTER EIGHT

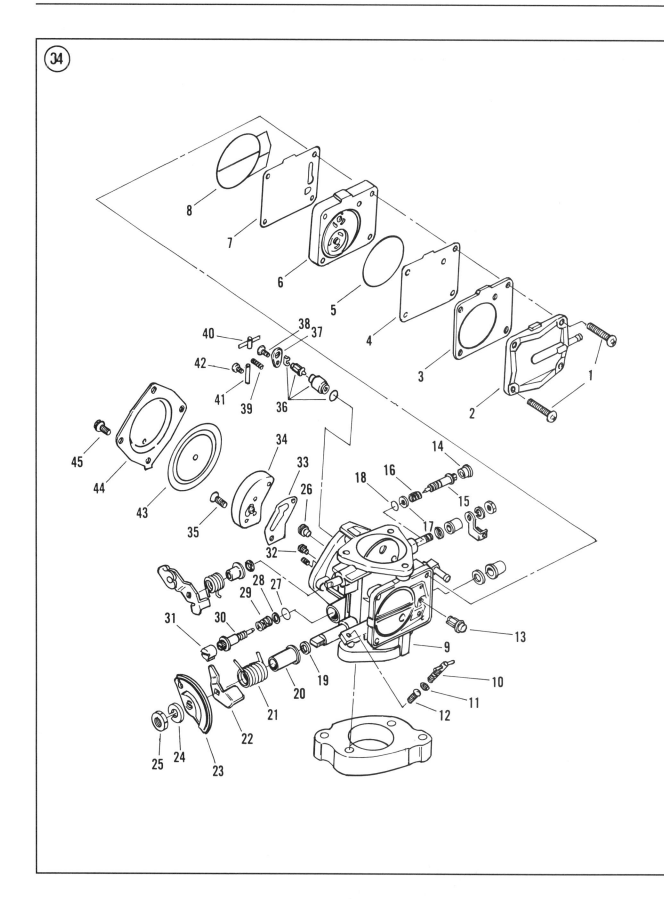

FUEL AND WATER BOX SYSTEMS

CARBURETOR
(TYPE V—RA700, RA700A, RA1100, SJ700A, WB700A, WVT700 AND WVT1100)

1. Screw
2. Pump Cover
3. Gasket
4. Diaphragm
5. O-ring
6. Body
7. Diaphragm
8. O-ring
9. Carburetor body
10. Idle adjust screw
11. Washer
12. Spring
13. Filter
14. Cap
15. High-speed mixture screw
16. Spring
17. Washer
18. O-ring
19. Washer
20. Collar
21. Spring
22. Bracket
23. Throttle assembly
24. Washer
25. Nut
26. Main jet
27. O-ring
28. Washer
29. Spring
30. Low-speed mixture screw
31. Limiter cap
32. Pilot jet
33. Gasket
34. Body
35. Screw
36. Needle valve assembly
37. Holder
38. Screw
39. Spring
40. Control arm
41. Pivot pin
42. Screw
43. Diaphragm
44. Carburetor cover
45. Screw

Fuel Pump
Disassembly

The fuel pump (**Figure 33** or **Figure 34**) is a pulse-operated diaphragm type that is integral with the carburetor housing.

NOTE
Do not disassemble the fuel pump body unless necessary. There are many gas-

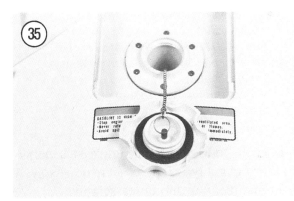

kets that must be replaced whenever the fuel pump is disassembled.

1. Remove the carburetor(s) as described in this chapter.

2. Remove the screws securing the pump cover (**Figure 38**) to the carburetor housing. Remove the pump cover.

CAUTION
Because some of the components may be stuck together, disassembly of the diaphragms and gaskets must be performed carefully. The diaphragms are very thin and can be easily damaged from improper handling and storage. Do not set the diaphragms on an oily, rough or dirty surface.

3. Referring to **Figure 33** or **Figure 34**, separate the pump assembly. Lay parts out in order so you do not mix them up during reassembly.

FUEL AND WATER BOX SYSTEMS

4. Remove the fuel filter (**Figure 39**) from the carburetor body.

Carburetor Disassembly

Refer to **Figure 33** or **Figure 34** when performing this procedure.

1. Remove the fuel pump as described in this chapter.
2. Remove the carburetor cover (**Figure 40**) then carefully remove the diaphragm (**Figure 41**) from the carburetor.
3. Remove the control arm pivot pin screw (**Figure 42**). Then remove the control arm, pivot pin (**Figure 43**) and spring (**Figure 44**).
4. Remove the needle valve holder screw and the holder (**Figure 45**). Pull the needle valve assembly out of the carburetor body (**Figure 46**).
5. Remove the 2 body Phillips screws (A, **Figure 47**) and remove the body (B, **Figure 47**).
6. Remove the pilot (**Figure 48**) and main (**Figure 49**) jets.

7. Lightly seat the high-speed mixture screw (**Figure 50**), counting the number of turns required for reassembly reference. Back out the screw and remove it from the carburetor with the cap, spring, washer and O-ring.

8. Lightly seat the low-speed mixture screw (**Figure 51**), counting the number of turns required for reassembly reference. Back out the screw and remove it from the carburetor along with the limiter cap, spring, washer and O-ring.

Cleaning and Inspection

CAUTION
*If the fuel pump was not removed from the carburetor body, do **not** use compressed air to clean the carburetor body in the following steps. Compressed air will damage the pump's diaphragms and check valves.*

1. Thoroughly clean and dry all parts. Because seals are used at the choke and throttle shaft positions, and these shafts should not be removed, a special carburetor cleaning solvent should be used. Select a cleaning solution which will not damage rubber or plastic parts. Metal and alloy parts which are removed from the carburetor housing can be soaked in a carburetor cleaner. After soaking parts in a cleaner, wash these parts with hot soapy water then rinse with water. If you do not have compressed air, place all of the carburetor parts on a clean lint free cloth and allow to air dry.

2. Remove any old gasket material and check that all fuel and vent passages are clear. Blow them clean with compressed air, if necessary. Do not use a wire to clean any of the orifices; wire will enlarge them and change fuel flow rates.

3. Clean the internal fuel filter (**Figure 52**) in solvent and check the filter element for tearing or debris that will not clean out. Replace the filter if it cannot be thoroughly cleaned or if it is torn or otherwise damaged.

FUEL AND WATER BOX SYSTEMS

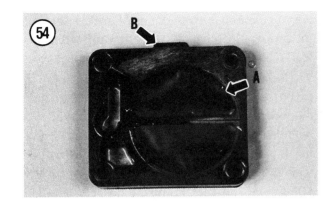

54

55

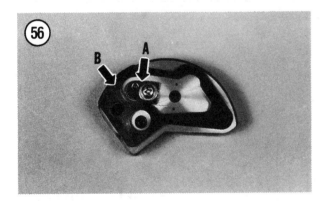

56

57

4. Check the carburetor (**Figure 53**) and fuel pump (**Figure 54**) diaphragms for tears, pin holes or deterioration. Replace as required.

5. Inspect the check valves (**Figure 55**) installed in the fuel pump body for warpage and spring tension. If any valve appears worn or damaged, replace the fuel pump body.

6. Inspect the carburetor control arm and replace if excessively worn. Replace the spring if it is damaged or weakened.

7. Check the taper on the fuel valve seat and fuel valve needle for wear, scratches or other damage. Replace the seat and needle as a set if either part is defective.

8. Check the tapered ends of the high- and low-speed mixture screws for grooves or roughness and replace if any is found. Inspect the O-rings and replace if damaged.

9. Check the body assembly check valve (A, **Figure 56**) for cracks, weakness or damage. Replace the check valve as required.

10. Operate the choke and throttle shafts (A, **Figure 57**) and check shaft movement and operation. The shafts should move smoothly. Check each shaft for side play; excessive play indicates worn O-rings. If the shafts show excessive play, you may have to replace the entire carburetor body. Separate components for the choke and throttle shafts are not available from Yamaha.

CAUTION
*The Phillips screws (B, **Figure 57**) securing the throttle plate are in a vulnerable position. A loose screw will back out and fall into the engine, causing expensive damage. Remove loose screws and reinstall them with Loctite 242 (blue). When loosening or tightening these screws, a screwdriver that fits the end of the screw perfectly must be used. These screws are made of soft material and will round out easily if the screwdriver size is incorrect.*

Carburetor Assembly

Refer to **Figure 33** or **Figure 34** when reassembling the carburetor.

1. Install the high- and low-speed mixture screws with their springs, washers and O-rings. Install the limiter cap on the low-speed mixture screw.
2. *Lightly* seat each screw, then back each screw out the number of turns recorded during removal.

CAUTION
Do not seat the mixture screws hard or the tips will be damaged and require replacement.

3. Install the main (**Figure 49**) and pilot (**Figure 48**) jets.
4. Install the body assembly with a new gasket (B, **Figure 56**) into the carburetor. Torque the 2 Phillips screws (A, **Figure 47**) to 2 N·m (18 in.-lb.).
5. Check that the needle valve and its seat are clean. Then insert the needle valve assembly—with its O-ring—into the carburetor until it bottoms (**Figure 46**). Install the needle valve holder and secure it with its Phillips screw (**Figure 45**). Torque the screw to 1 N·m (9 in.-lb.).
6. Install the control arm spring (**Figure 44**) into its mounting hole in the carburetor housing.
7. Assemble the fuel valve needle onto the control arm with the arm pin (**Figure 43**) and lower the fuel valve needle into the fuel valve seat. Place the control arm into position with the rounded projection down in the top of the spring. Insert the control arm pivot pin through the arm and install its retaining screw (**Figure 42**). Torque the retaining screw to 1 N·m (9 in.-lb.).
8. Measure the distance from the carburetor body surface to the top control arm as shown in **Figure 58**. The distance should be 0-0.2 mm (0-0.008 in.) when measured with a feeler gauge or vernier caliper. If the distance is incorrect, take out the arm and bend it carefully to obtain the correct clearance. If the control arm is not set at the proper clearance, the fuel mixture will be too lean or too rich throughout the rpm range.
9. Fit the diaphragm (**Figure 41**) into the carburetor housing, making sure the outer diaphragm lip fits into the housing groove. Then install the cover (**Figure 40**) and secure it with its mounting screws. Torque the mounting screws to 4 N·m (35 in.-lb.).
10. Install the fuel pump as described in this chapter.

Fuel Pump Assembly

Refer to **Figure 33** or **Figure 34** when performing this procedure.

NOTE
The fuel pump is assembled by stacking each component on the carburetor housing. To ensure proper assembly, the fuel

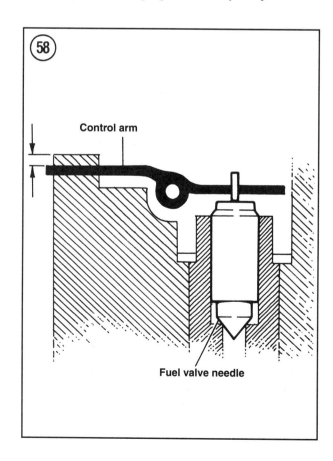

FUEL AND WATER BOX SYSTEMS

pump body pieces have locating tabs (B, Figure 54) that align with a slot (A, Figure 59) machined in the carburetor housing. If these parts are not aligned correctly, the fuel pump will not function properly.

1. Install the fuel filter (B, **Figure 59**).

2. Assemble the O-ring, diaphragm, body and O-ring as shown in **Figure 33** or **Figure 34**. Then place the assembly onto the carburetor so the body check valves face up as shown in **Figure 60**.

3. Assemble the diaphragm and gasket onto the fuel pump cover and install the cover onto the carburetor. See **Figure 61**.

CAUTION
Be careful not to cross thread or strip the fuel pump cover screws. The aluminum threads in the carburetor are easily damaged, resulting in air or fuel leaks.

4. Install the fuel pump cover so the nozzle faces in the direction shown in **Figure 62**. Install the cover screws and torque the screws to 5 N•m (44 in.-lb.).

Remote Fuel Pump (RA1100 and WVT1100)

RA1100 and WVT1100 models have a remote fuel pump on the rear of the carburetor assembly. You can remove the remote fuel pump without removing the carburetors or flame arrestor.

NOTE
Do not disassemble the remote fuel pump unless necessary. There are several gaskets that should be replaced whenever the fuel pump is disassembled.

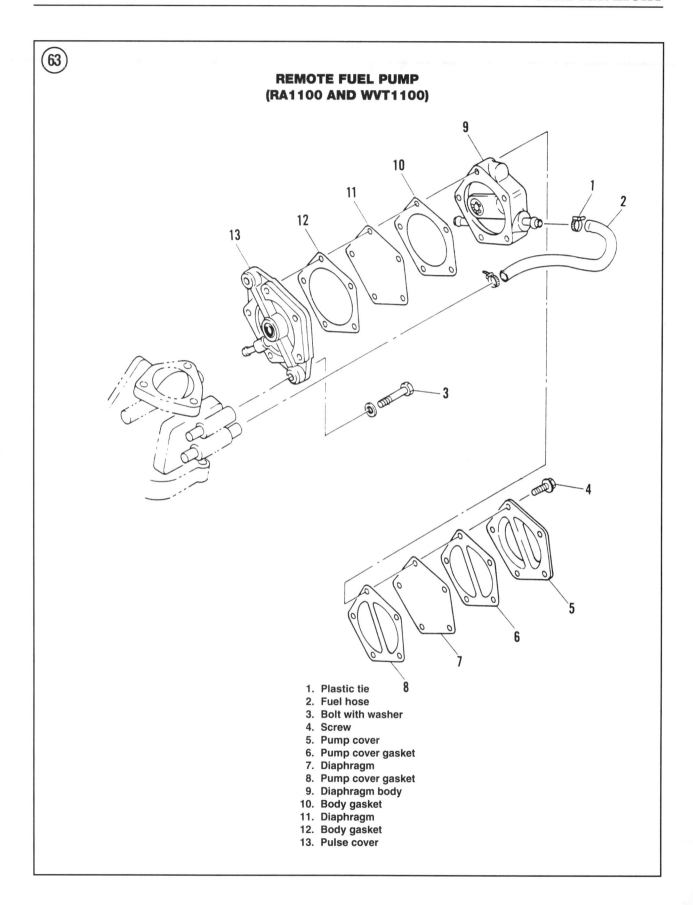

FUEL AND WATER BOX SYSTEMS

Refer to **Figure 63** when performing this procedure.

1. If necessary, remove the carburetors as described earlier in this chapter.
2. Remove all hoses from the fuel pump body.
3. Remove the 2 pump mounting bolts and remove the fuel pump from the carburetor assembly. Be sure to remove the collar along with the upper mounting bolt (A, **Figure 64**).
4. Remove the pump cover screws and remove the cover from the pump body (B, **Figure 64**).
5. Remove the cover gasket, diaphragm and the second cover gasket.
6. Remove the diaphragm body, then remove the body gasket, diaphragm and the second body gasket from the pulse cover.

Cleaning and Inspection

1. Thoroughly clean and dry all parts. Metal and alloy can be soaked in a carburetor cleaner. After soaking in a cleaner, wash these parts with hot soapy water then rinse with water. If you do not have compressed air, place all of the fuel pump parts on a clean lint free cloth and allow to air dry.
2. Remove any old gasket material and check that all fuel and vent passages are clear. Blow them clean with compressed air, if necessary. Do not use wire to clean any of the orifices; wire will enlarge them and change fuel flow rates.
3. Check the diaphragms and diaphragm body for tears, pin holes or deterioration. Replace as required.
4. Inspect the check valve installed in the diaphragm body for warpage and spring tension. If the valve appears worn or damaged, replace the diaphragm body.

Reassembly/Reinstallation

1. Stack the body gasket, diaphragm and second body gasket onto the pulse cover, then set the diaphragm body in place on the gasket. Be sure the hose fitting on the diaphragm body is opposite the hose fitting on the pulse cover as shown in **Figure 63**.
2. Stack the cover gasket, diaphragm and second cover gasket onto the pump cover. Be sure the cutouts in both gaskets align with the chambers in the pump cover. See **Figure 63**.
3. Set the cover onto the diaphragm body. Be sure the diagonal in the pump cover aligns with the diagonal on the pulse cover. Install the 5 cover screws. Torque the screws to 3 N•m (26 in.-lb.).
4. Secure the remote fuel pump to the carburetor assembly with the 2 mounting bolts. Remember to include the collar when securing the fuel pump to the flame arrestor bracket (A, **Figure 64**). Apply Loctite 242 to the bolt threads.
5. Reattach the fuel and pulse lines.
6. Reinstall the carburetors, if removed.

FUEL TANK

Refer to the illustration for your model when servicing the fuel tank:

a. **Figure 65**—SJ650, SJ700 and SJ700A.
b. **Figure 66**—FX700.
c. **Figure 67**—WB700 and WB700A.
d. **Figure 68**—WRA650, WRA650A and WRA700.
e. **Figure 69**—WRB650, WRB650A and WRB700.
f. **Figure 70**—RA700, RA700A, RA700B, RA760, RA1100, WB760, WVT700 and WVT1100.
g. **Figure 71**—WR500 and WR650

Removal/Inspection/Installation

Refer to **Figures 65-71**, as appropriate for your model, when performing this procedure.

WARNING
Some fuel may spill during these procedures. Work in a well-ventilated area at least 50 ft. (15 m) from any sparks or

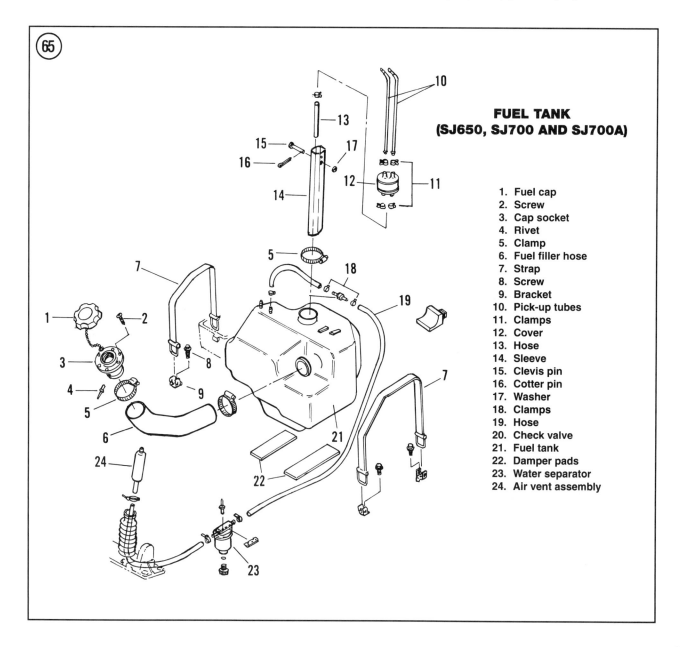

FUEL TANK (SJ650, SJ700 AND SJ700A)

1. Fuel cap
2. Screw
3. Cap socket
4. Rivet
5. Clamp
6. Fuel filler hose
7. Strap
8. Screw
9. Bracket
10. Pick-up tubes
11. Clamps
12. Cover
13. Hose
14. Sleeve
15. Clevis pin
16. Cotter pin
17. Washer
18. Clamps
19. Hose
20. Check valve
21. Fuel tank
22. Damper pads
23. Water separator
24. Air vent assembly

FUEL AND WATER BOX SYSTEMS

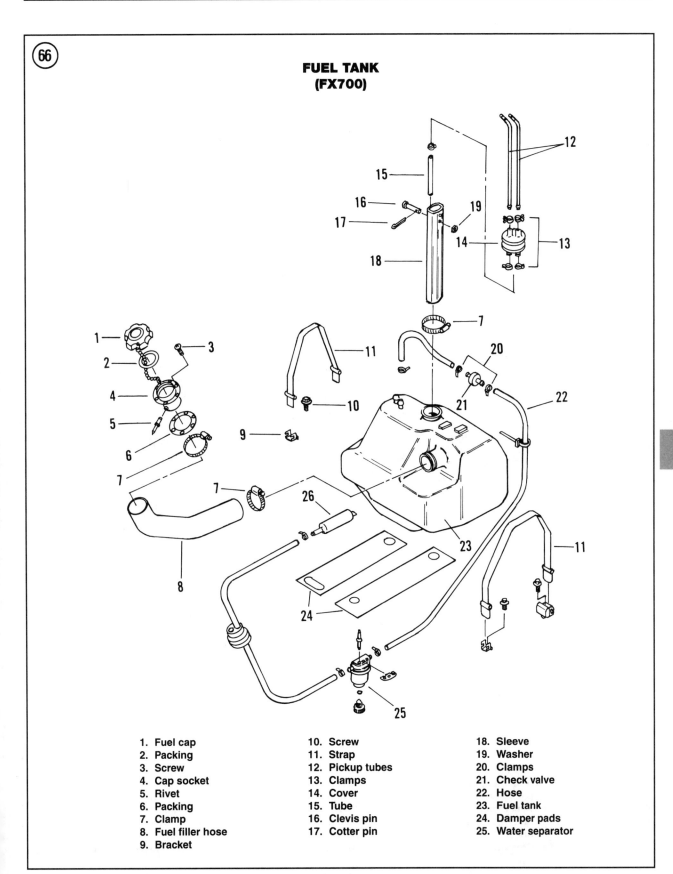

FUEL TANK (FX700)

1. Fuel cap
2. Packing
3. Screw
4. Cap socket
5. Rivet
6. Packing
7. Clamp
8. Fuel filler hose
9. Bracket
10. Screw
11. Strap
12. Pickup tubes
13. Clamps
14. Cover
15. Tube
16. Clevis pin
17. Cotter pin
18. Sleeve
19. Washer
20. Clamps
21. Check valve
22. Hose
23. Fuel tank
24. Damper pads
25. Water separator

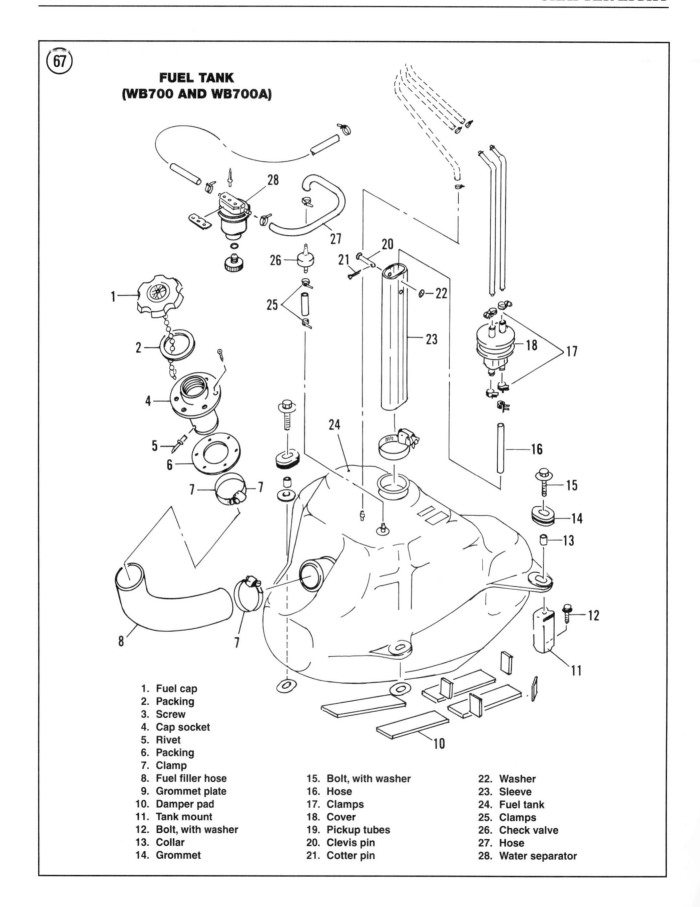

FUEL AND WATER BOX SYSTEMS

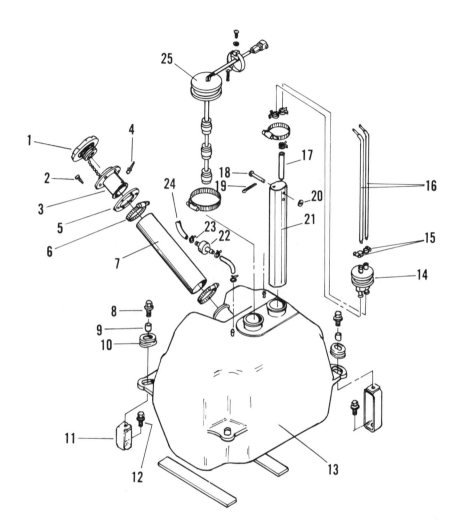

FUEL TANK (WRA650, WRA650A AND WRA700)

1. Fuel cap
2. Screw
3. Cap socket
4. Rivet
5. Packing
6. Clamp
7. Fuel filler hose
8. Bolt, with washer
9. Collar
10. Grommet
11. Tank mount
12. Bolt, with washer
13. Fuel tank
14. Cover
15. Clamps
16. Pickup tubes
17. Hose
18. Clevis pin
19. Cotter pin
20. Washer
21. Sleeve
22. Check valve
23. Clamps
24. Hose
25. Fuel sender

CHAPTER EIGHT

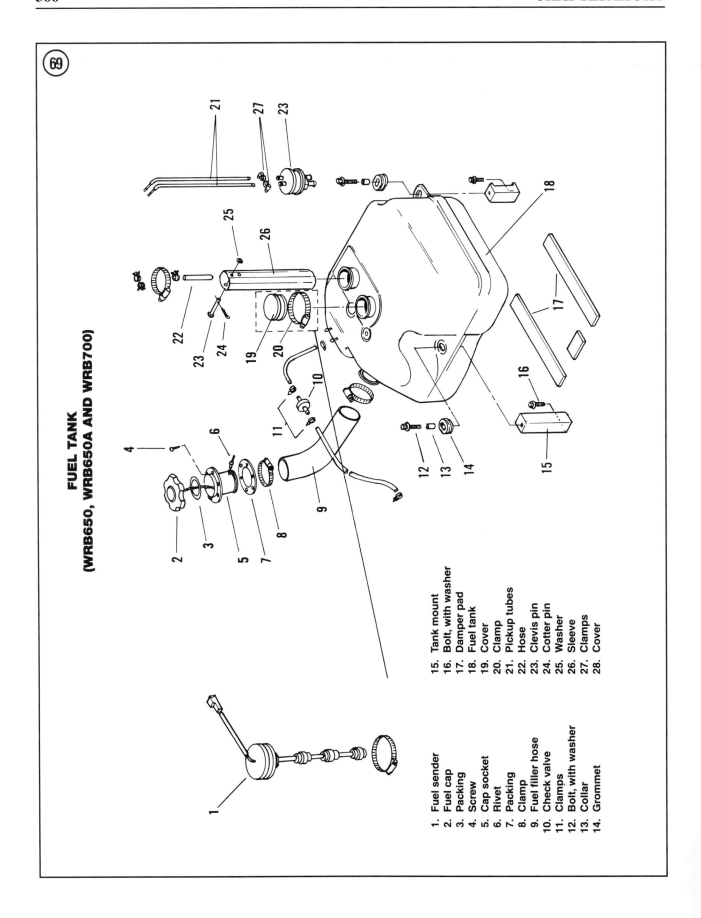

FUEL AND WATER BOX SYSTEMS

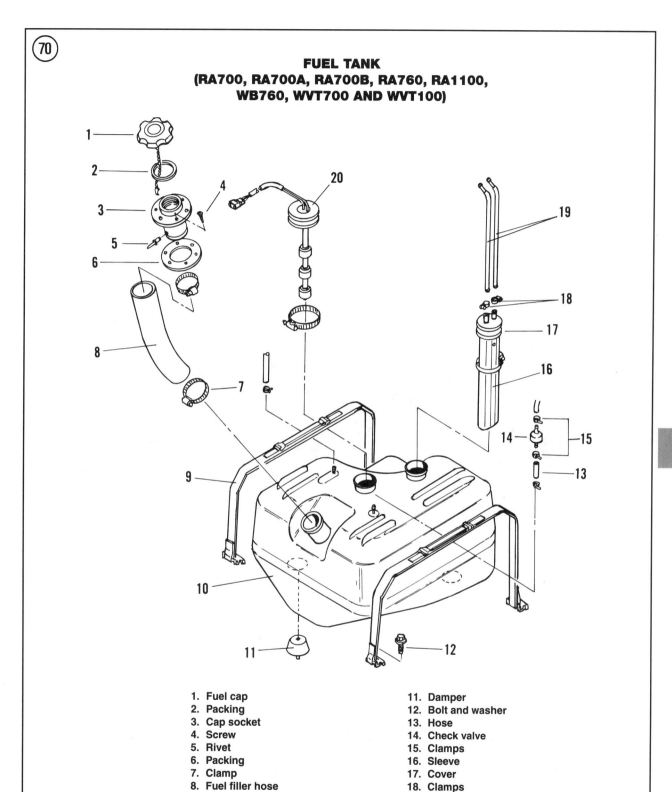

FUEL TANK (RA700, RA700A, RA700B, RA760, RA1100, WB760, WVT700 AND WVT100)

1. Fuel cap
2. Packing
3. Cap socket
4. Screw
5. Rivet
6. Packing
7. Clamp
8. Fuel filler hose
9. Strap
10. Fuel tank
11. Damper
12. Bolt and washer
13. Hose
14. Check valve
15. Clamps
16. Sleeve
17. Cover
18. Clamps
19. Pickup tubes
20. Fuel sender

368 CHAPTER EIGHT

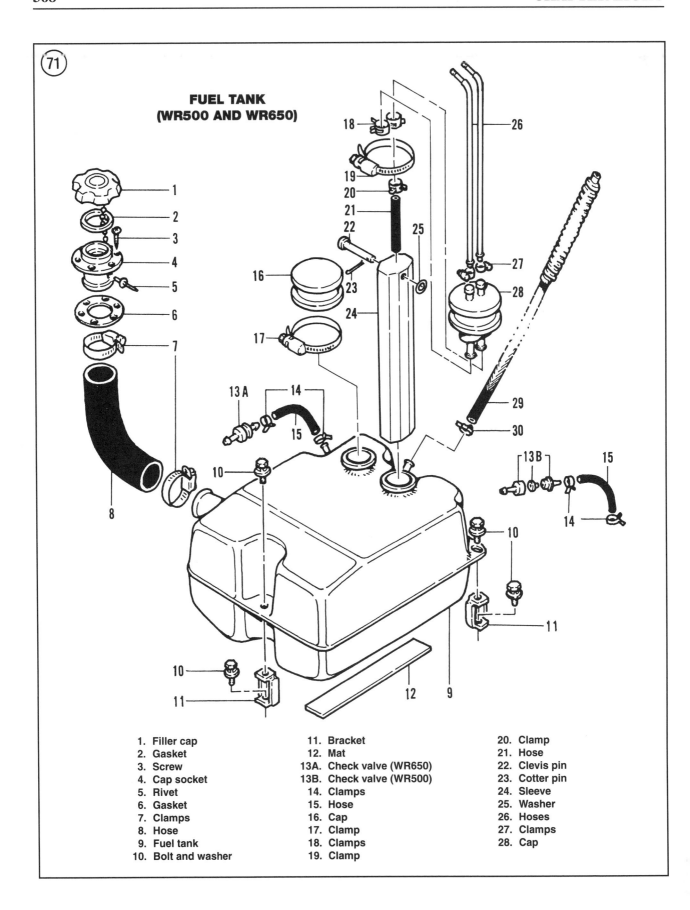

FUEL AND WATER BOX SYSTEMS

flames, including gas appliance pilot lights. Do not smoke in the area. Keep a B:C rated fire extinguisher handy.

WARNING
Before disconnecting any fuel lines, loosen the fuel filler cap to relieve any pressure in the fuel tank.

1. Review the *WARNING* at the beginning of this chapter before removing the fuel tank.
2. Remove the engine hood.
3. Disconnect the negative battery cable. Remove the battery if it is located next to the fuel tank. See Chapter Nine.

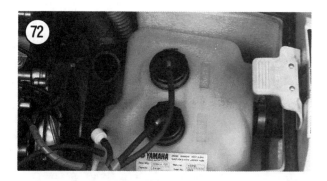

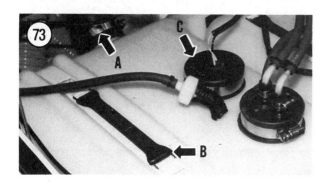

4. Disconnect the fuel sender electrical connector on models so equipped. The following models do not use a fuel sender: SJ650, SJ700, SJ700A, FX700, WB700, WB700A, WRB650 and WRB650A.
5. Siphon the contents of the tank into a container approved for gasoline storage.
6. Disconnect the fuel filler hose from the fuel tank. See **Figure 72** or A, **Figure 73**.
7. Label and disconnect all remaining fuel hoses from the fuel tank. See **Figure 73** or **Figure 74**.
8. Unhook the fuel tank's rubber straps from the clips on the hull. See B, **Figure 73** or **Figure 74**.
9. Remove the fuel sender from models so equipped (C, **Figure 73**).
10. Remove any bolts securing the tank to the hull and remove the fuel tank.
11. To install, reverse the removal steps. Note the following:
   a. Check that the fuel tank damper pads are in good condition. Install new pads, if necessary, using a waterproof contact cement.
   b. When installing the tank retainer straps, position the strap ends with their tabs toward the tank to keep the hooks from chafing the tank.
   c. Make sure the filler hose clamp and tank outlet connections are tight.

Inspection

1. Inspect the fuel tank (**Figure 75**) for cracks or leakage. Especially check all hose or tube

joints and areas where straps are positioned against the tank. Replace the tank if required.

2. Check all of the fuel tank covers, hose clamps and tubing for deterioration, cracks or damage. Replace worn or damaged parts as required. Fuel filter replacement is described in Chapter Three.

3. Inspect the pickup tubes for cracks or damage. Measure the pickup tube for both the main line and reserve line as shown **Figure 76**. Be sure the main and reserve pickup tubes are within the specifications given in **Table 3**.

4. Inspect the fuel filler tube (**Figure 77**) for cracks or damage. Check the filler tube hose clamps for weakness or damage. Replace parts as required.

5. Discard any fuel in the tank and pour about a pint of clean fuel into the tank. Install the cap, slosh the fuel around for about a minute and pour it into a safe container.

FUEL PETCOCK

A nonserviceable petcock controls fuel flow. If the petcock leaks or becomes inoperative, it must be replaced as an assembly. Refer to the appropriate illustration for your model:

a. **Figure 78**—WR500.
b. **Figure 79**—WR650.
c. **Figure 80**—SJ650.
d. **Figure 81**—SJ700 and SJ700A.
e. **Figure 82**—FX700.
f. **Figure 83**—WRA650, WRA650A and WRA700.
g. **Figure 84**—WRB650, WRB650A and WRB700.
h. **Figure 85**—WB700, WB700A and WB760.
i. **Figure 86**—RA700, RA700A and RA1100.
j. **Figure 87**—RA700B and RA760.
k. **Figure 88**—WVT700 and WVT1100.

Removal/Installation

WARNING
*Before disconnecting any fuel lines, loosen the fuel filler cap (**Figure 89**) to relieve any built-up pressure in the fuel tank.*

1. Read the *WARNING* at the beginning of this chapter before performing the following.

2. Remove the engine hood.

3. Disconnect the negative battery cable.

4. Locate the fuel petcock mounting position for your model. Remove any components that interfere with its removal. When you have access to the petcock hoses, label then disconnect each hose. Plug each hose to prevent fuel leakage and/or contamination.

5. Remove the petcock lever (**Figure 90**) and remove the ring nut holding the petcock to the hull. Remove the petcock.

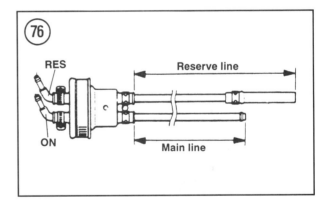

FUEL AND WATER BOX SYSTEMS

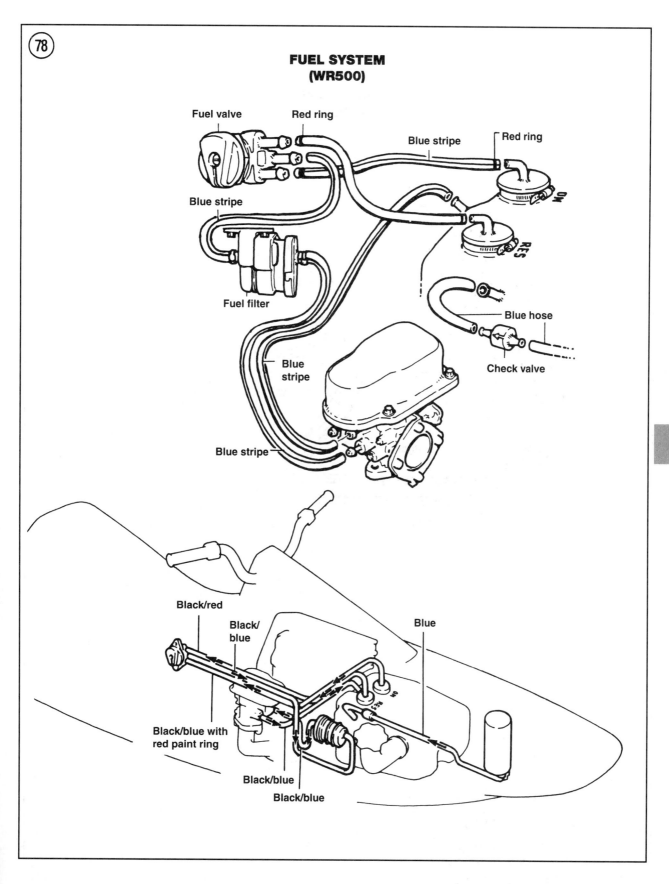

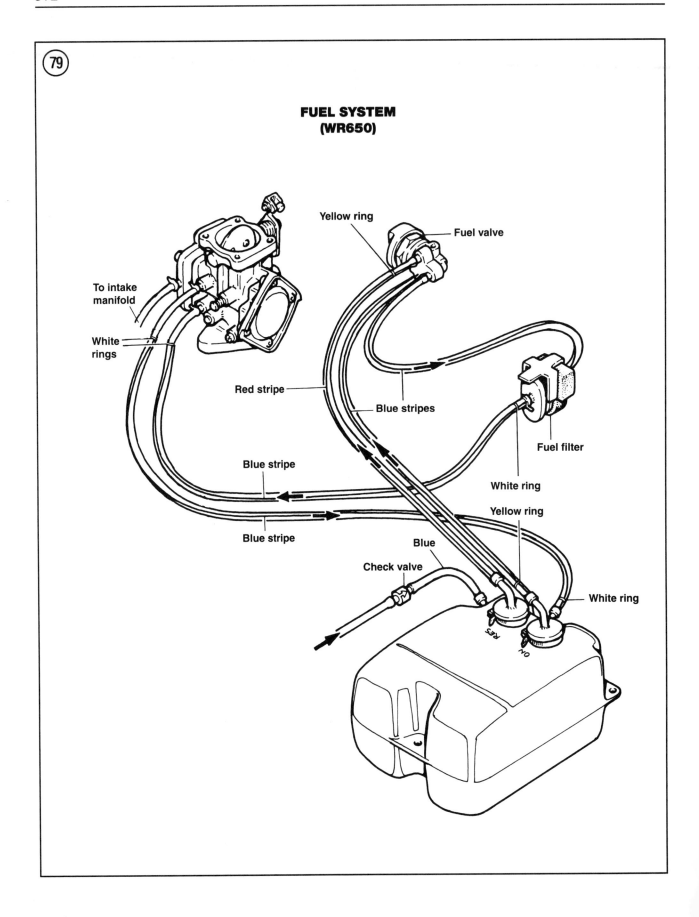

FUEL AND WATER BOX SYSTEMS

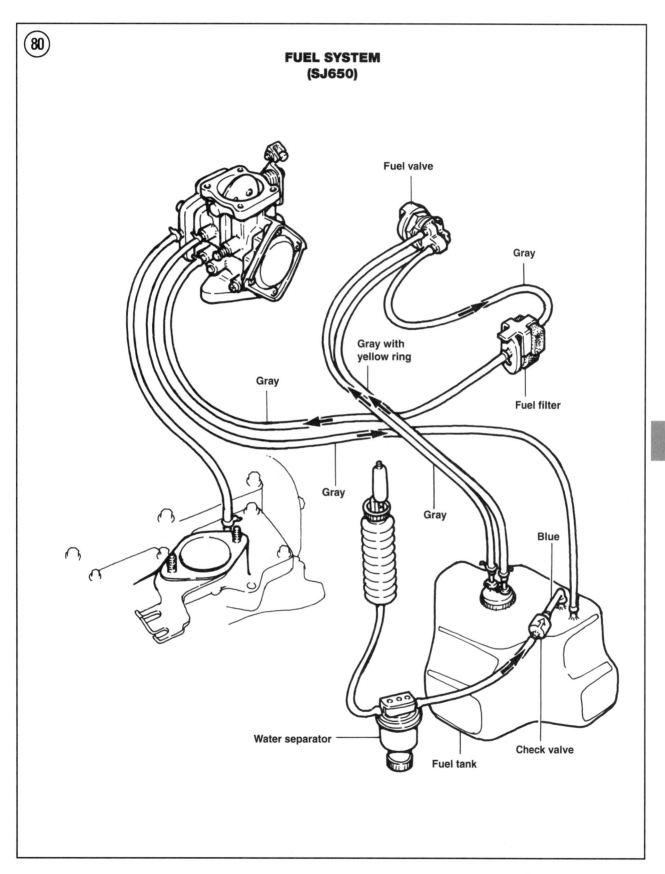

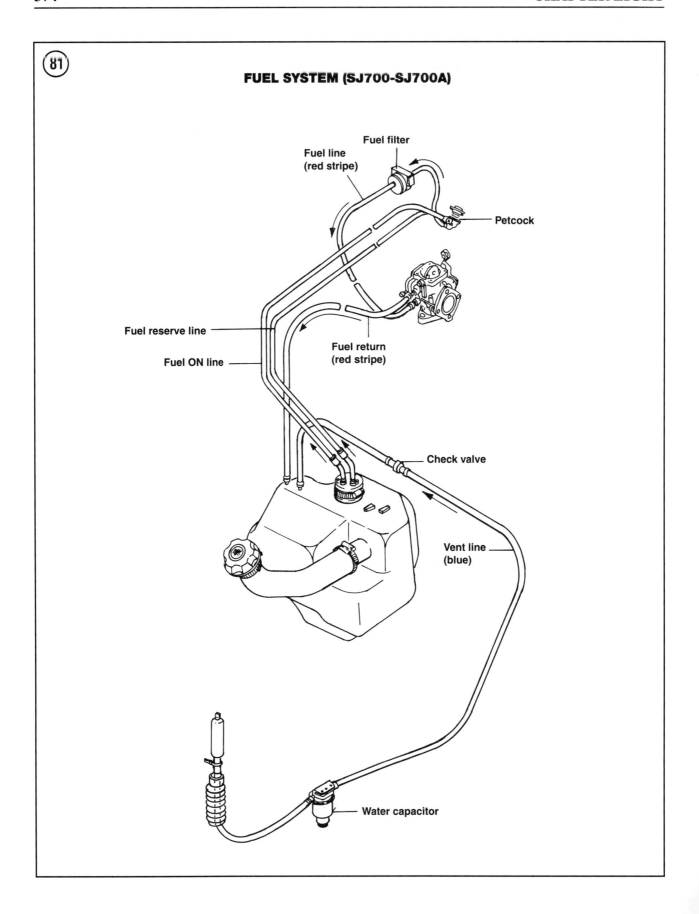

FUEL AND WATER BOX SYSTEMS

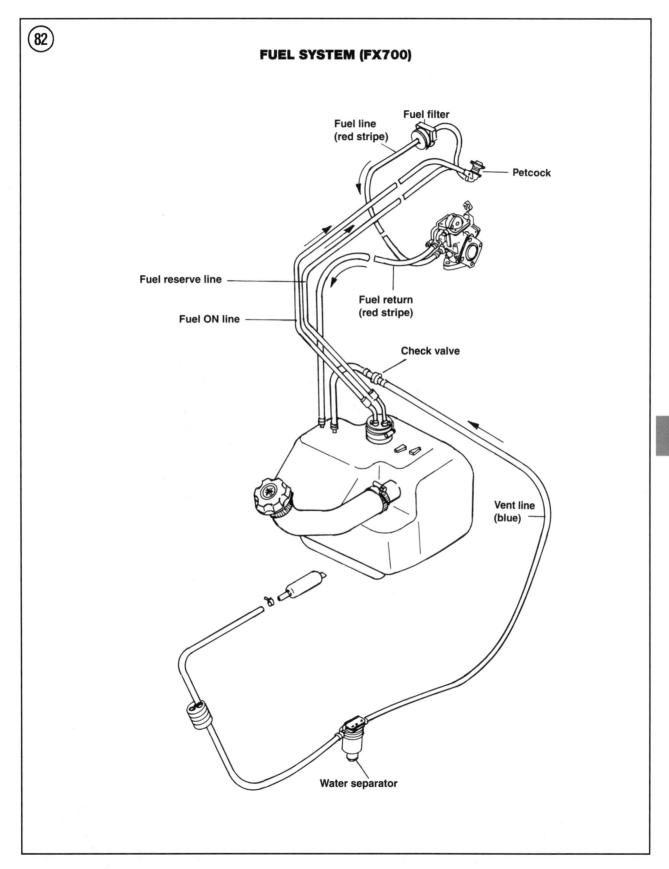

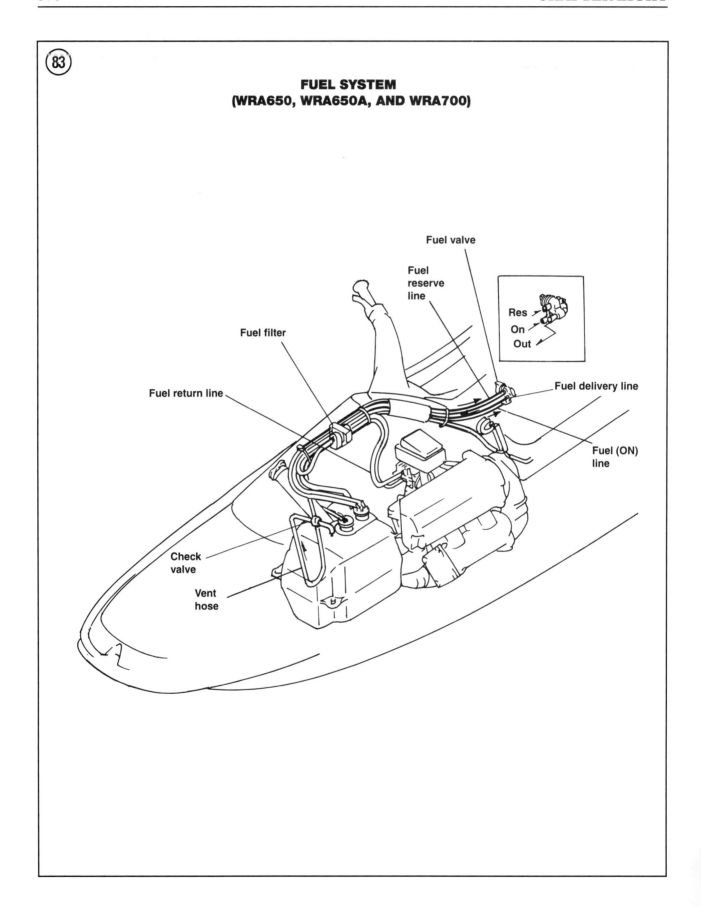

FUEL AND WATER BOX SYSTEMS

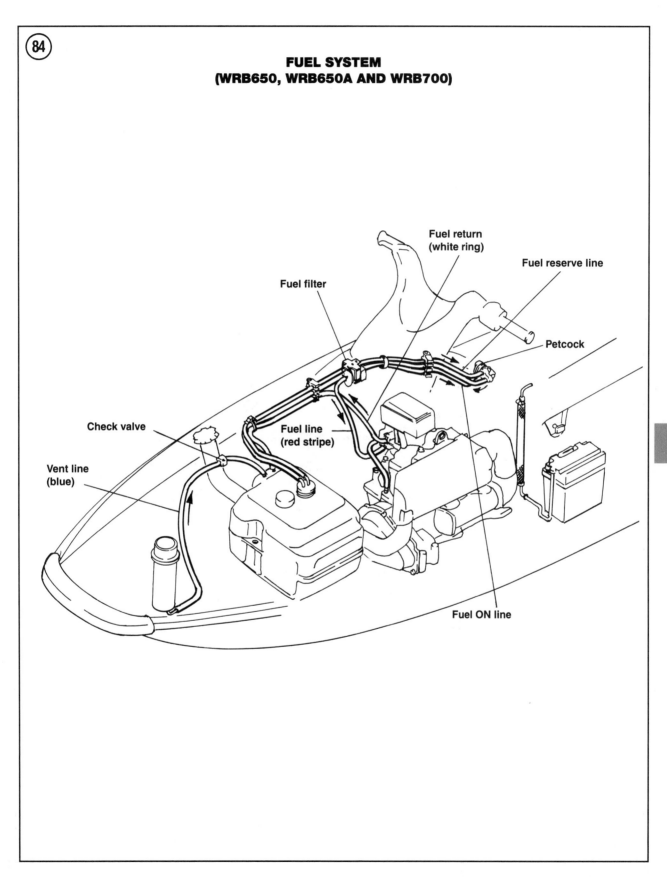

⑧⁴ **FUEL SYSTEM (WRB650, WRB650A AND WRB700)**

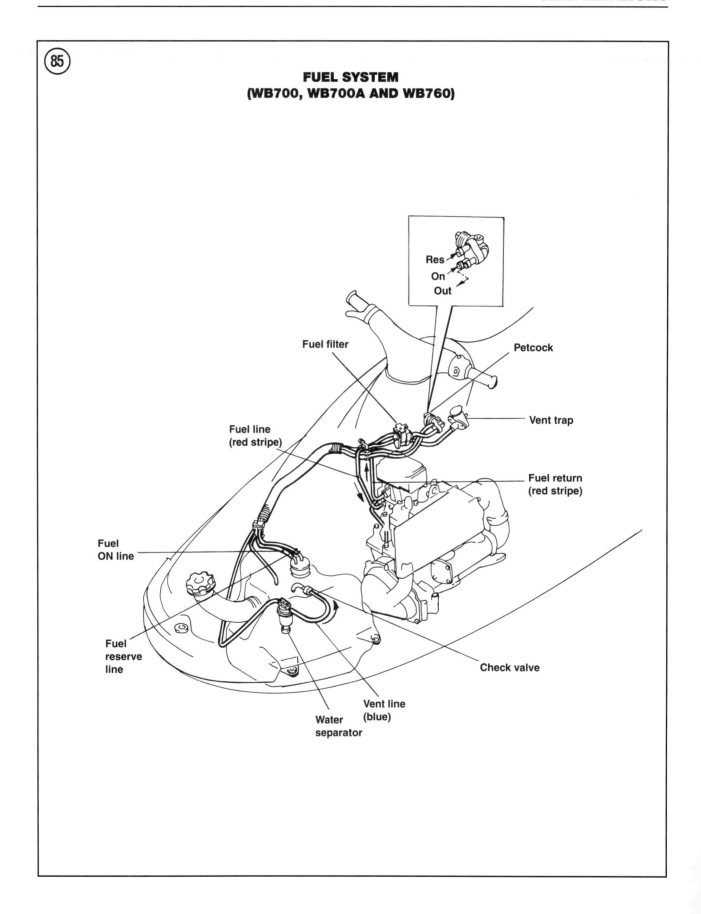

85 FUEL SYSTEM (WB700, WB700A AND WB760)

FUEL AND WATER BOX SYSTEMS

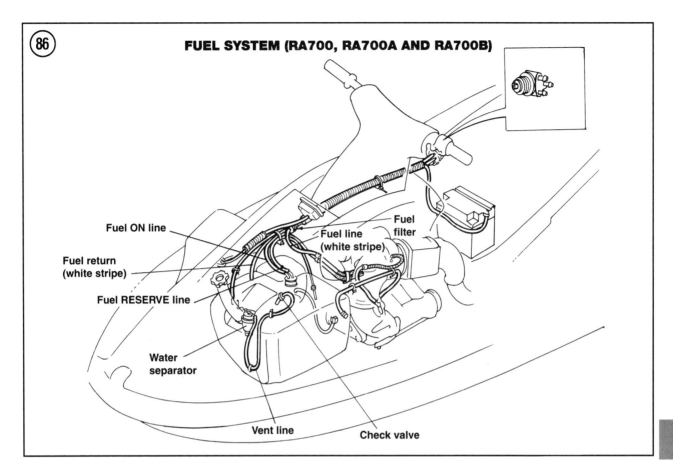

⑧⑥ **FUEL SYSTEM (RA700, RA700A AND RA700B)**

- Fuel ON line
- Fuel return (white stripe)
- Fuel RESERVE line
- Fuel line (white stripe)
- Fuel filter
- Water separator
- Vent line
- Check valve

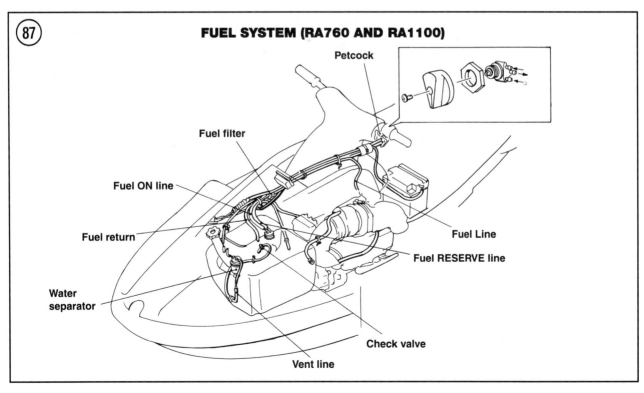

⑧⑦ **FUEL SYSTEM (RA760 AND RA1100)**

- Petcock
- Fuel filter
- Fuel ON line
- Fuel return
- Water separator
- Fuel Line
- Fuel RESERVE line
- Check valve
- Vent line

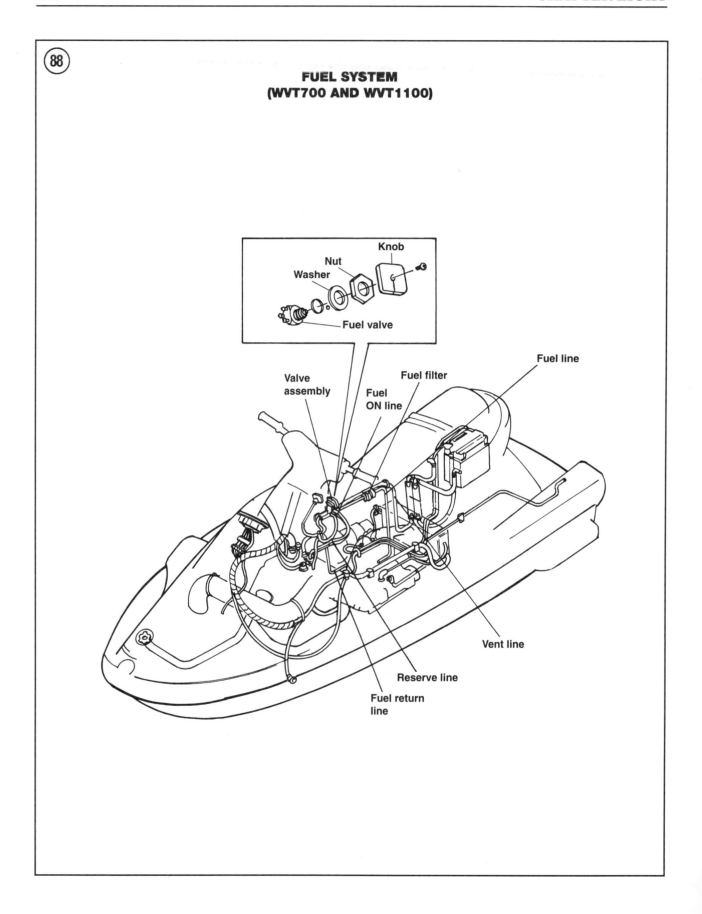

FUEL AND WATER BOX SYSTEMS

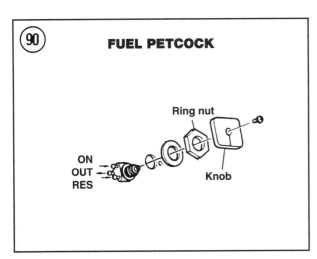

6. Installation is the reverse of these steps. Check for leaks after reinstalling the petcock and reconnecting the hoses.

Inspection

1. Inspect the petcock for proper operation by turning the valve through all of its operating positions. The valve should move smoothly without excessive play.

2. If petcock operation is sluggish or if it appears contaminated or clogged, replace it.

VENT TRAP SYSTEM
(FX700, RA700, RA700A, RA700B, RA760, RA1100, WB700, WB700A AND WB760)

WaveBlaster, WaveRaider and FX700 models have 2 ventilation traps: one vents the fuel tank, the other vents the battery. When servicing or replacing a vent trap, install it so the TOP is up as shown in **Figure 91**.

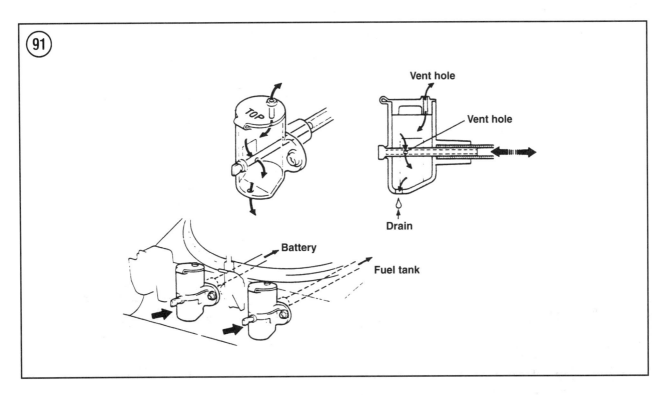

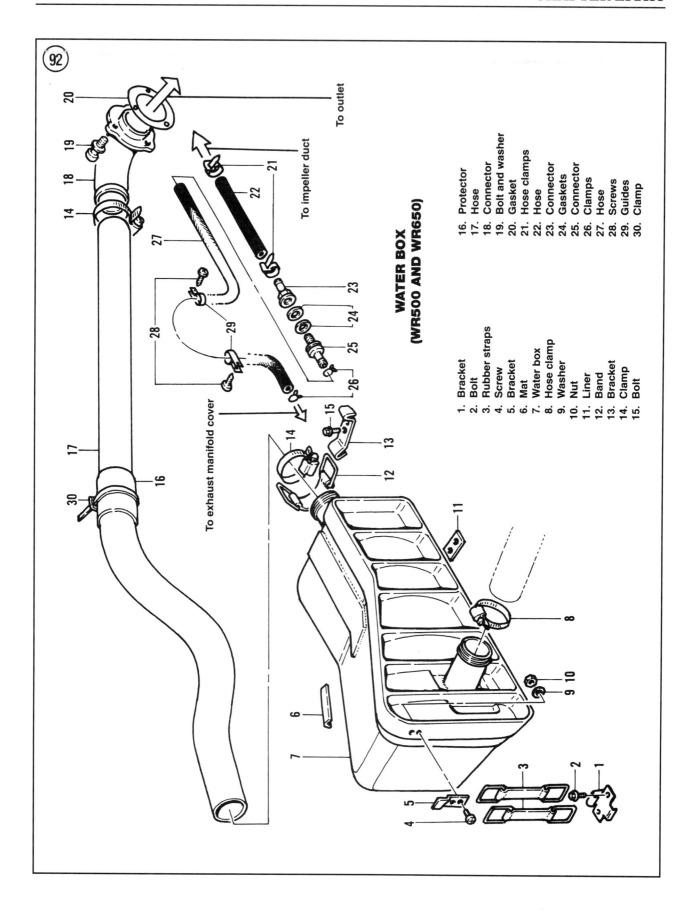

FUEL AND WATER BOX SYSTEMS

WATER BOX

Removal/Installation (WR500 and WR650)

Refer to **Figure 92** for this procedure.
1. Remove the engine as described in Chapter Four or Chapter Five.
2. Remove the seat.
3. Remove the storage compartment box.
4. Disconnect the hoses from the water box.
5. Remove the water box mounting bolt and remove the water box.
6. Inspect the water box as described in this chapter.
7. Installation is the reverse of these steps.

Removal/Installation (SJ650, SJ700, SJ700A and FX700)

Refer to **Figure 93** for this procedure.
1. Remove the battery as described in Chapter Nine.
2. Remove the engine as described in Chapter Five.
3. Remove the fuel tank as described in this chapter.
4. Disconnect the hoses from the water box.
5. Release the mounting straps at the water box and remove the water box from the hull. See **Figure 94**.
6. Inspect the water box as described in this chapter.
7. Installation is the reverse of these steps.

Removal/Installation (WRA650, WRA650A and WRA700)

Refer to **Figure 95** for this procedure.
1. Remove the oil tank as described in Chapter Ten.
2. Remove the seat and storage compartment.
3. Disconnect the hoses from the water box.
4. Release the mounting straps at the water box and remove the water box from the hull.
5. Inspect the water box as described in this chapter.
6. Installation is the reverse of these steps.

Removal/Installation (WRB650, WRB650A and WRB700)

Refer to **Figure 96** for this procedure.
1. Open and properly support the engine compartment hood.
2. Remove the oil tank as described in Chapter Ten.
3. Remove the fuel tank as described in this chapter.
4. Disconnect the hoses from the water box.
5. Release the mounting straps at the water box and remove the water box from the hull.
6. Inspect the water box as described in this chapter.
7. Installation is the reverse of these steps.

Removal/Installation (WB700 and WB700A)

Refer to **Figure 97** for this procedure.
1. Turn the petcock OFF.
2. Open and properly support the engine compartment hood.
3. Disconnect the hoses at the water box.
4. Release the mounting straps from the water box and remove the water box from the hull.
5. Inspect the water box as described in this chapter.
6. Installation is the reverse of these steps.

Removal/Installation (WB760)

Refer to **Figure 98** for this procedure.
1. Remove the battery as described in Chapter Nine.

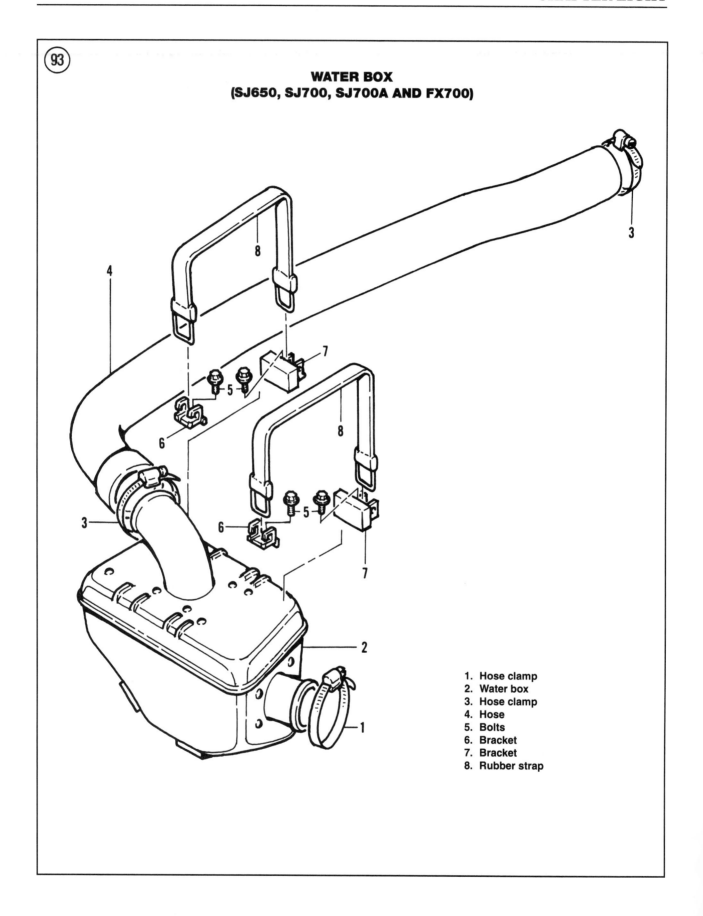

FUEL AND WATER BOX SYSTEMS

2. Remove the seat and seat lock (**Figure 99**).

3. Remove the screws securing the battery case and remove the battery case. During reinstallation, be sure the ends of the battery case packing butt together at the front centerline of the case as shown in **Figure 99**. Seal the ends with silicon sealant. Torque the battery case screws to 3 N•m (26 in.-lb.).

4. Disconnect the hoses from the water box.

5. Release the mounting straps at the water box and remove the water box from the hull.

6. Inspect the water box as described in this chapter.

7. Installation is the reverse of these steps. Be sure the forward indicator on the water box faces the front as shown in **Figure 98**.

Removal/Installation (RA700, RA700A, RA700B, RA760, RA1100, WVT700 and WVT1100)

Refer to **Figures 100-103** for this procedure.

1A. *RA700, RA700A, RA700B, RA760, RA1100*—Remove the fire extinguisher box (**Figure 104**).
   a. Remove the self-tapping screws and remove the storage box.
   b. Remove the 2 bolts securing the fire extinguisher box, and remove the fire extinguisher box. During reinstallation, apply Loctite 271 (red) to the threads of the extinguisher-box bolts, and torque the bolts to 13 N•m (115 in.-lb.).

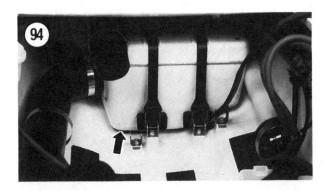

1B. *WVT700 and WVT100*—Remove the storage box (**Figure 105**).
   a. Lift the seat and remove the storage box from the hull.
   b. During installation, apply silicon sealant to the front lip of the storage box as shown in **Figure 105**. Be sure the ends of the packing butt against one another at the centerline at the storage box. Seal the ends with the silicon sealant.

2. Disconnect the hoses from the water box.

3. Release the mounting straps at the water box and remove the water box from the hull.

4. Inspect the water box as described in this chapter.

5. Installation is the reverse of these steps.

6. *RA700 and RA700B*—To secure the straps, hook the strap frame into the cutout in the bracket, and then twist the frame over the bracket and into place (**Figure 106**).

Inspection (All Models)

A leaking or damaged water box will allow the engine compartment to fill with exhaust gasses. This condition will result in poor engine operation.

1. Discard any water in the box.

2. Check all of the hose clamps for weakness or damage and replace as required.

3. Check all of the mounting brackets for looseness. If a bracket is loose, remove its mounting bolt and reinstall after applying Loctite 242 (blue) to its bolt threads. Tighten the bolt securely.

4. Check all of the rubber straps (if used) for fatigue, fraying or other damage. Replace all questionable straps.

5. Check the water box inlet for heat damage. If heat damage is present, check the cooling system for proper operation.

6. Check the water box seams for leaks. Replace the box if it is leaking or damaged.

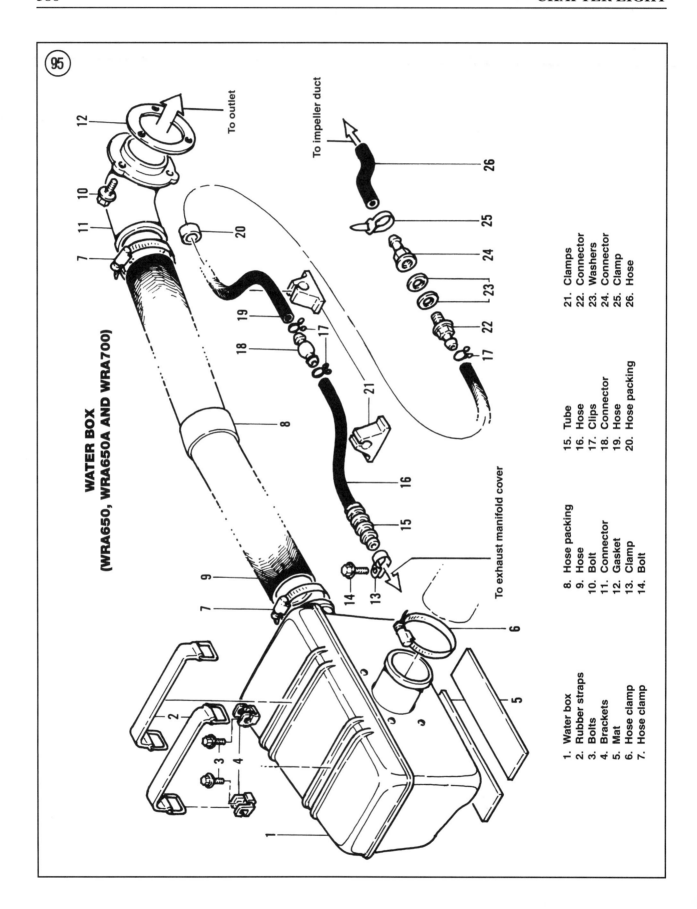

FUEL AND WATER BOX SYSTEMS

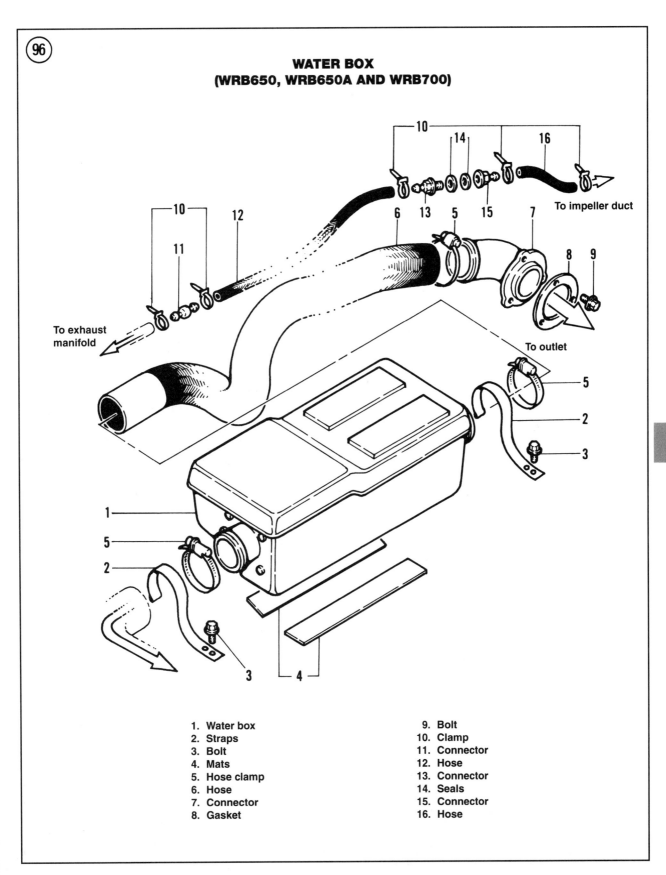

WATER BOX (WRB650, WRB650A AND WRB700)

1. Water box
2. Straps
3. Bolt
4. Mats
5. Hose clamp
6. Hose
7. Connector
8. Gasket
9. Bolt
10. Clamp
11. Connector
12. Hose
13. Connector
14. Seals
15. Connector
16. Hose

**WATER BOX
(WB700 AND WB700A)**

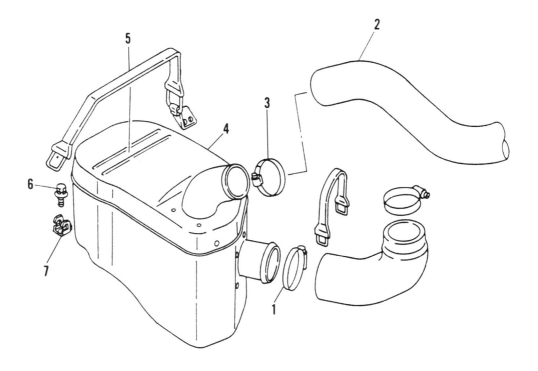

1. Clamp
2. Exhaust hose
3. Clamp
4. Water box
5. Rubber strap
6. Bolt
7. Bracket

FUEL AND WATER BOX SYSTEMS

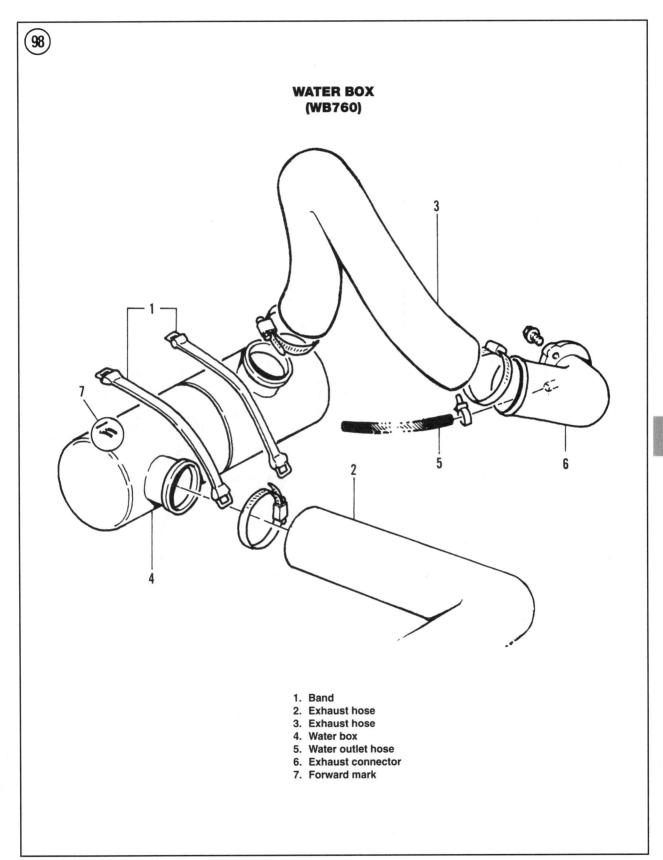

WATER BOX (WB760)

1. Band
2. Exhaust hose
3. Exhaust hose
4. Water box
5. Water outlet hose
6. Exhaust connector
7. Forward mark

CHAPTER EIGHT

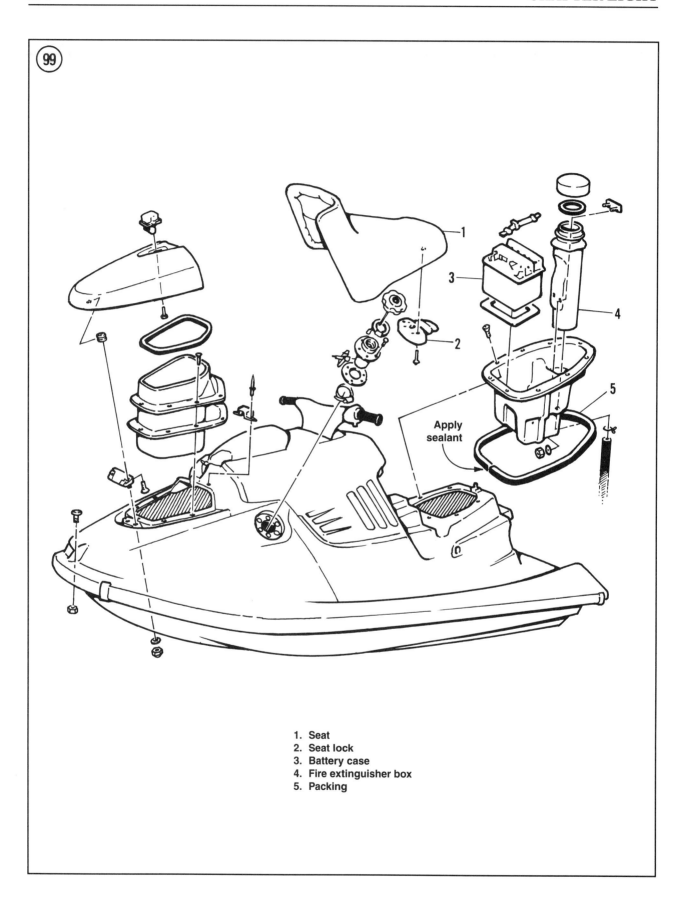

1. Seat
2. Seat lock
3. Battery case
4. Fire extinguisher box
5. Packing

FUEL AND WATER BOX SYSTEMS

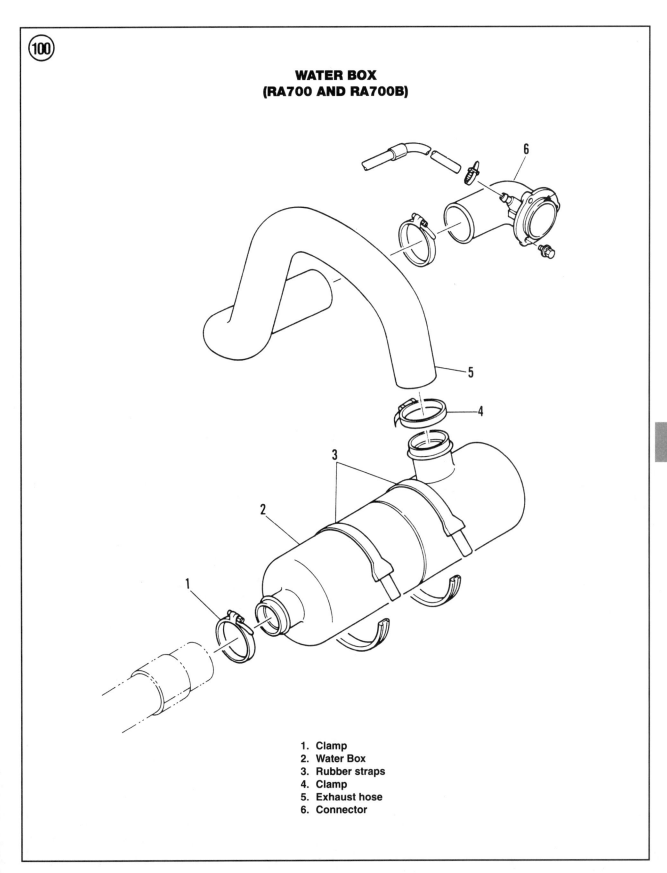

WATER BOX
(RA700 AND RA700B)

1. Clamp
2. Water Box
3. Rubber straps
4. Clamp
5. Exhaust hose
6. Connector

CHAPTER EIGHT

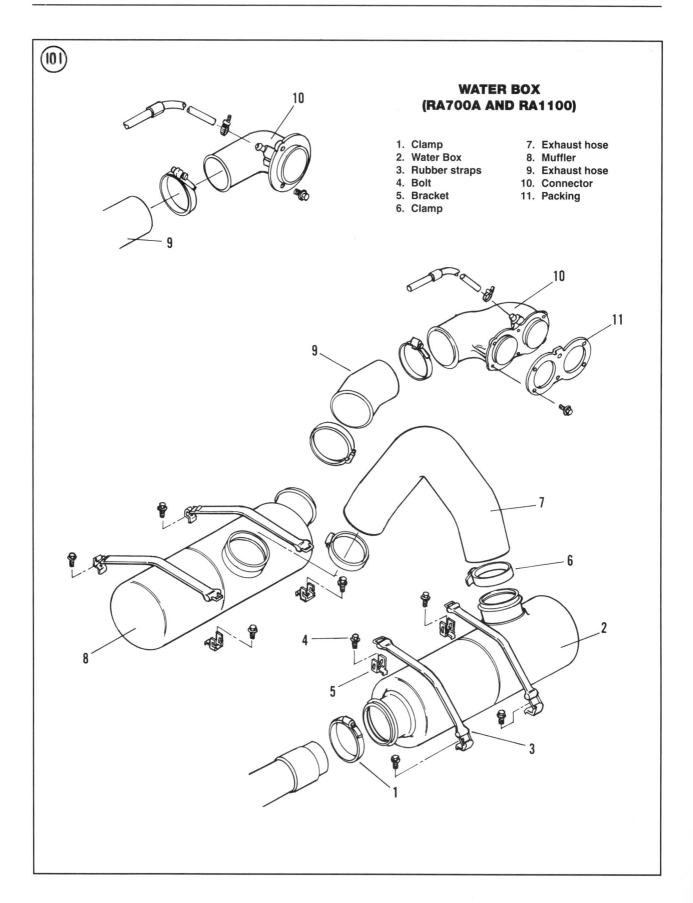

FUEL AND WATER BOX SYSTEMS

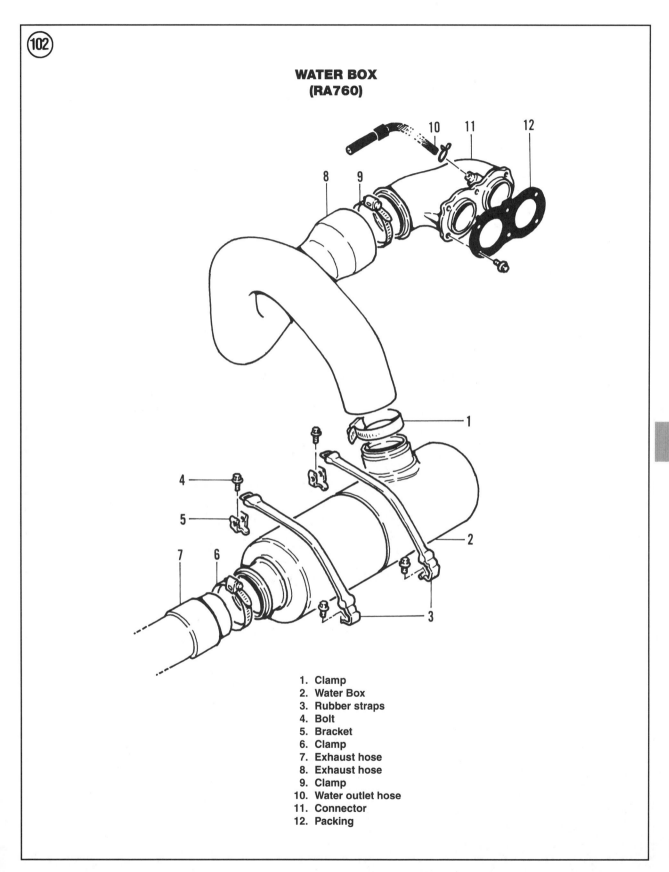

WATER BOX (RA760)

1. Clamp
2. Water Box
3. Rubber straps
4. Bolt
5. Bracket
6. Clamp
7. Exhaust hose
8. Exhaust hose
9. Clamp
10. Water outlet hose
11. Connector
12. Packing

394

CHAPTER EIGHT

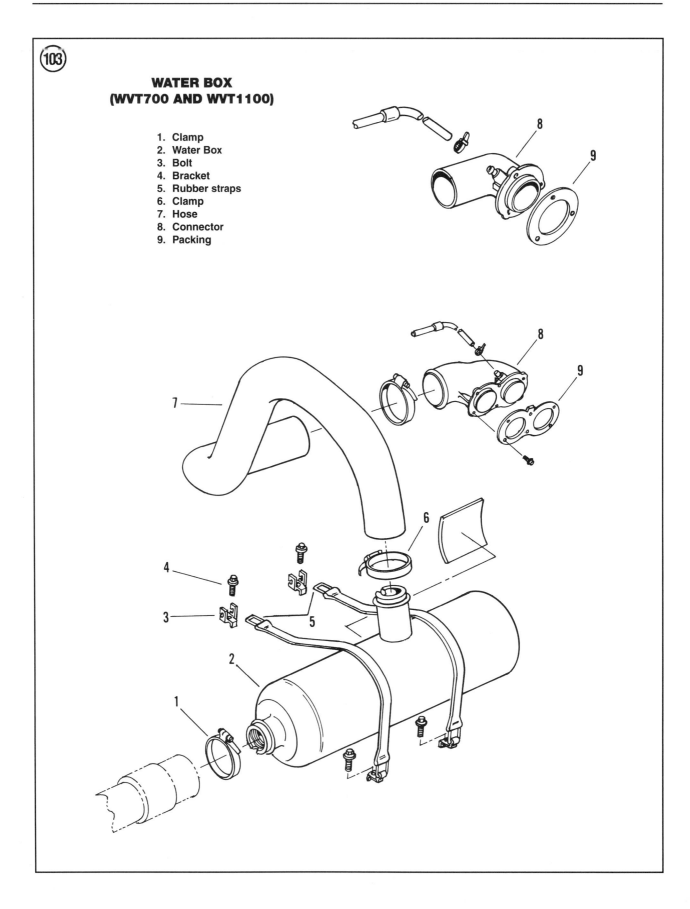

FUEL AND WATER BOX SYSTEMS

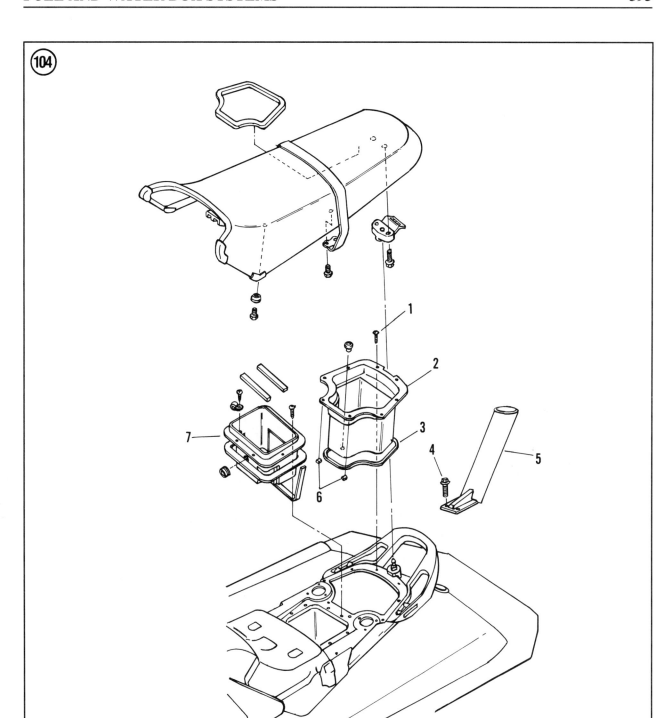

1. Self-tapping screws
2. Storage box
3. Packing
4. Bolts
5. Fire extinguisher box
6. Packing (RA700 only)
7. Battery case (Except RA700)

396 CHAPTER EIGHT

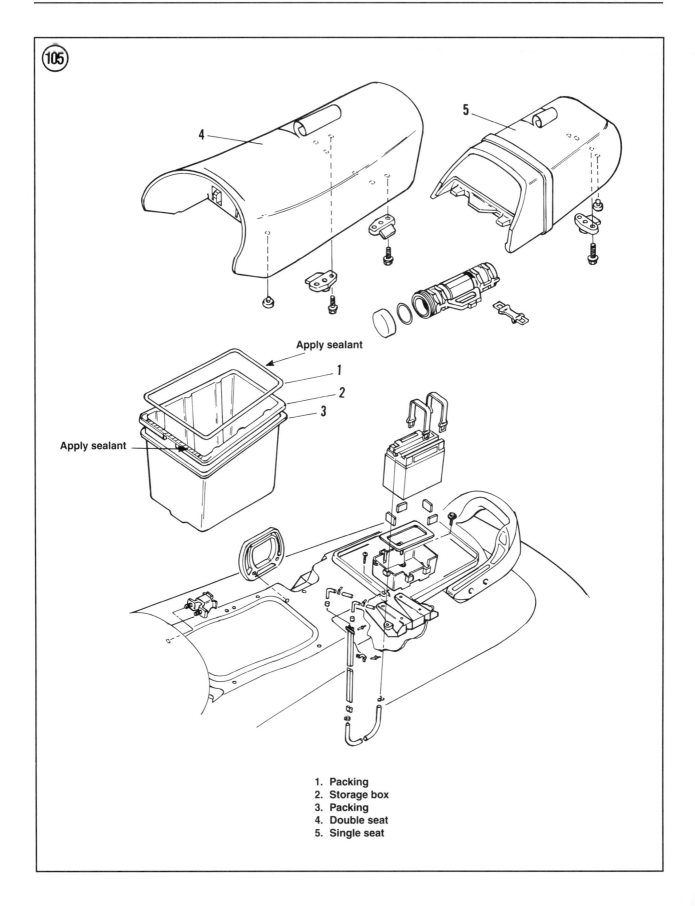

FUEL AND WATER BOX SYSTEMS

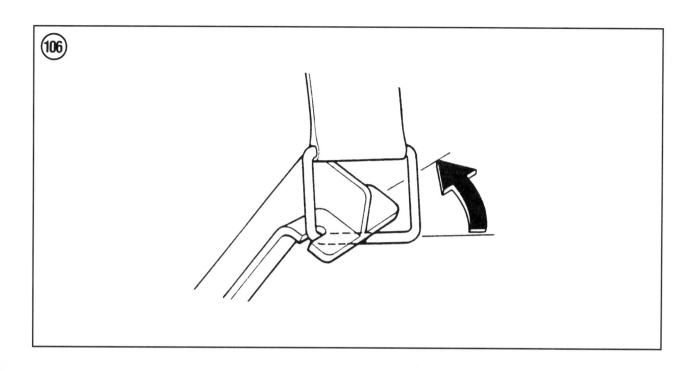

Table 1 CARBURETOR SPECIFICATIONS

Model	Carburetor identification	Main jet mm	Pilot jet mm
FX700, SJ700	61X00-01	135	115
SJ650	6R702	115	120
SJ700A	64U00F/R	130	70
RA700, RA700A			
Front carb	62T00F	120	67.5
Rear carb	62T00	130	67.5
RA700	61X01	135	115
RA760			
Front carb	64X01	135	1.5
Rear carb	64X02	137.5	1.5
RA1100			
Front carb	63M00F	107.5	75
Middle carb	63M00C	95	75
Rear carb	63M00R	107.5	75
WB700, WB700A	61X00	135	115
WB760			
Front carb	64Y01	135	1.5
Rear carb	64Y02	137.5	1.5
WR500	6K811	80	77.5
WR650	6M600-11	115	115
WRA650, WRA650A	6R801	115	120
WRA700	61X01	135	115
WRB650, WRB650A	61L00-L02	125	120
WRB700	61X00	135	115
WVT700			
Front carb	62T01F	120	67.5
Rear carb	62T01R	130	67.5
(continued)			

Table 1 CARBURETOR SPECIFICATIONS (continued)

Model	Carburetor identification	Main jet mm	Pilot jet mm
WVT1100			
Front carb	64T00F	107.5	75
Middle carb	64T00C	95	75
Rear carb	64T00R	107.5	75

Table 2 REED VALVE SERVICE SPECIFICATIONS

	mm	in.
Thickness		
RA1100, WVT1100	0.42	0.017
RA760, WB760	0.4	0.016
All other models	0.2	0.008
Valve stopper height		
SJ650, WRA650, WRA650A	10.5 ± 0.2	0.415 ± 0.008
SJ700A	9.0 ± 0.2	0.35 ± 0.01
WR500	6.0	0.024
WR650	9.6	0.38
WRB650, WRB650A	9.9 ± 0.2	0.390 ± 0.008
All other models	10.9 ± 0.2	4.29 ± 0.008
Bending limit		
WR500	0.9	0.04
All other models	0.2	0.008

Table 3 PICKUP TUBE SPECIFICATIONS

	Main line mm (in.)	Reserve line mm (in.)
FX700	148.5 ± 2 (5.85 ± 0.08)	234 ± 2 (9.21 ± 0.08)
RA700, RA700A, RA700B, RA760, WB700, WB700A, WB760	180 ± 2 (7.09 ± 0.08)	266 ± 2 (10.7 ± 0.08)
RA1100, WVT700, WVT1100	245 ± 2 (9.65 ± 0.08)	320 ± 2 (12.6 ± 0.08)
SJ650, SJ700	165 ± 2 (6.50 ± 0.08)	247 ± 2 (9.72 ± 0.08)
SJ700A	165 ± 2 (6.50 ± 0.08)	242 ± 2 (9.53 ± 0.08)
WR500, WR650	101 ± 2 (3.98 ± 0.08)	165 ± 2 (6.50 ± 0.08)
WRA650, WRA650A, WRA700	245 ± 2 (9.65 ± 0.08)	305 ± 2 (12.0 ± 0.08)
WRB650, WRB650A, WRB700	180 ± 2 (7.09 ± 0.08)	240 ± 2 (9.45 ± 0.08)

Chapter Nine

Electrical System

This chapter provides service procedures for the battery, starting system, charging system, ignition system and signal system. Electrical troubleshooting procedures are described in Chapter Two. Wiring diagrams are at the end of the book. **Tables 1-4** are at the end of the chapter.

BATTERY

The electric starting system requires a fully charged battery to provide the large amount of current required to operate the starter motor. Because there are no lights on the water vehicle, the battery's sole purpose is to provide power to the starter motor. A lighting coil (mounted on the stator plate) and a voltage regulator, connected in circuit with the battery, keep the battery charged while the engine is running. The battery can also be charged externally.

Care and Inspection

The battery is the heart of the electrical system. Most electrical system trouble can be attributed to neglect of this vital component.

In order to service the electrolyte level correctly, it is necessary to remove the battery from the hull. Maintain the electrolyte level between the 2 marks on the battery case. If the electrolyte level is low, remove the battery completely so it can be thoroughly cleaned, serviced and checked.

On all models covered in this manual, the negative side is grounded. When removing the battery, disconnect the negative (−) ground cable first, then the positive (+) cable. This minimizes the chance of a tool shorting to ground when disconnecting the hot positive cable.

WARNING
When performing the following procedure, protect your eyes, skin and clothing. If electrolyte gets into your eyes, flush your eyes thoroughly with clean water and get prompt medical attention.

Battery location

On WRA650, WRA650A and WRA700 models, the battery is in the battery compartment with its own hatch.

On RA700, RA700A, RA700B, RA760, RA1100, WB700, WB700A, WB760, WVT700 and WVT1100 models, the battery is in the battery compartment beneath the seat.

On SJ650, SJ700, SJ700A, FX700, WRB650, WRB650A and WRA700 models, the battery is located in the engine compartment.

1. Open the engine compartment or the battery compartment to access the battery on your model. (See **Figure 1** or **Figure 2**).
2. Unlatch the battery strap or straps and carefully position the battery to access the negative battery cable (**Figure 3**).
3. Disconnect the negative battery cable from the battery and pull the cable away from the battery so it cannot accidentally fall back against the negative battery terminal.
4. Disconnect the positive battery cable.
5. Disconnect the battery vent tube (**Figure 4**) from the battery and remove the battery.

CAUTION
Be careful not to spill battery electrolyte on painted or polished surfaces. The liquid is highly corrosive and will damage the finish. If it is spilled, wash it off immediately with soapy water and thoroughly rinse with clean water.

6. Check the entire battery case for cracks. Replace the battery if it is cracked.
7. Inspect the battery tray (**Figure 5**) or the battery compartment for corrosion. Clean if necessary with a solution of baking soda and water. Check that the mounting screws or straps are tight.
8. Check the battery hold-down straps for deterioration, cracks or other damage. Replace the straps as required.

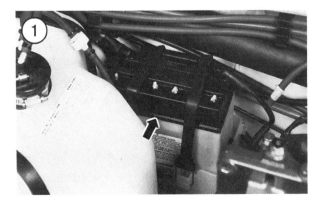

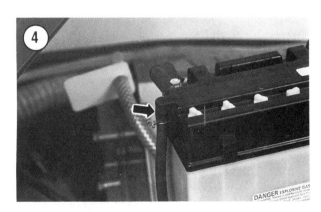

ELECTRICAL SYSTEM

NOTE
Keep cleaning solution out of the battery cells in Step 9 or the electrolyte level will be seriously weakened.

9. Clean the top of the battery with a stiff bristle brush using the baking soda and water solution. Rinse the battery case with clean water and wipe it dry with a clean cloth or paper towel.

10. Check the battery cable clamps for corrosion and damage. If corrosion is minor, clean the battery cable clamps with a stiff wire brush. Replace severely worn or damaged cables.

NOTE
Do not overfill the battery cells in Step 11. The electrolyte expands due to heat from charging and will overflow if the level is above the upper level line.

11. Remove the caps (**Figure 6**) from the battery cells and check the electrolyte level. Add distilled water, if necessary, to bring the level within the upper and lower level lines on the battery case (**Figure 7**).
12. Reconnect the positive battery cable, then the negative cable.

CAUTION
Be sure the battery cables are connected to their proper terminals. Connecting the battery backward will reverse the polarity and damage the rectifier and ignition system.

WARNING
*After installing the battery, make sure the vent tube (**Figure 4**) is not pinched. A pinched or kinked tube would allow high pressure to accumulate in the battery and cause the battery electrolyte to overflow. If the tube is pinched or otherwise damaged, install a new tube.*

13. Place the battery in its tray or compartment and secure it with the mounting strap or straps.
14. Tighten the battery connections and coat them a water resistant grease.

Testing

Hydrometer testing is the best way to check battery condition. Use a hydrometer with numbered graduations from 1.100 to 1.300 rather than one with just color-coded bands. To use the hydrometer, squeeze the rubber ball, insert the tip into the cell and release the ball (**Figure 8**).

NOTE
Do not attempt to test a battery with a hydrometer immediately after adding water to the cells. Charge the battery for 15-20 minutes at a rate high enough to cause vigorous gassing.

Draw enough electrolyte to float the weighted float inside the hydrometer. When using a temperature-compensated hydrometer, release the electrolyte and repeat this process several times to make sure the hydrometer has adjusted to the electrolyte temperature before taking the reading.

Hold the hydrometer vertically and note the number aligned with the surface of the electrolyte (**Figure 9**). This is the specific gravity for this cell. Return the electrolyte to the cell from which it came.

The specific gravity of a cell is the indicator of the cell's state of charge. A fully charged cell will read 1.260 or more at 80° F (26.7° C). A cell that is 75% charged will read from 1.220-1.230 while a cell with a 50% charge will read from 1.170-1.180. Any cell reading 1.160 or less should be considered discharged. All cells should be within 30 points specific gravity of each other. If over 30 pints variation is noted, the battery's condition is questionable. Charge the battery and recheck the specific gravity. If 30 points or more variation remains between cells after charging, the battery has failed and must be replaced.

NOTE
If a temperature-compensated hydrometer is not used, add 0.004 points to the specific gravity reading for every 10° above 80° F (26.7° C). For every 10° below 80° F (26.7° C), subtract 0.004 points.

Charging

A good state of charge must be maintained in batteries used for starting. When charging the battery, note the following:

a. During charging, the cells will show signs of gas bubbling. If one cell has no gas

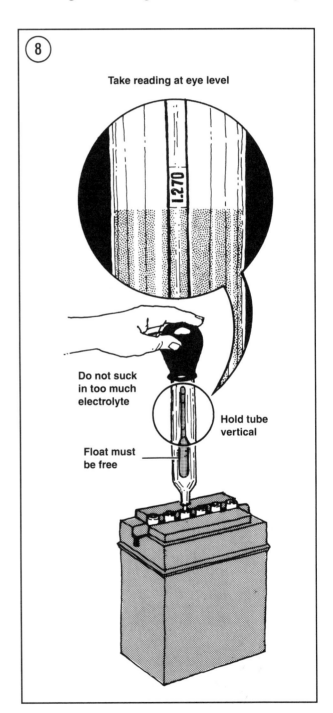

ELECTRICAL SYSTEM

bubbles or if its specific gravity is low, the cell is probably defective.

b. If a battery that is not in use looses its charge within a week after charging or if the specific gravity drops quickly, the battery is defective. A good battery should self-discharge approximately 1% each day.

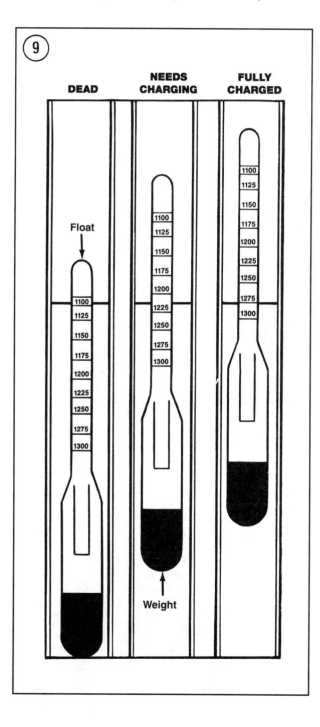

CAUTION
Always remove the battery from the hull before connecting charging equipment.

WARNING
During charging, highly explosive hydrogen gas is released from the battery. Charge the battery only in a well-ventilated area. Open flames and cigarettes should be kept away. Never check the charge of the battery by arcing across the terminals. The resulting spark can ignite the hydrogen gas.

1. Remove the battery from the hull as described in this chapter.

2. Connect the positive (+) charger lead to the positive battery terminal, and the negative (−) charger lead to the negative battery terminal.

3. Remove all vent caps from the battery, set the charger to 12 volts and switch it on. Normally, a battery should be charged at a slow charge rate of 1/10 its given capacity. See **Table 1** for battery capacity.

CAUTION
Maintain the electrolyte level at the upper level during the charging cycle. Check and refill with distilled water as necessary.

4. The charging time depends on the discharged condition of the battery. Use the chart in **Figure 10** to determine approximate charging times at

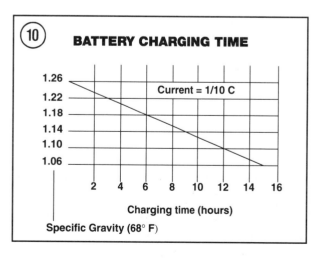

different specific gravity readings. For example, if the specific gravity of your battery is 1.180, the approximate charging time is 6 hours.

5. After the battery has charged for the recommended number of hours, turn the charger off, disconnect the leads and check the specific gravity.

New Battery Installation

When replacing the old battery, be sure to charge it completely (specific gravity, 1.260-1.280) before installing it in the hull. Failure to do so or using the battery with a low electrolyte level will permanently damage the battery.

Jump Starting

If the battery becomes severely discharged, it is possible to start and run an engine by jump starting it from another battery. If the proper procedure is not followed, however, jump starting can be dangerous. Check the electrolyte level before jump starting any battery. If it is not visible or if it appears to be frozen, *do not* attempt to jump start the battery.

The booster battery must be a 12-volt battery.

WARNING
Use extreme caution when connecting a booster battery to one that is discharged to avoid personal injury or damage to the system. Do not lean over the battery when making the connections. Wear safety glasses when performing the following procedure.

1. Secure the vehicle with the dead battery with a strong rope (minimum 500 lb. test) connected to the rear of the water vehicle and to a stationary object (**Figure 11**). Be sure to choose an object that is strong enough to withstand the full thrust of the water vehicle.
2. Remove the engine cover or open the battery compartment.

3. Connect the jumper cables in the following order:
   a. Connect the positive (+) jumper cable between the 2 battery positive terminals.
   b. Connect one end of the negative (−) jumper cable to the booster battery negative terminal. Connect the opposite end to an unpainted cylinder head bolt. *Do not* connect the jumper cable to the negative battery terminal on the dead battery.

WARNING
An electrical arc may occur when the final connection is made. This could cause an explosion if it occurs near the battery. For this reason, make the final connection to a good ground away from the battery and not to the battery itself.

4. Check that all jumper cables are out of the way of moving engine parts.

ELECTRICAL SYSTEM

NOTE
When attempting to start the engine in Step 5, do not operate the starter longer than 5 seconds. Excessive starter operation will overheat the starter and cause damage. Allow 15 seconds between starting attempts.

5. Start the engine. Once it starts, run it at a moderate speed.

CAUTION
Running the engine at wide-open throttle may cause damage to the electrical system.

6. Remove the jumper cables in the exact reverse order of connection.
7. Install the engine cover, battery compartment hatch, or the seat and remove the test rope.

CHARGING SYSTEM

The charging system consists of a battery, lighting coil (**Figure 12**), flywheel with attached magnets and a regulator/rectifier.

As the flywheel turns, the magnets located in the rim of the flywheel (**Figure 13**), rotate past the lighting coil, create alternating current. This current is sent to the rectifier where it is converted into direct current to charge the battery. The regulator is a solid-state device that prevents overcharging of the battery.

A malfunction in the charging system generally causes the battery to remain undercharged.

Perform the following visual inspection to determine the cause of the problem. If the visual inspection proves satisfactory, test the charging system as described in Chapter Two.

1. Make sure the battery cables are connected properly. The red cable must be connected to the positive battery terminal. If polarity is reversed, check for a damaged rectifier.
2. Inspect the terminals for loose or corroded connections. Tighten or clean as required.
3. Inspect the physical condition of the battery. Look for bulges or cracks in the case, leaking electrolyte or corrosion buildup.
4. Carefully check the wiring between the lighting coil and battery for chafing, deterioration or other damage.
5. Check the circuit wiring for corroded or loose connections. Clean, tighten or connect as required.

ELECTRIC STARTING SYSTEM

The starting system consists of the engine stop switch, start switch, starter solenoid, battery and starter motor. Starting system operation and troubleshooting are described in Chapter Two.

Starter Motor

The starter motor produces a very high torque but only for a brief period of time, due to heat buildup. Never operate the starter motor continuously for more than 5 seconds. Let the motor cool for at least 15 seconds before operating it again.

If the starter motor does not turn, check the battery and all connecting wiring for loose or corroded connections. If this does not solve the problem, refer to Chapter Two.

Removal/Installation

1. Remove the engine cover.
2. Disconnect the battery ground lead.

3. Remove the exhaust system as described in Chapter Four, Chapter Five or Chapter Six.
4. Disconnect the starter relay lead from the starter motor.
5. Remove the bolts holding the starter motor to the engine (**Figure 14**). Remove the starter motor. **Figure 15** and **Figure 16** show 2 typical configurations used on Yamaha watercraft.
6. Installation is the reverse of these steps. Note the following:
   a. Clean the mounting lug on the starter and the flywheel cover where the starter is grounded.
   b. Check the condition of the starter O-ring and replace if worn or damaged. See **Figure 17** or **Figure 18**. Apply water-resistant grease to the O-ring before installation.
   c. Apply Loctite 242 (blue) to the starter mounting bolts.

Starter Motor Identification

The starter motors described in this section are basically the same with minor differences. It is important to pay particular attention to the location and order of parts during service. The starter motors are identified as follows:

The Type I starter motor is used on the following model:
  a. WR500.

The Type II starter motor is used on the following models:
  b. SJ650, SJ700 and FX700.
  c. WR650.
  d. WRA650, WRA650A and WRA700.
  e. WRB650, WRB650A and WRB700.
  f. 1994 RA700.

The Type III starter motor is used on the following models:
  a. SJ700A
  b. 1995-On RA700, RA700A, RA700B, RA760 and RA1100
  c. WB760.
  d. WVT700 and WVT1100.

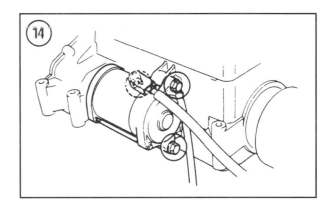

ELECTRICAL SYSTEM

Disassembly/Reassembly
(Type I—WR500 Models)

Refer to **Figure 19** for this procedure.

1. Place alignment marks on the front cover, housing and rear cover to aid reassembly. See **Figure 20**.

NOTE
An impact driver is required to loosen the starter housing through-bolts (A, Figure 21) and the brush holder screws (B, Figure 21). Attempting to loosen the

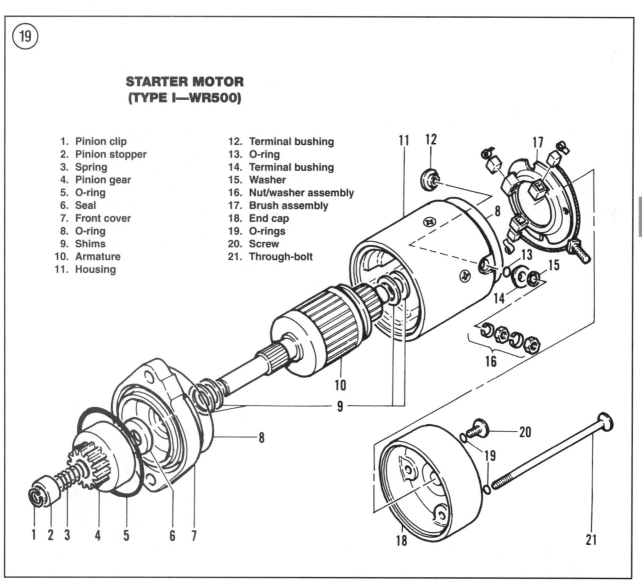

STARTER MOTOR (TYPE I—WR500)

1. Pinion clip
2. Pinion stopper
3. Spring
4. Pinion gear
5. O-ring
6. Seal
7. Front cover
8. O-ring
9. Shims
10. Armature
11. Housing
12. Terminal bushing
13. O-ring
14. Terminal bushing
15. Washer
16. Nut/washer assembly
17. Brush assembly
18. End cap
19. O-rings
20. Screw
21. Through-bolt

screws or bolts with a screwdriver may ruin the screw heads.

2. Loosen and remove the 2 through-bolts (A, **Figure 21**). Then slide the armature and front cover out of the center housing.

3. Remove any shims installed on the armature shaft next to the commutator (**Figure 22**).

4. Remove the end cover as follows:
   a. Remove the 2 brush holder screws (B, **Figure 21**).
   b. Tap the end cover off of the housing.

5. If it is necessary to remove the pinion gear assembly (**Figure 19**), perform the following:
   a. Mount the armature in a vise with soft jaws.
   b. Pry the pinion clip out of the groove in the end of the armature shaft (**Figure 23**).
   c. Remove the pinion clip, pinion stopper, spring, pinion gear and front cover.
   d. Remove the shims from the end of the armature.

6. Pry the 2 positive (+) springs out of the brush holder and remove the brush holder.

7. Inspect the starter assembly and perform the electrical test procedure described under *Inspection* in this chapter.

8. Lightly grease both ends of the armature shaft before installation.

9. If the pinion gear was removed, perform the following:
   a. Install the shims onto the end of the armature shaft as shown in **Figure 19**.
   b. Slide the front cover onto the armature shaft.
   c. Install the pinion gear, spring and pinion stopper.
   d. Compress the spring and install a new pinion clip into the groove in the armature shaft. See **Figure 23**.
   e. Compress the spring and check that the pinion clip seats in the groove completely. See **Figure 24**.

10. Install the brush holder and springs. Then push the brushes against the springs to allow armature installation.

11. Make sure an O-ring is installed on each side of the front cover.

12. Apply a thin coating of silicone sealant to the front cover.

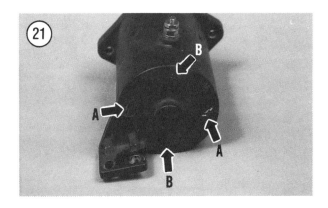

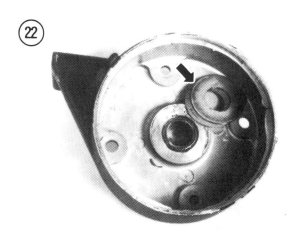

ELECTRICAL SYSTEM

13. Align the mark (made prior to disassembly) on the front cover with the housing mark and insert the housing over the armature.

14. Align the brush plate so the long bolts will pass through the housing and screw into the front end cover.

15. Install the shims (**Figure 22**) onto the end of the armature next to the commutator.

16. Make sure the O-ring is installed on the end of the housing. Then apply a thin coating of silicone sealant to the mating surface and install the end cover. Rotate the end cover and align the marks made prior to disassembly.

17. Install an O-ring on each brush holder screw. Then install the screws and tighten securely.

18. Install the 2 long through-bolts and tighten securely.

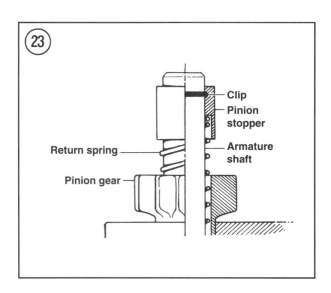

**Disassembly/Reassembly
(TYPE II—1994 RA700, SJ650, SJ700, FX700, WR650, WRA650, WRA650A, WRA700, WRB650, WRB650A and WRB700) (TYPE III—1995-ON RA700, RA700A, RA700B, RA760, RA1100, SJ700A, WB760, WVT700 and WVT1100)**

Refer to **Figure 25** or **Figure 26** for this procedure. 1. Place alignment marks on the front cover, housing and rear cover to aid reassembly.

2. Loosen the 2 case through-bolts. Then remove the bolts and washers.

3. Remove the front and rear end covers. Remove the washer from the front cover.

*NOTE
Record the number of washers (and thickness) used on both ends of the armature. Be sure to install the same number when reassembling the starter.*

4. Remove and identify the washers from both ends of the armature (**Figure 27**).

5. Slide the armature (**Figure 27**) out of the housing.

6. Pull the spring back and remove the brushes from their mounting position in the holder (**Figure 28**).

7. If it is necessary to remove the brush plate and brush yoke, perform the following:

   a. Lift the brush plate off of the center housing (**Figure 29**).

   b. Remove the nut and washer assembly (A, **Figure 30**) and remove the yoke and brush assembly (B, **Figure 30**).

8. Inspect the starter assembly and perform the electrical test procedure described under *Inspection* in this chapter.

9. If the brush plate and yoke assemblies were removed, perform the following:

   a. Install the brush yoke assembly (B, **Figure 30**) into the housing. Then install the

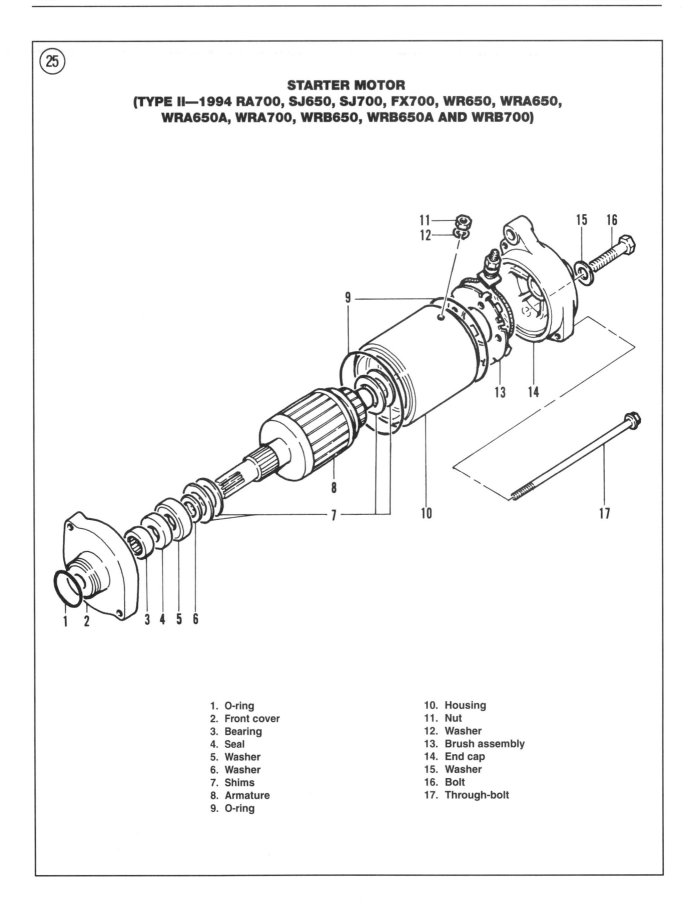

ELECTRICAL SYSTEM

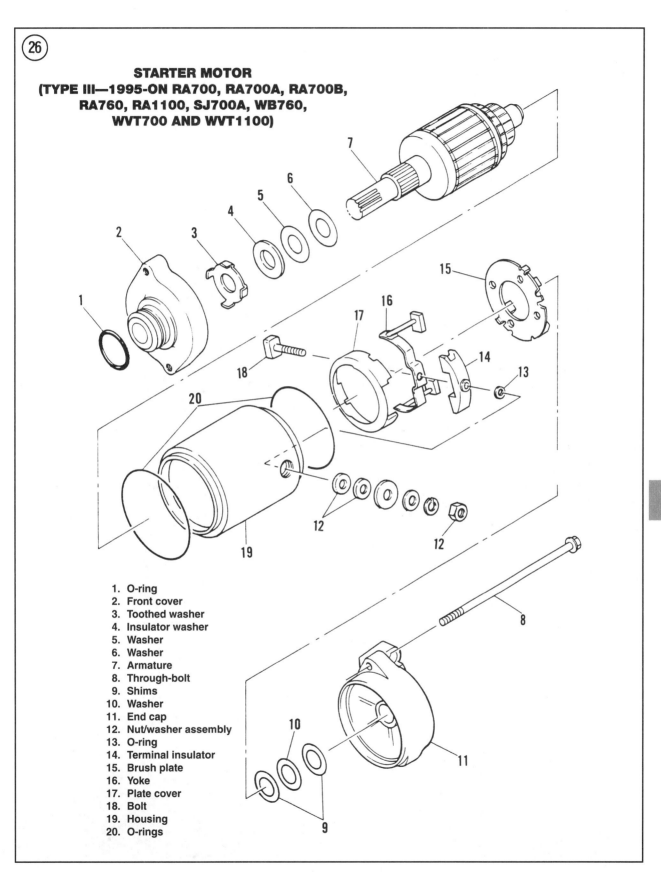

㉖ STARTER MOTOR
(TYPE III—1995-ON RA700, RA700A, RA700B, RA760, RA1100, SJ700A, WB760, WVT700 AND WVT1100)

1. O-ring
2. Front cover
3. Toothed washer
4. Insulator washer
5. Washer
6. Washer
7. Armature
8. Through-bolt
9. Shims
10. Washer
11. End cap
12. Nut/washer assembly
13. O-ring
14. Terminal insulator
15. Brush plate
16. Yoke
17. Plate cover
18. Bolt
19. Housing
20. O-rings

washer and nut assembly in the order shown in **Figure 25** or **Figure 26**.

b. Align the brush plate with the center housing and install it (**Figure 31**) into the housing.

10. Insert the brushes (**Figure 32**) into their holders and secure the brushes with the springs. Then place a piece of folded paper or washer between each brush and spring assembly as shown in **Figure 33**. This reduces spring tension on the brush face to allow easier armature installation.

11. Insert the armature into the center housing. Carefully work the commutator past the brushes until it bottoms. Remove the 4 pieces of paper or washers. Make sure each brush snaps against the commutator. See **Figure 34**.

12. Install the correct number of washers on the armature shaft next to the commutator (**Figure 27**).

13. Install the correct number of shims on the opposite end of the armature shaft (**Figure 27**).

14. Apply waterproof grease to the front cover seal.

15. On starters with a toothed washer, place the toothed washer into the front cover (**Figure 35**). There is only one position in which the washer will align correctly. Make sure the washer seats flush in the end cover.

16. Install both end covers, making sure to align the marks made during disassembly.

17. Install the 2 through-bolts and tighten them securely.

ELECTRICAL SYSTEM

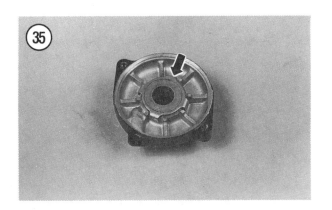

Inspection
(All Models)

Refer to **Figure 19** (Type I starters), **Figure 25** (Type II starters) or **Figure 26** (Type III starters) when performing this procedure.

1. Clean all grease, dirt and carbon from the armature, housing, and end covers with 600 grit abrasive paper.

CAUTION
Do not immerse the brushes or wire windings in solvent as the insulation may be damaged. Wipe the windings with a cloth lightly moistened with solvent and dry thoroughly.

2. Check the seal (**Figure 36**) in the front cover for tearing or excessive wear.

3. Check the cover bushings (**Figure 37**) for cracks or excessive wear. If the bushing is worn, replace the cover assembly.

4. Check all of the O-rings for flatness, cracks, hardness or other abnormal conditions. Replace the O-ring(s) as required.

5. Check the fiber washer(s) and replace if worn, cracked or damaged in any way. If this washer is not available from a Yamaha dealer, check with an automotive electrical repair shop.

6. Check the metal thrust washers for damage. Replace if necessary.

7. Measure the length of each brush with a vernier caliper (**Figure 38**). If the length is worn to the wear limit specified in **Table 3**, replace the brush holder assembly. The brushes cannot be replaced individually.

8. Inspect the commutator (**Figure 39**). The mica in a good commutator is below the surface of the copper bars. On a worn commutator, the mica and copper bars may be worn to the same level (**Figure 40**). See **Table 3** for mica undercut specifications. If necessary, have the commutator serviced by a dealer or electrical repair shop.

9. Inspect the commutator copper bars (**Figure 39**) for discoloration. A pair of discolored bars indicates grounded armature coils.

10. Measure the commutator outside diameter with a vernier caliper (**Figure 41**) and compare to specifications listed in **Table 3**. Replace the commutator if its outside diameter meets or exceeds wear limit.

NOTE
An ohmmeter is required to perform the following checks.

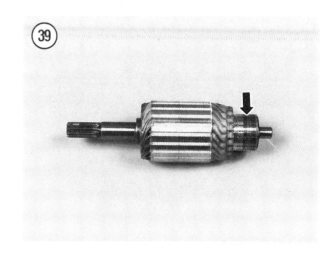

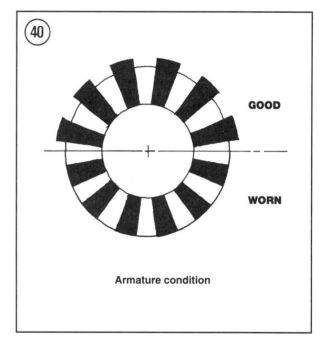

Armature condition

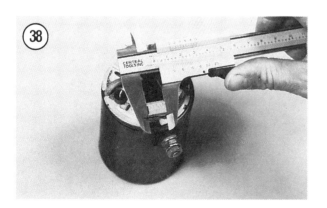

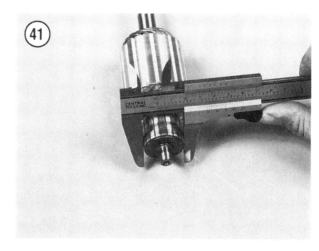

ELECTRICAL SYSTEM

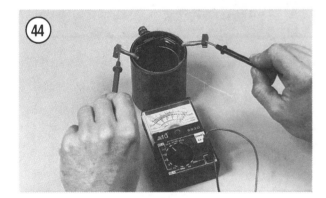

11. Use an ohmmeter and check for continuity between the commutator bars (**Figure 42**); there should be continuity between pairs of bars. Also check for continuity between the commutator bars and the shaft (**Figure 43**); there should be no continuity. If the unit fails either of these tests, the armature is faulty and must be replaced.

12. With an ohmmeter on set on the R × 1 scale, check for continuity between the 2 field coil brushes (**Figure 44**); there should be continuity. If there is high resistance or if the meter reads no continuity, there is an open circuit in one of the brush leads. Replace the field coil assembly.

13. Measure the resistance between the positive field coil terminal and the housing (**Figure 45**); there should be no continuity. If continuity is shown, replace the field coil and retest.

14. Measure the resistance between the insulated brush holders and the brush plate (**Figure 46**); there should be no continuity. If continuity is shown, replace the brush plate.

FLYWHEEL AND STATOR PLATE (500 CC MODELS)

NOTE
Refer to Chapter Two for troubleshooting and test procedures.

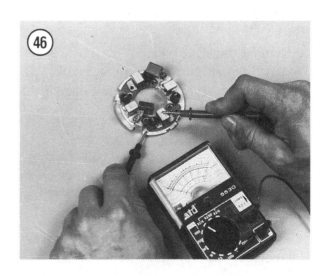

Flywheel
Removal/Installation

NOTE
Some of the following procedures are shown with the engine removed for clarity. The flywheel can be removed with the engine mounted in the hull.

1. Remove the fuel tank as described in Chapter Eight.
2. Disconnect the starter ground cable from the flywheel cover (**Figure 47**).
3. Remove the bolts holding the flywheel cover to the engine. Remove the cover (**Figure 48**) and gasket.
4. Hold the flywheel with a universal flywheel holding tool or the Yamaha flywheel holder (part No. YB-6139) and loosen the flywheel nut. Remove the nut and washer.

CAUTION
Do not use heat or hammer on the flywheel to remove it in Step 5. Heat may cause the flywheel to seize on the crankshaft, while hammering can damage the flywheel or bearings.

5. Install the Yamaha puller (part No. YB-6117) or equivalent onto the face of the flywheel (**Figure 49**) and break the flywheel free of the crankshaft. You may have to alternate tapping on the center puller bolt sharply with a hammer and tightening the bolt some more, but do not hit the flywheel.
6. Remove the flywheel puller, flywheel and Woodruff key (**Figure 50**) from the crankshaft.
7. Inspect flywheel carefully as described in this chapter.

NOTE
*Before installing the flywheel, check the magnets (**Figure 51**) for metal trash they may have picked up. Debris stuck to the magnets can cause coil damage.*

8. Place the Woodruff key in the crankshaft key slot (**Figure 50**). Position the flywheel over

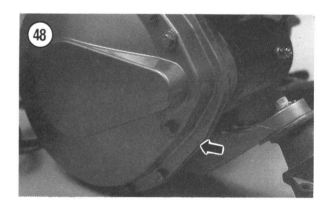

ELECTRICAL SYSTEM

crankshaft with the key slot in flywheel hub aligned with key in crankshaft. See **Figure 52**.

9. Oil the crankshaft threads and install the washer and nut. Hold the flywheel with the same tool used during removal and tighten the bolt to 140 N•m (103 ft.-lb.). See **Figure 53**.

10. Remove all gasket residue from the flywheel cover mating surface.

11. Install a new flywheel cover gasket. Install gasket so the side with sealer faces out (**Figure 54**).

12. Install the flywheel cover and its mounting bolts. Tighten bolts securely.

13. Reverse Steps 1 and 2 to complete installation.

Inspection

1. Check the flywheel (**Figure 55**) carefully for cracks or breaks.

WARNING
A cracked or chipped flywheel must be replaced. A damaged flywheel may fly apart at high speed, throwing metal fragments over a large area. Do not attempt to repair a damaged flywheel.

2. Check the tapered bore of the flywheel (**Figure 55**) and the crankshaft taper (**Figure 50**) for scoring, cracks or other damage.

3. Check the flywheel teeth (**Figure 55**) for excessive wear or damage.

4. Check the crankshaft and flywheel nut threads for wear or damage.

5. Check the crankshaft and flywheel lockwasher for weakness or damage.

6. Replace flywheel, crankshaft and/or flywheel nut and lockwasher as required.

Stator Plate
Removal/Installation

1. Remove the flywheel as described in this chapter.

2. Remove the stator plate wire harness cover (**Figure 56**) screws and remove the cover. Note the ground wire connection at the back of the plate at the upper screw position (**Figure 57**).

3. Disconnect the stator plate bullet connectors (A, **Figure 58**). Then remove the nuts (B, **Figure 58**) securing the stator plate to the seal housing and remove the stator plate (C).

4. Installation is the reverse of these steps. Note the following:

   a. Clean all of the electrical connectors with electrical contact cleaner.

   b. After reconnecting the bullet connectors, fit the connectors into the seal housing as shown in **Figure 59**.

   c. Reinstall the wire harness cover, making sure to reconnect the ground wire at the upper screw position. See **Figure 57**.

Coil Replacement

The stator plate coils can be replaced individually. When replacing an individual coil, it is necessary to heat the wire connection at the damaged coil with a soldering iron before disconnecting the wire. When the solder has melted, pull the wire away from the connection. This step should give you enough wire to work with when resoldering the new coil. If the wire is cut at the connection, it could cause the wire to fall short of the coils connecting tab. During reassembly, rosin core solder must be used to reconnect the

ELECTRICAL SYSTEM

wire. Never use acid core solder on electrical connections.

When replacing a defective coil, be sure that you have properly identified the leads to be unsoldered and removed.

FLYWHEEL, IDLER GEAR AND STATOR PLATE (650, 700, AND 760 CC MODELS)

NOTE
Refer to Chapter Two for troubleshooting and test procedures.

Flywheel/Idler Gear Removal/Installation

NOTE
Some of the following procedures are shown with the engine removed for clarity. The flywheel can be removed with the engine mounted in the hull.

1. Disconnect the negative battery cable.
2. Remove all components as required to gain access to the flywheel cover. Be sure to remove the oil pump on oil injection models.
3. *760 cc models*—Remove the grease hose from the flywheel cover (**Figure 60**).
4. Remove the bolts holding the flywheel cover to the engine. Remove the cover (**Figure 61**) and gasket. Remove the 2 cover dowel pins (A, **Figure 62**).
5. Locate and remove the 2 idler gear washers and spring (**Figure 63**). These parts can either be

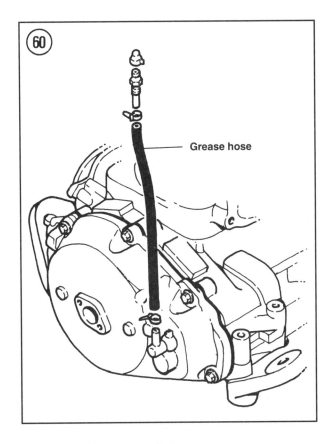

found in the front cover or mounted on the idler shaft.

6A. *650 cc and 700 cc models*—Hold the flywheel with a universal flywheel holding tool or the Yamaha flywheel holder (part No. YB-6139) and loosen the flywheel bolt. Remove the bolt (B, **Figure 62**) and washer.

6B. *760 cc models*—Bolt the Yamaha flywheel holder (part No. YW-06547) to the engine case (A, **Figure 64**), and loosen the flywheel bolt. Remove the bolt (B, **Figure 64**) and washer.

CAUTION
Do not use heat or hammer on the flywheel to remove it in Step 7. Heat may cause the flywheel to seize on the crankshaft, while hammering can damage the flywheel or bearings.

NOTE
*760 cc models uses a different shaped flywheel than the one shown in **Figure 65**.*

7. Install the Yamaha puller (part No. YB-6117) or equivalent onto the face of the flywheel (**Figure 65**) and break the flywheel free of the crankshaft. You may have to alternate tapping on the center puller bolt sharply with a hammer and tightening the bolt, but do not hit the flywheel.

8. Remove the flywheel puller and flywheel.

9. Remove the Woodruff key (A, **Figure 66**) from the crankshaft, if necessary.

10. Remove the idler gear (B, **Figure 66**).

11. Inspect flywheel carefully as described in this chapter.

12. Inspect the idler gear and idler gear bushing as described in this chapter.

NOTE
*Before installing the flywheel, check the magnets (**Figure 51**) for metal trash they may have picked up. Debris stuck to the magnets can cause coil damage.*

13. Place the Woodruff key in the crankshaft key slot (A, **Figure 66**).

14. Wipe the ends of the idler gear with marine grease and insert the idler gear into its bushing (B, **Figure 66**).

15. Position the flywheel over the crankshaft with the key slot in flywheel hub aligned with the key in the crankshaft.

16. Oil the crankshaft bolt threads and install the washer and bolt. Hold the flywheel with the same tool used during removal and tighten the bolt to 70 N•m (51 ft.-lb.).

17. Remove all gasket residue from the flywheel cover mating surface.

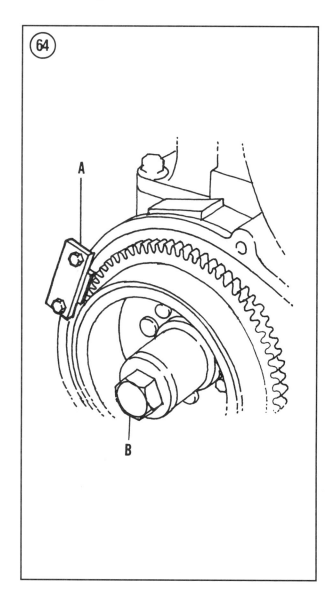

ELECTRICAL SYSTEM

18. Install the 2 flywheel cover dowel pins (A, **Figure 62**) and place a new cover gasket onto the crankcase.

19. Apply marine grease to the front cover idler gear bushing (**Figure 63**). Then insert a washer, spring and washer into the bushing.

20. Install the flywheel cover onto the crankcase, making sure to align the dowel pins with the cover. Also be sure that the idler gear spring and washers engage the end of the idler gear shaft correctly.

21. Reverse Steps 1 through 3 to complete installation.

Flywheel Inspection

1. Check the flywheel (**Figure 55**) carefully for cracks or breaks.

> *WARNING*
> *A cracked or chipped flywheel must be replaced. A damaged flywheel may fly apart at high speed, throwing metal fragments over a large area. Do not attempt to repair a damaged flywheel*

2. Check the tapered bore of the flywheel (**Figure 55** and **Figure 51**) and the crankshaft taper (**Figure 50**) for signs of scoring, cracks or other damage.

3. Check the flywheel teeth for excessive wear or damage.

4. Check the crankshaft and flywheel nut threads for wear or damage.

5. Check the crankshaft and flywheel lockwasher for weakness or damage.

6. Replace the flywheel, crankshaft and/or flywheel nut and lockwasher as required.

Idler Gear and Bushing Inspection/Replacement

1. Clean the pinion gear in solvent and dry thoroughly.

2. Check the pinion gear teeth (**Figure 67**) for cracks, deep scoring or excessive wear. If the pinion gear teeth are worn, check the flywheel and starter gear teeth for damage.

3. Check pinion gear movement by sliding it along the pinion shaft. It must move smoothly without sticking or dragging.

> *NOTE*
> *The stator on 760 cc models is mounted inside the flywheel cover. If you are*

working on a 760 cc model, refer to **Figure 68** to locate the crankcase housing bushing, but not the stator.

4. Check the flywheel cover (**Figure 63**) and crankcase housing bushings (A, **Figure 68**). Check for cracks, deep scoring or excessive wear. If a bushing is worn, replace it as follows:
   a. Remove the worn bushing with a blind bearing remover.
   b. Clean the bushing bore with solvent, then check for cracks or severe wear.
   c. Wipe the outside of the new bushing with marine grease and carefully drive the bushing into the housing with a suitable bearing driver.

Stator Plate Removal/Installation (650 and 700 cc)

1. Remove the electric box as described in this chapter. Set it aside so it can be removed together with the stator plate.
2. Remove the flywheel as described in this chapter.

NOTE
*The stator plate and crankcase each have an index mark, that when aligned sets the ignition timing. Note the position of the stator plate index mark before removing the stator plate in Step 3. Refer to C, **Figure 68**.*

3. Remove the stator plate mounting screws and remove the stator plate (B, **Figure 68**) together with the electric box. See **Figure 69**.
4. Installation is the reverse of these steps. Note the following:
   a. When installing the stator plate, align the index mark on the stator plate with the crankcase index mark (C, **Figure 68**); aligning these marks sets ignition timing.
   b. Coat the stator plate rubber seal with silicone sealant before installing it.

Stator Assembly Removal/Installation (760 cc)

Refer to **Figure 70** for this procedure.
1. Remove the electric box as described in this chapter. Set it aside so it can be removed together with the stator assembly.
2. Remove the flywheel cover as described in this chapter.
3. Unbolt the ground wire from the flywheel cover.
4. Remove the stator bracket mounting screws and remove the stator brackets.
5. Remove the pulser coil mounting bolts and remove the stator assembly.
6. Installation is the reverse of these steps. Note the following:

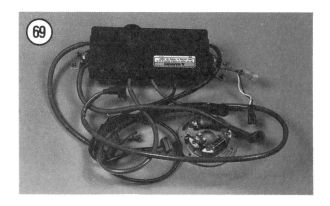

ELECTRICAL SYSTEM

a. Apply Loctite 242 to all mounting bolts when reinstalling the stator assembly.
b. Be sure to reconnect the stator ground when reinstalling the ground wire.

Coil Replacement

The stator coils can be replaced individually. When replacing an individual coil, it is necessary to heat the wire connection at the damaged coil with a soldering iron before disconnecting the wire. When the solder has melted, pull the wire away from the connection. This step should give you enough wire to work with when resoldering the new coil. If the wire is cut at the connection, it could cause the wire to fall short of the coils connecting tab. During reassembly, rosin core solder must be used-never use acid core solder on electrical connections-to reconnect the wire.

When replacing a defective coil, be sure that you have properly identified the leads to be unsoldered and removed. During reassembly, route and secure all wires properly and reconnect the ground wire (**Figure 70** or **Figure 71**).

FLYWHEEL, IDLER GEAR AND STATOR PLATE (1100 CC)

NOTE
Refer to Chapter Two for troubleshooting and test procedures.

Flywheel/Idler Gear Removal/Installation

Refer to **Figure 72** for this procedure.

1. Disconnect the negative battery cable, then the positive battery cable.
2. Remove all components as required to gain access to the flywheel cover.

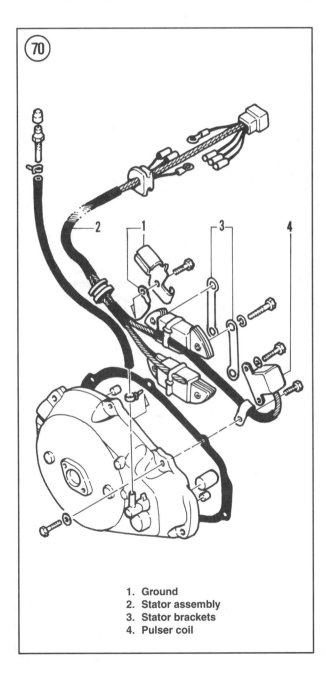

1. Ground
2. Stator assembly
3. Stator brackets
4. Pulser coil

CHAPTER NINE

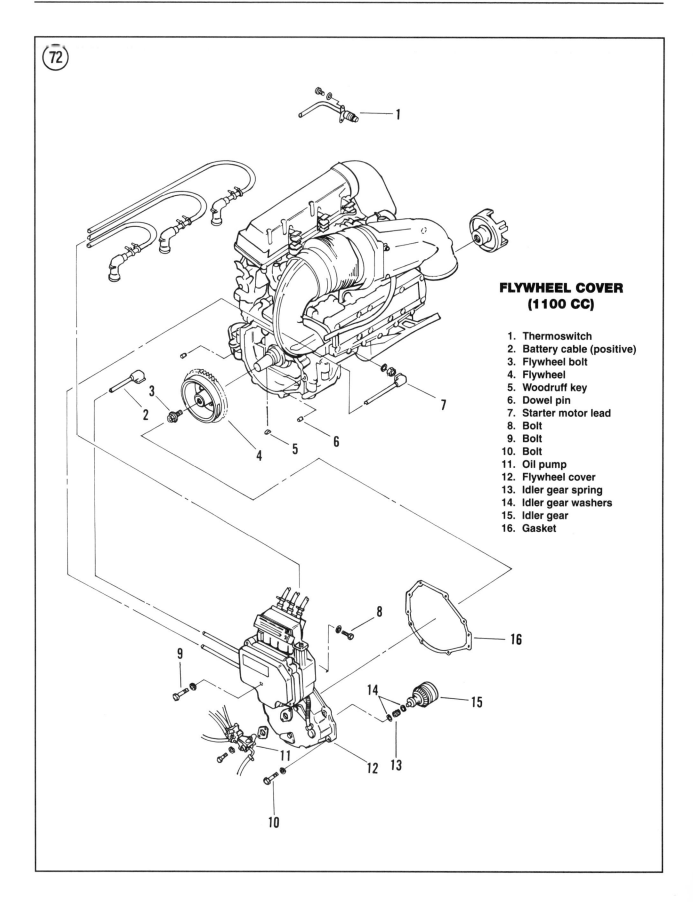

ELECTRICAL SYSTEM

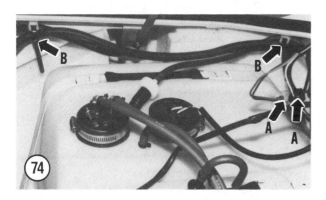

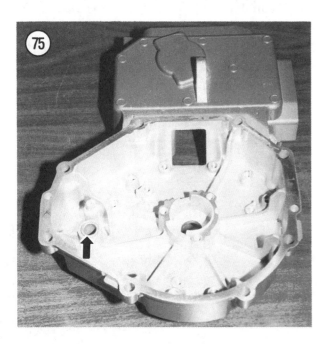

3. Remove the spark plug cap from each spark plug and remove the thermoswitch from the rear boss on the exhaust housing (**Figure 73**).

4. Disconnect the starter/stop switch connectors, and then disconnect the meter connector (A, **Figure 74**). Clip the 2 cable ties that secure the wiring harness to the hull so the harness is free. (B. **Figure 74**).

5. Remove the starter relay lead from the starter motor.

6. Remove the oil pump from the flywheel cover.

7. Remove the bolts holding the flywheel cover to the engine. (**Figure 72**) Remove the cover and gasket. Remove the 2 cover dowel pins.

8. Locate and remove the 2 idler gear washers and spring. These parts can either be found in the front cover (**Figure 75**) or mounted on the idler shaft.

9. Bolt the Yamaha flywheel holder (part No. YW-41528) to the engine case (A, **Figure 76**) and loosen the flywheel bolt. Remove the bolt (B, **Figure 76**) and washer.

CAUTION
Do not use heat or a hammer on the flywheel to remove it in Step 6. Heat may cause the flywheel to seize on the crankshaft, while hammering can damage the flywheel or bearings.

10. Install the Yamaha puller (part No. YB-6117) or equivalent onto the face of the flywheel

(**Figure 77**) and break the flywheel free of the crankshaft. You may have to alternate tapping on the center puller bolt sharply with a hammer and tightening the bolt some more, but do not hit the flywheel.

11. Remove the flywheel puller and flywheel.
12. Remove the Woodruff key from the crankshaft, if necessary.
13. Remove the idler gear (**Figure 67**).
14. Inspect the flywheel carefully as described in this chapter.
15. Inspect the idler gear and idler gear bushing as described in this chapter.

NOTE
Before installing the flywheel, check the magnets for metal trash they may have picked up. Debris stuck to the magnets can cause coil damage.

16. Place the Woodruff key in the crankshaft key slot (**Figure 72**).
17. Wipe the ends of the idler gear with marine grease and insert the idler gear into its bushing.
18. Position the flywheel over the crankshaft with the key slot in the flywheel hub aligned with the key in the crankshaft.
19. Oil the crankshaft bolt threads and install the washer and bolt. Hold the flywheel with the same tool used during removal and tighten the bolt to 70 N•m (51 ft.-lb.).
20. Remove all gasket residue from the flywheel cover mating surface.
21. Install the 2 flywheel cover dowel pins and then place a new cover gasket onto the crankcase.
22. Apply marine grease to the front cover idler gear bushing (**Figure 75**). Then insert a washer, spring and washer into the bushing.
23. Install the flywheel cover onto the crankcase, making sure to align the dowel pins with the cover. Also be sure the idler gear spring and washers engage the end of the idler gear shaft correctly.
24. Reverse Steps 1 through 5 to complete installation.

Flywheel Inspection

1. Check the flywheel carefully for cracks or breaks.

WARNING
A cracked or chipped flywheel must be replaced. A damaged flywheel may fly apart at high speed, throwing metal fragments over a large area. Do not attempt to repair a damaged flywheel.

2. Check the tapered bore of the flywheel and check the crankshaft for scoring, cracks or other damage.
3. Check the flywheel teeth for excessive wear or damage.
4. Check the crankshaft and flywheel nut threads for wear or damage.
5. Check the crankshaft and flywheel lockwasher for weakness or damage.
6. Replace the flywheel, crankshaft and/or flywheel nut and lockwasher as required.

Idler Gear and Bushing Inspection/Replacement

1. Clean the pinion gear in solvent and dry thoroughly.
2. Check the pinion gear teeth (**Figure 67**) for cracks, deep scoring or excessive wear. If the pinion gear teeth are worn, check the flywheel and starter gear teeth for damage.

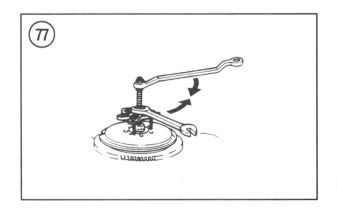

ELECTRICAL SYSTEM

3. Check pinion gear movement by sliding it along the pinion shaft. It should move smoothly without sticking or dragging.

4. Check the flywheel cover (**Figure 75**) and crankcase housing. Check for cracks, deep scoring or excessive wear. If a bushing is worn, replace it as follows:

   a. Remove the worn bushing with a blind bearing remover.

   b. Clean the bushing bore with solvent, then check for cracks or severe wear.

   c. Wipe the outside of the new bushing with marine grease and carefully drive the bushing into the housing with a suitable bearing driver.

Stator and Pulser Assemblies Removal/Installation

On 1100 cc models, the stator and pulser assemblies are mounted inside the flywheel cover. Refer to **Figure 78** for this procedure.

1. Remove the flywheel cover as described in this chapter.

2. Disconnect the stator assembly connectors and the pulser coil assembly connector.

3. Remove the stator assembly bolts and remove the stator assembly from the flywheel cover.

4. Remove the pulser coil lead bolts and the pulser coil bolts. Remove the pulser coil assembly from the flywheel cover. Note the position of each coil before disassembly. You must reinstall each pulser coil in its original position.

5. Installation is the reverse of these steps. Apply Loctite 242 to the threads of each Allen bolt, and torque the bolt to 11 N•m (97 in.-lb.).

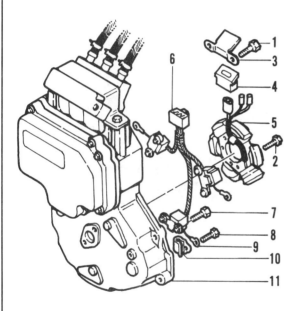

STATOR AND PULSER ASSEMBLIES (1100 CC)

1. Allen bolt
2. Allen bolt
3. Clamp
4. Grommet
5. Stator assembly
6. Pulser coil assembly
7. Allen bolt
8. Allen bolt
9. Pulser coil ground lead
10. Clamp
11. Flywheel cover

ELECTRIC BOX (500 CC MODELS)

The electric box installed on the WR500 contains the following electrical components (**Figure 79**):

   a. Voltage regulator/rectifier (mounted on outside of box).

   b. CDI unit.

   c. Starter relay.

   d. Ignition coil.

   e. Fuse.

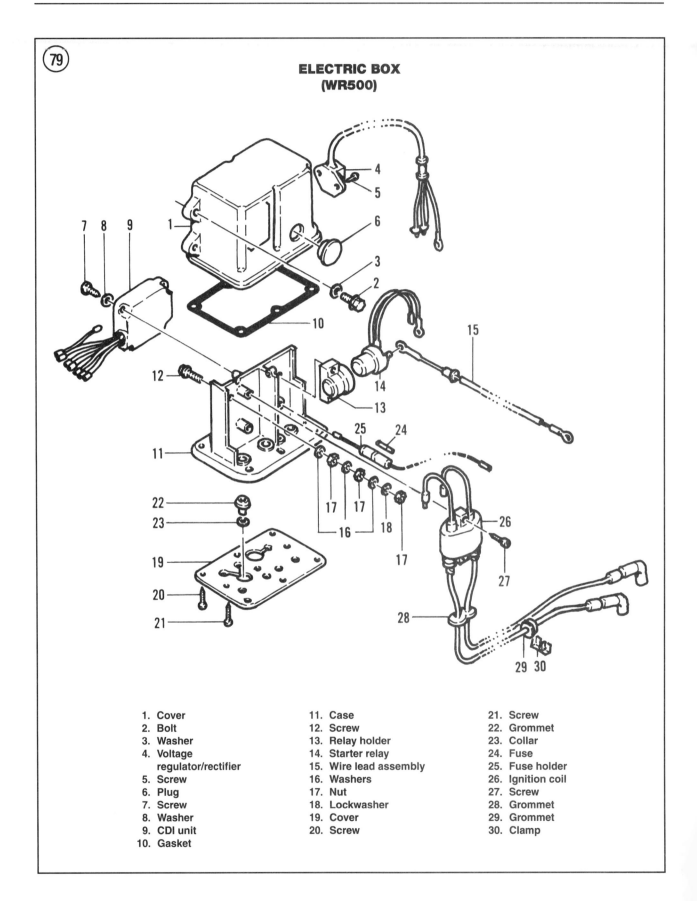

ELECTRICAL SYSTEM

Removal/Installation

1. Remove the engine cover.
2. Disconnect the negative battery cable.

NOTE
Label all wire connectors disconnected during this procedure.

3. Disconnect the stator coil-to-electric box electrical connectors as described in *Stator Coil Removal/Installation* in this chapter. Then pull the electric box wire harness out of the seal housing (**Figure 80**).
4. Disconnect all of the electrical connectors leading from the electric box.
5. Disconnect the spark plug caps from the plugs.
6. Remove the bolt and clamp securing the thermoswitch to the water passage in the cylinder head (**Figure 81**). Pull the sensor out of the cylinder head and lay it aside to be removed with the electric box.
7. Remove the bolts holding the electric box to the hull. Remove the electric box (**Figure 82**).
8. Installation is the reverse of these steps. Note the following:
    a. Apply marine grease to the thermoswitch (**Figure 83**) before reinstalling it into the cylinder head.
    b. Spray all of the exposed electrical connectors (**Figure 84**) with an electrical contact cleaner before reconnecting them.

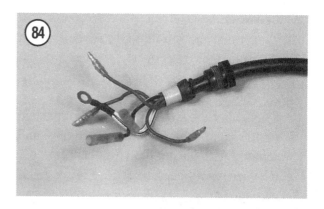

c. Apply Loctite 242 (blue) to each of the electric box mounting bolts. Tighten the bolts securely.

d. After installing the electrical box and reconnecting the electrical connectors, check the routing of all wires. Secure them with plastic ties as required.

Disassembly/Reassembly

Refer to **Figure 79** for this procedure.

1. Remove the electric box as described in this chapter.
2. Remove the screws securing the voltage regulator/rectifier (**Figure 85**) to the outside of the electric box and remove it.
3. Remove the Phillips screws holding the front and rear covers together and separate them (**Figure 86**). Note the position of the rubber gasket (A, **Figure 87**).
4. Remove the damaged components from the box as required:
   a. Starter relay (B, **Figure 87**).
   b. Ignition coil (C, **Figure 87**).
   c. CDI unit (**Figure 88**).
5. When installing new components, note the following:
   a. Route all wires carefully in the box to avoid pinching or damaging them. See **Figure 89** and **Figure 90**.
   b. Route the wires carefully through the front cover (**Figure 91**).
   c. Check the Phillips screws (where used) for tightness.
   d. Check that the ignition coil rubber mount (C, **Figure 87**) is properly secured.
   e. Clean the electrical connectors with electrical contact cleaner.
   f. Check that the spark plug caps are pushed tightly onto their high tension leads and secured with plastic ties (**Figure 92**).
   g. Replace the rubber gasket (A, **Figure 87**) if damaged.

ELECTRICAL SYSTEM

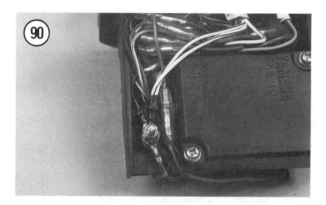

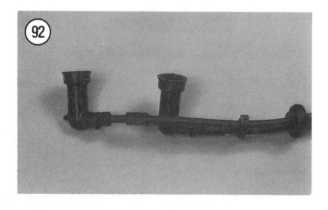

6. When reassembling the electric box, note the following:
   a. Check that all of the wire connectors are clean and tight.
   b. Install the rubber gasket (A, **Figure 87**).
   c. Coat the outside of all grommets with a waterproof grease.
   d. Make sure that none of the wires are pinched between the electric box halves.
   e. Install the Phillips screws securely (**Figure 86**).
   f. Remount the voltage regulator/rectifier to the outside of the box (**Figure 85**).

ELECTRIC BOX (650, 700 AND 760 CC MODELS)

The electric box installed on the 650, 700 and 760 cc models contains the following electrical components (**Figure 93** or **Figure 94**):
   a. Voltage regulator/rectifier.
   b. CDI unit.
   c. Starter relay.
   d. Ignition coil.
   e. Fuse.

Removal/Installation

1. Remove the engine cover.
2. Disconnect the negative battery cable.

NOTE
Label all wire connectors disconnected during this procedure.

3. *600 cc and 700 cc models*—Remove the stator plate or stator assembly as described in this chapter. Set it aside to be removed with the electric box.
4. Disconnect all of the exposed electrical connectors leading to the electric box.
5. Disconnect the spark plug caps from the plugs.

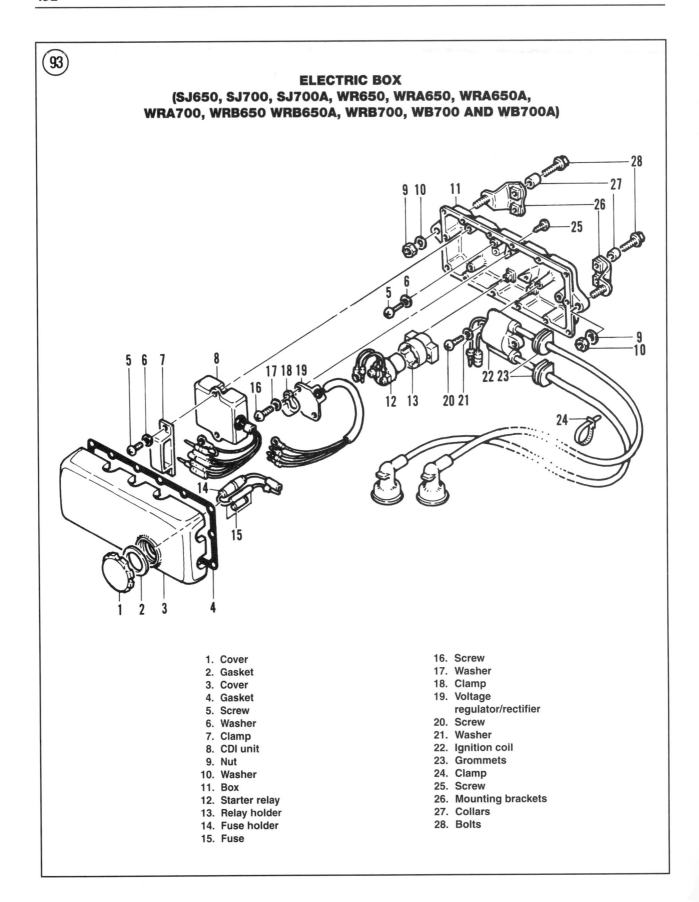

ELECTRICAL SYSTEM

**ELECTRIC BOX
(FX700, RA700, RA700A, RA700B, RA760, WB760 AND WVT700)**

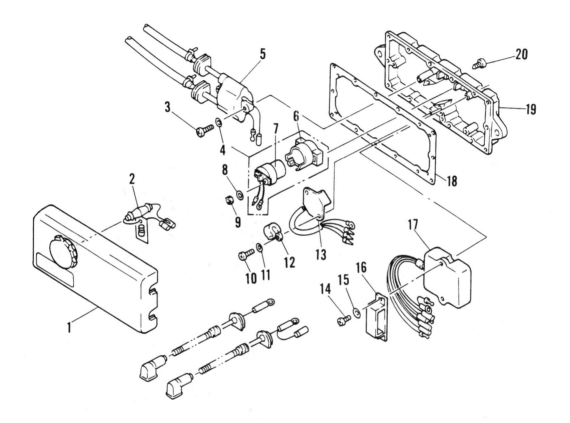

1. Cover
2. Fuse holder
3. Screw
4. Washer
5. Ignition coil
6. Relay holder
7. Starter relay
8. Washer
9. Nut
10. Screw
11. Washer
12. Clamp
13. Voltage rectifier/regulator
14. Screw
15. Washer
16. Clamp
17. CDI unit
18. Gasket
19. Box
20. Screw

6A. Remove the bolt and clamp securing the thermoswitch to the water galley in the cylinder head (**Figure 95**). Pull the thermoswitch out of the cylinder head and lay it aside to be removed with the electric box.

6B. *On RA760 and WB760 models*—Remove the thermoswitch from the muffler housing (**Figure 73**).

7. *FX700, SJ650, SJ700 and SJ700A*—Disconnect the fire extinguisher compartment strap (A, **Figure 96**) and remove the compartment (B, **Figure 96**).

8. Remove the bolts or nuts holding the electric box to the hull.

9A. *650cc and 700 cc models*—Remove the electric box (**Figure 97**) along with the stator plate (**Figure 98**) or the stator assembly.

9B. *RA760 and WB760 models*—Open the electric box and disconnect the stator from the CDI as follows:
   a. Remove the Phillips screws (B, **Figure 99**) holding the front and rear covers together.
   b. Separate the covers, and note the position of the rubber gasket.
   c. Disconnect the CDI connector and remove the electric box from the watercraft.

10. Installation is the reverse of these steps. Note the following:
   a. On SJ650 and SJ700 models, the electric box mounting brackets (A, **Figure 99**) align with the mounting bolt holes in the hull.
   b. Wipe the thermoswitch mounting hole with marine grease before reinstalling the thermoswitch.
   c. Spray all of the exposed electrical connectors (**Figure 100**) with an electrical contact cleaner before reconnecting them.
   d. Apply Loctite 242 (blue) to each of the electric box mounting bolts. Tighten the bolts securely.
   e. After installing the electric box and reconnecting the electrical connectors, check the routing of all wires and secure with plastic ties as required.

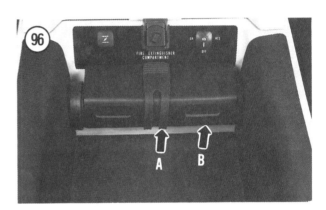

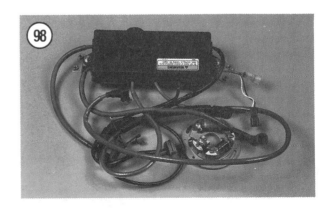

ELECTRICAL SYSTEM

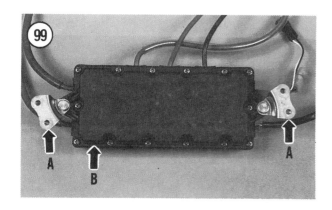

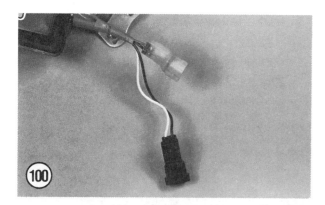

f. *FX700, SJ650, SJ700 and SJ700A*—Make sure the fire extinguisher compartment is installed and secured properly.

Disassembly/Reassembly

Refer to **Figure 93** or **Figure 94** for this procedure.

1. Remove the electric box as described in this chapter.
2. Remove the Phillips screws (B, **Figure 99**) holding the front and rear covers together and separate them. Note the position of the rubber gasket.
3. Remove the damaged components from the electric box as required:
    a. Ignition coil (A, **Figure 101**).
    b. Voltage regulator/rectifier (B, **Figure 101**).
    c. Starter relay (**Figure 102**).
    d. CDI unit (**Figure 103**).
4. When installing new components, note the following:
    a. Route all wires carefully in the box to avoid pinching or damaging them. See **Figure 103**.
    b. Route the wires carefully through their grommets during assembly.
    c. Check the Phillips screws for tightness.
    d. Check that the ignition coil rubber mount is properly secured (A, **Figure 101**).
    e. Clean the electrical connectors with electrical contact cleaner.

f. Check that the spark plug caps are pushed tightly onto their high tension leads and secured with plastic ties (**Figure 104**).

g. Replace the rubber gasket if damaged.

5. When reassembling the electric box, note the following:

　a. Check that all of the wire connectors are clean and tight.

　b. Install the rubber gasket, if removed.

　c. Coat the outside of all grommets with a waterproof grease.

　d. Make sure that none of the wires are pinched between the electric box halves.

　e. Make sure the rubber grommets fit into the box halves correctly. See **Figure 105**.

　f. Install the Phillips screws and secure the box securely (B, **Figure 99**).

ELECTRIC BOX (1100 CC MODELS)

The electric box installed on the RA1100 and WVT1100 is cast into the flywheel cover. You can disassemble the electric box without removing the flywheel cover from the engine. The electric box includes the following electrical components (**Figure 106**):

　a. CDI unit.

　b. Starter relay.

　c. Voltage regulator/rectifier, on the side of the electric box.

　d. Ignition coils, in the damper located on top of the electric box.

　e. Fuse, also in the damper.

Disassembly/Reassembly

Refer to **Figure 106** for this procedure.

NOTE
Label all wire connectors disconnected during this procedure.

1. If necessary, remove the flywheel cover as described in this chapter.

2. Remove the negative lead from the battery.

3. Remove the cover bolts along with their washers and collars. Remove the cover and gasket from the electric box.

4. Remove the damaged components from the electric box as required:

　a. *CDI unit*—Remove the CDI mounting bolts along with the washers, collars, and grommets shown in **Figure 106**). Disconnect the CDI unit leads, and remove the CDI unit.

　b. *Voltage regulator/rectifier*—Disconnect the voltage regulator leads in the electric box. Remove the regulator mounting bolts and

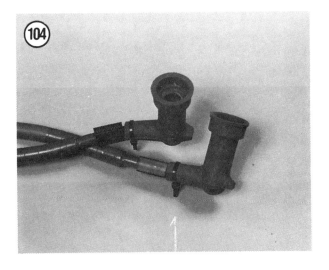

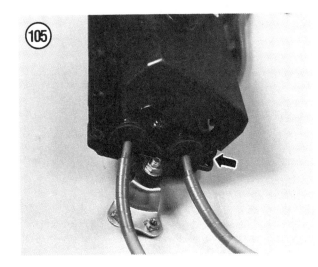

ELECTRICAL SYSTEM

437

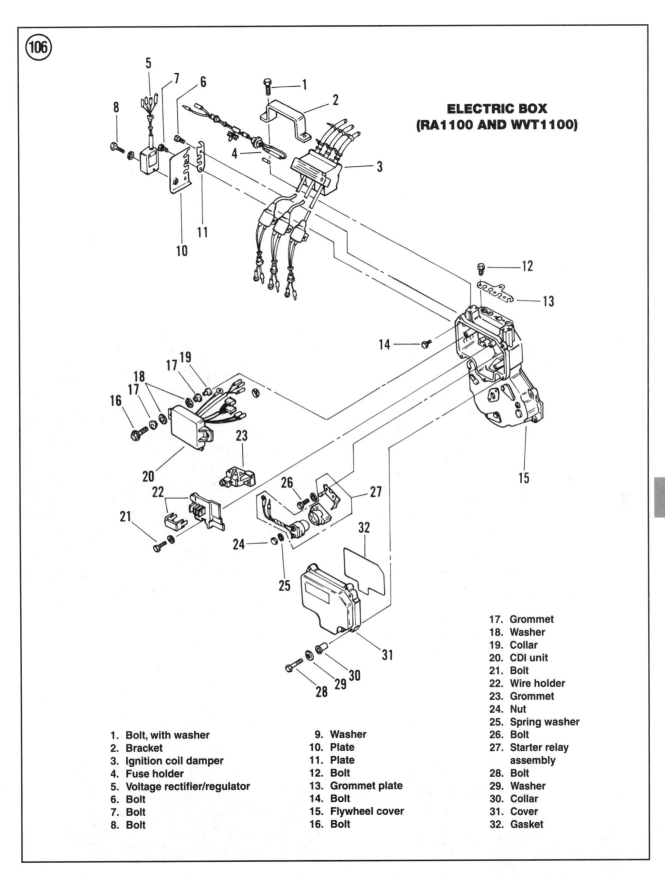

ELECTRIC BOX (RA1100 AND WVT1100)

1. Bolt, with washer
2. Bracket
3. Ignition coil damper
4. Fuse holder
5. Voltage rectifier/regulator
6. Bolt
7. Bolt
8. Bolt
9. Washer
10. Plate
11. Plate
12. Bolt
13. Grommet plate
14. Bolt
15. Flywheel cover
16. Bolt
17. Grommet
18. Washer
19. Collar
20. CDI unit
21. Bolt
22. Wire holder
23. Grommet
24. Nut
25. Spring washer
26. Bolt
27. Starter relay assembly
28. Bolt
29. Washer
30. Collar
31. Cover
32. Gasket

remove the regulator from the side of the electric box (A, **Figure 107**).

c. *Starter relay*—Disconnect the relay leads. Remove the securing nut and spring washer. Remove the relay.

d. *Ignition coil*—Remove the bracket securing the ignition damper to the top of the electric box (A, **Figure 108**). Remove the grommet plate from the top of electric box (B, **Figure 108**). Disconnect the ignition coil leads from the inside the electric box and remove the ignition coil.

5. When installing new components, note the following:
   a. Route all wires carefully in the box to avoid pinching or damaging them.
   b. Route the wires carefully through their grommets during assembly.
   c. Check the screws and bolts for tightness.
   d. Check that the fuse holder grommet is properly secured in the ignition damper (B, **Figure 107**).
   e. Be sure the ignition coil and thermoswitch grommets are properly secured by the plate.
   f. Clean the electrical connectors with electrical contact cleaner.
   g. Check that the ignition coil high tension leads are secured to their grommets with plastic ties (C, **Figure 108**).
   h. Check that the spark plug caps are pushed tightly onto their high tension leads and secured with plastic ties (**Figure 104**).
   i. Replace the cover gasket if it is damaged.

6. When reassembling the electric box, note the following:
   a. Check that all of the wire connectors are clean and tight.
   b. Install the rubber gasket, if removed.
   c. Coat the outside of all grommets with a waterproof grease.
   d. Make sure that none of the wires are pinched by the cover.
   e. Make sure the rubber grommets fit into the box correctly.
   f. Apply Loctite 242 (blue) to the bolt threads when installing the ignition damper bracket. Torque the bolt in 2 steps to 15 N•m (11 ft.-lb.) and 28 N•m (20 ft.-lb.).

SWITCHES

Start/Stop Switch Testing

1. Locate the start and stop switch electrical connectors (**Figure 109**). The start switch connector is white; the stop switch is black.

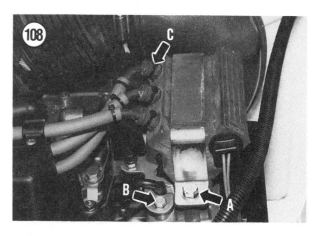

ELECTRICAL SYSTEM

2. Test the start switch (B, **Figure 110**) as follows:

   a. Disconnect the start switch electrical connector.

   b. Install the lanyard lock plate (A, **Figure 110**) into the engine stop switch.

   c. Switch an ohmmeter to the R × 1 scale. Connect the positive ohmmeter lead to the red starter switch lead and the negative ohmmeter lead to the starter switch brown lead.

   d. Check resistance with the start button depressed then test with the button released. When the starter button is depressed, the meter should read 0 ohms. When the start button is released, the meter should read infinity.

   e. Disconnect the ohmmeter leads and reconnect the start switch electrical connectors.

3. Test the stop switch (C, **Figure 110**) as follows:

   a. Disconnect the stop switch electrical connector.

   b. Install the lanyard lock plate into the engine stop switch.

   c. Switch an ohmmeter to the R × 1 scale. Connect the positive ohmmeter lead to the white stop switch lead and the negative ohmmeter lead to the stop switch black lead.

   d. With the lock plate installed, the ohmmeter should read infinity.

   e. With the lock plate removed, the ohmmeter should read 0 ohms.

   f. With the lock plate installed and the stop button depressed, the ohmmeter should read 0 ohms.

   g. Disconnect the ohmmeter leads and reconnect the stop switch electrical connectors.

4. Replace the switch assembly if it failed any of the tests in Steps 2 and 3.

Stop/Start Switch Replacement

The stop and start switches (**Figure 111**) are combined into one housing. If any one switch is faulty, the switch housing must be replaced.

1. Disconnect the switch housing electrical connectors.

2. Remove the screws securing the switch housing to the handlebar and separate the housing.

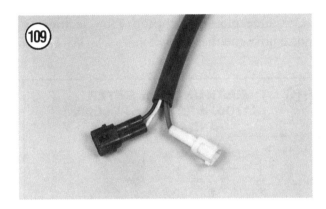

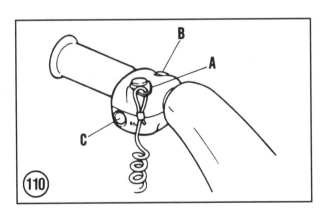

3. Remove the switch assembly while noting the routing of the switch wiring harness.

4. Installation is the reverse of these steps. Test the switches as described in this chapter before starting the engine.

Thermoswitch Testing

All models are equipped with a thermoswitch that monitors engine temperature and protects the engine from overheating by limiting engine speed. If the engine temperature meets or exceeds a predetermined temperature, the thermoswitch sends a signal through the CDI speed control circuit, and engine speed is limited to 2500-3000 rpm. The engine will not run above this speed until engine temperature is reduced.

On most models, the thermoswitch is mounted in the cylinder head (**Figure 95**). On RA760, RA1100, WB760 and WVT1100 models, however, the thermoswitch is mounted in the rear boss on the muffler housing (**Figure 73**).

1. Remove the thermoswitch from the cylinder head or muffler housing.

2A. *Except RA1100 and WVT 1100 models*— Remove the electric box from its mounting position in the hull as described in this chapter. Open the electric box, and disconnect the thermoswitch wires from the CDI unit. Remove the thermoswitch.

2B. *RA1100 and WVT1100 models*—Remove the electric box cover and disconnect the thermoswitch wires from the CDI unit. Remove the grommet plate from the top of the electric box and remove the thermoswitch.

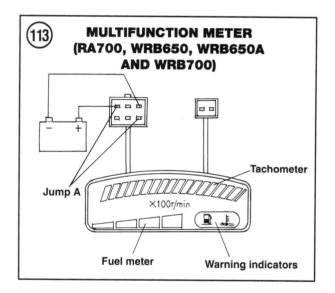

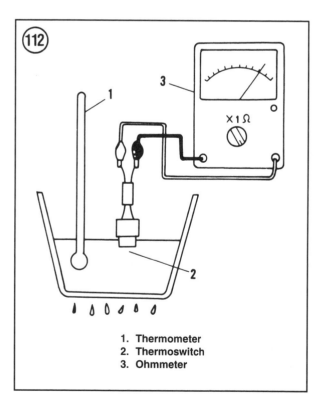

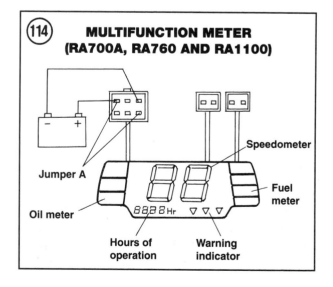

ELECTRICAL SYSTEM

441

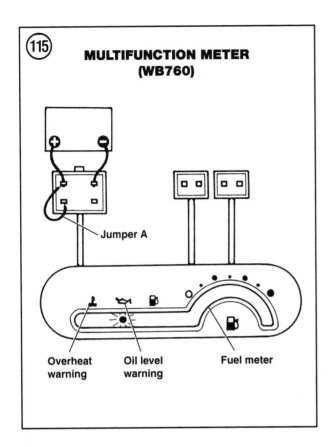

115 MULTIFUNCTION METER (WB760)

Overhead warning — Oil level warning — Fuel meter

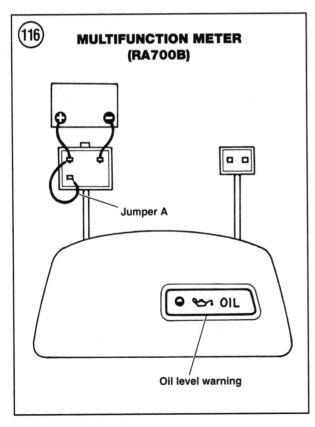

116 MULTIFUNCTION METER (RA700B)

Oil level warning

3. Pour some water in a container that can be heated. Suspend a thermometer in the container so it does not touch the sides or bottom.

4. Connect an ohmmeter to the thermoswitch leads and suspend the tip of the thermoswitch in the water as it is being heated. See **Figure 112**. Do not submerge the thermoswitch in the water as the readings will be incorrect.

5. No continuity should be shown until the water temperature reaches the switch activation temperature indicated in **Table 4**. For example, on WB700 models, the switch activation temperature is 169-183° F (76-84° C).

6. When the water reaches the switch activation temperature, the meter should show continuity (0 ohms). Allow the water to reach the boiling point, then remove the heat and allow it to cool. The meter should show continuity until the water temperature cools to switch deactivation temperature indicated in **Table 4**. For example, on WB700 models, the switch deactivation temperature is 145-171° F (63-77° C).

7. When the water cools to the switch deactivation temperature, the meter should show no continuity (infinity).

8. If the thermoswitch does not provide the specified readings, replace it.

9. Reinstall the thermoswitch by reversing steps 1 and 2. When reinstalling the thermoswitch, note the following:

   a. Wipe the thermoswitch with marine grease before reinstalling it.
   b. On 1100 cc models, be sure the grommet on the thermoswitch lead is secured by the grommet plate.

MULTIFUNCTION METERS

Several models are equipped with a multifunction meter that monitor various functions on the watercraft. **Figures 113-117** show the meters and the functions they monitor. Refer to **Figures 113-117** when checking the display as described below.

To check the display:

1. Disconnect the meter connector.

2. Disconnect the fuel sender connector and the oil sensor connector on models so equipped.

3. Connect the positive battery terminal to the red lead in the meter connector and connect the negative terminal to the black lead in the connector.

4A. *RA700, RA700A, RA700B, RA760, RA1100, WB760, WVT700 and WVT1100*—Connect jumper A from the red to green leads in the meter connector.

4B. *WRB650, WRB650A, and WRB700*—Connect jumper A from the red to brown leads in the meter connector.

5. All LCDs and displays in the meter should light up when you complete the above connections.

6. Disconnect jumper A. All displays should go out, except the fuel meter, which should operate for approximately 30 seconds and then go out.

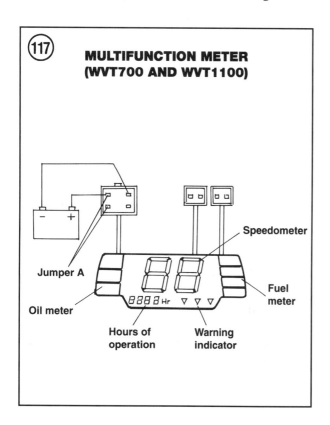

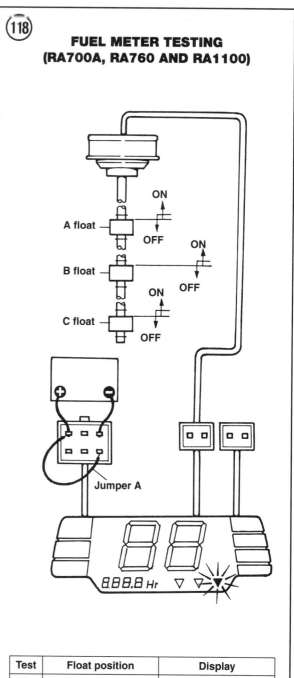

Test	Float position	Display
1	A, B, C: On	F0, F1, F2, F3: On
2	A: Off B, C: On	F0, F1, F2: On
3	A, B: Off C: On	F0, F1: On
4	A, B, C: Off	F0, F-AL: Blinking

ELECTRICAL SYSTEM

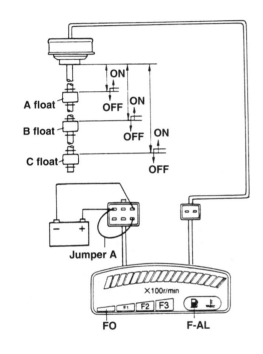

7. Replace the multifunction meter if its display functions are not operating.

FUEL METER
(RA700, RA700A, RA760, RA1100, WB760, WRA650, WRA650A, WRA700, WRB650, WRB650A, WRB700, WVT700 AND WVT1100)

The RA700, RA700A, RA760, RA1100, WB760, WRA650, WRA650A, WRA700, WVT700 and WVT1100 models are equipped with a 3-step LCD fuel meter that displays fuel level when the engine is running. **Figures 118-122** identify the various fuel meters and their float operating positions.

During operation, floats in the fuel sender turn magnetic switches on and off. When a float moves to the OFF position, its corresponding LCD on the meter turns off. When the fuel in the tank nears the reserve level, the EMP symbol flashes on and off. You will need fuel soon.

Please note that an LCD does not immediately turn off when its float descends to the OFF position. The system has a 5-10 second delay to prevent erratic readouts due to fuel sloshing. Keep this in mind when troubleshooting the fuel meter.

Because the fuel meter only operates when the engine is running, visually check the fuel level before starting the engine. Use the fuel meter to monitor fuel level when riding on the water only.

Fuel Meter Testing

Refer to **Figures 118-122** when performing this procedure.

1. Disconnect the multifunction meter connector.

2. Connect the positive battery terminal to the red lead in the meter connector. Connect the negative battery terminal to the black lead in the connector. See the figure for your model.

CHAPTER NINE

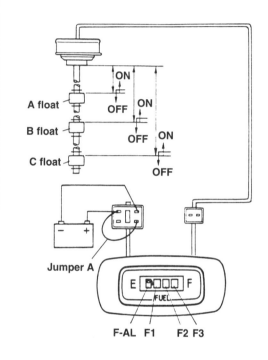

120 FUEL METER TESTING (WRA650, WRA650A AND WRA700)

Test	Float position	Display
1	A, B, C: On	F1, F2, F3: On
2	A: Off B, C: On	F1, F2: On
3	A,B: Off C: On	F1: On
4	A, B, C: Off	F-AL: Blinking

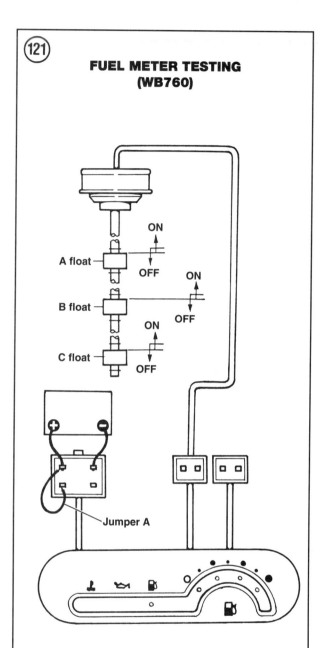

121 FUEL METER TESTING (WB760)

Test	Float position	Display
1	A, B, C: On	F0, F1, F2, F3: On
2	A: Off B, C: On	F0, F1, F2: On
3	A,B: Off C: On	F0, F1: On
4	A, B, C: Off	F0, F-AL: Blinking

ELECTRICAL SYSTEM

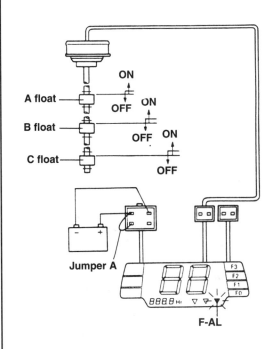

3A. *RA700, RA700A, RA760, RA1100, WB700, WVT700 and WVT1100*—Connect jumper A across the red and green leads in the meter connector as shown in the figure for your model.

3B. *WRA650, WRA650A, WRA700, WRB650, WRB650A and WRB700*—Connect jumper A across the red and brown leads in the meter connector as shown in **Figure 119** or **Figure 120**.

4. Loosen the fuel sender hose clamp and pull it out of the fuel tank. Hold a rag underneath the sender to prevent fuel from dripping onto the engine or other components. Cover the fuel tank opening while performing the following steps.

5. Position the fuel sensor so you can move the floats by hand. Check the fuel meter in the 4 positions shown in the figure for your model. The indicated LCDs should come on approximately 20 seconds after you move a float.

6. Replace the meter if it fails any of the above tests.

Fuel Sender Testing/Replacement

Functional testing of the fuel sender will determine the accuracy of the sender in each of its 3 operating positions. Check the fuel meter before checking the fuel sender.

1. Disconnect the negative battery cable.

2. Disconnect the fuel sender electrical connector. Loosen the fuel sender hose clamp and pull the fuel sender out of the fuel tank. Hold a rag underneath the fuel sender to prevent fuel from dripping onto the engine or other components. Cover the fuel tank opening while performing the following steps.

3. Support the fuel sender so that you can reposition each of the 3 floats by hand. Then connect an ohmmeter to the fuel sender switch leads. Set the ohmmeter on the $R \times 1$ or $R \times 100$ scale as required by the test procedure. Touch the meter leads and zero the meter when switching between resistance scales.

Test	Float position	Display
1	A, B, C: On	F0, F1, F2, F3: On
2	A: Off B, C: On	F0, F1, F2: On
3	A, B: Off C: On	F0, F1: On
4	A, B, C: Off	F0, F-AL: Blinking

4. Refer to **Figure 123** or **Figure 124** for test connections and values. Check the fuel sender in the 4 different test positions by positioning the floats as indicated in **Figure 123** and **Figure 124**. For example, to run test No. 1, move floats A, B and C to their respective ON (upper) positions and note the reading on the ohmmeter.

5. If any of the meter readings differ from the stated values, replace the fuel sender.

6. Reverse to install the fuel sender.

OIL METER (RA700A, RA760, RA1100, WVT700 AND WVT1100)

The RA700A, RA760, RA1100, WVT700 and WVT1100 models use an oil meter to indicate the oil level in the oil tank. LCDs in the display turn on or off as the floats in the oil sensor open and close magnetic switches.

Testing

1. Disconnect the multifunction meter connector.

2. Connect the battery to the red and black leads of the meter connector as shown in the **Figure 125** or **Figure 126**.

3. Connect jumper A across the red and green leads in the meter connector as shown in the **Figure 125** or **Figure 126**.

4. Loosen the oil sensor hose clamp and pull it out of the oil tank. Hold a rag underneath the sensor to prevent oil from dripping onto the engine or other components. Cover the oil tank opening while performing the following steps.

5. Position the oil sensor so you can move the floats by hand. Check the oil meter in the 3 positions shown in **Figure 125** or **Figure 126**. The indicated LCDs should come on approximately 20 seconds after you perform each test.

6. Connect jumper B across the blue and black terminals in the oil sensor connector. The display

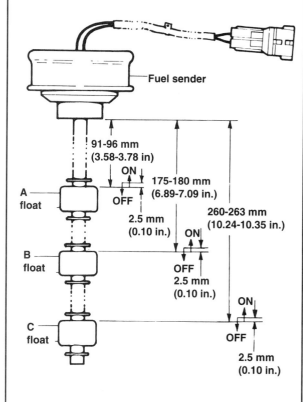

FUEL SENSOR TESTING (RA700A, RA760, RA1100, WRA650, WRA650A, WRA700, WVT700 AND WVT1100)

Test	Float position	Resistance
1	A, B, C: On	0-2 ohms
2	A: Off B, C: On	97-103 ohms
3	A, B: Off C: On	292-308 ohms
4	A, B, C: Off	667-713 ohms

ELECTRICAL SYSTEM

FUEL SENSOR TESTING
(RA700, RA700B, WB760, WRB650, WRB650A AND WRB700)

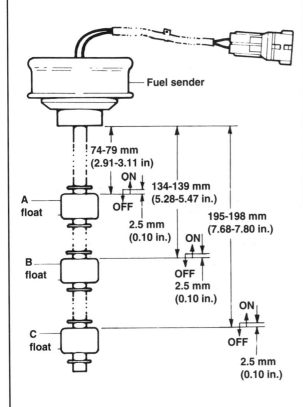

Test	Float position	Resistance
1	A, B, C: On	0-2 ohms
2	A: Off B, C: On	97-103 ohms
3	A, B: Off C: On	292-308 ohms
4	A, B, C: Off	667-713 ohms

OIL METER TESTING
(RA700A, RA760 AND RA1100)

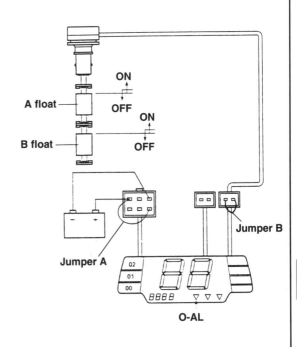

Test	Float position	Display
1	A, B: On	00, 01, 02: On
2	A: Off B: On	00, 01: On
3	A, B: Off	00, 0-AL: Blinking

**OIL METER TESTING
(WVT700 AND WVT1100)**

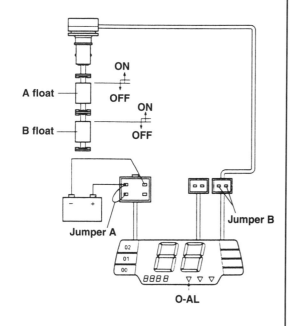

Test	Float position	Display
1	A, B: On	00, 01, 02: On
2	A: Off B: On	00, 01: On
3	A, B: Off	00, 0-AL: Blinking

**OIL LEVEL SENSOR TESTING
(RA700A, RA760, RA1100,
WVT700 AND WVT1100)**

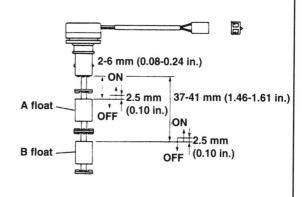

Test	Float position	Resistance
1	A, B: On	0-2 ohms
2	A: Off B: On	97-103 ohms
3	A, B: Off	292-308 ohms

ELECTRICAL SYSTEM

should stop blinking approximately 20 seconds after you make this connection.

7. Replace the meter if it fails any of the above tests.

Oil Sensor
Testing/Replacement

Functional testing of the oil sensor will determine the accuracy of the sender in each of its 3 operating positions. Always test the oil meter as described above before testing the sensor.

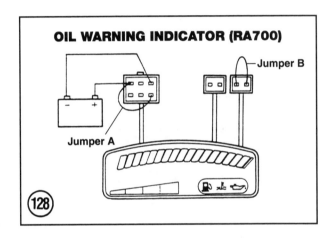

128

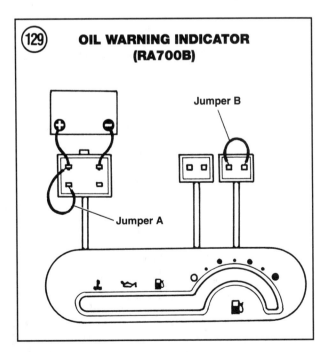

129

1. Disconnect the negative battery cable.
2. Disconnect the oil sensor electrical connector. Loosen the oil sensor hose clamp and pull it out of the oil tank. Hold a rag underneath the sensor to prevent oil from dripping onto the engine or other components. Cover the oil tank opening while performing the following steps.
3. Support the oil sensor so you can reposition each of the 2 floats by hand. Then connect an ohmmeter to the oil sensor leads. Set the ohmmeter on the R × 1 or R × 100 scale as required by the test procedure. Touch the meter leads and zero the meter when switching between resistance scales.
4. Refer to **Figure 127** for test values. Check the fuel sender in the 3 different test positions by positioning the floats as indicated in **Figure 127**. For example, to run test No. 1, move floats A and B to their respective ON (upper) positions and note the reading on the ohmmeter.
5. If any of the meter readings differ from the indicated values, replace the oil sensor.
6. Reverse to install the oil sensor.

OIL WARNING INDICATOR
(RA700, RA700B, WB760)

The RA700, RA700B, RA760 and WB760 models use an oil level warning indicator in the multifunction meter. The oil level indicator blinks when the oil level in the tank drops below a critical level. The indicator is activated when the float in the oil sensor opens the magnetic switch in the oil sensor.

Testing
(RA700, RA700B and WB760)

Refer to **Figures 128-130** when performing this procedure.

Check the operation of the oil warning indicator in the multifunction meter as follows:

1. Disconnect the multifunction meter connector and the oil level sensor connector.

2. Connect the positive battery terminal to the red lead in the connector. Connect the negative battery terminal to the black lead in the connector.

3. Connect jumper A across the red and green leads in the meter connector as shown in the figure for your model. The oil warning indicator should blink when you make this connection.

4. Connect jumper B across the 2 leads in the oil level sensor connector as shown in the figure for your model. The oil warning indicator should stop blinking 15-30 seconds after you make this connection.

Oil Sensor Testing/Replacement

Functional testing of the oil sensor will determine the accuracy of the sender in each of its 2 operating positions. Check the oil warning indicator before you check the sensor.

1. Disconnect the negative battery cable.

2. Disconnect the oil sensor electrical connector. Loosen the oil sensor hose clamp and pull it out of the oil tank. Hold a rag underneath the sensor to prevent oil from dripping onto the engine or other components. Cover the oil tank opening while performing the following steps.

3. Support the oil sensor so that you can reposition the float by hand. Then connect an ohmmeter to the oil sensor leads (blue and black). Set the ohmmeter on the R × 1, and check the continuity when the float is in the ON and OFF positions shown in **Figure 131** or **Figure 132**.

4. The meter should show continuity (0 ohms) when the float is in the ON position. The meter should indicate no continuity (infinity) when the float is in the OFF position.

5. If either meter reading differs from these values, replace the oil sensor.

6. Reverse to install the oil sensor.

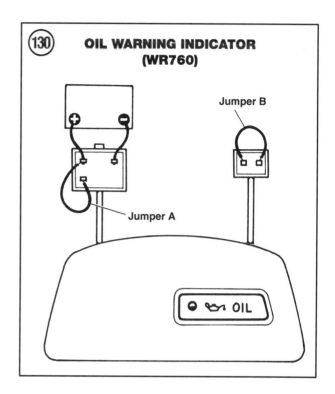

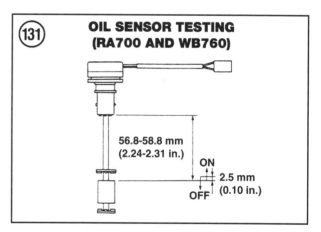

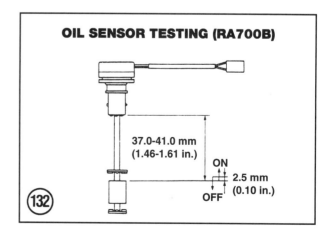

ELECTRICAL SYSTEM

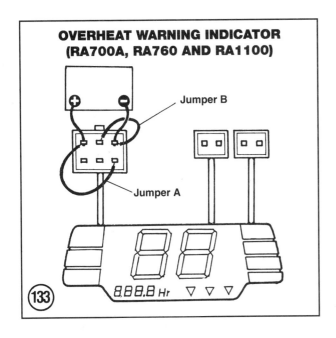

133

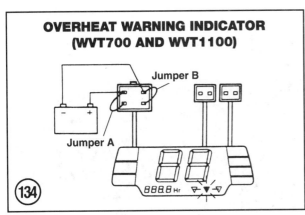

134

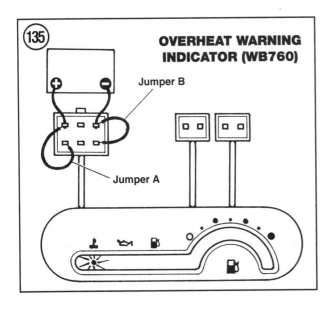

135

OVERHEAT WARNING INDICATOR (RA700, RA700A, RA760, RA1100, WB760, WRB650, WRB650A, WRB700, WVT700 AND WVT1100)

Refer to **Figures 133-136** when performing this procedure.

If the engine begins to overheat, the overheat warning indicator blinks, and engine speed decreases to below 3000 rpm. Engine speed is held below 3000 rpm until operating temperature drops to a safe level.

Testing

1. Disconnect the multifunction meter connector and the oil level sensor connector.

2. Connect the positive battery terminal to the red lead in the connector. Connect the negative battery terminal to the black lead in the connector.

3A. *RA700, RA700A, RA760, RA1100, WB760, WVT700 and WVT1100*—Connect jumper A across the red and green leads in the meter connector as shown in the **Figure 133**, **Figure 134** or **Figure 135**.

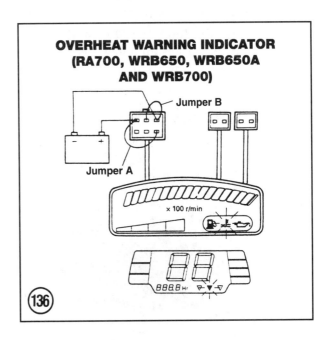

136

3B. *WRB650, WRB650A and WRB700* Connect jumper A across the red and brown leads in the meter connector as shown in the **Figure 136**.

4. Connect jumper B across the pink and black leads in the meter connector as shown in the figure for your model. The temperature warning indicator should start blinking when you make this connection.

5. Replace the meter if the temperature warning indicator is not operating properly.

FUSE

All models are equipped with a single fuse. See **Table 1** for fuse capacity for each model.

On most models, the fuse is located in the electric box. If the fuse blows, remove the electric box fuse cover and pull the fuse out of the box. See **Figure 137** or **Figure 138**. If a fuse is blown, install a new fuse with the same amperage rating. After replacing the fuse, insert the fuse holder back into the electric box and reinstall the fuse cover. On 650 cc models, make sure to install the fuse cover together with its rubber gasket (**Figure 139**).

On 1100 cc models, the fuse is located inside the ignition coil damper. Pull the fuse holder grommet from the ignition coil damper (B, **Figure 107**). The fuse holder is behind the grommet. If the fuse is blown, install a new fuse with the same amperage rating, and reinstall the fuse holder in the ignition coil damper. Be sure the grommet is properly seated in the damper.

NOTE
If the fuse blows, find out the reason for the failure before replacing the fuse. Usually the trouble is a short circuit in the wiring. A blown fuse may be caused by worn-through wiring or a disconnected wire shorting to ground.

CAUTION
Never substitute tin foil or wire for a fuse. Never use a higher amperage fuse than specified. An overload could result in fire and complete loss of the craft.

WIRING DIAGRAMS

Complete wiring diagrams are located at the end of this book.

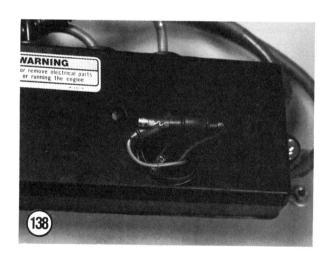

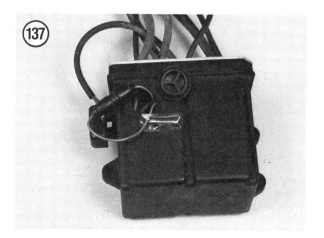

ELECTRICAL SYSTEM

Table 1 ELECTRICAL SPECIFICATIONS

Fuse Capacity	10 amp
Battery capacity	12 volt/19 amp hours

Table 2 BATTERY STATE OF CHARGE

Specific gravity	State of charge
1.1101-1.130	Discharged
1.140-1.160	Almost discharged
1.170-1.190	One-quarter discharged
1.200-1.220	One half discharged
1.230-1.250	Three-quarters discharged
1.260-1.280	Fully charged

Table 3 STARTER SPECIFICATIONS

	New mm (in.)	Wear limit mm (in.)
Brush length		
WR500, SJ650	17 (0.67)	13 (0.51)
WR650	12.5 (0.49)	7.5 (0.295)
WRA650, WRA650A	NS	13 (0.51)
RA700, RA700A	NS	5.0 (0.20)
All other models	NS	6.5 (0.26)
Commutator diameter		
WR500, SJ650, WRA650, WRA650A	33 (1.30)	31 (1.22)
WR650	28 (1.10)	27 (1.06)
WRA650, WRA650A	33 (1.30)	31 (1.22)
RA700, RA700A, RA700B RA760, RA1100, WB760	NS	27 (1.06)
WVT700, WVT1100	28.0 (1.10)	27 (1.06)
All other models	27-28 (1.06-1.10)	27 (1.06)
Mica undercut		
WR500, WRA650, WRA650A	0.5-0.8 (0.02-0.03)	0.2 (0.008)
WR650	0.7 (0.028)	0.15 (0.005)
RA700, RA700A, RA700B, RA760, RA1100, WB760	NS	0.2 (0.008)
WVT700, WVT1100	0.7 (0.03)	0.15 (0.006)
All other models	0.2-0.7 (0.008-0.028)	0.2 (0.008)

NS–Not specified

Table 4 THERMOSWITCH SPECIFICATIONS

Model	Switch Activated	Switch Deactivated
RA700, RA700A, RA700B, WVT700	100.4-125.6° F (66-74°C)	78.8-93.2° F (43-57° C)
RA760, RA1100	199.4 F (93° C)	181.4° F (83° C)
FX700, SJ700, SJ700A, WB700, WB700A, WRA650, WRB650A, WRA700, WRB650, WRB650A, WRB700	169-183° F (76-84° C)	145-171° F (63-77° C)
SJ650, WR500, WR650	139-213° F (60-100° C)	121-195° F (50-90° C)
WB760, WVT1100	194-204.8° F (90-96° C)	68.8-194° F (76-90° C)

Chapter Ten

Oil Injection System

A mechanical oil injection system using a crankshaft-driven injection pump and external oil tank is used on all RA, WB, WRA, WRB and WVT models. The pump draws oil from the oil tank and supplies it under pressure to the flame arrestor holder where nozzles spray it into the air/fuel mixture.

WB700, WB700A, WRA650, WRA650A, WRA700, WRB650, WRB650A and WRB700, models use self-bleeding oil pumps. RA700, RA700A, RA700B, RA760, RA1100, WB760, WVT700 and WVT1100 models use manual bleed pumps. Refer to Chapter Three for oil bleeding procedures as well as other regular oil injection system maintenance procedures. **Figures 1-5** show the oil injection systems used on these models.

OIL TANK

Removal/Installation

Refer to **Figures 6-10** for this procedure.

1. Open the engine cover.
2. Cut all plastic ties used to secure hoses at the oil tank as required. Spring type clamps can be opened and then slid along the hose.
3. Tag and disconnect all hoses at the oil tank. Plug all hoses and tank fittings to prevent oil leakage and contamination.
4A. *WRA650, WRA650A, WRA700, WRB650, WRB650A, WRB700, WB700, WB700A*—Remove the oil tank mounting bolts and remove the oil tank.
4B. *RA700, RA700A, RA700B, RA760, RA1100, WB760, WVT700 and WVT1100*—On these models, the oil tank sits atop the fuel tank and is held in place by one of the fuel tank straps. Remove the strap and remove the oil tank.
5. Wipe up all spilled oil inside the hull.
6. Installation is the reverse of these steps. Install each hose onto the oil tank and secure the hose with the spring clamp or a new plastic tie as required.
7. Bleed the oil injection system as described in Chapter Three before operating the engine.

OIL INJECTION SYSTEM

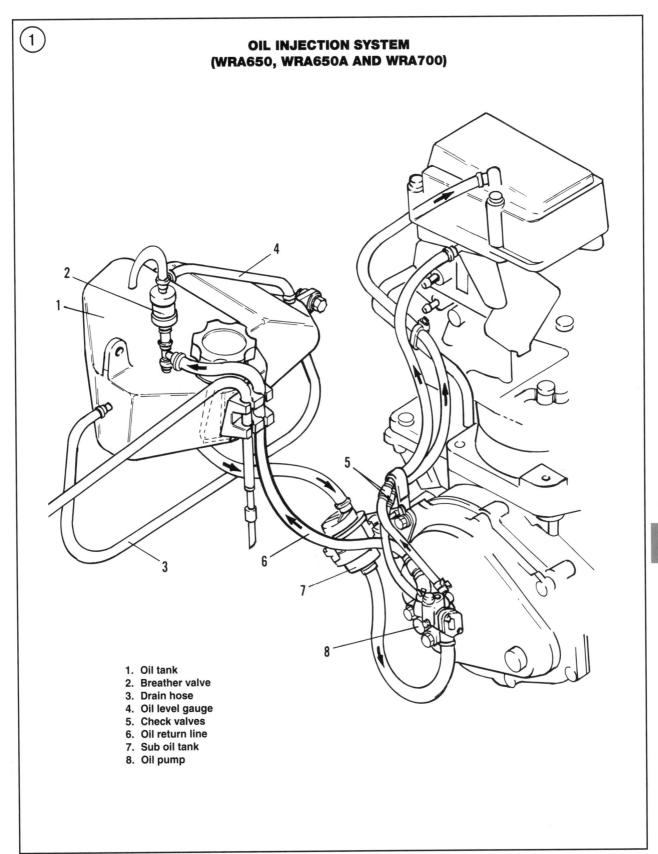

**OIL INJECTION SYSTEM
(WRA650, WRA650A AND WRA700)**

1. Oil tank
2. Breather valve
3. Drain hose
4. Oil level gauge
5. Check valves
6. Oil return line
7. Sub oil tank
8. Oil pump

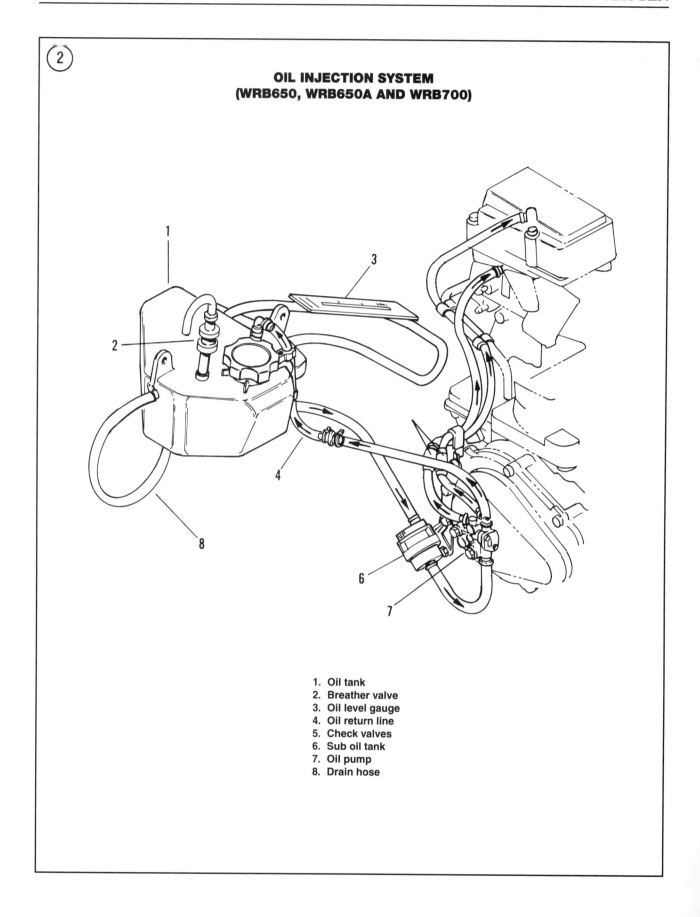

② OIL INJECTION SYSTEM
(WRB650, WRB650A AND WRB700)

1. Oil tank
2. Breather valve
3. Oil level gauge
4. Oil return line
5. Check valves
6. Sub oil tank
7. Oil pump
8. Drain hose

OIL INJECTION SYSTEM

457

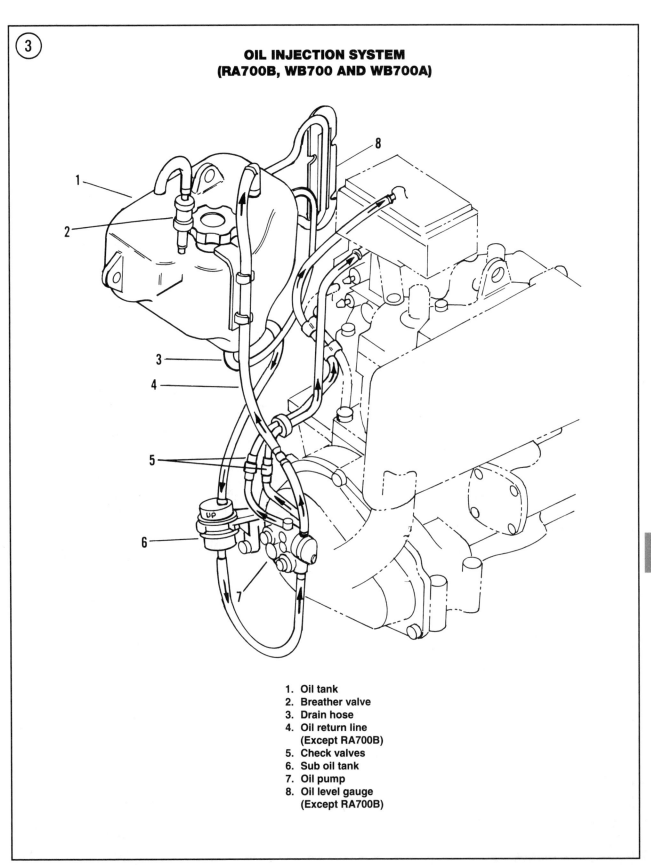

1. Oil tank
2. Breather valve
3. Drain hose
4. Oil return line
   (Except RA700B)
5. Check valves
6. Sub oil tank
7. Oil pump
8. Oil level gauge
   (Except RA700B)

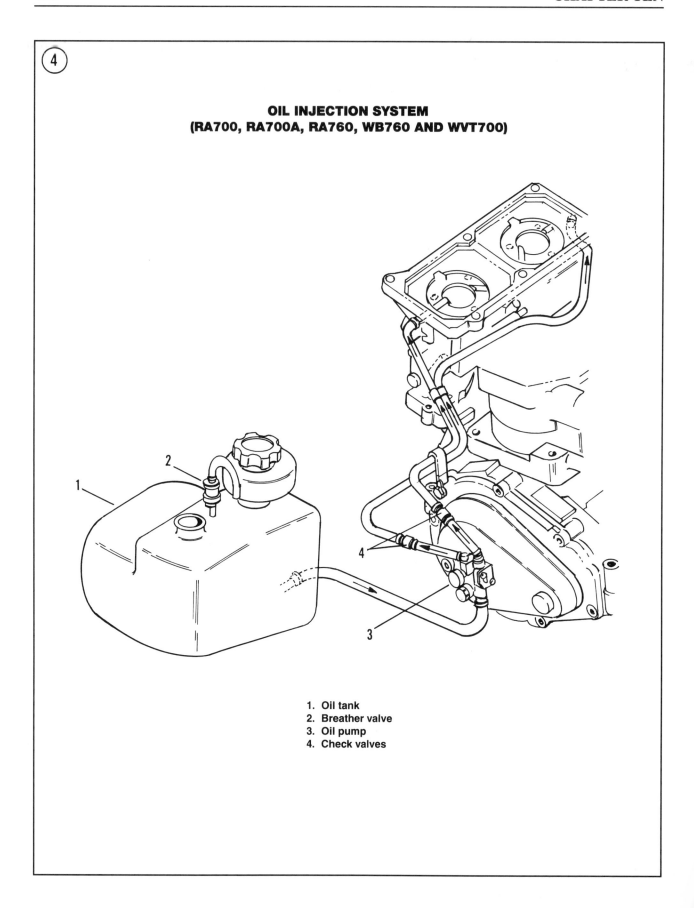

1. Oil tank
2. Breather valve
3. Oil pump
4. Check valves

OIL INJECTION SYSTEM

459

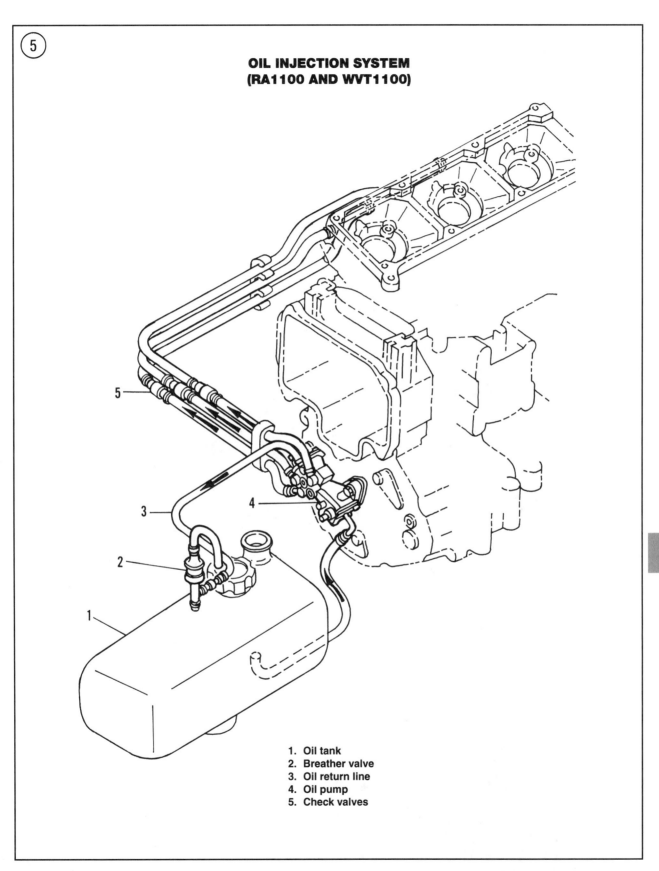

OIL INJECTION SYSTEM (RA1100 AND WVT1100)

1. Oil tank
2. Breather valve
3. Oil return line
4. Oil pump
5. Check valves

460

CHAPTER TEN

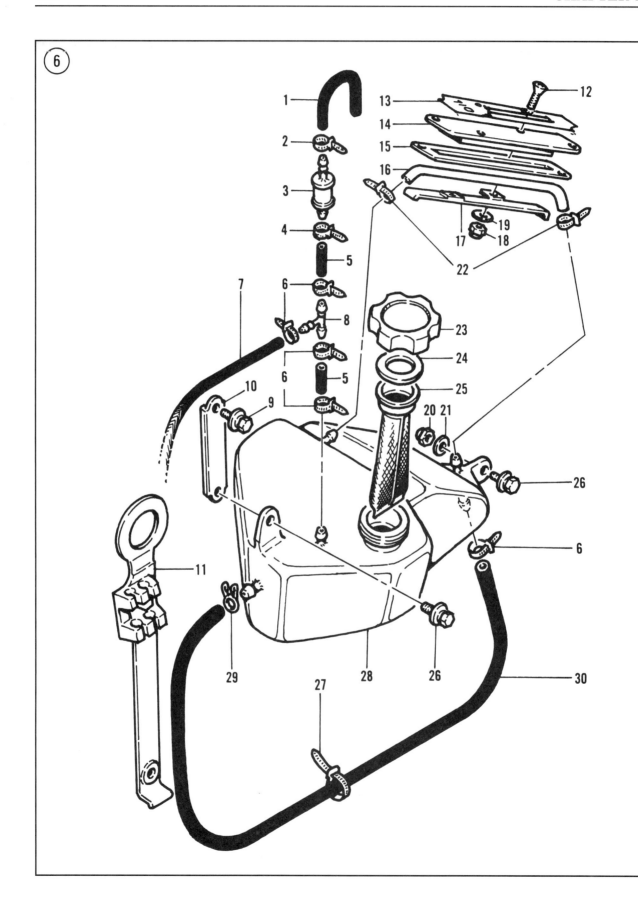

OIL INJECTION SYSTEM

**OIL TANK
(WRA650, WRA650A, WRA700, WRB650,
WRB650A AND WRB700)**

1. Hose
2. Hose clamp
3. Breather valve
4. Hose clamp
5. Hose
6. Hose clamps
7. Hose*
8. T-fitting*
9. Bolt
10. Bracket
11. Holder
12. Screw
13. Label
14. Oil level gauge
15. Gasket
16. Hose
17. Base
18. Nut
19. Washer
20. Nut
21. Washer
22. Hose clamps
23. Cap
24. Gasket
25. Oil filter
26. Bolt and washer
27. Clamp
28. Oil tank
29. Clamp
30. Hose

*Not used on WRB650, WRB650A and WRB700 models. Return oil from oil injection pump is routed to fitting on oil tank.

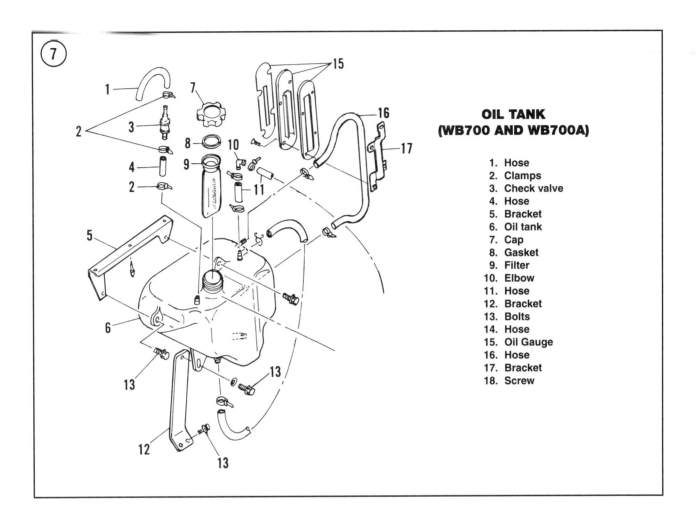

OIL TANK (WB700 AND WB700A)

1. Hose
2. Clamps
3. Check valve
4. Hose
5. Bracket
6. Oil tank
7. Cap
8. Gasket
9. Filter
10. Elbow
11. Hose
12. Bracket
13. Bolts
14. Hose
15. Oil Gauge
16. Hose
17. Bracket
18. Screw

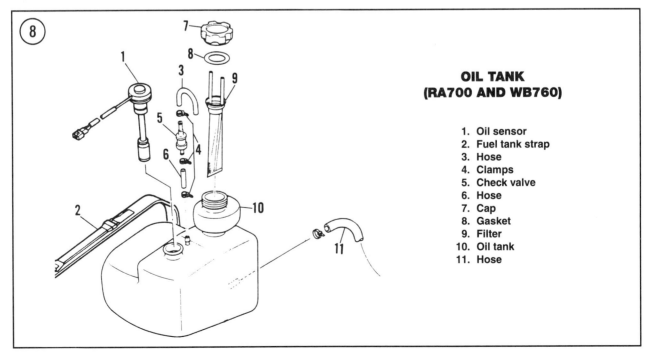

OIL TANK (RA700 AND WB760)

1. Oil sensor
2. Fuel tank strap
3. Hose
4. Clamps
5. Check valve
6. Hose
7. Cap
8. Gasket
9. Filter
10. Oil tank
11. Hose

OIL INJECTION SYSTEM

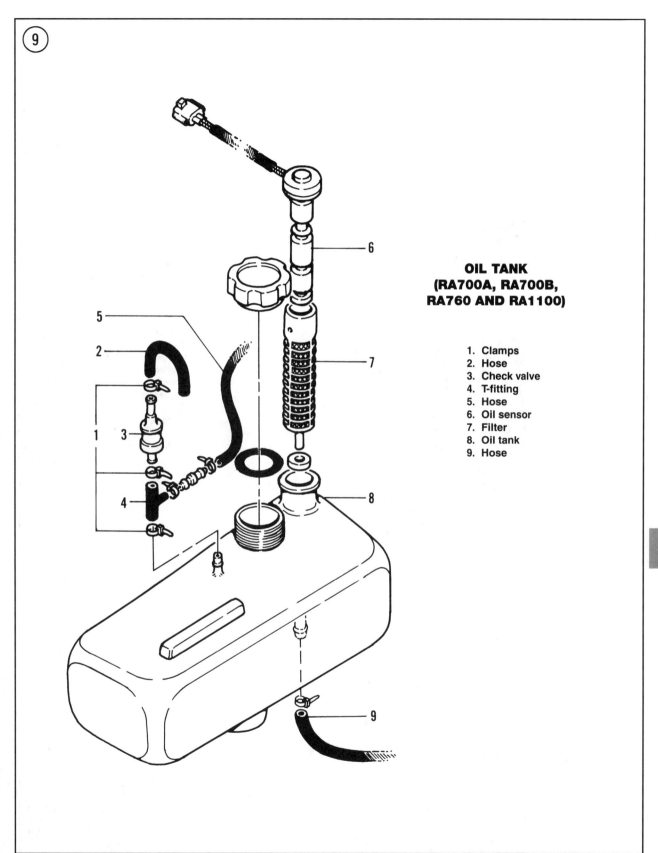

OIL TANK (RA700A, RA700B, RA760 AND RA1100)

1. Clamps
2. Hose
3. Check valve
4. T-fitting
5. Hose
6. Oil sensor
7. Filter
8. Oil tank
9. Hose

CHAPTER TEN

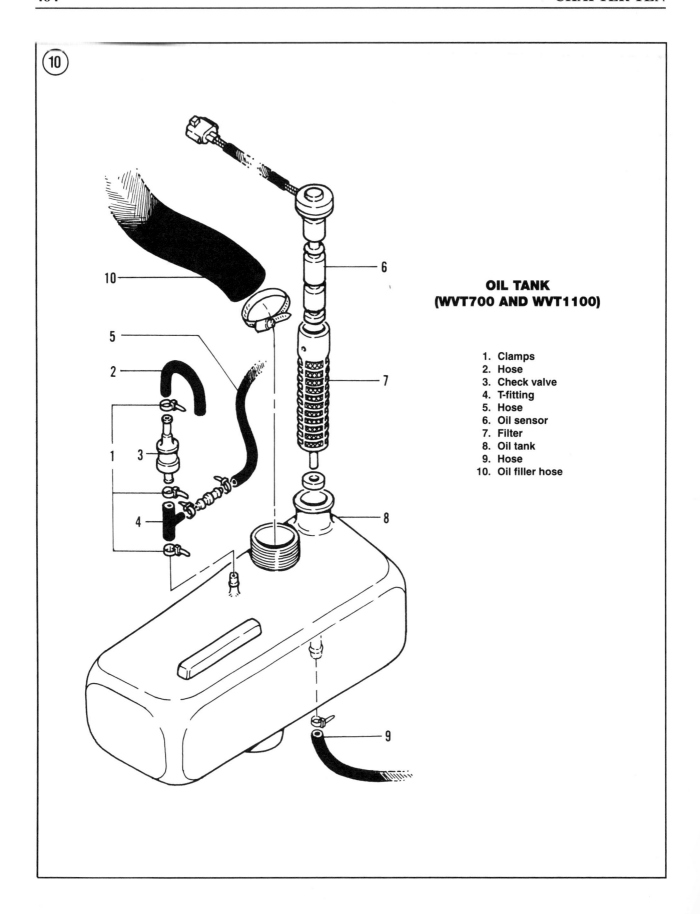

OIL TANK (WVT700 AND WVT1100)

1. Clamps
2. Hose
3. Check valve
4. T-fitting
5. Hose
6. Oil sensor
7. Filter
8. Oil tank
9. Hose
10. Oil filler hose

OIL INJECTION SYSTEM

SUB OIL TANK

Removal/Installation

(WRA650, WRA650A, WRA700, WRB650, WRB650A, WRB700, WB700 and WB700A)

Refer to **Figure 11** for this procedure.
1. Open the engine cover.
2. Pull the sub oil tank out of its holder.
3. Cut the plastic ties used to secure the hoses to the sub oil tank. Then disconnect and plug the hoses to prevent oil leakage and contamination. Remove the sub oil tank.
4. Installation is the reverse of these steps. Note the following:
   a. Reconnect the oil hoses to the sub oil tank so the UP mark on the sub oil tank faces toward the main oil tank (**Figure 11**).
   b. When inserting the sub oil tank into its holder, be sure the tab in the holder engages the cutout in sub oil tank. See **Figure 11**.
   c. If the sub oil tank holder was removed, position the clamp so it is within the 30° angle shown in **Figure 12**.
   d. Bleed the oil injection system as described in Chapter Three before operating the engine.

CHECK VALVES

Removal/Installation

Check valves are installed in each of the oil pump delivery lines (**Figure 13**).

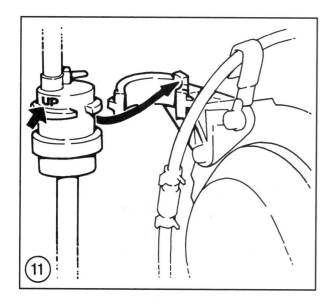

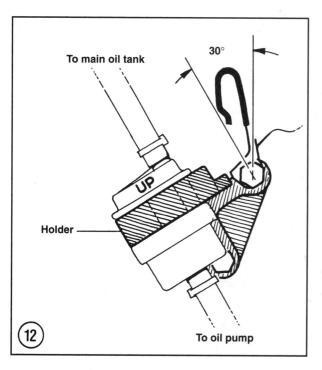

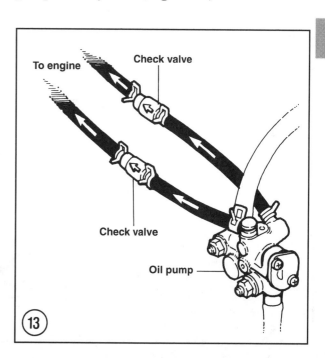

1. Slide the oil line clamps off of the check valve hose fittings.
2. Disconnect the 2 oil lines from each check valve. Plug the oil lines to prevent leakage or contamination.
3. Installation is the reverse of these steps. Note the following:
    a. Install the check valve so the arrow on each valve points toward the intake manifold. See **Figure 13**.
    b. Push the oil lines onto the check valve hose fittings until they bottom. Then secure each hose with its hose clamp.
    c. Replace weak or questionable hose clamps.

OIL TANK BREATHER VALVE

Testing/Replacement

Refer to **Figures 6-10** for this procedure.
1. Disconnect the 2 hoses from the breather valve and remove the valve.
2. Wipe the ends of the breather valve with a clean rag.
3. Test the breather valve by blowing through it from both ends. Air should only blow through the breather valve one way. If air blows through both ways or not at all, replace the breather valve.

4. Install the breather valve so its green hose connection (**Figure 14**) faces toward the oil tank.
5. Use new plastic ties when reconnecting the hoses at the valve.

OIL PUMP

Removal/Installation

Refer to **Figures 15-18** for this procedure.
1. Label and disconnect the inlet and outlet hoses from the oil pump. Plug the end of the hoses.
2. Remove the bolts and washers securing the oil pump to the flywheel cover. Remove the oil pump and gasket. Discard the gasket.
3. Installation is the reverse of these steps. Note the following:
    a. Install a new oil pump gasket.
    b. Align the oil pump shaft with the slot in the end of the crankshaft. See **Figure 19**.
    c. Apply Loctite 242 to the threads of the oil pump bolts and tighten the bolts securely.
    d. Wipe the oil pump hose fittings clean before reconnecting the hoses.
    e. Bleed the oil injection system as described in Chapter Three before operating the engine.

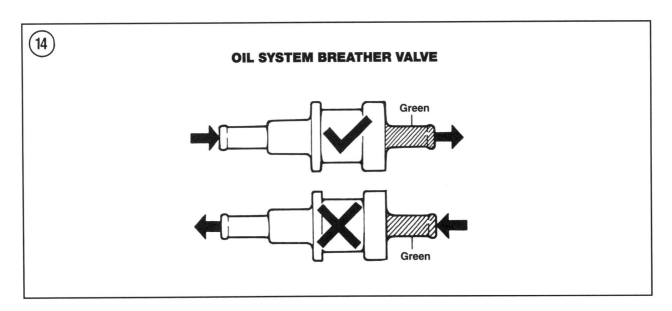

OIL INJECTION SYSTEM

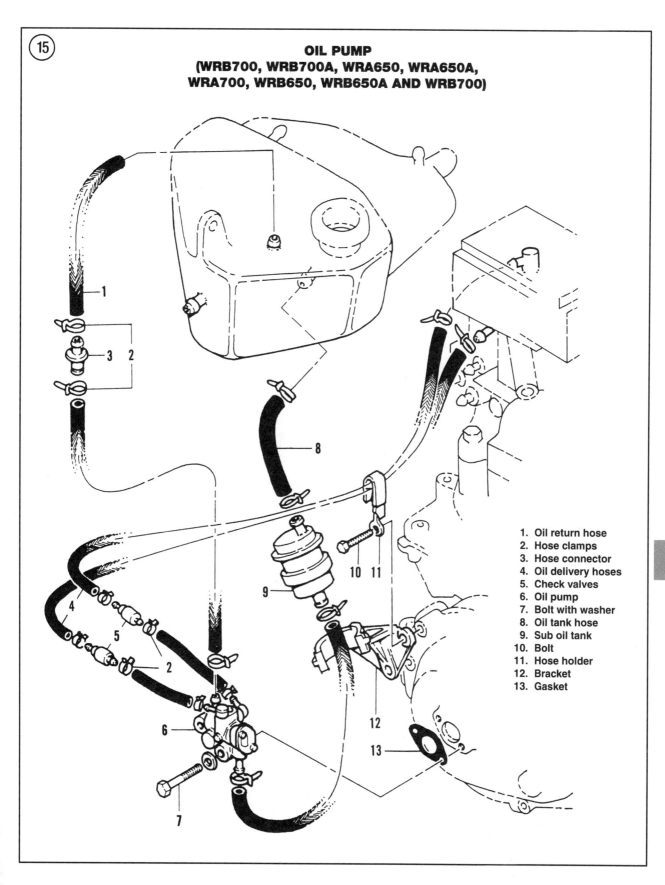

⑮ **OIL PUMP**
(WRB700, WRB700A, WRA650, WRA650A, WRA700, WRB650, WRB650A AND WRB700)

1. Oil return hose
2. Hose clamps
3. Hose connector
4. Oil delivery hoses
5. Check valves
6. Oil pump
7. Bolt with washer
8. Oil tank hose
9. Sub oil tank
10. Bolt
11. Hose holder
12. Bracket
13. Gasket

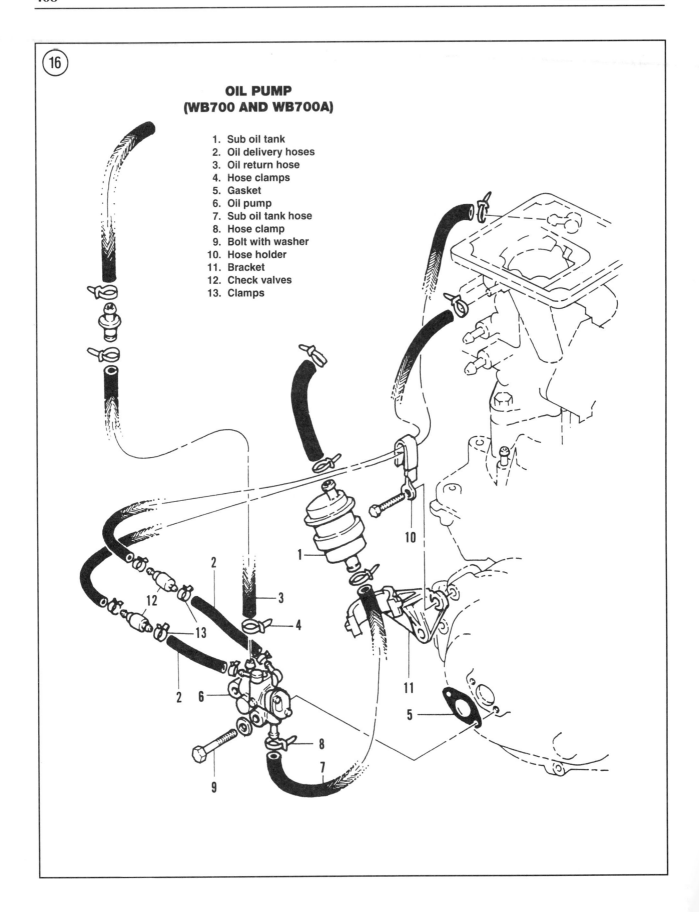

OIL INJECTION SYSTEM

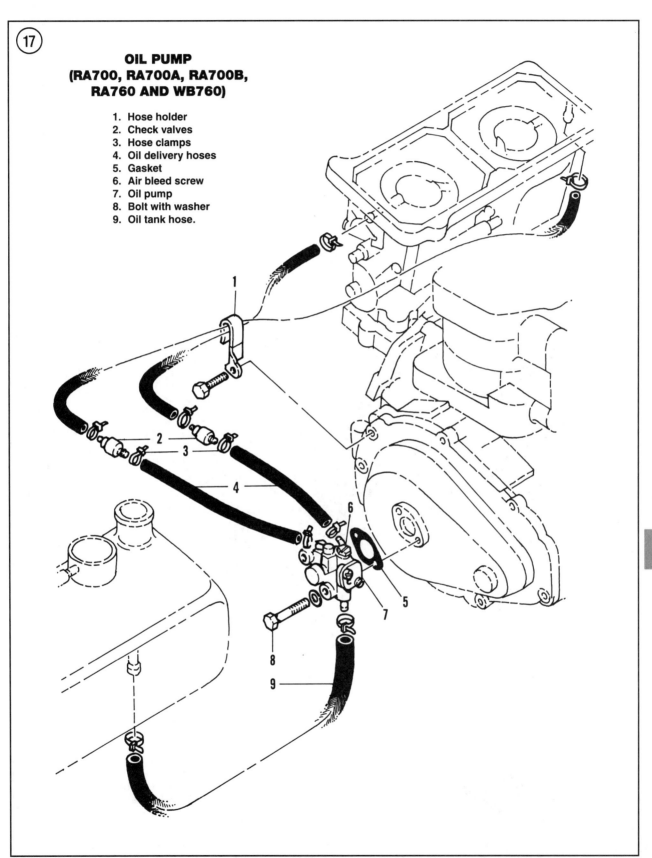

⑰ **OIL PUMP (RA700, RA700A, RA700B, RA760 AND WB760)**

1. Hose holder
2. Check valves
3. Hose clamps
4. Oil delivery hoses
5. Gasket
6. Air bleed screw
7. Oil pump
8. Bolt with washer
9. Oil tank hose.

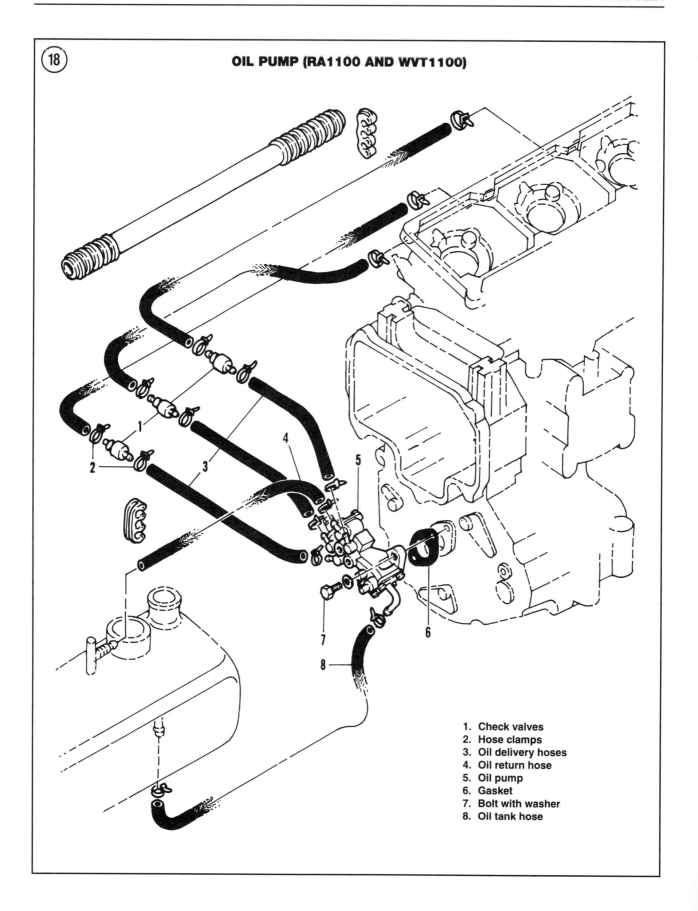

OIL INJECTION SYSTEM

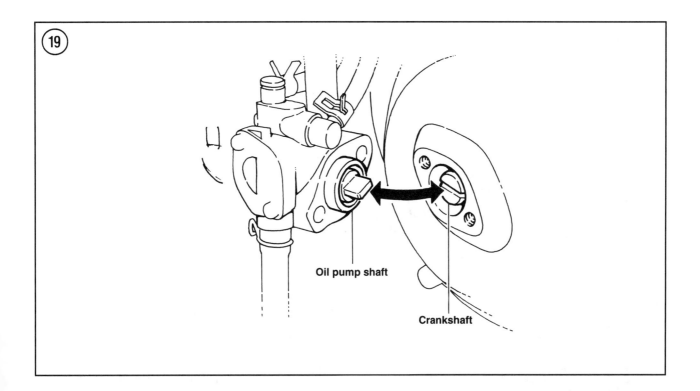

19

Chapter Eleven

Bilge System

Despite the fact that water vehicles are designed to prevent water from entering the engine compartment, some always manages to get in. The purpose of the bilge system is to remove this unwanted water.

The bilge system has no moving parts. It consists of a bilge pickup filter (**Figure 1**) mounted in the bottom of the hull, an antisiphon loop and bilge suction fitting.

When the jet pump is operating, water is forced out of the pump outlet at high speed. At the same time, a small portion of water bypasses the pump outlet and exits through the bilge venturi. This creates a low pressure area or vacuum at the bilge suction fitting on the venturi. This vacuum draws bilge water up from the bilge filter, through the antisiphon valve and out of the bilge venturi (**Figure 2**).

Since the bilge system depends on a fast flow of water through the jet pump to cause suction, the bilge system does not work when the engine is idling. Water is emptied from the engine compartment when the craft is being ridden normally. The purpose of the antisiphon valve is to prevent water from siphoning back into the craft when the engine is not running.

The bilge systems for the different models are identified as follows:
   a. WR500 and WR650 (**Figure 3**).
   b. WRA650, WRA650A, WRA700 and WVT700 (**Figure 4**).
   c. SJ650, SJ700 and SJ700A (**Figure 5**).
   d. FX700 (**Figure 6**).
   e. WRB650, WRB650A and WRB700 (**Figure 7**).
   f. WB700 and WB700A (**Figure 8**).
   g. RA700, RA700A and RA700B (**Figure 9**).
   h. RA760, RA1100, WB760 and WVT1100 (**Figure 10**).

BILGE SYSTEM

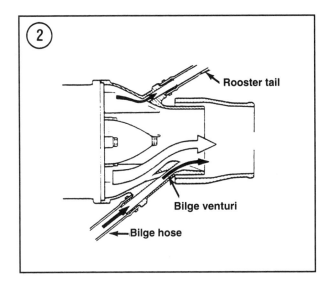

Bilge Filter
Removal/Installation

1. Locate the bilge filter (**Figure 1**) in the engine compartment. Disconnect the hose from the filter.

2. Remove the brace from the hull and remove the bilge filter.

3. Installation is the reverse of the above.

CAUTION
When installing a through-the-hull connector during service, tighten the connector from both sides of the hull.

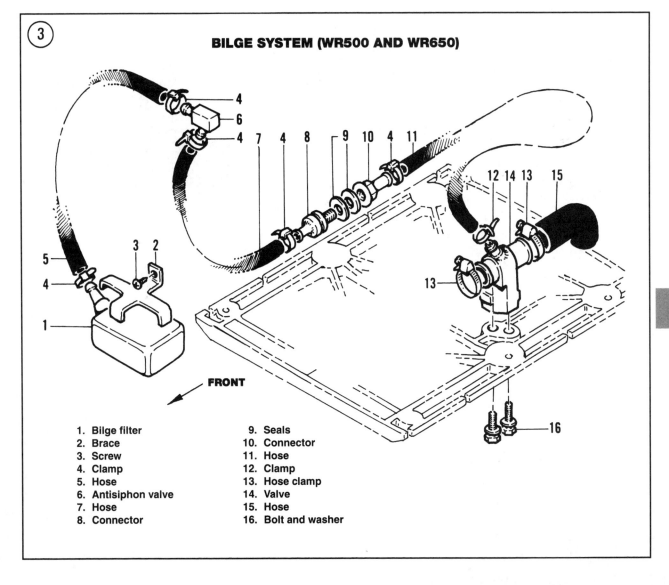

1. Bilge filter
2. Brace
3. Screw
4. Clamp
5. Hose
6. Antisiphon valve
7. Hose
8. Connector
9. Seals
10. Connector
11. Hose
12. Clamp
13. Hose clamp
14. Valve
15. Hose
16. Bolt and washer

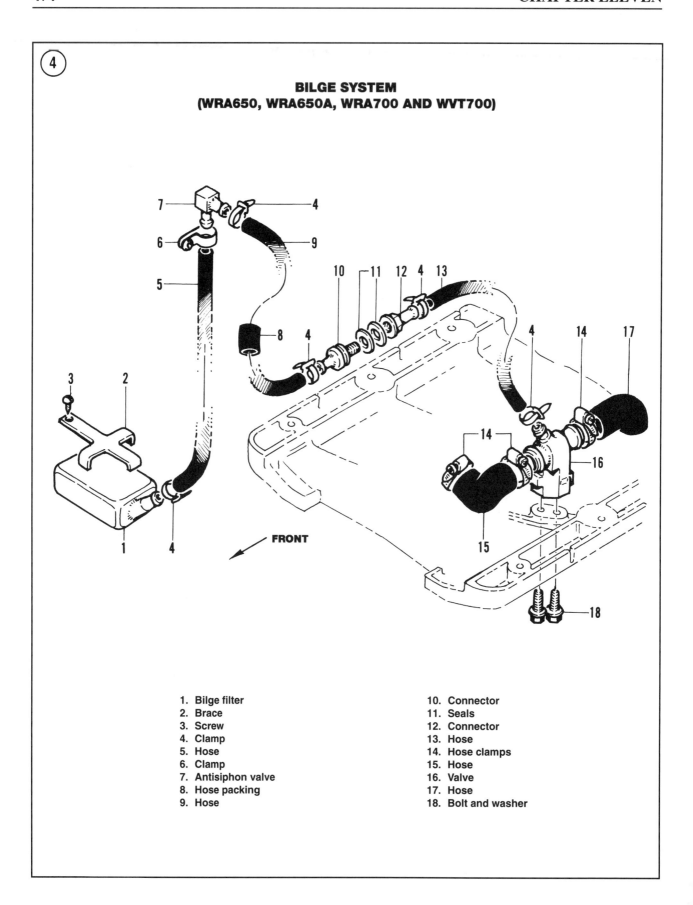

④ BILGE SYSTEM
(WRA650, WRA650A, WRA700 AND WVT700)

1. Bilge filter
2. Brace
3. Screw
4. Clamp
5. Hose
6. Clamp
7. Antisiphon valve
8. Hose packing
9. Hose
10. Connector
11. Seals
12. Connector
13. Hose
14. Hose clamps
15. Hose
16. Valve
17. Hose
18. Bolt and washer

BILGE SYSTEM

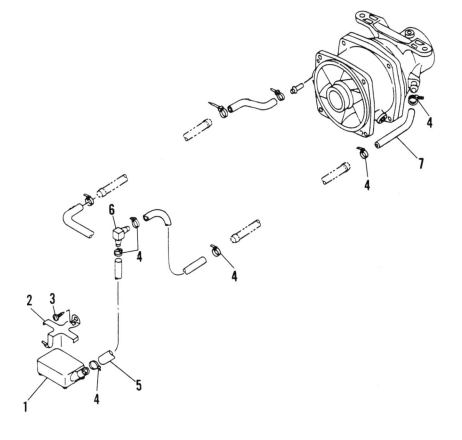

**BILGE SYSTEM
(SJ650, SJ700 AND SJ700A)**

1. Bilge filter
2. Brace
3. Screw
4. Clamp
5. Hose
6. Antisiphon valve
7. Hose

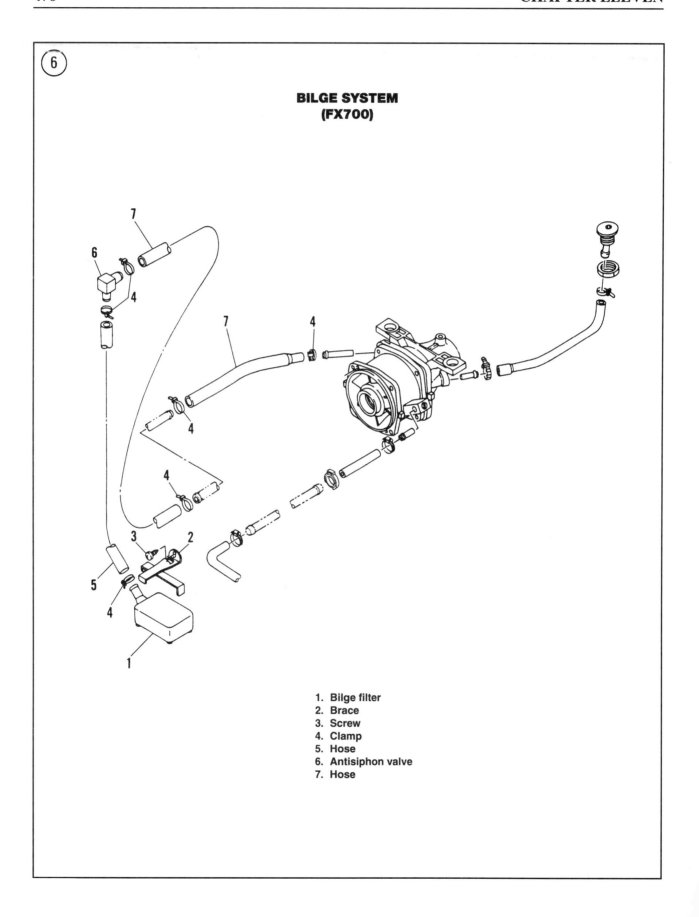

BILGE SYSTEM 477

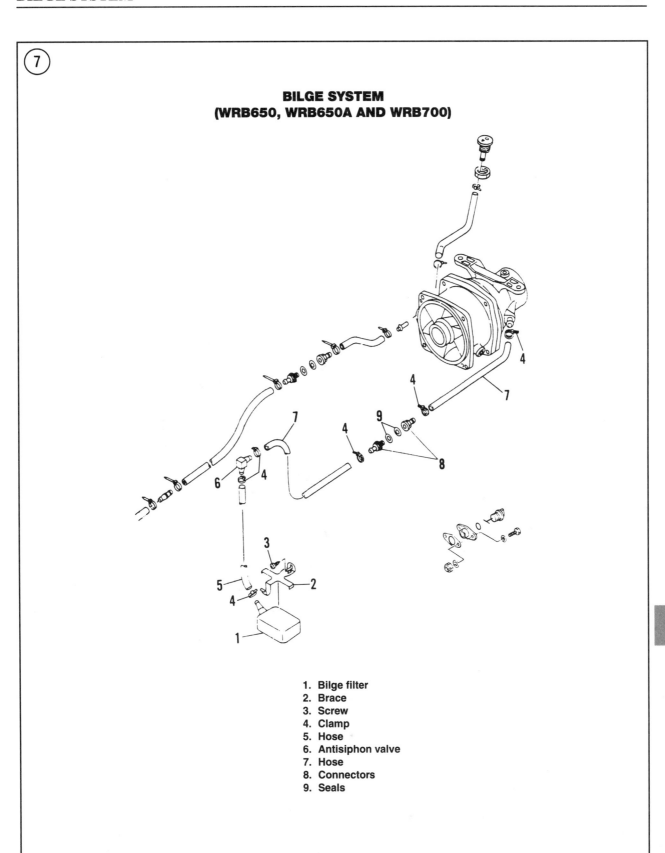

⑦

**BILGE SYSTEM
(WRB650, WRB650A AND WRB700)**

1. Bilge filter
2. Brace
3. Screw
4. Clamp
5. Hose
6. Antisiphon valve
7. Hose
8. Connectors
9. Seals

11

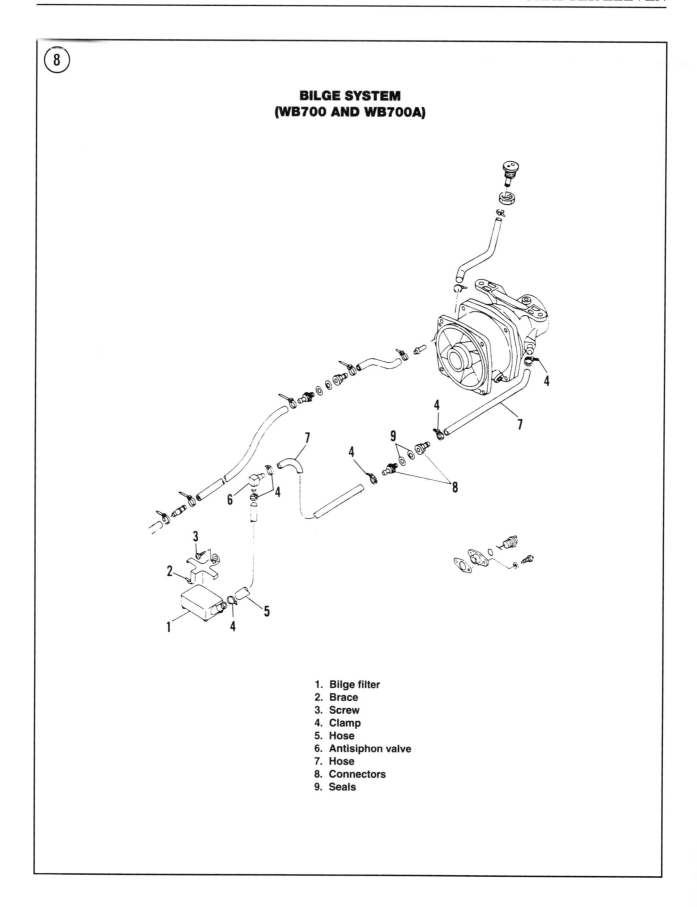

BILGE SYSTEM

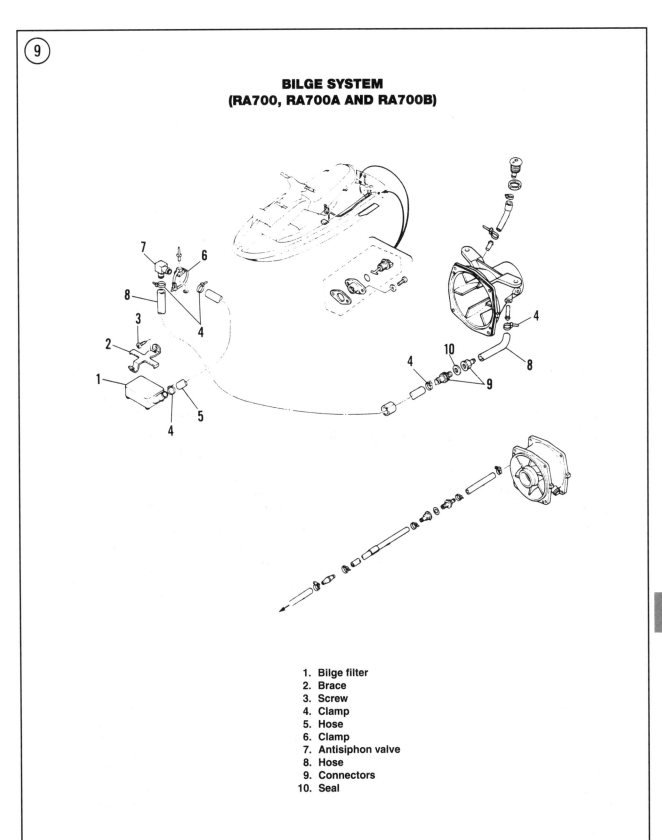

**BILGE SYSTEM
(RA700, RA700A AND RA700B)**

1. Bilge filter
2. Brace
3. Screw
4. Clamp
5. Hose
6. Clamp
7. Antisiphon valve
8. Hose
9. Connectors
10. Seal

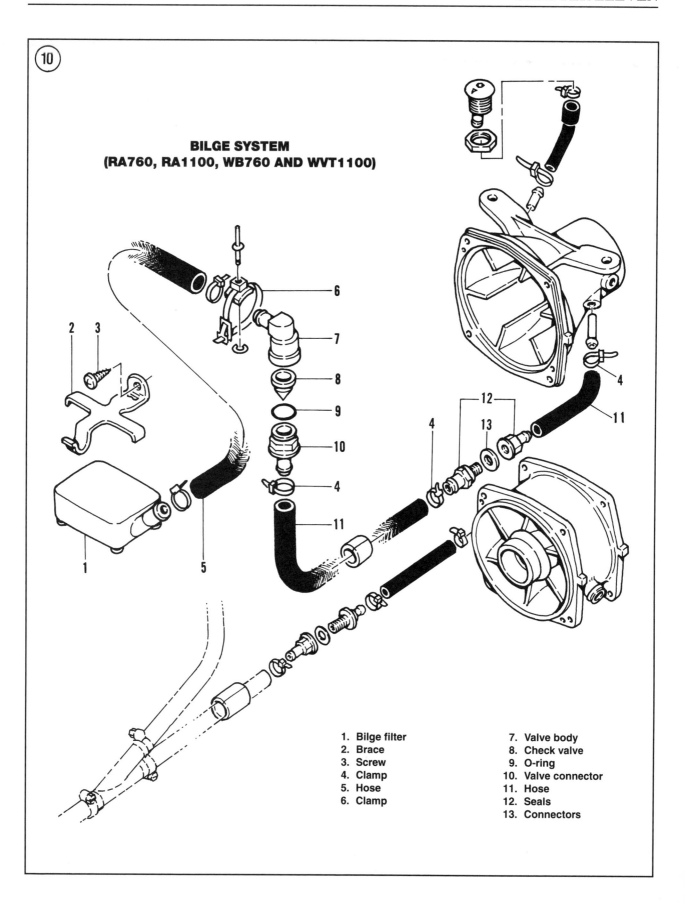

Chapter Twelve

Steering Systems

This chapter describes service to the handlebar and steering assemblies. **Table 1** and **Table 2** are at the end of the chapter.

STEERING ASSEMBLY
(RA700, RA700A, RA700B, RA760, RA1100, WB700, WB700A, WB760, WR500, WR650, WRA650, WRA650A, WR700, WRB650, WRB650A, WRB700, WVT700 AND WVT1100)

Refer to **Figures 1-7** when performing the procedures in this section.

Removal

1. To avoid short circuits while disconnecting and moving wires, disconnect the negative battery cable.
2. Remove the handlebar pad.
3. Disconnect the start/stop switch electrical connectors.
4. Disconnect the throttle cable from the throttle lever.
5. Disconnect the steering cable from the ball joint on the steering bracket (A, **Figure 8**).
6. If necessary, remove the starter/stop switch housing screws and remove the switch from the handlebar assembly.
7. On models with a trim adjuster, remove the trim adjuster control from the handlebar if necessary.
8. Loosen the handlebar mounting bolts, then remove the bolts, holders and handlebar.

NOTE
The wiring harness and throttle cable pass through packing installed in the center of the steering column. Note the position and alignment of the wiring harness, cable and packing before removing them in Step 8.

CHAPTER TWELVE

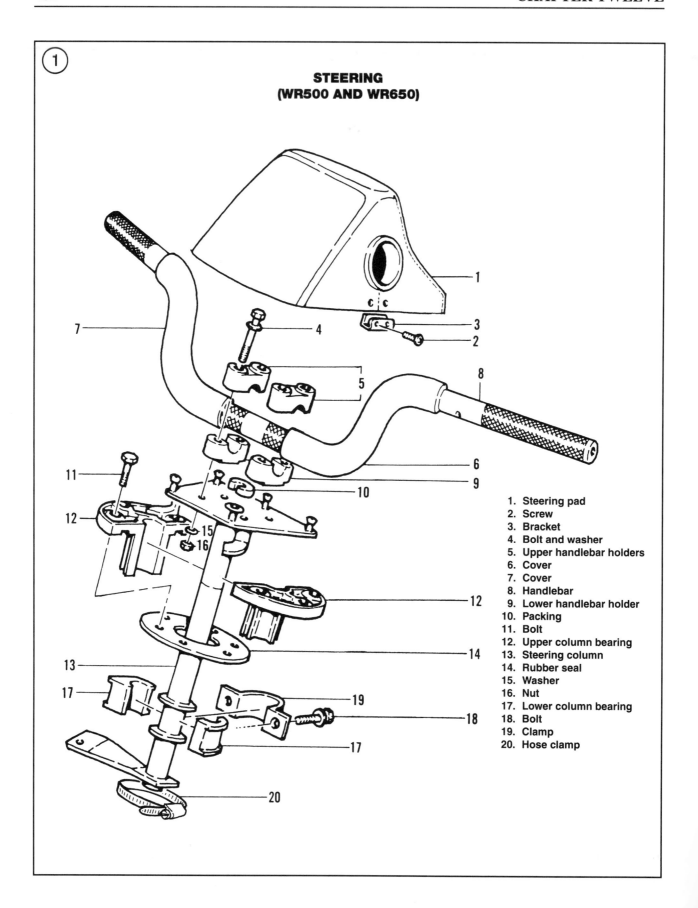

STEERING SYSTEM

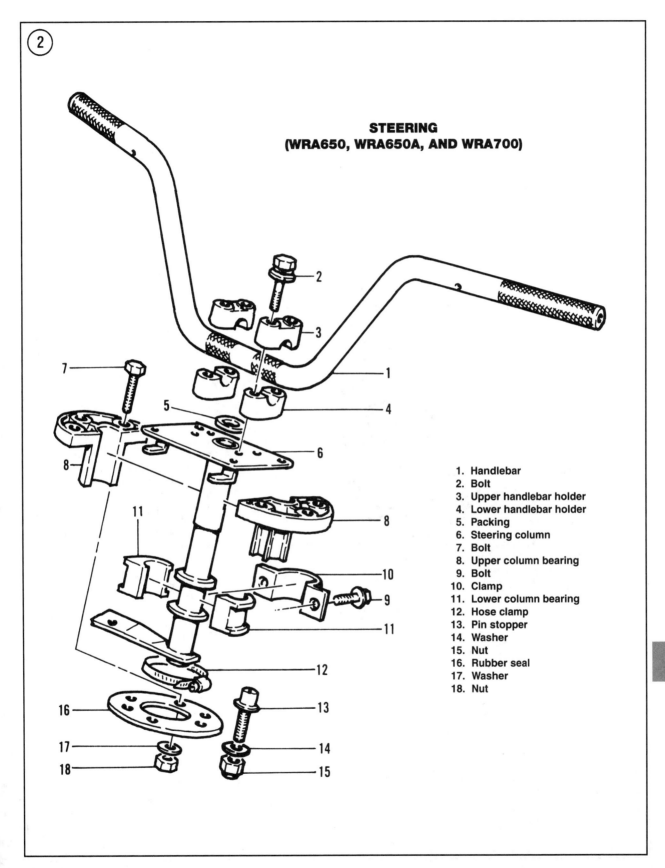

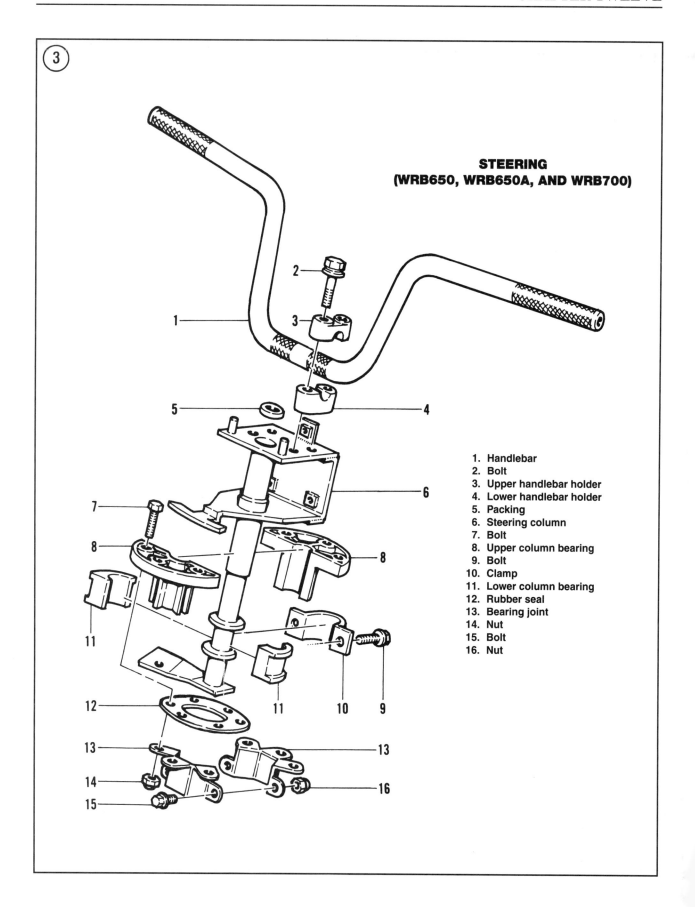

STEERING SYSTEM

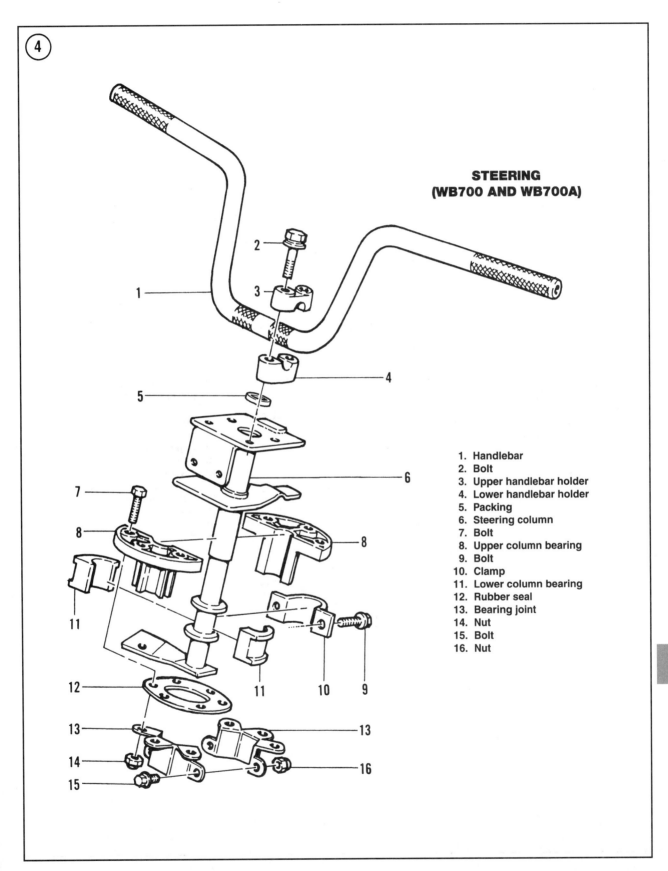

STEERING (WB700 AND WB700A)

1. Handlebar
2. Bolt
3. Upper handlebar holder
4. Lower handlebar holder
5. Packing
6. Steering column
7. Bolt
8. Upper column bearing
9. Bolt
10. Clamp
11. Lower column bearing
12. Rubber seal
13. Bearing joint
14. Nut
15. Bolt
16. Nut

CHAPTER TWELVE

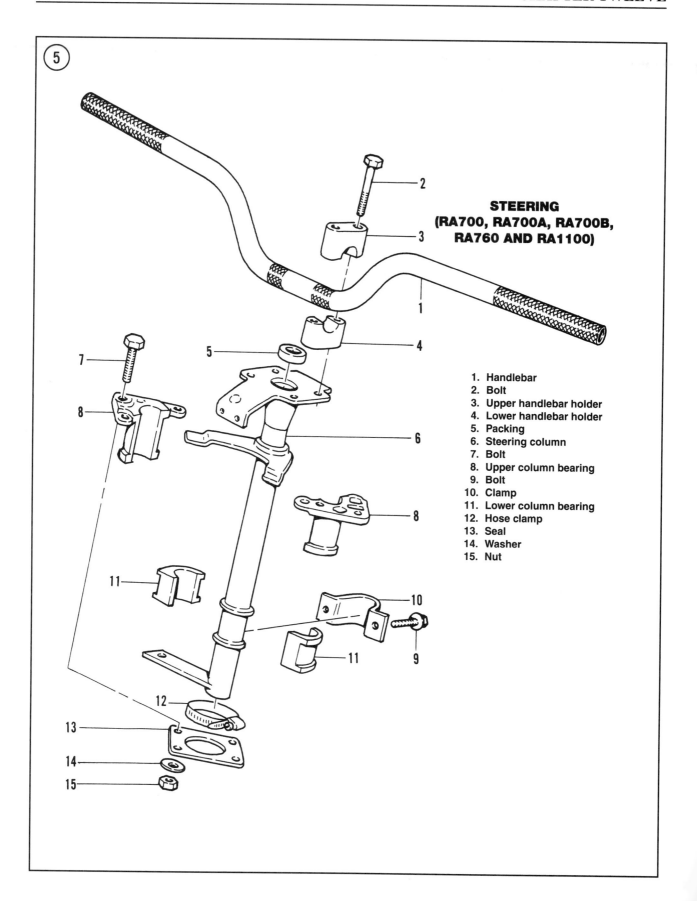

STEERING SYSTEM

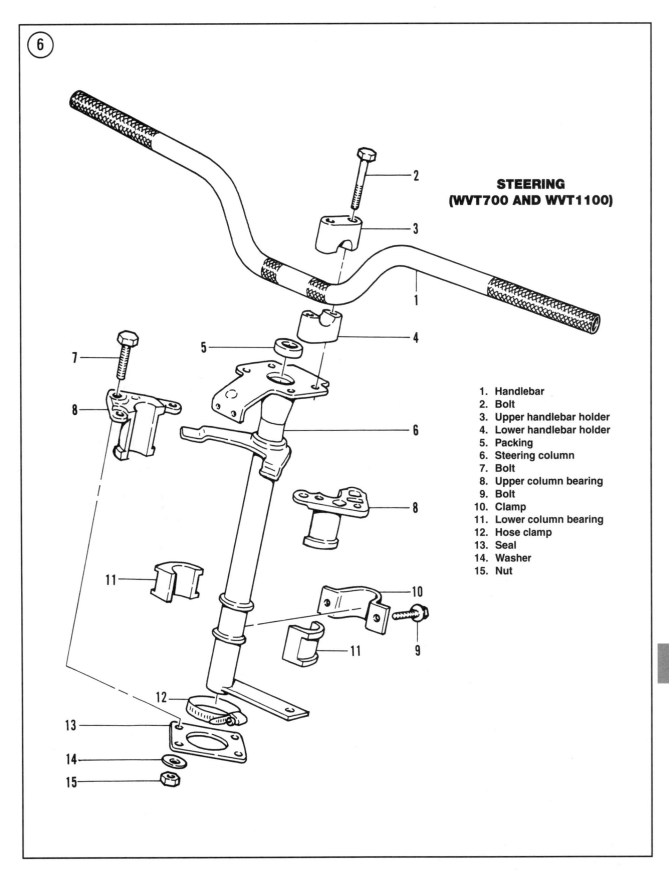

STEERING
(WVT700 AND WVT1100)

1. Handlebar
2. Bolt
3. Upper handlebar holder
4. Lower handlebar holder
5. Packing
6. Steering column
7. Bolt
8. Upper column bearing
9. Bolt
10. Clamp
11. Lower column bearing
12. Hose clamp
13. Seal
14. Washer
15. Nut

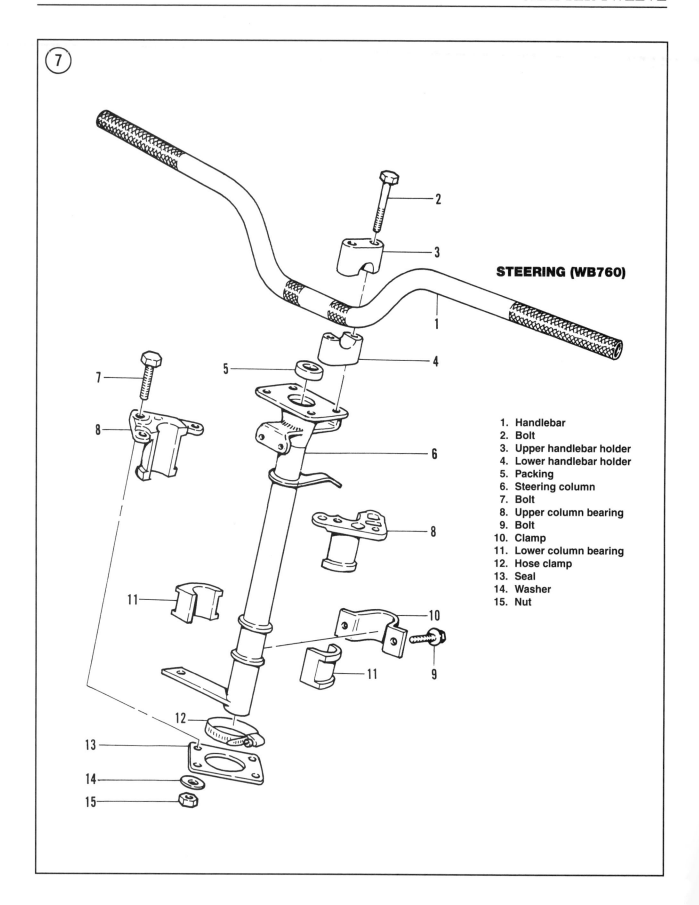

STEERING SYSTEM

9. Pull the wiring harness and throttle cable through the steering column. Remove the packing.

10. Remove the clamp (B, **Figure 8**) from the bracket on the engine cover and remove the 2 lower column bearings (C, **Figure 8**).

11A. *WRB650, WRB760A, WRB700, WB700 and WB700A models*—Perform the following:

   a. Remove the 2 bolts clamping the bearing joint to the upper column bearings (**Figure 9**).
   b. Remove the nuts and bolts holding the upper bearings to the engine cover.
   c. Remove the bearing joint from inside the engine cover (**Figure 9**).
   d. Remove the 2 upper column bearings and remove the steering column.

11B. *WR500, WR650, WB760, all WRA models, all RA models and all WVT models*—Perform the following:

   a. Loosen the hose clamp from around the upper column bearings (D, **Figure 8**).
   b. Remove the nuts and bolts holding the upper bearings to the engine cover.
   c. Remove the 2 upper column bearings and remove the steering column.

Inspection

1. Check the steering column for cracks, deep scoring or excessive wear.
2. Check the handlebars for cracks or damage.
3. Inspect the bearings for scoring or wear.
4. Replace worn or damaged parts as required.

Assembly

1. Wipe the steering and bearing columns with marine grease before assembly.
2. Apply Loctite 242 to all threads during assembly.

3A. *WR500, WR650, WRA650, WRA650A, WRA700, RA700, RA700A, RA700B, RA760, RA1100, WVT700, WVT1100 and WB760*—Set the steering column in place in the engine cover and install the upper bearings by performing the following:

   a. Set the upper bearings in place on the steering column. Secure the bearings to the engine cover with the mounting nuts and bolts. Torque the mounting nuts to the specification given in **Table 1**.
   b. Install the hose clamp around the upper bearing. Torque the clamp to the specification given in **Table 1**.

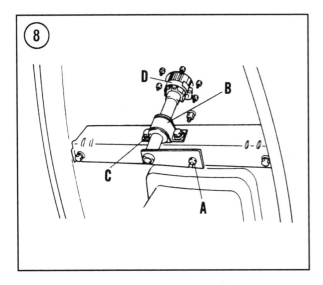

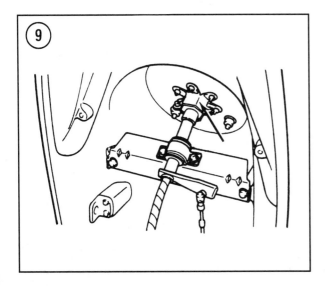

3B. *WRB650, WRB650A, WRB700, WB700 and WB700A*—Set the steering column in place in the engine cover and install the upper bearings by performing the following:

   a. Apply RTV sealant to the mating surfaces of the upper bearings and set the upper bearings in place on the steering column (**Figure 10**).

   b. Place the bearing joint around the bearings and secure the assembly to the engine cover (**Figure 9**). Torque the upper bearing mounting bolts to the specification given in **Table 1**.

   c. Install the bearing joint clamp bolts. Torque them to the specification given in **Table 1**.

4. *WR500, WR650, WRA650, WRA650A, WRA700, WRB650, WRB650A, WRB700, WB700 and WB700A*—Once the upper bearings are installed, apply RTV sealant into the gaps formed by the 2 upper bearing mating surfaces (**Figure 11**).

5. Assemble the lower bearing assembly by performing the following:

   a. Set the lower bearings in place on the steering column.

   b. Fit the clamp around the bearings and secure the clamp to the bracket on the engine cover (B, **Figure 8**).

   c. Torque the lower bearing clamp bolts to the specification listed in **Table 1**.

6A. *WR500 and WR650*—Perform the following:

   a. Route the wire harness and throttle cable through the steering column. Be sure the spiral tubing extends 50 mm (2.0 in.) into the column shaft as shown in **Figure 12**.

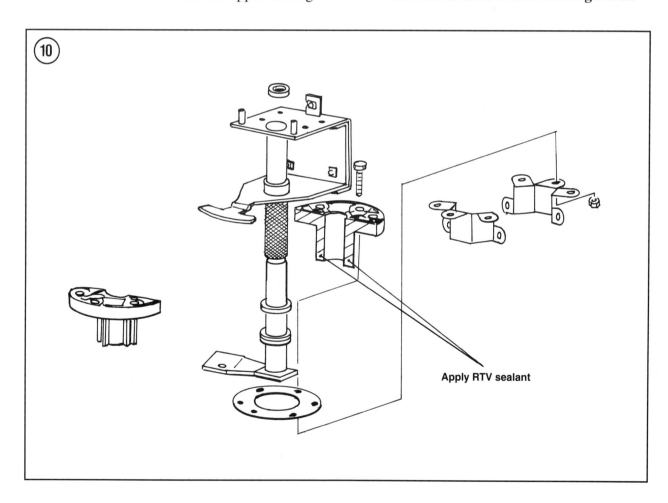

Apply RTV sealant

STEERING SYSTEM

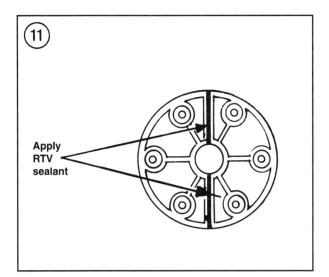

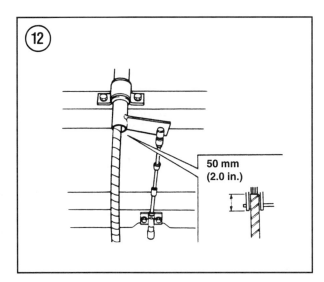

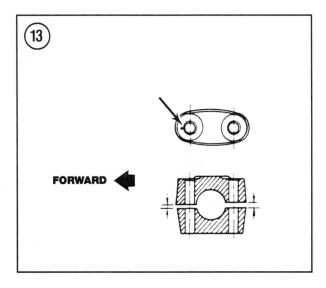

b. Place the handlebar into position and install the upper holders and bolts. Tighten the bolts hand tight at this time. Position the handlebar in the holders so a distance of 580 mm (22.8 in.) is maintained from the center of the handlebar grip to the floor mat. Also position the handlebar so the alignment dot on the handlebar is centered in the gap between the 2 holders.

c. Be sure the punch mark on the upper holder faces forward, toward the bow, as shown in **Figure 13**. When tightening the holder bolts, be sure the gap between the holders on the front (bow) side of the holders is less than the gap on the rear (stern) side as shown in **Figure 13**. Torque the handlebar holder bolts to the specification given in **Table 1**.

NOTE
When routing the throttle cable and the start/stop switch wiring harness in the following steps, a radius of 60 mm (2.36 in.) must be maintained at the center of the handlebar. See Figure 14.

WARNING
If the throttle cable is not routed properly, the inner cable will not move smoothly. This may cause the throttle to stick wide open.

d. Loop the throttle cable as shown in **Figure 14** and reconnect the cable to the throttle lever.

e. Route the start/stop switch wiring harness as shown in **Figure 14**. Then reconnect the electrical connectors.

f. When the wire harness and throttle cable are properly routed, install the packing into the steering column as shown in **Figure 14**. Make sure the wire harness and throttle cable are positioned in the packing as shown. Also be sure that none of the packing sticks out of the steering column.

g. Install the steering pad. Align the pad with the 4 pins on the steering column and push the pad into position. Maintain a clearance of 7-10 mm (0.28-0.39 in.) at the front of the pad.

6B. *WRA650, WRA650A and WRA700*—Perform the following:

a. Route the wire harness and throttle cable through the steering column. Be sure the spiral tubing extends 30-50 mm (1.2-2.0 in) into the column shaft as shown in **Figure 15**.

b. Install the left- and right-hand steering pads.

c. Place handlebar into position and install upper holders and bolts. Tighten bolts finger tight at this time. Then position the handlebar so the alignment dot on the handlebar is centered between the gap in the 2 holders.

d. Be sure the punch mark on the upper holder faces forward, toward the bow, as shown in **Figure 13**. When tightening the holder bolts, be sure the gap between the holders on the front (bow) side of the holders is less than the gap on the rear (stern) side as shown in **Figure 13**. Torque the handlebar holder bolts to the specification given in **Table 1**.

NOTE
When positioning the handlebar, check that the handlebar does not cut into the left- and right-hand steering pads. Reposition the handlebar as required.

NOTE
*When routing the throttle cable and the start/stop switch wiring harness in the following steps, maintain a radius of 60 mm (2.36 in.) at the center of the handlebar. See **Figure 16**.*

WARNING
If the throttle cable is not routed properly (see previous NOTE and following step), the inner cable will not move smoothly. This may cause the throttle to stick wide open.

e. Loop the throttle cable as shown in **Figure 16** and reconnect the cable to the throttle lever.

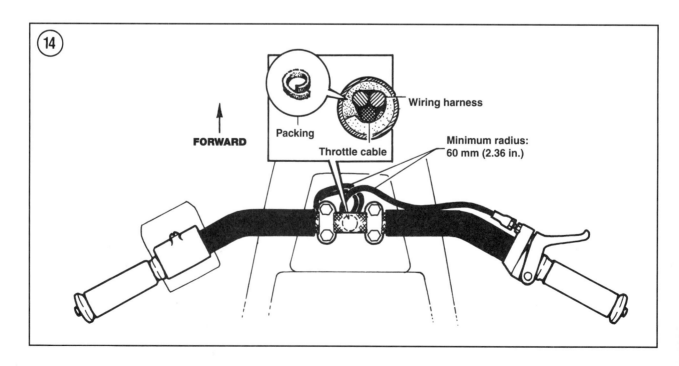

STEERING SYSTEM

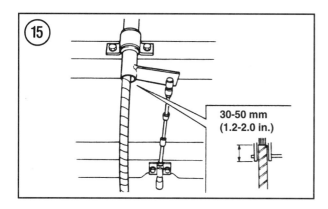

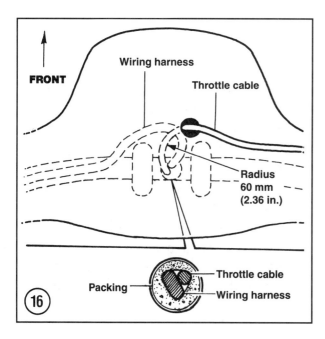

f. Route the start/stop switch wiring harness as shown in **Figure 16**. Then reconnect the electrical connectors.

g. When the wire harness and throttle cable are properly routed, install the packing into the steering column as shown in **Figure 16**. Make sure the wire harness and throttle cable are positioned in the packing as shown. Also be sure that none of the packing sticks out of the steering column.

h. Install the handlebar pad.

6C. *WRB650, WRB650A, and WRB700*—Perform the following:

a. Route the throttle cable and wiring harness through the steering column as shown in **Figure 17**. Be sure the spiral tubing extends 30-50 mm (1.2-2.0 in.) into the column shaft as shown in **Figure 15**.

b. Install the rear steering pad.

c. Place the handlebar into position and install the upper holders and bolts. Be sure the punch mark on the upper holder faces forward, toward the bow, as shown in **Figure 13**. Tighten the bolts finger-tight at this time. Pivot the handlebar so it will align with the rear steering pad. Tighten the front holder bolts first then the rear holder bolts

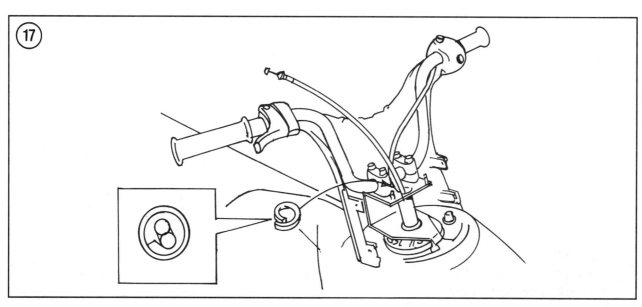

to the torque specified in **Table 1**. Be sure the gap between the holders on the front (bow) side of the holders is less than the gap on the rear (stern) side as shown in **Figure 13**.

d. Reconnect the throttle cable to the throttle lever.

e. If removed, install the start/stop switch assembly.

f. Reconnect the electrical connectors.

g. Install the packing into the steering column. Be sure the throttle cable and wiring harness are positioned as shown in **Figure 17**. Also make sure that none of the packing sticks out of the steering column.

h. Install the front steering pad.

6D. *WB700 and WB700A*—Perform the following:

a. Route the throttle cable and wiring harness through the steering column as shown in **Figure 17**. Be sure the spiral tubing extends 30-50 mm (1.2-2.0 in.) into the column shaft as shown in **Figure 15**.

b. Place the handlebar into position and install the upper holders and bolts. Be sure the punch mark on the upper holder faces forward, toward the bow, as shown in **Figure 13**. Tighten the front holder bolts first then the rear holder bolts to the torque specified in **Table 1**. Be sure the gap between the holders on the front (bow) side of the holders is less than the gap on the rear (stern) side as shown in **Figure 13**.

c. Reconnect the throttle cable to the throttle lever.

d. If removed, install the start/stop switch assembly.

e. Reconnect the electrical connectors.

f. Install the packing into the steering column. Be sure the throttle cable and wiring harness are positioned as shown in **Figure 17**. Also make sure that none of the packing sticks out of the steering column.

g. Install the steering pad.

6E. *RA700, RA700A, RA700B, RA760, RA1100, WB760, WVT700 and WVT1100*—Perform the following:

a. Route the throttle cable and wiring harness through the steering column. Be sure that 20.1 cm (7.9 in.) of the throttle cable and wiring harness protrudes from the steering column as shown in **Figure 18**. Also be sure that the specified amount of spiral tubing extends into the steering column. For WVT700 and WVT1100 models, three turns of the windings should be inside the column.

For RA1100 models, 25 mm (1.0 in.) of tubing should extend into the column.

For RA700, RA700A, RA700B, RA760 and WB760 models, 50 mm (2.0 in.) of tubing should extend into the column. See **Figure 12**.

b. Place the handlebar into position and install the upper holders and bolts. Be sure the punch mark on the upper holder faces forward, toward the bow, as shown in **Figure 13**. Also be sure that the alignment dot on the handlebar is aligned with the top edge of the lower handlebar holder as shown in **Figure 19**.

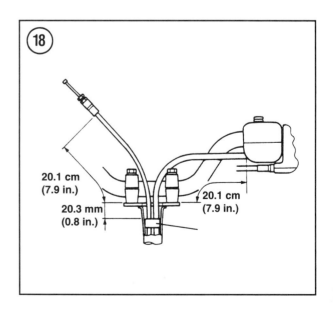

STEERING SYSTEM

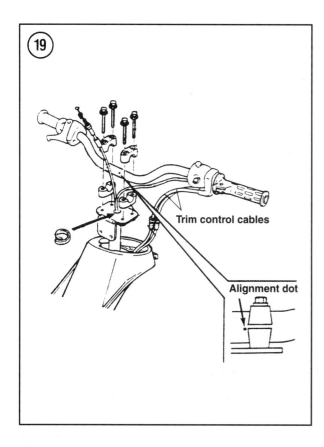

⑲ Trim control cables / Alignment dot

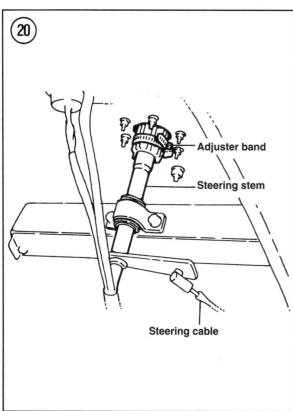

⑳ Adjuster band / Steering stem / Steering cable

c. Tighten the front holder bolts first and then the rear holder bolts to the torque specified in **Table 1**. Be sure the gap between the holders on the front (bow) side of the holders is less than the gap on the rear (stern) side as shown in **Figure 13**.

d. On RA700, RA700A, RA760 and RA1100 models, be sure the trim control cables are routed as shown in **Figure 19**.

e. Reconnect the throttle cable to the throttle lever.

f. If removed, install the start/stop switch assembly.

g. Reconnect the electrical connectors.

h. Install the packing into the steering column. Be sure the packing is positioned as shown in **Figure 18**, so no packing sticks out of the steering column.

h. Install the steering pad.

7. Reconnect the steering cable to the steering plate (**Figure 20**).

8. Adjust the following as described in Chapter Three.

   a. Steering cable.
   b. Throttle cable.
   c. Choke cable.
   d. Trim adjusters (RA700, RA700A, RA760 and RA1100).

9. Reconnect the negative battery cable.

STEERING ASSEMBLY (SJ650, SJ700, SJ700A AND FX700)

The steering assembly consists of the steering pole and steering pivot assembly. While the following section describes complete removal and installation of the steering assembly, the pole and pivot assembles can be serviced separately. If steering pivot service is required, you can remove the steering pole without removing the handlebar or steering column assembly. Read through the *Removal* procedure to determine the service required.

CHAPTER TWELVE

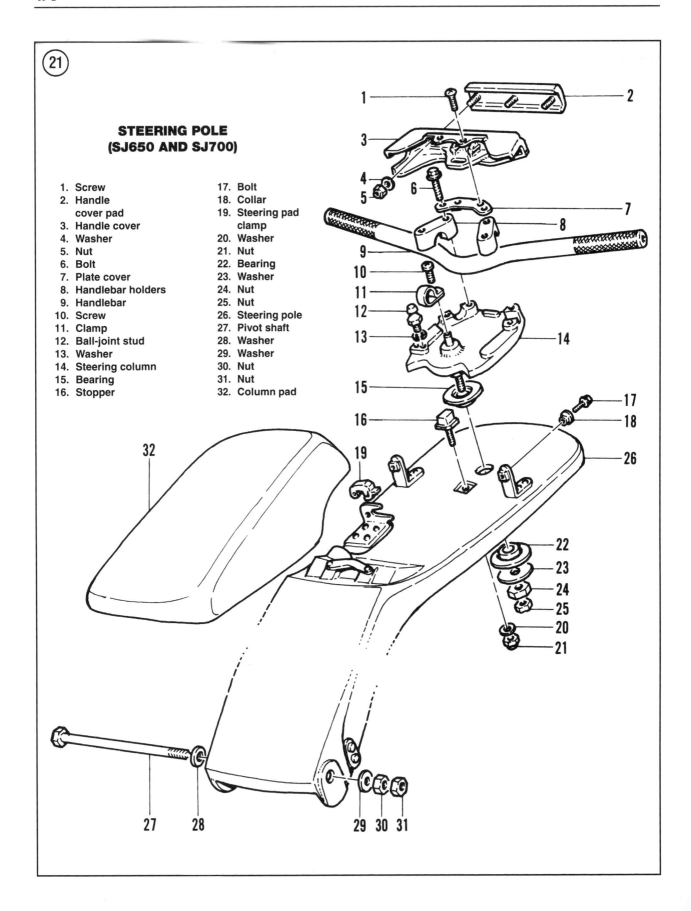

STEERING SYSTEM

Refer to **Figures 21-24** when performing procedures in this section.

Removal

1. Raise and secure the steering pole.
2. Remove the engine hood.
3. To avoid short circuits while disconnecting and moving wires around, disconnect the negative battery cable.
4. Remove the column pad (A, **Figure 25**) from the steering pole.
5. Remove the handlebar as follows:
   a. Remove the handle cover (A, **Figure 26**).
   b. Open the throttle lever and disconnect the throttle cable (B, **Figure 26**) from the lever.
   c. Remove the screws securing the start/stop switch housing to the left-hand handlebar (C, **Figure 26**). Then remove the switch housing from the handlebar and lay it aside or remove it, as required.
   d. Unbolt and remove the handlebar holders and remove the handlebars from the steering pole. See D, **Figure 26**.
6. Remove the steering column (A, **Figure 27**) as follows:
   a. Disconnect and remove the cable clamp (B, **Figure 27**) from the steering column.
   b. Disconnect the steering cable (C, **Figure 27**) from the steering column. Pull the steering cable sleeve backwards to uncover the ball joint, then lift the cable end from the ball joint.
   c. The steering column is secured to the steering pole with a nut and locknut (B, **Figure 25**). Loosen and remove the locknut and nut. Then remove the large washer and bearing (**Figure 21**). Lift the steering column (A, **Figure 27**) off of the steering pole.
7A. *SJ650 and SJ700*—Remove the screws securing the bow cover (B, **Figure 28**) to the hull and remove the bow cover. Cut the plastic tie securing the throttle cable and wiring harness to the cable bracket (**Figure 29**). Then, disconnect the steering cable from the cable bracket.
7B. *FX700 and SJ700A*—Remove the cable bracket (A, **Figure 30**) from the steering pole.
8. Cut the plastic tie at the lower boot (A, **Figure 31** and pull the packing out of the boot.

NOTE
Remove the wiring harness slowly and carefully to avoid cutting or otherwise damaging the harness. If the harness should become tight when removing it, stop and determine the problem.

9. Locate the start/stop switch wiring harness connectors inside the hull (**Figure 32**). Then trace the wiring harness from the connector to the steering pole, cutting or disconnecting each plastic tie. When the harness is free of all plastic ties and other clamps, pull the wiring harness out of the steering pole and remove the start/stop switch housing assembly.

NOTE
An assistant is required to remove the steering pole in the following steps.

10. Loosen and remove the 2 steering pole pivot shaft nuts (A, **Figure 33**). Note that only 1 nut is used on FX700 and SJ700A models.

NOTE
Have an assistant hold the steering pole as the pivot shaft is tapped out of the steering pivot assembly. Then, when the shaft is free, remove the steering pole in such a way that it allows free removal of the steering and throttle cables.

11. Using a brass or aluminum drift, carefully drive the pivot shaft (B, **Figure 33**) out of the steering pivot assembly. Catch the shaft to prevent it from falling to the floor. Then carefully remove the steering pole assembly. Remove the 2 pivot shaft springs and washers. Only 1 pivot shaft spring is used on FX700 and SJ700A models.

498 CHAPTER TWELVE

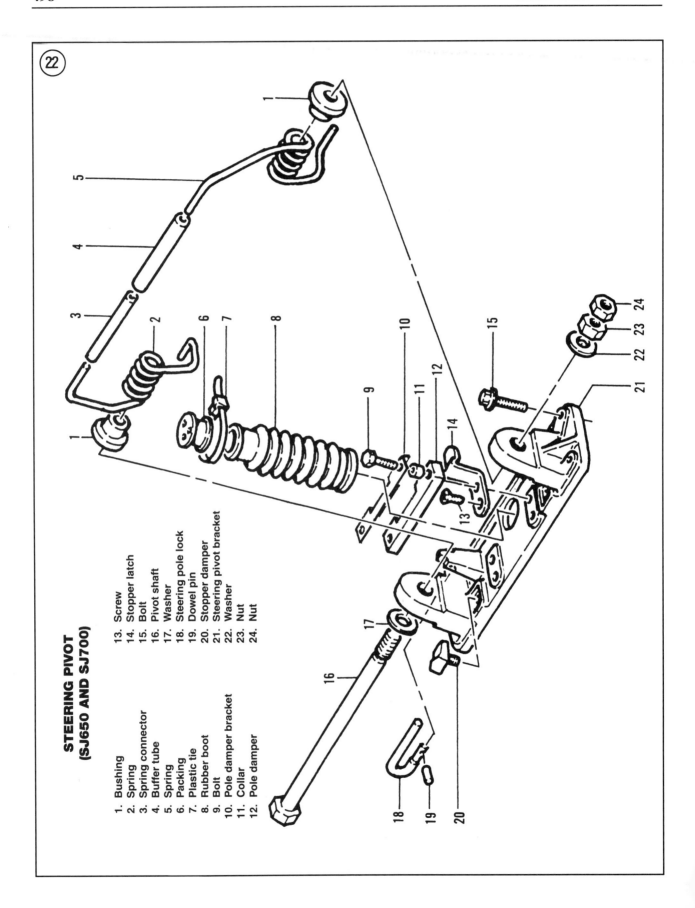

STEERING PIVOT (SJ650 AND SJ700)

1. Bushing
2. Spring
3. Spring connector
4. Buffer tube
5. Spring
6. Packing
7. Plastic tie
8. Rubber boot
9. Bolt
10. Pole damper bracket
11. Collar
12. Pole damper
13. Screw
14. Stopper latch
15. Bolt
16. Pivot shaft
17. Washer
18. Steering pole lock
19. Dowel pin
20. Stopper damper
21. Steering pivot bracket
22. Washer
23. Nut
24. Nut

STEERING SYSTEM

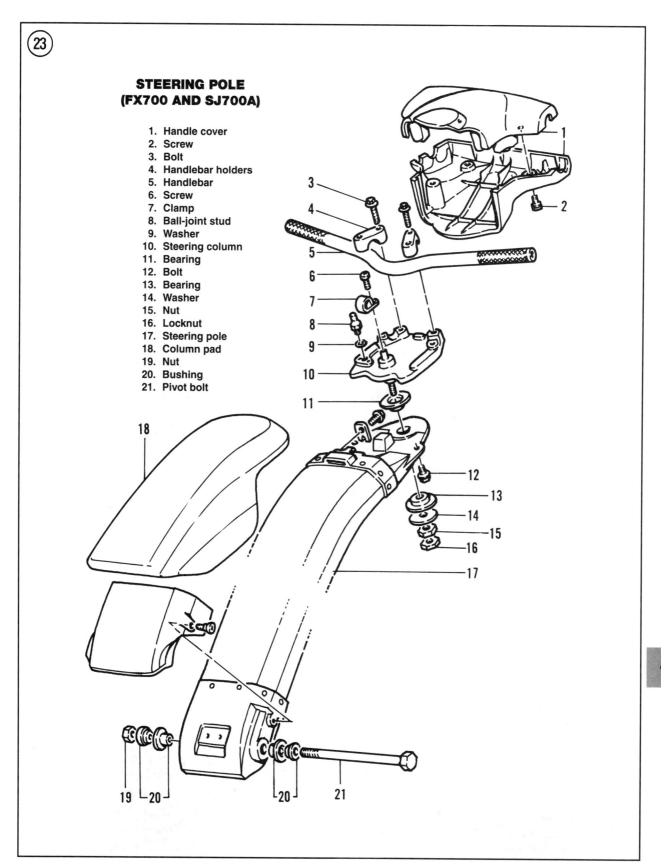

STEERING POLE (FX700 AND SJ700A)

1. Handle cover
2. Screw
3. Bolt
4. Handlebar holders
5. Handlebar
6. Screw
7. Clamp
8. Ball-joint stud
9. Washer
10. Steering column
11. Bearing
12. Bolt
13. Bearing
14. Washer
15. Nut
16. Locknut
17. Steering pole
18. Column pad
19. Nut
20. Bushing
21. Pivot bolt

CHAPTER TWELVE

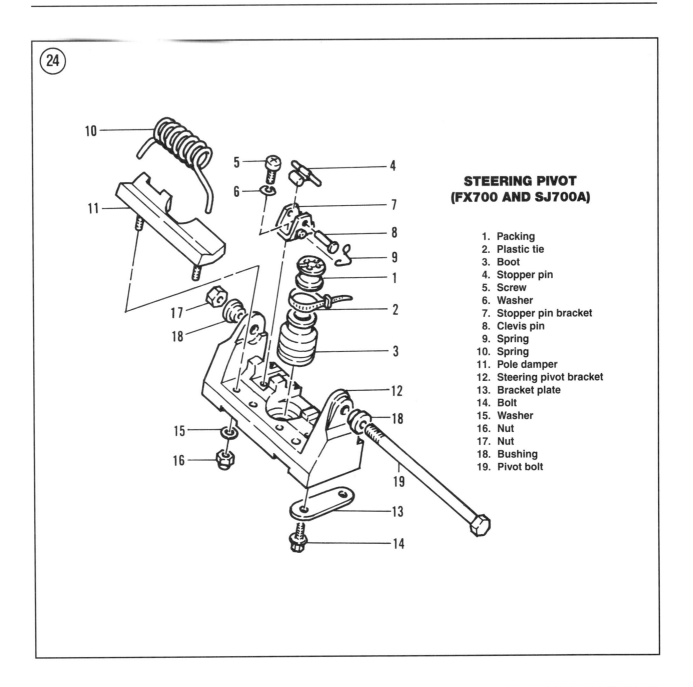

STEERING PIVOT (FX700 AND SJ700A)

1. Packing
2. Plastic tie
3. Boot
4. Stopper pin
5. Screw
6. Washer
7. Stopper pin bracket
8. Clevis pin
9. Spring
10. Spring
11. Pole damper
12. Steering pivot bracket
13. Bracket plate
14. Bolt
15. Washer
16. Nut
17. Nut
18. Bushing
19. Pivot bolt

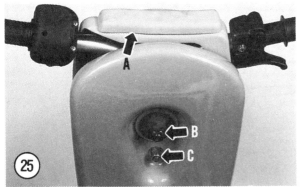

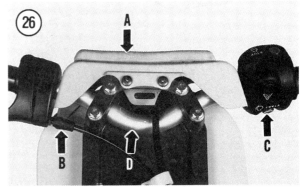

STEERING SYSTEM

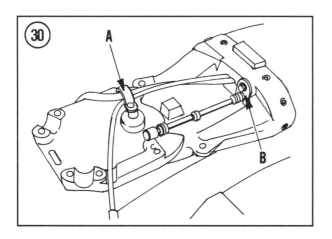

12. Position the steering and throttle cables to avoid damaging them.

13A. *SJ650 and SJ700*—Remove the dowel pin from the steering pole lock (A, **Figure 28**) and remove the pole lock.

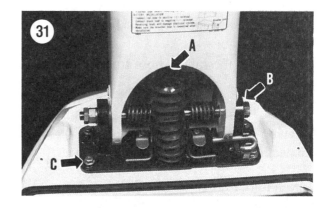

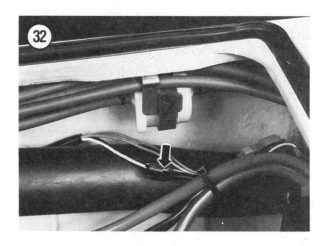

13B. *FX700 and SJ700A*—Remove the stopper pin assembly (A, **Figure 34**) from the steering pivot bracket (B, **Figure 34**).

14. Remove the nuts holding the pole damper (C, **Figure 34**) and steering pivot bracket to the hull. Then remove the steering pivot bracket by sliding it off the steering and throttle cables. Do not lose the mounting plates from under the steering pivot bracket.

15. Remove the rubber boot from the steering pivot.

**Steering Pivot
Inspection/Disassembly/Reassembly**

Refer to **Figure 22** or **Figure 24** for this procedure.

1. Clean the steering pivot assembly in solvent and dry it thoroughly.

2A. *SJ650 and SJ700*—Inspect the 2 rubber stopper dampers (**Figure 35**) for deterioration, wear or other damage. If necessary, remove stopper dampers by pulling them out of the steering bracket. Install the stopper dampers by pushing them into the steering bracket so they face in the direction shown in **Figure 36**. When installing the dampers, make sure the plug seats in the bracket completely. Replace stopper dampers as a set.

2B. *FX700 and SJ700A*—Inspect the stopper damper for deterioration, wear or other damage (B, **Figure 34**). Replace the damper if it is worn or damaged.

3. *SJ650 and SJ700*—Remove and inspect the pole damper (B, **Figure 36**), pole damper bracket (A, **Figure 36**), and stopper hatch (C, **Figure 36**). Replace any part that is worn or damaged. Apply Loctite 242 to the bolt threads during reassembly.

4. Inspect the steering pivot bracket for cracks or damage.

5. Check the pivot shaft bushings for excessive or abnormal wear. If a bushing is damaged, replace both bushings as a set:

   a. Support the steering pivot bracket in a vise with soft jaws.
   b. Using an aluminum or brass drift, carefully tap the bushings out of the bracket.
   c. Check the bracket bushing bore for cracks or damage.

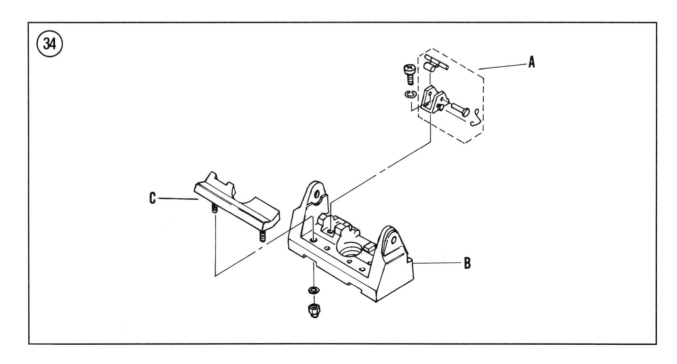

STEERING SYSTEM

d. Align the new bushing with the bracket and drive it into the bracket with a suitable bearing driver.

e. Repeat for the opposite bushing.

6. *FX700 and SJ700A*—Inspect the stopper pin assembly. Pay particular attention to the stopper pin (A, **Figure 37**) and the spring (B, **Figure 37**). Replace either if cracked, worn, or bent.

7. Visually check the pivot shaft surface for cracks, deep scoring or excessive wear. Check the shaft and 2 nut threads (only 1 nut on FX700 and SJ700A models) for damage or contamination. Clean threads thoroughly.

8. Check the rubber boot for tears or other damage.

9. Check the bracket threads in the hull for corrosion. Use a tap to true the threads and remove any deposits.

Steering Pole Cleaning/Inspection

1. Thoroughly check the steering pole for cracks or other damage.

2. Check the pivot shaft bore holes in the bottom of the steering pole for elongation or other damage.

3. Check all of the brackets mounted on the steering pole for damage or looseness. Retighten brackets by first removing bolts or screws and then applying Loctite 242 (blue) to the screw threads. Reinstall screws and tighten securely.

4. *FX700 and SJ700A*—Inspect the bushings for excessive wear or damage. If either bushing is worn, replace them both.

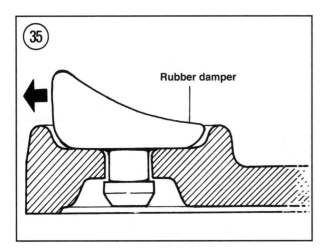

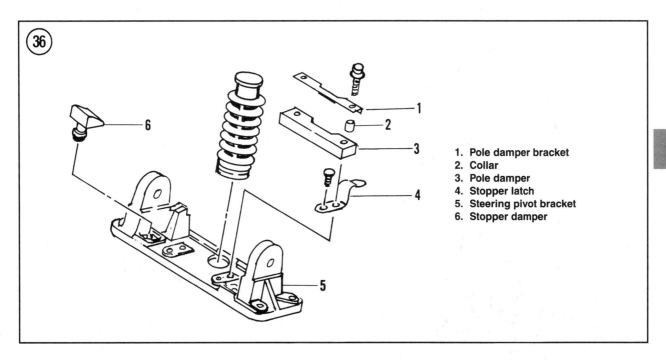

1. Pole damper bracket
2. Collar
3. Pole damper
4. Stopper latch
5. Steering pivot bracket
6. Stopper damper

Steering Column/Handlebar Inspection

1. Clean the steering column assembly in solvent and thoroughly dry. Place all parts on a clean lint free paper towel.
2. Check the steering column for cracks or damage. Check the pivot shaft on the column for cracks, deep scoring or excessive wear. Check the threads for contamination or damage.
3. Check the bearings for cracks, deep scoring or excessive wear. Replace the bearings as a set.
4. Check the steering column washer and 2 nuts for damage.
5. Check the handlebar for damage.
6. Replace worn or damaged parts as required.

Installation (SJ650 and SJ700)

Refer to **Figures 21** and **Figure 22** when performing this procedure.

1. Assembly the steering pivot bracket as follows:
   a. Install the stopper latch (C, **Figure 36**) to the steering pivot bracket. Apply Loctite 242 (blue) to the threads of the stopper hatch screw and torque the screw to the specification given in **Table 2**.
   b. Install the pole damper and the pole damper bracket onto the steering pivot bracket. Apply Loctite 242 (blue) to the threads of the pole damper bolt and torque the bolt to the specification given in **Table 2**. Be sure the collar is in place as shown in **Figure 36**.
2. Install the steering pivot bracket as follows:
   a. Slide steering pivot bracket and rubber boot over the steering and throttle cables.
   b. Fit the rubber boot into the steering pivot bracket. Then align the mounting plates with the steering pivot bracket and position them onto the hull. Align all of the bracket and plate mounting bolt holes with the threads in the hull.
   c. Wipe the steering pivot mounting bolts (A, **Figure 38**) with Loctite 271 (red) and install the bolts and washers. Tighten the bolts to the torque specification listed in Table 2. Make sure to fit the steering pole lock (B, **Figure 38**) into position before installing the right-rear bracket mounting bolt.
   d. Insert the steering pole lock (B, **Figure 38**) into the steering pivot bracket so it faces in the direction shown in A, **Figure 28**. Secure the lock with a new dowel pin.
3. Install the steering pole as follows:
   a. Wipe the steering pole pivot shaft and bushings with marine grease.
   b. Place the pivot pole springs into position and join them with their connector spring and buffer tube (**Figure 39**).

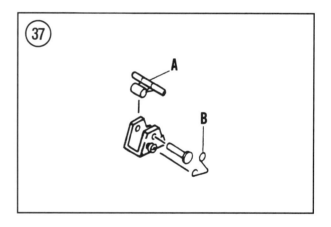

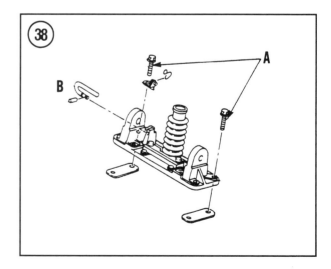

STEERING SYSTEM

c. Temporarily install the boot packing into the boot. Be sure the packing engages the steering and throttle cables.

d. Run the steering and throttle cables through the steering pole, following the original path, and place the pole into position on the steering pivot.

e. Install the pivot shaft from the right-hand side, making sure to install the 2 left- and right-hand washers between the steering pole and bracket. When slipping the steering pole over the springs, make sure the springs seat against the steering bracket and that the buffer tube is positioned inside the steering pole housing.

f. Lock the steering pole with the lock (A, **Figure 28**).

g. Wipe the pivot shaft threads with Loctite 271 (red) and install the large hex nut (A, **Figure 33**). Tighten the nut to 15 N·m (11 ft.-lb.). Unlock the steering pole, and pivot it back and forth. It should pivot smoothly. If the pole is tight, check the hex nut's break away torque. If this torque is excessive, retighten the nut to its correct torque specification. If the torque is correct, but pole movement is still tight, check for damaged or misaligned components.

h. Relock the steering pole (A, **Figure 28**).

i. Install the pivot shaft locknut (B, **Figure 33**). Then hold the hex nut (A, **Figure 33**) with a wrench and tighten the locknut to 70 N·m (51 ft.-lb.). Recheck steering pole movement (substep g).

j. Insert the start/stop switch wiring harness through the steering pole (**Figure 29**) and then through the rubber boot. When the connectors are free of the boot, pull them into the engine compartment and route the wiring harness along the side of the hull, following original routing path. Secure the wiring harness with new plastic ties where required, then reconnect the 2 electrical connectors.

k. Install the air vent hose through the rubber boot, making sure the hose is not kinked or pinched.

l. Fit the packing around the steering cable, throttle cable, wiring harness and air vent hose where they run through the rubber boot. When the packing is properly fitted around the cables, install a new plastic tie around the top of the boot. Tighten the tie securely.

m. Reinstall the bow cover (B, **Figure 28**) and secure it with its mounting screws.

4. Install the steering column as follows:

a. Wipe the steering column pivot shaft and the 2 bearings with marine grease. When lubricating the 2 bearings, apply marine grease to the bearing's flat surfaces as well as the bearing bore where it rides against the steering column pivot shaft.

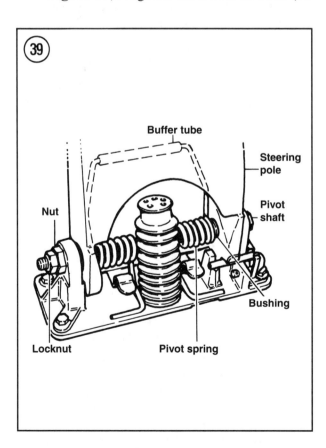

b. Install the upper bearing onto the steering column (**Figure 21**), then install the steering column and bearing onto the steering pole. Install the lower bearing and washer and hold them in position by hand. Install the large hex nut (B, **Figure 25**), and torque it to 1 N•m (9 in.-lb.). Pivot the steering column from side to side. It must pivot smoothly. Remove and reinstall the nut if pole movement is too tight. When the steering column pivots correctly, hold the large hex nut with a wrench, then install and torque the locknut to 29 N•m (21 ft.-lb.).

c. Secure the steering cable bracket to the steering pole then fasten the throttle cable and wiring harness to the bracket with a new plastic tie. Cut the tie arm to length after tightening. See **Figure 29**.

d. Reconnect the steering cable (C, **Figure 27**) to the steering column. Apply marine grease to the ball joint on the steering column. Pull the shift cable sleeve backward to open its fitting and push the fitting onto the ball joint. Release the cable sleeve to lock the cable.

4. Install the handlebar as follows:

a. Place the handlebar (D, **Figure 26**) onto the steering column, then lay the wiring harness over the handlebar as shown in **Figure 26**. Install the 2 handlebar holders and plate cover, then secure into position with the 4 bolts; run the bolts down finger-tight. Make sure the wiring harness is routed over the handlebar and under the plate cover as show in **Figure 26**.

b. Position the handlebar in the holders so the punch mark in the middle of the handlebars is centered between the left-and right-hand handlebar holders. Mount the machine and rotate the handlebar up or down until its position suits your riding preference. Then tighten the 4 handlebar holder bolts—starting with the 2 rear bolts—to the torque specification listed in **Table 2**. After tightening the bolts, check the gap between holders and steering column. It should be toward the front of the vehicle as shown in **Figure 40**. If the gap is at the rear, loosen and then retighten handlebar bolts to reposition the gap. Make sure the handlebar punch mark is centered between the handlebar holders.

c. Reinstall the start/stop switch housing assembly onto the left-hand handlebar. See C, **Figure 26**. Tighten the screw on the passenger side of the housing first, then tighten the screw on the engine side of the housing (**Figure 41**). Torque the screws to the specification given in **Table 2**.

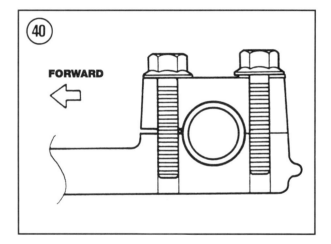

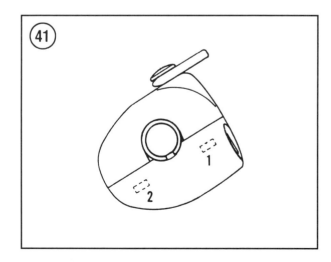

STEERING SYSTEM

d. Reinstall the handle cover. Apply Loctite 242 (blue) to the cover screws and torque the screws to the specification given in **Table 2**.

e. Reconnect the cable clamp (B, **Figure 27**) to the steering pole.

f. Reconnect the throttle cable (B, **Figure 26**) to the throttle lever.

g. Check the steering cable and wiring harness cable routing as shown in **Figure 26**.

5. Reinstall the steering pad (A, **Figure 25**). Apply Loctite 242 (blue) to the threads of the steering pad bolts and torque the bolts to the specification given in **Table 2**.

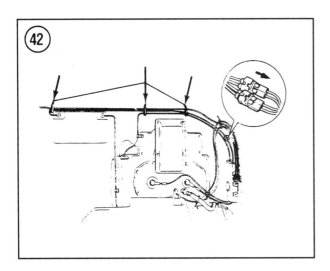

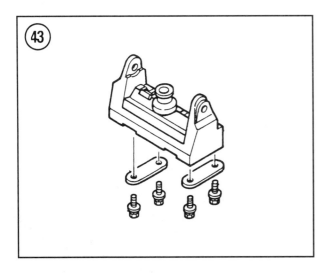

6. Unlock the steering pole and check its operation. It should move smoothly in all operating positions. Check throttle operation while moving the pole up and down and the handlebars from side to side. The lever should move smoothly.

7. Adjust the following as described in Chapter Three:

　a. Steering cable.

　b. Throttle cable.

8. Make sure all connections are secure and the wires are correctly routed and secured to the hull. (See **Figure 42**).

9. Reconnect the negative battery cable and install the engine hood.

Installation (FX700 and SJ700A)

Refer to **Figures 23** and **Figure 24** when performing this procedure.

1. Assemble the steering pivot bracket as follows:

　a. Install the stopper pin assembly to the steering pivot bracket. Apply Loctite 242 (blue) to the threads of the stopper pin assembly screw and torque the screw to the specification given in Table 2.

　b. Set the boot in place on the steering pivot bracket, then install the pole damper.

2. Install the steering pivot bracket as follows:

　a. Slide steering pivot bracket and rubber boot over the steering and throttle cables.

　b. Set the mounting plates in position on the steering pivot bracket and place the bracket onto the hull. Align all of the bracket and plate mounting bolt holes with the holes in the hull **Figure 43**.

　c. Wipe the steering pivot mounting bolts with Loctite 242 (blue) and install the bolts and washers. Tighten bolts to the torque specification listed in Table 2.

　d. Reinstall the bow cover if it was removed. Finger-tighten the screws at this time.

3. Install the steering pole as follows:
   a. Apply marine grease to the steering pole bushings and install the bushings to the steering pole (**Figure 44**). Be sure to apply grease to the inner circumference of the bushing as well as its flange.
   b. Apply marine grease to the steering pole pivot shaft and bushings.
   c. Place the pivot pole into position in the steering pivot bracket.
   d. Set the pivot spring in place in the steering pivot bracket. Be sure the ends of the spring engage the pole and the pivot bracket as shown in **Figure 45**.
   e. Adjust the bow cover as needed to align the cover holes with those in the steering pivot bracket. Install the pivot shaft from the left-hand side. Be sure the left bushing sits between the shaft head and the bow cover. Also be sure that the pivot shaft passes through the pivot spring as shown in **Figure 45**.
   f. Install the right bushing onto the pivot shaft. Apply Loctite 242 (blue) to the threads of the shaft and install the lock nut. Torque both the pivot shaft and the lock nut to the specifications given in **Table 2**.
   g. Check the movement of the steering pole. If the pole is tight, check the lock nut's break-away torque. If this torque is excessive, retighten the nut to the correct torque specification. If the torque is correct but pole movement is still tight, check for damaged or misaligned components.
   h. Run the steering and throttle cable through the steering pole, following the original path. Be sure the cables are routed on the

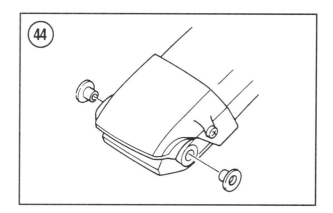

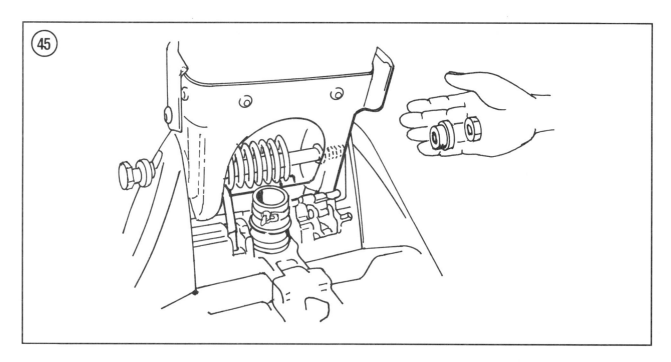

STEERING SYSTEM

rear side of the pivot spring as shown in **Figure 46**.

i. Insert the start/stop switch wiring harness through the steering pole (**Figure 29**) and then through the rubber boot (and packing). When the connectors are free of the boot, pull them into the engine compartment and route the wiring harness along the side of the hull, following original routing path. Secure the wiring harness with new plastic ties where required, then reconnect the 2 electrical connectors (**Figure 42**).

j. Install the air vent hose through the rubber boot, making sure the hose is not kinked or pinched.

k. Fit the packing around the steering cable, throttle cable, wiring harness and air vent hose where they run through the rubber boot. When the packing is properly fitted around the cables, install a new plastic tie around the top of the boot. Tighten the tie securely then cut tie arm.

l. Tighten the bow cover mounting screws.

4. Install the steering column as follows:

a. Wipe the steering column pivot shaft and the 2 bearings with marine grease. When lubricating the bearings, apply marine grease to each bearing's flat surfaces as well as the bearing bore where it rides against the steering column pivot shaft.

b. Install the upper bearing onto the steering column (**Figure 23**) and install the steering column and bearing onto the steering pole. Install the lower bearing and washer and hold them in position by hand. Install the large hex nut (B, **Figure 25**) and torque it to 1 N·m (9 in.-lb.). Pivot the steering column from side to side. It must pivot smoothly. Remove and reinstall the nut if pole movement is too tight. When the steering column pivots correctly, hold the large hex nut with a wrench, then install and torque the locknut to 29 N·m (21 ft.-lb.).

c. Secure the throttle cable and wiring harness to the steering column with the cable clamp (B, **Figure 30**).

d. Reconnect the steering cable (C, **Figure 27**) to the steering column. Apply marine grease to the ball joint on the steering column. Pull the shift cable sleeve backward to open its fitting and push the fitting onto the ball joint. Release the cable sleeve to lock the cable.

e. Reinstall the cable bracket (A, **Figure 30**) to the steering pole.

5. Install the handlebar as follows:

a. Place the handlebar onto the steering column, then lay the wiring harness over the handlebar as shown in **Figure 47**. Install the 2 handlebar holders and secure them in

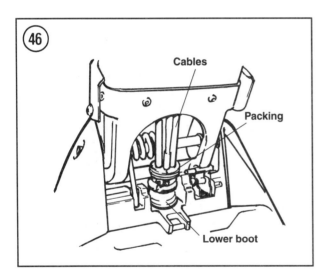

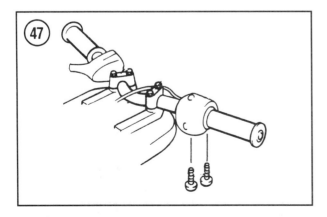

place with the 4 bolts. Apply Loctite 242 (blue) to the threads of the holder bolts and run the bolts down finger-tight.

b. Position the handlebar in the holders so the punch mark on the middle of the handlebars is centered between the left-and right-hand handlebar holders. Mount the machine and then rotate the handlebar up or down until its position suits your riding preference. Then tighten the 4 handlebar bolts, starting with the 2 rear bolts, to the torque specification listed in **Table 2**. After tightening the bolts, check the gap between holders and steering column; it should be toward the front of the vehicle as shown in **Figure 40**. If the gap is at the rear, loosen and then retighten the handlebar bolts to reposition the gap. Make sure the handlebar punch mark is centered.

c. Reinstall the start/stop switch housing assembly onto the left-hand handlebar. Tighten the screw on the passenger side of the housing first, then tighten the screw on the engine side of the housing (see **Figure 41**). Torque the screws to the specification given in **Table 2**.

d. Reconnect the throttle cable to the throttle lever (B, **Figure 26**).

e. Reinstall the handle cover. Apply Loctite 242 (blue) to the cover screws and torque the screws to the specification given in **Table 2**.

6. Reinstall the handle cover and the steering pad. (**Figure 48**).

7. Check the operation of the steering pole. It should move smoothly in all operating positions. Check throttle operation while moving the pole up and down and the handlebars from side to side; the lever should move smoothly.

8. Adjust the following as described in Chapter Three:
   a. Steering cable.
   b. Throttle cable.

9. Be sure all connections are secure and that the wires are correctly routed and secured to the hull. (See **Figure 42**).

10. Reconnect the negative battery cable and install the engine hood.

CABLES

Refer to **Figures 49-56** for exploded drawings of the cable system for each model.

Steering Cable (FX700, SJ650, SJ700 and SJ700A)

Refer to **Figure 49** and **Figure 50** when performing the following procedure.

Removal

NOTE
Note how the cable is routed through your vehicle for installation reference.

1. Drain the fuel from the tank.
2. Disconnect both battery leads from the battery.
3. Raise and secure the steering pole.
4. Remove the ride plate as described in Chapter Seven (**Figure 57**).
5. Disconnect the steering cable from the ball joint on the steering nozzle. Slide the spring-

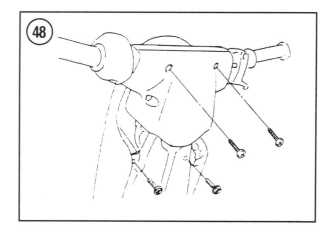

STEERING SYSTEM

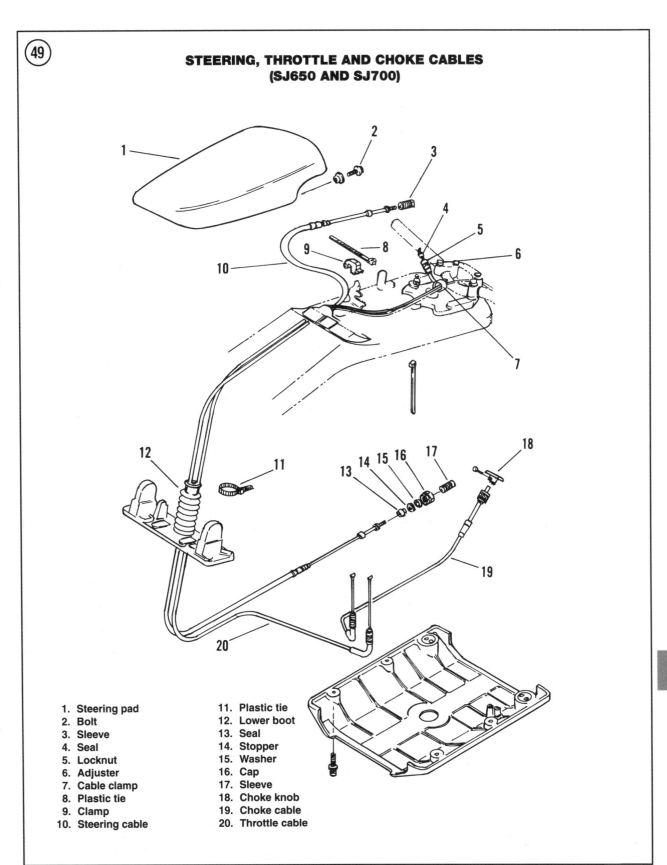

STEERING, THROTTLE AND CHOKE CABLES (SJ650 AND SJ700)

1. Steering pad
2. Bolt
3. Sleeve
4. Seal
5. Locknut
6. Adjuster
7. Cable clamp
8. Plastic tie
9. Clamp
10. Steering cable
11. Plastic tie
12. Lower boot
13. Seal
14. Stopper
15. Washer
16. Cap
17. Sleeve
18. Choke knob
19. Choke cable
20. Throttle cable

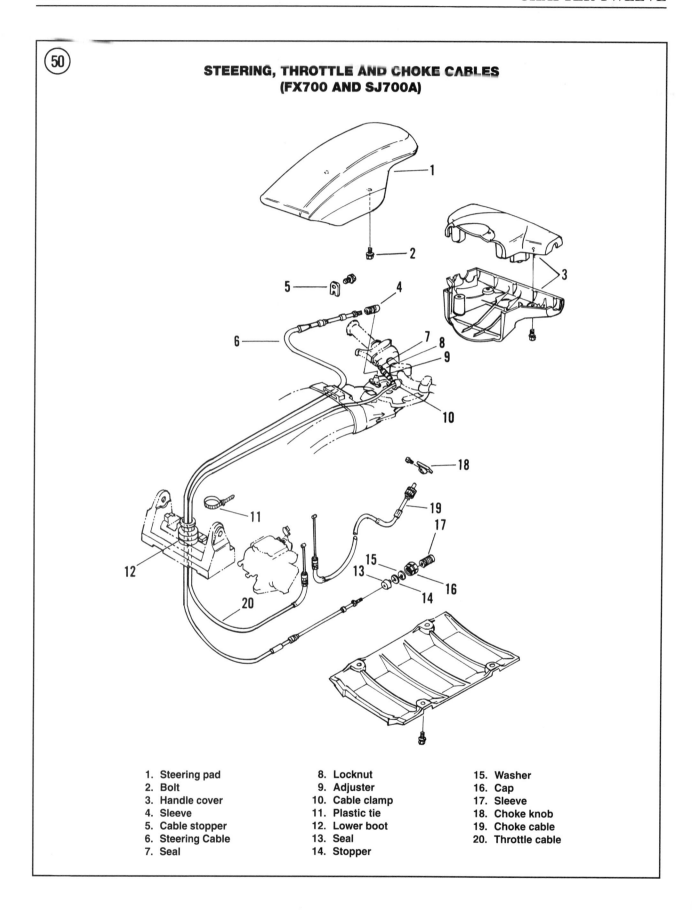

STEERING, THROTTLE AND CHOKE CABLES (FX700 AND SJ700A)

1. Steering pad
2. Bolt
3. Handle cover
4. Sleeve
5. Cable stopper
6. Steering Cable
7. Seal
8. Locknut
9. Adjuster
10. Cable clamp
11. Plastic tie
12. Lower boot
13. Seal
14. Stopper
15. Washer
16. Cap
17. Sleeve
18. Choke knob
19. Choke cable
20. Throttle cable

STEERING SYSTEM

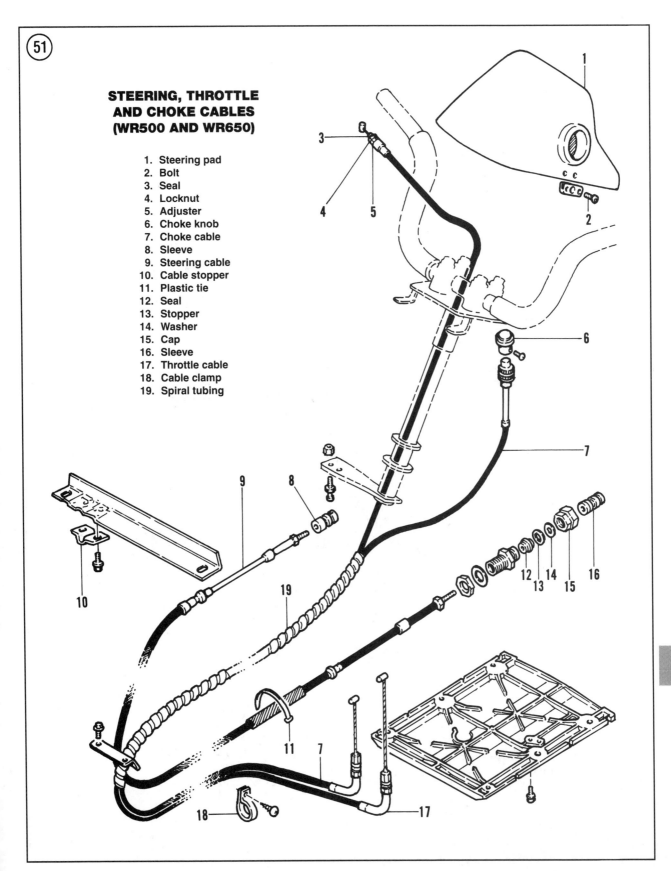

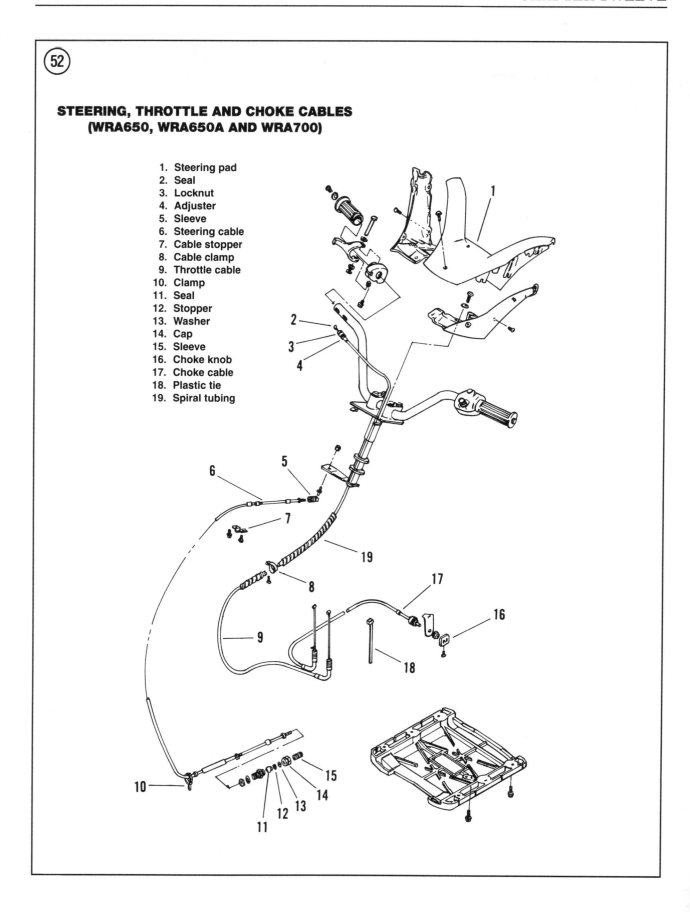

STEERING SYSTEM

515

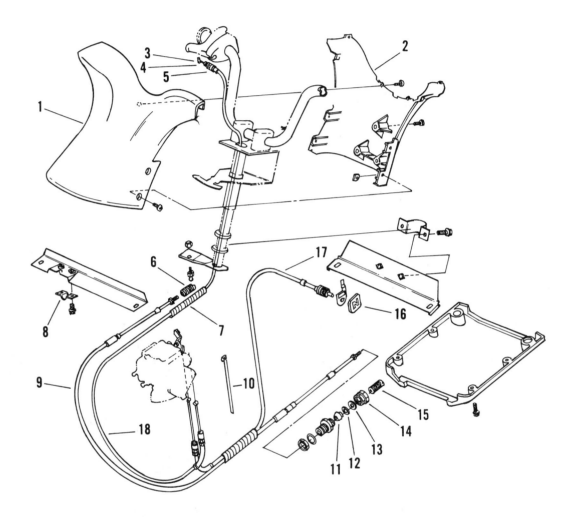

STEERING, THROTTLE AND CHOKE CABLES (WRB650, WRB650A AND WRB700)

1. Front steering pad
2. Rear steering pad
3. Seal
4. Locknut
5. Adjuster
6. Sleeve
7. Spiral tubing
8. Cable stopper
9. Steering cable
10. Plastic tie
11. Seal
12. Stopper
13. Washer
14. Cap
15. Sleeve
16. Choke knob
17. Choke cable
18. Throttle cable

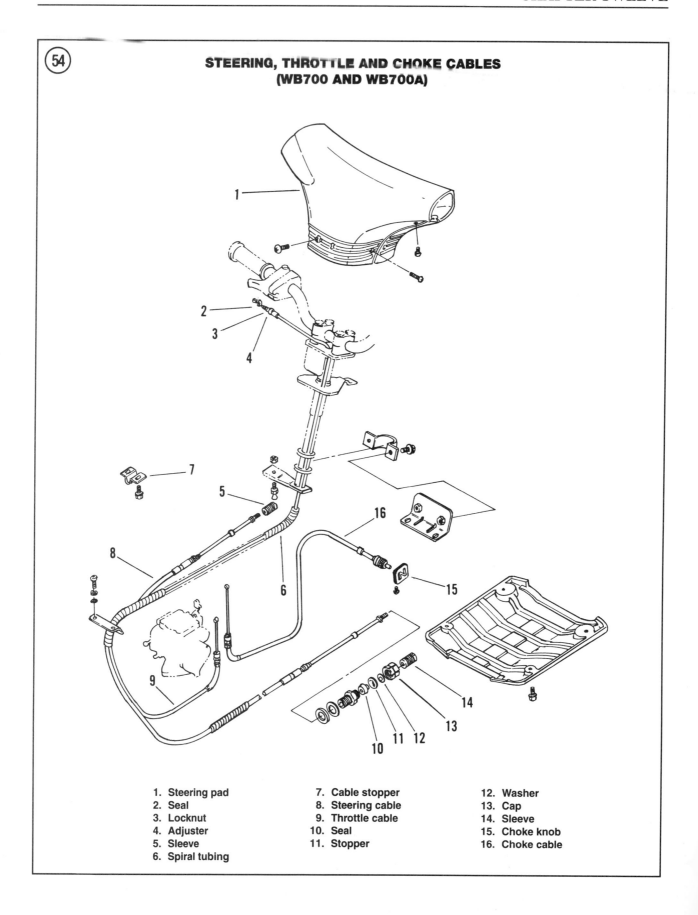

STEERING SYSTEM

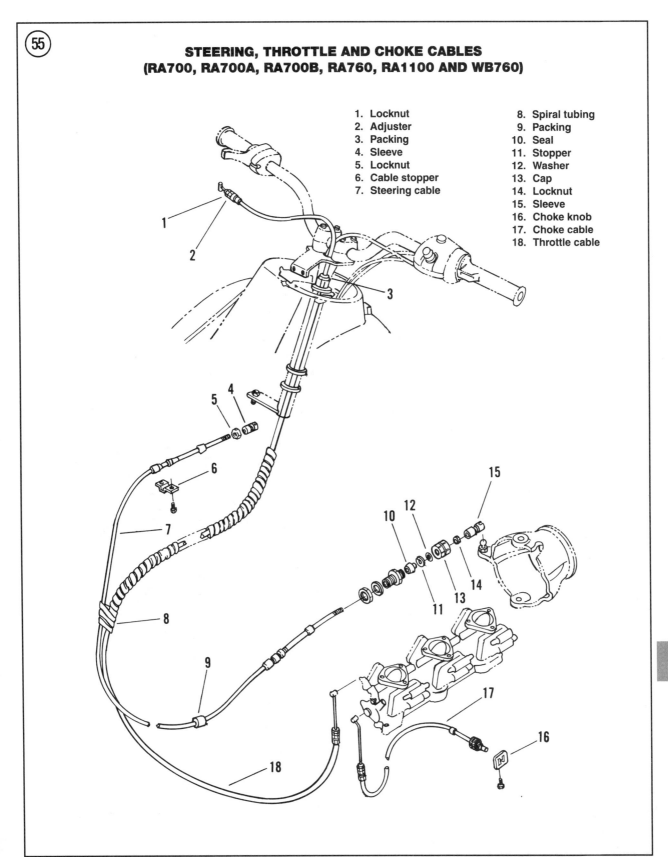

STEERING, THROTTLE AND CHOKE CABLES (RA700, RA700A, RA700B, RA760, RA1100 AND WB760)

1. Locknut
2. Adjuster
3. Packing
4. Sleeve
5. Locknut
6. Cable stopper
7. Steering cable
8. Spiral tubing
9. Packing
10. Seal
11. Stopper
12. Washer
13. Cap
14. Locknut
15. Sleeve
16. Choke knob
17. Choke cable
18. Throttle cable

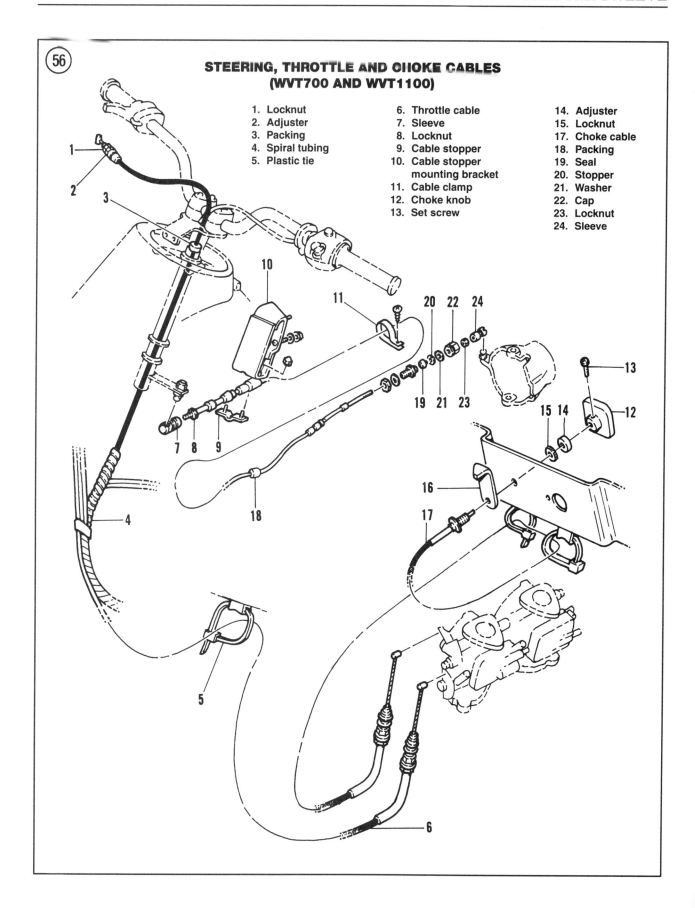

STEERING SYSTEM

loaded cable sleeve back from the ball joint, and remove the cable joint (**Figure 58**).

6. Disassemble the cable joint by removing the sleeve, locknut, cap, washer, stopper and seal (**Figure 59**).

7. Remove the steering pad from the top of the steering pole.

8. On SJ700A and FX700 models, remove the cover from the handlebars.

9. Disconnect and remove the cable clamp (B, **Figure 27** or B, **Figure 30**) from the steering column.

10. Disconnect the steering cable (C, **Figure 27**) from the steering column. Pull the steering cable sleeve backward to uncover the ball joint, then lift the cable joint from the ball joint.

11. On FX700 and SJ700A models, remove the cable stopper from the steering column (A, **Figure 30**).

12. Remove the plastic tie from the lower boot and pull the packing from the boot (**Figure 46**).

13. Trace the routing of the steering cable and remove the plastic ties that secure the cable to the hull (**Figure 60** or **Figure 61**).

14. Remove the steering cable.

15. Inspect the cable. Replace it if it is worn, kinked or frayed.

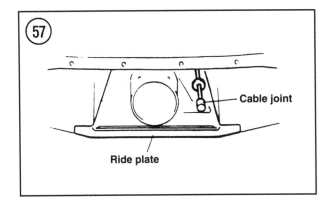

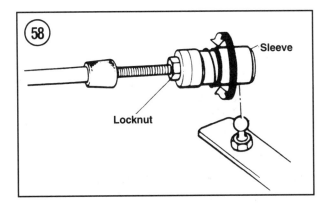

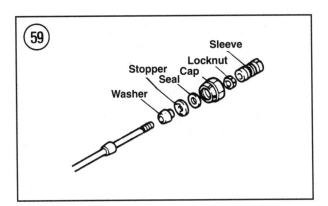

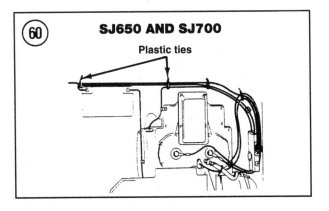

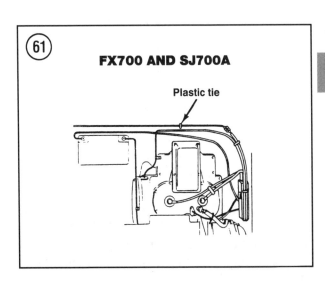

Installation

1. Route the cable through the vehicle as you noted before removal. Take care to reroute the cable in its original position through the steering pole. On FX700 and SJ700A models, make sure the cable is routed on the rear side of the pivot spring as shown in **Figure 46**.

WARNING
When assembling the cable joint on the steering nozzle end of the cable, the steering cable must be threaded into the sleeve so the minimum specified cable-to-sleeve engagement is maintained as shown in Figure 62.

2. Reassemble the cable joint at the steering nozzle end of the cable as follows:
    a. Install the seal, stopper, washer, cap, locknut and cable sleeve as shown in **Figure 59**.
    b. Apply Loctite 242 to the threads of the locknut and cable joint. Torque the cap to 6 N·m (53 in.-lb.), and torque the locknut to 3 N·m (26 in.-lb.)
    c. Be sure to maintain the minimum cable-to-sleeve engagement as shown in **Figure 62**.

3. Apply marine grease to both the cable joint and the ball joint, and attach the steering cable to the steering nozzle ball joint. Pull the sleeve back from the cable joint and fit the joint over the ball joint.

WARNING
When assembling the cable joint onto the steering pole end of the cable, the steering cable must be threaded into the sleeve so the minimum specified cable-to-sleeve engagement is maintained as shown in Figure 63.

4. Reassemble the cable joint to the steering pole end of the cable. Screw the sleeve onto the cable shaft so it extends approximately 13 mm (0.5 in.) into the sleeve. Be sure to maintain the minimum cable-to-sleeve engagement as shown in **Figure 63**.

5. Reattach the cable to the steering column ball joint. Apply marine grease to both the cable joint and the ball joint. Pull the sleeve back from the

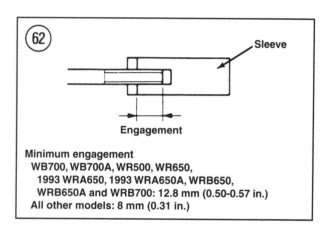

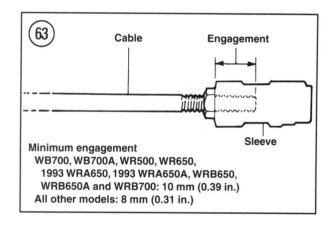

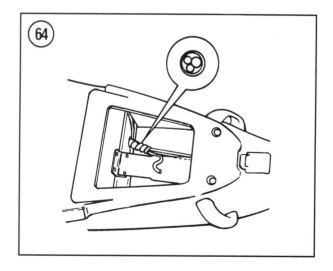

STEERING SYSTEM

cable joint and fit the cable joint over the ball joint.

6. Install the cable clamp (B, **Figure 27** or B, **Figure 30**) to the steering column.

7. On FX700 and SJ700A models, reinstall the cable stopper (A, **Figure 30**) to the steering column.

8. Reinstall the packing into the lower boot and install a new plastic tie around the top of the boot (**Figure 46**). After tightening the plastic tie on the boot, clip the tie arm to length.

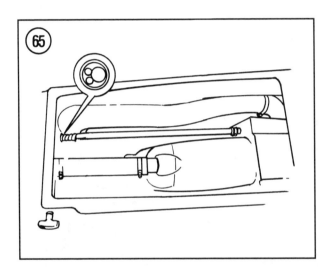

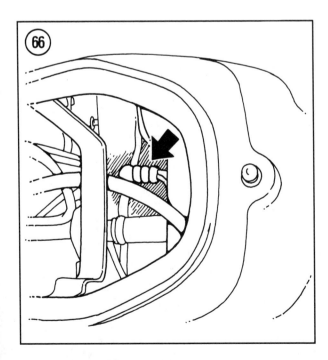

9. Secure the cable to the hull with new plastic ties (**Figure 60** or **Figure 61**). After tightening each plastic tie, clip the tie arm to length.

10. Install the steering pad.

11. Reinstall the ride plate as described in Chapter Seven.

12. Adjust the steering cable as described in Chapter Three.

13. Check the operation of the steering cable while moving the pole up and down. The steering nozzle should move smoothly.

14. Reconnect the battery leads to the battery.

Steering Cable (RA700, RA700A, RA700B, RA760, RA1100, WB700, WB700A, WB760, WR500, WR650, WRA650, WRA650A, WRA700, WRB650, WRB650A, WRB700, WVT700 and WVT1100)

Refer to **Figures 51-56** when performing the following procedures.

Removal

NOTE
Note how the cable is routed through your vehicle for installation.

1. Drain the fuel from the tank.

2. Disconnect both battery leads from the battery.

3. Remove the ride plate as described in Chapter Seven (**Figure 57**).

4A. *WB700, WB700A, WRB650, WRB650A, WRB700*—Remove the packing from the hull and remove the spiral tubing that holds the steering cable, water hose, and choke cable together. On WB models, see **Figure 64**. On WRB models, see **Figure 65**.

4B. *WB760*—Remove the spiral tubing that holds the steering cable, water inlet hose, and battery breather hose together (**Figure 66**).

5. Remove the spiral tubing that holds the steering cable to the throttle cable and the start/stop switch wiring harness.

6. Disconnect the steering cable from the ball joint on the steering nozzle. Slide the spring-loaded sleeve backward and remove the cable joint from the ball joint.

7. Disassemble the cable joint by removing the sleeve, lock nut, cap, washer, stopper and seal (**Figure 59**).

8. Disconnect the steering cable from the arm on the steering column (**Figure 58**).

9. Remove the sleeve and the lock nut from the steering column end of the cable shaft.

10A. Remove the cable stopper from the mounting bracket (**Figure 67**).

10B. *WVT700 and WVT1100*—Remove the cable stopper mounting bracket, then remove the cable stopper (**Figure 56**).

11. *WRA650, WRA650A and WR700*—Remove the cable clamp securing the cables near the fuel tank (**Figure 68**).

12. Remove the cable. Replace the cable if it is worn, kinked or frayed.

Installation

1. Route the cable through the vehicle as you noted before removal.

2A. *WB700, WB700A, WRB650, WRB650A and WRB700*—Use the spiral tubing to wrap at least 200 mm (7.9 in.) of the steering cable, water hose and choke cable together. When the cables are arranged as shown, reinstall the packing in the hull. On WB models, see **Figure 64**. On WRB models, see **Figure 65**.

2B. *WB760*—Reinstall the spiral tubing that holds the steering cable, water inlet hose, and the battery breather hose together (**Figure 66**).

3. Reinstall the packing in the hull.

WARNING
When assembling the cable joint on the steering nozzle end of the cable, the steering cable must be threaded into the sleeve so the minimum specified cable-to-sleeve engagement is maintained as shown in Figure 62.

4. Reassemble the cable joint to the steering nozzle end of the cable as follows:

a. Install the seal, stopper, washer, cap, locknut and cable sleeve as shown in b12-32.

b. Apply Loctite 242 to the threads of the locknut and cable joint. Torque the cap to 6 N•m (53 in.-lb.), and torque the locknut to 3 N•m (26 in.-lb.).

c. Be sure to maintain the minimum cable-to-sleeve engagement as shown in **Figure 62**.

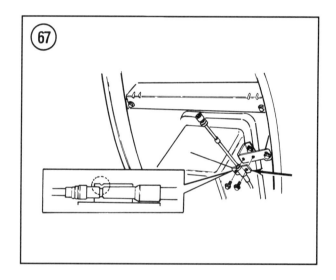

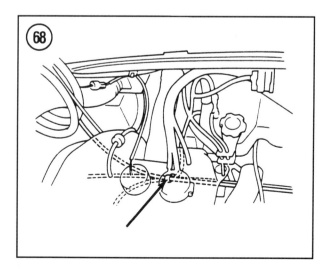

STEERING SYSTEM

WARNING
*When assembling the cable joint onto the steering handle end of the cable, the steering cable must be threaded into the sleeve so the minimum specified cable-to-sleeve engagement is maintained as shown in **Figure 63**.*

5. Reassemble the cable joint to the steering column end. Screw the sleeve onto the cable shaft so it extends approximately 13 mm (0.5 in.) into the sleeve. Be sure to maintain the minimum cable-to-sleeve engagement as shown in **Figure 63**.

6A. *WB700, WB700A, WRB650, WRB650A, and WRB700*—Fit the cable stopper over the steering cable and secure the stopper to the mounting bracket. Apply Loctite 242 (blue) to the cable stopper bolts. Be sure the index in the cable stopper engages the cutout in the cable as shown in **Figure 67**.

6B. *WVT700 and WVT1100 models*—Reinstall the mounting bracket and cable stopper.

   a. Apply Loctite 271 (red) to the mounting bracket bolt threads and torque the nut to 7 N•m (62 in.-lb.).

   b. Fit the cable stopper over the steering cable and secure the stopper to the mounting bracket. Apply Loctite 271 (red) to the cable stopper bolts. Be sure the index in the cable stopper engages the cutout in the cable. See the inset, **Figure 67**.)

7. Reconnect the steering cable to the ball joint on the steering handle and on the steering nozzle. Apply marine grease to both the ball joint and cable joint at each end. When connecting a cable, slide the sleeve backward and fit the cable joint onto the ball joint (**Figure 58**).

8A. Install the spiral tubing that holds the steering cable to the throttle cable and the start/stop switch wiring harness.

8B. *WRB650, WRB650A and WRB700*—Use a least 4 turns of the spiral tubing to secure the steering cable to the throttle cable and start/stop switch wiring harness. Start the tubing at a point 160 mm (6.3 in.) below the cable stopper as shown in **Figure 69**.

8C. *WB700 and WB700A*—Use spiral tubing to secure the steering cable to the throttle cable and start/stop switch wiring harness. Start the tubing at a point 150 mm (5.9 in.) below the cable stopper as shown in **Figure 70**.

9. *WRA650, WRA650A and WRA700*—Secure the cable with the cable clamp (**Figure 68**).

10. Adjust the steering cable as described in Chapter Three.

11. Check the operation of the steering by turning the handlebars. Be sure the steering nozzle moves smoothly throughout its entire range of motion.

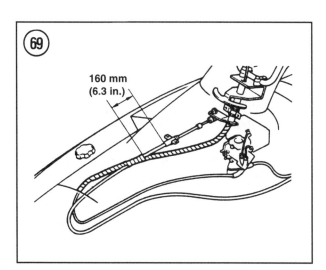

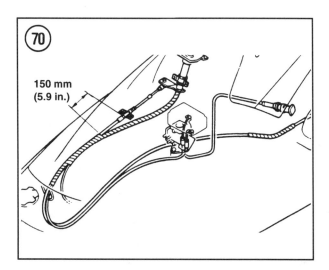

12. Reattach the ride plate as described in Chapter Seven.
13. Reattach the positive battery lead then the negative battery lead.

Throttle Cable
(FX700, SJ650, SJ700 and SJ700A)

Refer to **Figure 49** and **Figure 50** when performing the following procedure.

Removal

NOTE
Note how the cable is routed through your vehicle for installation reference.

1. Drain the fuel from the tank.
2. Disconnect both battery leads from the battery.
3. Raise and secure the steering pole.
4. Remove the column pad (A, **Figure 25**) and the handle cover (A, **Figure 26**).
5. Open the throttle lever and disconnect the throttle cable. Loosen the locknut, back out the cable adjuster and remove the throttle cable along with the seal (**Figure 71**).
6. Disconnect and remove the cable clamp (B, **Figure 27** or B, **Figure 30**) from the steering column.
7. *SJ650 and SJ700*—Cut the plastic tie securing the throttle cable and wiring harness to the cable bracket (**Figure 29**).
8. Remove the plastic tie from the lower boot, and pull the packing from the boot. (**Figure 46**).
9. At the carburetor, loosen the throttle cable adjuster locknut and loosen the adjuster. Remove the throttle cable inner wires from the carburetor throttle and remove the throttle cable from the cable bracket (**Figure 72**).
10. Trace the cable and clip any plastic ties securing the throttle cable to the hull (**Figure 60** or **Figure 61**). Note the location of the ties so you can install new ties in the same location.
11. Remove the throttle cable.
12. Inspect the cable. Replace it if it is worn, kinked or frayed.

Installation

1. Fit the end of the inner cable into the carburetor throttle and slide the adjuster in place in the cable bracket.
2. Turn the throttle cable adjuster so the end of the cable extends 17 mm (0.67 in.) above the cable bracket as shown in **Figure 72**. Tighten the locknut.

WARNING
If the throttle cable is not routed properly, the inner cable will not move smoothly. This may cause the throttle to stick wide open.

3. Route the cable through the vehicle as you noted before installation. On FX700 and SJ700A models, make sure the cable is routed on the rear side of the pivot spring as shown in **Figure 46**.
4. Reconnect the throttle cable to the throttle lever and insert the seal into the handle (**Figure 71**).
5. Secure the cable with the cable clamp (B, **Figure 27** or B, **Figure 30**) at the steering column.

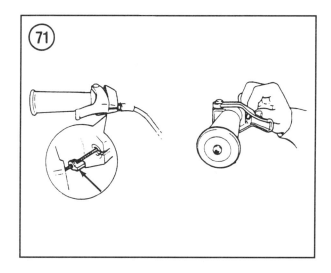

STEERING SYSTEM

6. *SJ650 and SJ700*—Secure the throttle cable and wiring harness to the cable bracket with a new plastic tie (**Figure 29**).

7. Reinstall the packing in the lower boot and install a new plastic tie around the top of the lower boot (**Figure 46**). Clip the arm of the plastic tie to size once the tie is secure.

8. Secure the cable to the hull with a new plastic tie (**Figure 60** or **Figure 61**). Clip the arm on each tie to length when the tie is secure.

9. Adjust the throttle lever freeplay as described in Chapter Three.

10. Check throttle operation while moving the pole up and down and side to side; the lever should move smoothly.

11. Reinstall the column pad (A, **Figure 25**) and the handle cover (A, **Figure 31**).

12. Reconnect both battery leads from the battery.

Throttle Cable
(RA700, RA700A, RA700B, RA760, RA1100, WB700, WB700A, WB760, WR500, WR650, WRA650, WRA650A, WRB650, WRB650A, WRB700, WVT700 and WVT1100)

Refer to **Figures 51-56** when performing the following procedures.

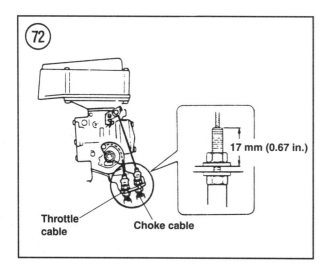

Removal

1. Drain the fuel from the tank.
2. Disconnect both battery leads from the battery.
3. Remove the steering pad from the handlebars.
4. Open the throttle lever and disconnect the throttle cable. Loosen the locknut, back out the cable adjuster and remove the throttle cable along with the seal (**Figure 71**).

NOTE
Note the position of the spiral tubing as you remove it. You must reinstall the tubing in the same position during installation.

5. Remove the spiral tubing that holds the steering cable to the throttle cable and the start/stop switch wiring harness.

6. Remove the spiral tubing that wraps the throttle cable and start/stop switch wiring harness together.

7. Remove the packing from the steering column, and remove the throttle cable from the steering column.

8. *WVT700 and WVT1100*—Remove the plastic tie that secures the cable. Note its location. You must reinstall a new tie in the same position during installation.

9. At the carburetor, loosen the throttle cable adjuster locknut and loosen the adjuster. Remove the throttle cable inner wires from the carburetor throttle and remove the throttle cable from the cable bracket (**Figure 72**).

10. Replace the cable if it is worn, kinked or frayed.

Installation

1. Fit the end of the inner cable into the carburetor throttle and slide the adjuster in place in the cable bracket.

2A. Turn the throttle cable adjuster so the end of the cable extends 17 mm (0.67 in.) above the cable bracket as shown in **Figure 72** or **Figure 73**. Tighten the locknut.

2B. *RA1100*—Turn the throttle cable adjuster so the end of the cable extends 14 mm (0.55 in.) above the cable bracket as shown in **Figure 74**. Tighten the locknut.

3A. *WR500 and WR650*—Perform the following:

NOTE
*When routing the throttle cable and the start/stop switch wiring harness in the following steps, maintain a radius of 60 mm (2.36 in.) at the center of the handlebar. See **Figure 75**.*

WARNING
If the throttle cable is not routed properly, the inner cable will not move smoothly. This may cause the throttle to stick wide open.

a. Route the throttle cable through the steering column. Loop the throttle cable as shown in

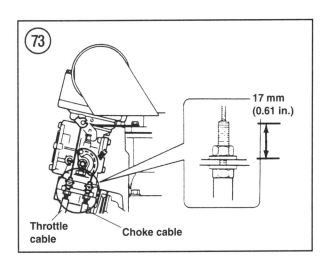

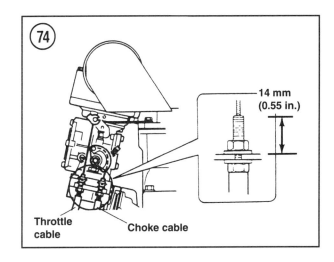

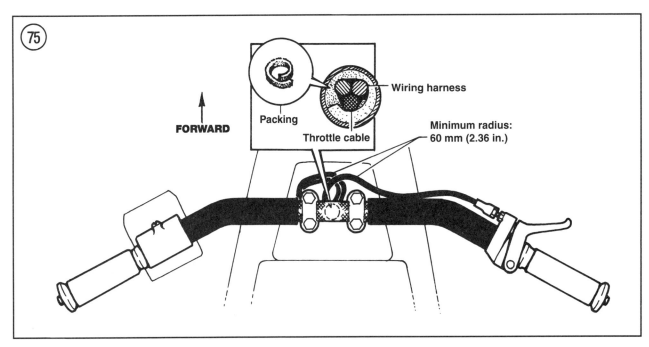

STEERING SYSTEM

Figure 75, and reconnect the cable at the throttle lever.

b. Reconnect the throttle cable to the throttle lever and insert the seal into the handle (**Figure 71**).

c. Wrap the spiral tubing around the throttle cable and the start/stop switch wiring harness. Insert 30-50 mm (1.2-2.0 in.) of spiral tubing into the steering column shaft as shown in **Figure 76**.

d. When the wiring harness and throttle cable are properly routed, install the packing into the steering column as shown in **Figure 75**. Make sure the wiring harness and throttle cable are positioned in the packing as shown. Also be sure that none of the packing sticks out of the steering column.

3B. *WRA650, WRA650A, and WRA700*—Perform the following:

NOTE
When routing the throttle cable and the start/stop switch wiring harness in the following steps, maintain a radius of 60 mm (2.36 in.) at the center of the handlebar. See Figure 77.

WARNING
If the throttle cable is not routed properly, the inner cable will not move smoothly. This may cause the throttle to stick wide open.

a. Route the throttle cable through the steering column. Loop the throttle cable as shown in **Figure 77**, and reconnect the cable at the throttle lever.

b. Reconnect the throttle cable to the throttle lever and insert the seal into the handle (**Figure 71**).

c. Wrap the spiral tubing around the throttle cable and the start/stop switch wiring harness. Insert 30-50 mm (1.2-2.0 in.) of spiral tubing into the steering column shaft as shown in **Figure 76**.

d. When the wire harness and throttle cable are properly routed, install the packing into the steering column as shown in **Figure 77**. Make sure the wire harness and throttle cable are positioned in the packing as shown. Also be sure that none of the packing sticks out of the steering column.

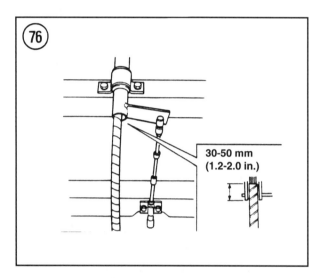

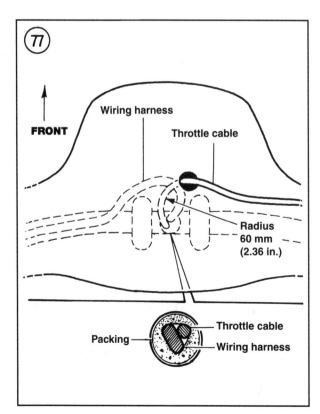

3C. *WB700, WB700A, WRB650, WRB650A, and WRB700*—Perform the following:

WARNING
If the throttle cable is not routed properly, the inner cable will not move smoothly. This may cause the throttle to stick wide open.

a. Route the throttle cable through the steering column. Loop the throttle cable as shown in **Figure 78**.
b. Reconnect the throttle cable to the throttle lever and insert the seal into the handle (**Figure 71**).
c. Wrap the spiral tubing around the throttle cable and the start/stop switch wiring harness. Insert 30-50 mm (1.2-2.0 in.) of spiral tubing into the steering column shaft as shown in **Figure 76**.
d. When the wire harness and throttle cable are properly routed, install the packing into the steering column as shown in **Figure 78**. Make sure the wire harness and throttle cable are positioned in the packing as shown. Also be sure that none of the packing sticks out of the steering column.

3D. *RA700, RA700A, RA700B, RA760, RA1100 and WB760*—Perform the following:

WARNING
If the throttle cable is not routed properly the inner cable will not move smoothly. This may cause the throttle to stick wide open.

a. Route the throttle cable through the steering column. Be sure that 200 mm (7.9 in.) of the throttle cable protrudes from the steering column as shown in **Figure 79**.
b. Reconnect the throttle cable to the throttle lever and insert the seal into the handle (**Figure 71**).
c. Wrap the spiral tubing around the throttle cable and the start/stop switch wiring harness. *On RA700, RA700B, WB760 models,* insert 50 mm (2.0 in.) of spiral tubing into the steering column shaft as shown in **Figure 80**. *On RA700A, RA760 and RA1100 models,* insert 25 mm (0.98 in.) of the tubing into the steering column.
d. When the wiring harness and throttle cable are properly routed, install the packing into the steering column as shown in **Figure 79**. Make sure the wiring harness and throttle cable are positioned in the packing as shown. Also be sure that none of the packing sticks out of the steering column.

3E. *WVT700 and WVT1100*—Perform the following:

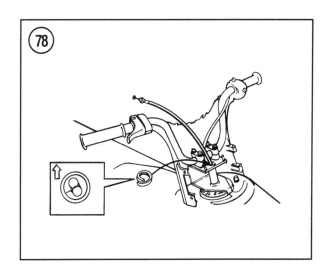

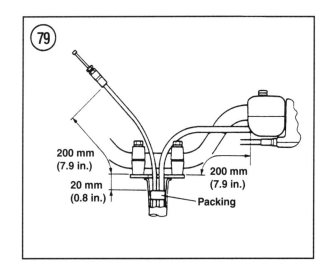

STEERING SYSTEM

WARNING
If the throttle cable is not routed properly, the inner cable will not move smoothly. This may cause the throttle to stick wide open.

a. Route the throttle cable through the steering column. Be sure that 200 mm (7.9 in.) of the throttle cable from the steering column as shown in **Figure 79**.

b. Reconnect the throttle cable to the throttle lever and insert the seal into the handle (**Figure 71**).

c. Wrap 7 windings of the spiral tubing around the throttle cable and the start/stop switch wiring harness. Insert 3 windings of the tubing into the steering column (**Figure 81**).

d. Continue securing the remaining cables and wires with the spiral tubing. Use the next 7 windings to include the fuel sensor and oil sensor leads. Then use the next winding to include the meter breather hose. Finally, wind remaining tubing around all the cables. However, leave the throttle cable out of this final pass.

e. When the wiring harness and throttle cable are properly routed, install the packing into the steering column as shown in **Figure 79**. Make sure the wiring harness and throttle cable are positioned in the packing as shown. Also be sure that none of the packing sticks out of the steering column.

4A. Install the spiral tubing that holds the steering cable to the throttle cable and the start/stop switch wiring harness.

4B. *WRB650, WRB650A and WRB700*—Use a least 4 turns of the spiral tubing to secure the steering cable to the throttle cable and start/stop switch wiring harness. Start the tubing at a point 160 mm (6.3 in.) below the cable stopper as shown in **Figure 69**.

4C. *WB700 and WB700A*—Use spiral tubing to secure the steering cable to the throttle cable and start/stop switch wiring harness. Start the tubing at a point 150 mm (5.9 in.) below the cable stopper as shown in **Figure 70**.

5. *WVT700 and WVT1100*—Secure the throttle cable with a new plastic tie.

6. Reattach the steering pad.

7. Adjust throttle lever freeplay as described in Chapter Three.

8. Check throttle operation while moving the handlebars from side to side; the lever should move smoothly.

9. Reconnect the leads to the battery.

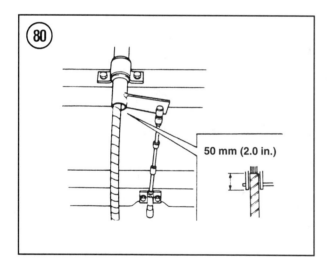

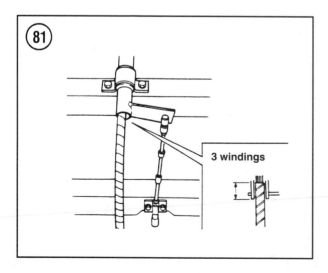

Choke Cable

Refer to **Figures 49-56** when performing the following procedure.

Removal

1. *WB700, WB700A, WRB650, WRB650A and WRB700*—Remove the packing from the hull and remove the spiral tubing that holds the steering cable, water hose, and choke cable together. On WB models, see **Figure 64**. On WRB models, see **Figure 65**.

2A. Disconnect the choke cable from the mounting bracket as follows (**Figure 82** or **Figure 83**):
   a. Remove the set screw and remove the choke knob.
   b. Remove the adjuster and locknut, and then remove the choke cable from the choke knob bracket.
   c. On WRA, WRB and WVT models, remove the choke plate.

3. At the carburetor end, loosen the choke cable adjuster locknut and loosen the adjuster (**Figure 72** or **Figure 73**).

4. Remove the end of the inner wire from the choke lever on the carburetor and remove the choke cable from the cable bracket.

5. *FX700, SJ650, SJ700 and SJ700A*—Clip the plastic ties securing the cable to the hull and remove the choke cable (**Figure 84** or **Figure 85**).

6. Inspect the cable. Replace the cable if it is worn, kinked, or frayed.

Installation

1. Fit the end of the inner cable into the choke lever on the carburetor and slide the adjuster in place in the cable bracket.

2. Turn the choke cable adjuster so the end of the cable extends 17 mm (0.67 in.) above the

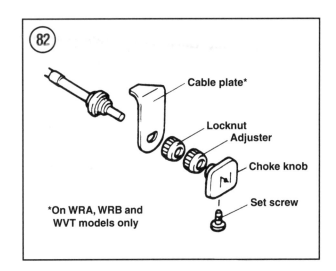

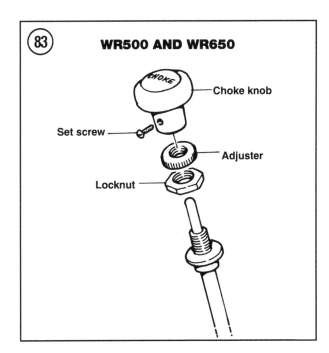

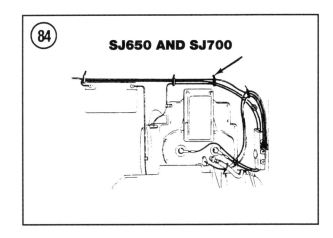

STEERING SYSTEM

cable bracket as shown in **Figure 72** or **Figure 73**. Tighten the locknut.

3. Fit the other end of the cable through the mounting bracket and reassemble the choke knob (**Figure 82** or **Figure 83**).

4. *FX700, SJ650, SJ700 and SJ700A*—Securing the cable to the hull with a new plastic tie (**Figure 84** or **Figure 85**).

5. Adjust the choke cable freeplay as described in Chapter Three.

6. Pull the knob out to its farthest point. If it moves back in, tighten the adjuster on the choke knob. Recheck and repeat this procedure until the choke knob holds at the fully-opened position (**Figure 86**).

Shift Cable (WRA650, WRA650A, WRA700, WVT700 and WVT1100)

Refer to **Figure 87** or **Figure 88** when performing this procedure.

NOTE
Note how the shift cable is routed through the hull for installation reference.

1. Remove the cable joint from reverse gate. Slide the spring-loaded sleeve back and remove the joint from the ball on the reverse gate (**Figure 89**).

2. Disassemble the cable joint by removing the sleeve, locknut, cap, washer, stopper and seal (**Figure 90**).

3. Remove the cable joint from shift lever as follows.
   a. On WVT700 and WVT1100 models, slide the spring loaded sleeve back from the joint and remove the cable joint from the ball joint on the shift lever arm.
   b. On WRA650, WRA650A, and WRA700 models, remove the cotter pin from the pivot, and remove the pivot and washer from the shift lever arm (**Figure 91**).

4. *WVT700 and WVT1100*—Remove the sleeve and lock nut from the lever end of the shift cable.

5. Remove the cable stopper, and remove the shift cable.

6. Inspect the cable. Replace it if it is worn, kinked or frayed.

Installation

1. Route the shift cable through the hull and reinstall the packing in the hull (**Figure 92**).

2. Fit the cable stopper over the reverse cable and secure the stopper to the mounting bracket.

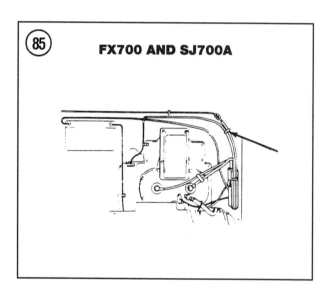

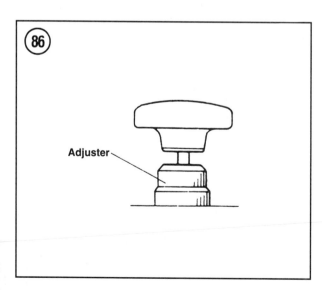

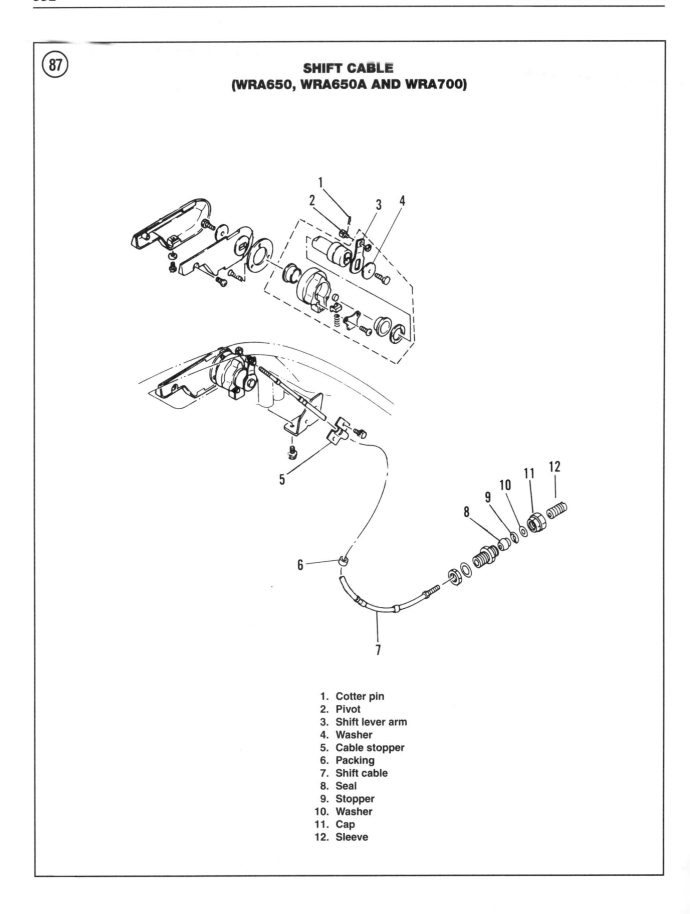

87 SHIFT CABLE (WRA650, WRA650A AND WRA700)

1. Cotter pin
2. Pivot
3. Shift lever arm
4. Washer
5. Cable stopper
6. Packing
7. Shift cable
8. Seal
9. Stopper
10. Washer
11. Cap
12. Sleeve

STEERING SYSTEM

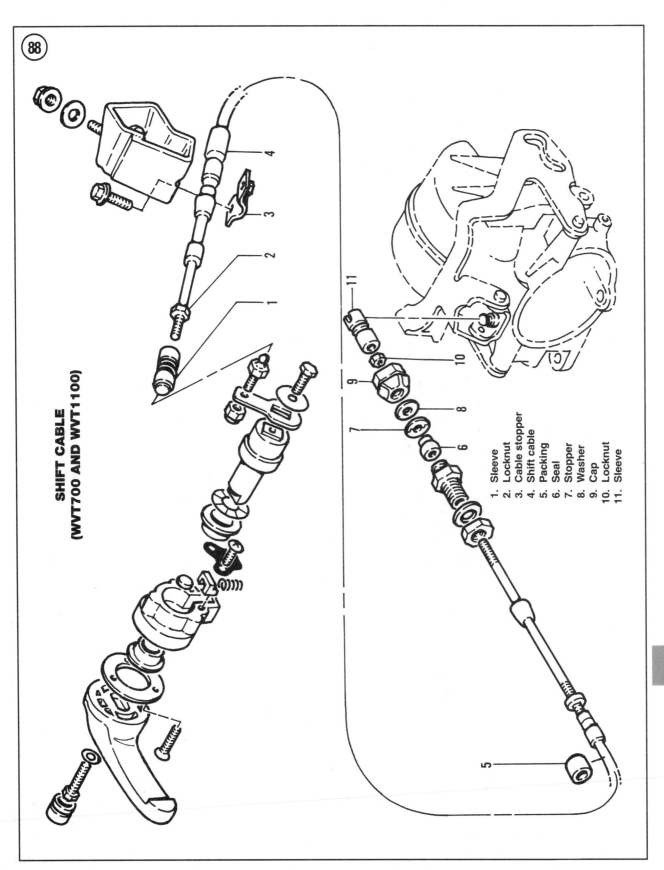

Be sure the index in the cable stopper engages the cutout in the cable as shown in **Figure 93**.

a. On WRA650, WRA650A, and WRA700 models, apply Loctite 242 (blue) to the threads of the cable stopper bolt and torque the bolt to 5 N•m (44 in.-lb.).

b. On WVT700 and WVT1100 models, apply Loctite 271 (red) to the threads of the cable stopper bolt and torque the bolt to 7 N•m (62 in.-lb.).

WARNING
When assembling the cable joint at the shift gate end of the cable, at least 8 mm (0.31 in.) of cable must be threaded into the sleeve. **Figure 94**.

3. Reassemble the cable joint to the reverse gate end of the cable as follows:

a. Install the seal, stopper, washer, cap, locknut and cable sleeve as shown in **Figure 90**.

b. Apply Loctite 242 to the threads of the locknut and cable joint. Torque the cap to 6 N•m (4.3 in.-lb.), and torque the locknut to 3 N•m (26 in.-lb.)

c. Be sure at least 8 mm (0.31) of cable engages the sleeve as shown in **Figure 94**.

4. Apply marine grease to the ball on the shift gate and to the cable joint. Slide the sleeve back

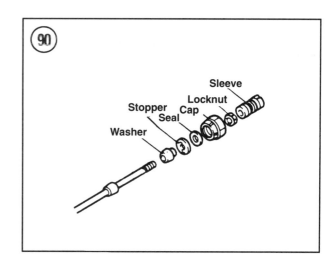

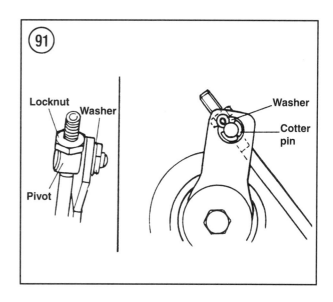

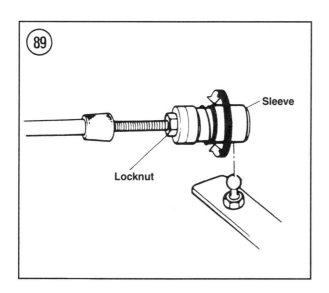

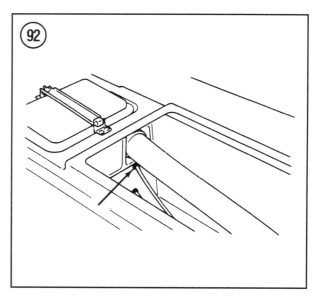

STEERING SYSTEM

from the cable joint and fit the cable joint over the ball joint.

5A. *WVT700 and WVT1100*—Reassemble the cable joint at the reverse lever as follows:

  a. Thread the locknut and cable sleeve onto the cable shaft.

  b. Be sure at least 8 mm (0.31 in.) of cable engages the sleeve as shown in **Figure 94**.

  c. Apply marine grease to the ball on the shift lever and to the cable joint. Slide the sleeve back from the joint and fit the cable joint over the ball joint.

5B. *WRA65, WRA650A and WRA700*—Reassemble the cable joint at the reverse lever as follows (**Figure 91**):

  a. Slide the pivot over the cable shaft and fit the pivot through the shift lever arm.

  b. Install the washer over the pivot and secure the pivot to the arm with a new cotter pin.

  c. Install the lock nut onto the cable shaft.

6. Adjust the shift cable as described in Chapter Three.

7. Check the operation of the shift cable. The shift lever and reverse gate should operate smoothly.

Trim Control Cables (RA700, RA700A, RA760 and RA1100)

Refer to **Figure 95** when performing this procedure.

The trim control system consists of 2 pull cables that connect the grip assembly to the trim wheel and a nozzle control cable that connects the steering nozzle to the trim wheel. The pull cables and nozzle control cable can be serviced separately.

Nozzle Control Cable
Removal

NOTE
Note how the nozzle control cable is routed through the hull for installation reference.

1. Remove the ride plate as described in Chapter 7.

2. Remove the cable joint from the ball joint on the steering nozzle. Pull the sleeve back from the cable joint and remove the cable joint from the ball joint (**Figure 89**).

3. Disassemble the cable joint by removing the cable sleeve, locknut, cap, washer, stopper and seal (**Figure 90**).

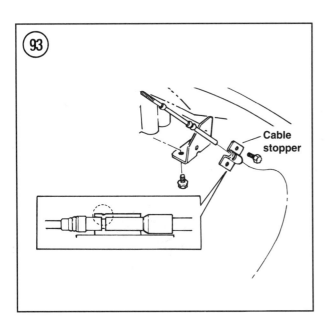

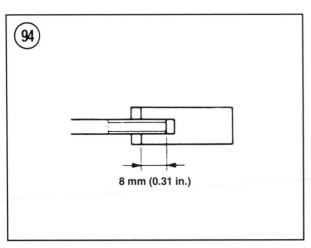

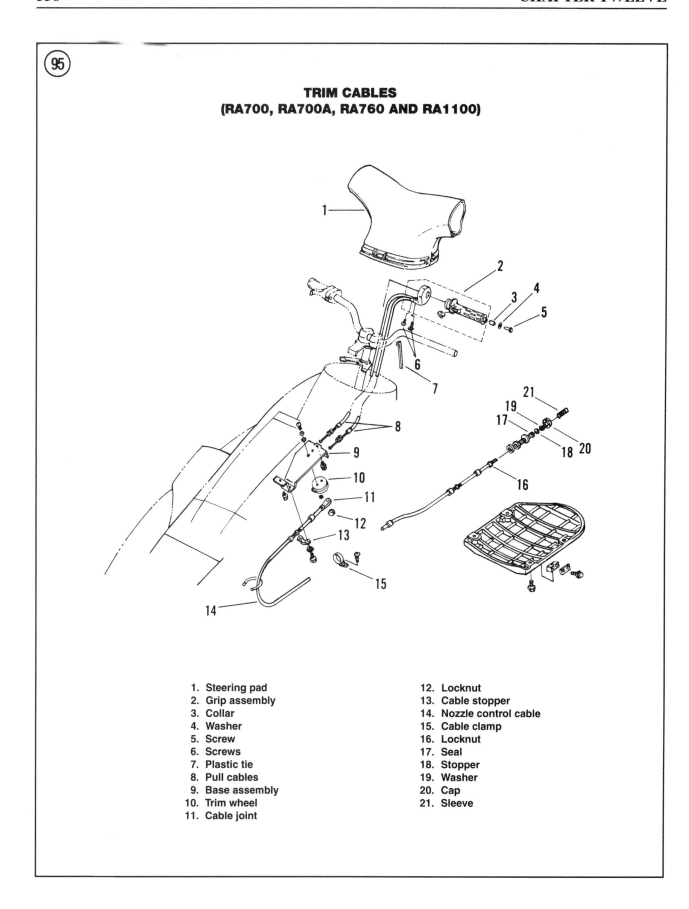

STEERING SYSTEM

4. Remove the cable clamp securing the nozzle control cable.

5. Remove the cable stopper from the base assembly.

6. Remove the locknut and remove the cable joint from the trim wheel (**Figure 96**).

7. Remove the cable joint and locknut from the nozzle control cable.

8. Remove the packing from the hull and remove the nozzle control cable from the watercraft (**Figure 97**).

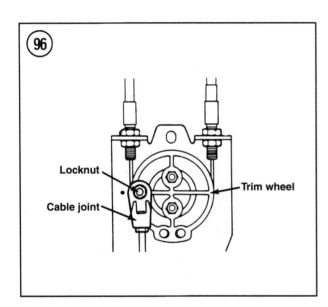

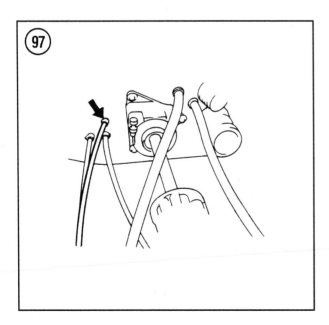

9. Inspect the cable. If it is worn, kinked or frayed, replace it.

Installation

1. Route the cable through the hull, and install the packing (**Figure 97**).

> *WARNING*
> *When assembling the cable joint at the steering nozzle end of the cable, at least 8 mm (0.31 in.) of cable must be threaded into the sleeve (**Figure 94**).*

2. Reassemble the cable joint to the steering nozzle end of the cable as follows:
   a. Install the seal, stopper, washer, cap, locknut and cable sleeve as shown in **Figure 90**.
   b. Apply Loctite 242 to the threads of the locknut and cable joint. Torque the cap to 6 N•m (53 in.-lb.).
   c. Be sure at least 8 mm (0.31) of cable engages the sleeve as shown in **Figure 94**.

> *WARNING*
> *When assembling the cable joint at the trim wheel end of the cable, thread the cable joint at least 8 mm (0.31 in.) down the cable.*

3. Reassemble the cable joint at the trim wheel as follows:
   a. Thread the locknut and cable joint onto the trim control cable. Be sure to thread the cable joint at least 8 mm (0.31 in.) onto the cable.
   b. Secure the cable joint to the trim wheel with the locknut (**Figure 96**). Torque the nut to 4 N•m (35 in.-lb.).

4. Fit the cable stopper over the trim control cable and secure the stopper to the base assembly. Be sure the index in the cable stopper

538

engages the cutout in the cable as shown in **Figure 98**.

5. Secure the trim control with the cable clamp.
6. Adjust the nozzle control cable as described in Chapter Three.

Pull Cables

NOTE
Note how the pull cables are routed through the engine cover for installation reference.

Removal

1. Disconnect each pull cable from the trim wheel as follows: (**Figure 99**).
    a. Loosen a pull cable locknut and loosen the adjuster.
    b. Remove the inner cable from the trim wheel and remove the pull cable from the base assembly.
2. Remove the steering pad from the handlebars.
3. Clip the plastic tie from the handlebars (**Figure 100**).
4. Remove the screw (A, **Figure 101**) from the end of the grip assembly.
5. Remove the screw (B, **Figure 101**) from the housing cover and remove the grip assembly from the handlebar. The pull cables will come out from the engine cover with the grip assembly. Do not lose the collar and washer from the end of the grip assembly.
6. Remove the remaining screw (C, **Figure 101**) from the housing cover and remove the cover and housing from the grip assembly.
7. Inspect the cables. Replace them if they are worn, frayed or kinked.
8. Replace the housing if it is worn or damaged.

Installation

1. Reassemble the grip assembly as follows (**Figure 102**):

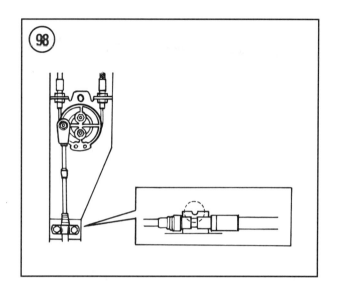

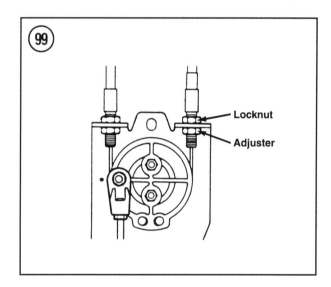

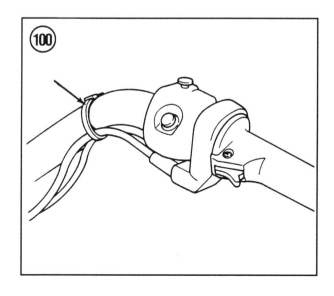

STEERING SYSTEM

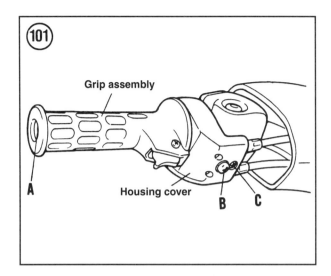

101

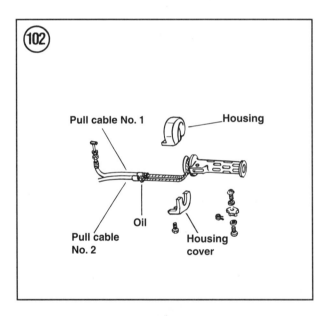

102

a. Reassemble the release button if necessary.
b. Be sure pull cable No. 1 and pull cable No. 2 are positioned on the grip assembly as shown in **Figure 102**. Pull cable No. 1 can be identified by the white tape.
c. Fit the housing over the grip assembly. Put the cover in place on the housing and secure the cover with the screw (C, **Figure 101**).
d. Apply oil to the cables where shown in **Figure 102**.

2. Route both pull cables through the engine cover and slide the grip assembly onto the handlebar.
3. Insert the collar and washer into the end of the grip assembly. Apply Loctite 242 (blue) to the threads of the screw and secure the grip assembly to the handlebar with the screw (**Figure 103**).
4. Install the washer and screw into the housing (B, **Figure 101**). Apply Loctite 242 (blue) to the threads of the screw, and securely tighten the screw.
5. Check that the cables are correctly routed through the engine cover, and seal the opening with packing.
6. Secure the cables to the handlebar with a new plastic tie (**Figure 100**).

CAUTION
*Pull cable No. 1 must be installed on the side of the base assembly marked with the D (**Figure 104**). Pull cable No. 1 is the cable with the white tape.*

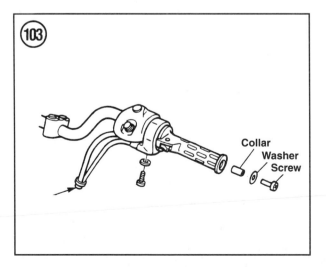

103

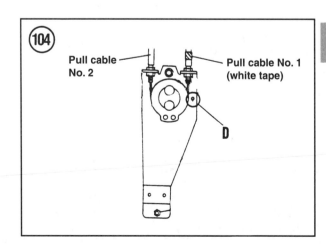

104

7. Install each pull cable to the trim wheel as described below. Install pull cable No. 1 on the side of the base assembly marked D as shown in **Figure 104**.
   a. Apply marine grease to the inner cable.
   b. Attach the end of the inner cable to the trim wheel.
   c. Fit the cable around the wheel, and secure the cable to the base assembly with the locknut.

8. Set the handle grip to the neutral position. Also be sure that the handlebars are steering straight ahead.

9. Adjust the adjuster on each cable so the top of the adjuster is 75 mm (2.95 in.) from the end of the inner cable (**Figure 105**).

10. Adjust the pull cables as described in Chapter Three.

11. Check the operation of the trim adjuster. Movement should be smooth at both the grip assembly and at the steering nozzle.

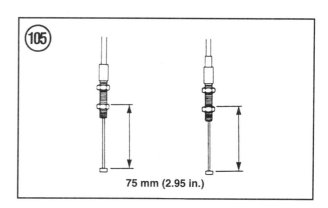

Table 1 STEERING TIGHTENING TORQUE—RA700, RA700A, RA700B, RA760, RA1100, WB700, WB700A, WB760, WR500, WR650, WRA650, WRA650A, WRA700, WRB650, WRB650A, WRB700, WVT700 AND WVT1100

	N·m	in.-lb.	ft.-lb.
Lower bearing clamp bolt	17	–	12
Upper bearing mounting nut			
WR500, WR650	5.5	49	–
WB700, WB700A	6.0	53	–
WRA650, WRA650A, WRA700, WRB650, WRB650A, WRB700 RA700, RA700A, RA700B, RA760 RA1100, WVT700, WVT1100, WB760	5.0	44	–
Bearing joint clamp bolt			
WRB650, WRB650A, WRB700, WB700, WB700A	6	53	–
Upper bearing hose clamp			
RA700, RA700A, RA700B, RA760, RA1100, WB760	2.0	18	–
WVT700, WVT1100	1.5	13	–
Handlebar holder bolt	17	–	12

STEERING SYSTEM

Table 2 STEERING TIGHTENING TORQUE—FX700, SJ650, SJ700 AND SJ700A

	N·m	ft.-lb.
Handlebar holder bolts		
SJ650, SJ700	17	12
FX700, SJ700A	16	11
Steering column		
Locknut	29	21
Pivot shaft		
SJ650, SJ700		
First nut (Large hex nut)	15	11
Locknut		
FX700, SJ700A	70	50
Pivot shaft head	15	11
Locknut	70	50
Steering pivot mounting bolt	16	11
Stopper hatch screw	6	4.3
Stopper pin assembly screw	3	2.2
Pole damper bolt	6	4.3
Start/stop switch housing screw	4	2.9
Steering pad bolt	3	2.2
Handle cover screw		
SJ650, SJ700	3	2.2

Index

B

Battery 399-405
Bearing replacement........... 27-30
Bilge system................ 472-473
Break-in procedure............. 85

C

Cable
   choke 530-531
   nozzle control............ 535-538
   pull 538-540
   shift, WRA650, WRA650A,
     WRA700, WVT700
     and WVT1100 531-535
   steering
     RA700, RA700A, RA700B,
     RA760, RA1100, WB700,
     WB700A, WB760, WR500,
     WR650, WRA650, WRA650A,
     WR700, WRB650, WRB650A,
     WRB700, WVT700 and
     WVT1100............ 521-524
     FX700, SJ650, SJ700
     and SJ700A 510-521
   trim control, RA700, RA700A,
     RA760 and RA1100......... 535
Carburetor.................. 351-361
   synchronization 117-118
Charging system............... 405
   troubleshooting.............. 47-49
Check valves, oil 465-466
Choke
   cable 530-531
   synchronization 118-119

D

Clearing a submerged
   water vehicle................ 30-31
Cooling system, flushing 107-108
Crankcase and crankshaft
   650, 700 and
     760 cc engines 200-205
   1100 cc engine 246-258
Cylinder
   650, 700 and
     760 cc engines 189-192
   1100 cc engine 236-239

D

Drive train
   intermediate shaft and housing
     RA700, RA700A, RA700B,
     RA760, RA1100, WVT700
     and WVT1100 279-284
     FX700, SJ650, SJ700, SJ700A,
     WB700, WB700A, WB760,
     WRB650, WRB650A and
     WRB700............. 273-279
     WRA650, WRA650A
     and WRA700 267-273
     WR500 and WR650 261-267
   jet pump
     FX700, SJ650, SJ700, SJ700A,
     WB700, WB700A, WB760,
     WRB650, WRB650A
     and WRB700 319-332
     RA700, RA700A, RA700B,
     RA760, RA1100, WRA650,
     WRA650A, WRA700,
     WVT700 and
     WVT1100............ 300-319
     WR500 and WR650 285-300
   ride plate adjustment........ 333-334

E

Electrical
   battery 399-405
   box
     500 cc models 427-431
     650, 700 and
     760 cc models 431-436
     1100 cc models 436-438
   charging system 405
   flywheel and stator plate,
     500 cc models 415-419
   flywheel, idler gear and stator plate
     650, 700, and
     760 cc models 419-423
     1100 cc 423-427
   fuel meter, RA700, RA700A, RA760,
     RA1100, WB760, WRA650,
     WRA650A, WRA700, WRB650,
     WRB650A, WRB700, WVT700
     and WVT1100.......... 443-446
   fuse 452
   multifunction meters....... 441-443
   oil meter, RA700A, RA760,
     RA1100, WVT700
     and WVT1100.......... 446-449
   oil warning indicator
     RA700, RA700B
     and WB760 449-450
   overheat warning indicator
     RA700, RA700A, RA760,
     RA1100, WB760, WRB650,
     WRB650A, WRB700, WVT700
     and WVT1100.......... 451-452
   starting system 405-415
   switches.................. 438-441
   wiring diagrams 545-559

INDEX

Emergency troubleshooting 41-43
Engine
  500 cc 139-166
    lubrication. 126-127
    muffler housing and
      exhaust guide 133-139
    removal. 129-133
    serial numbers 128
    service precautions 127-128
    servicing in hull 128
    special tools 129
  650, 700 and 760 cc
    crankcase and
      crankshaft 200-205
    cylinder. 189-192
    lubrication. 168
    muffler housing 177-183
    piston, piston pin
      and piston rings 192-200
    precautions 171
    removal. 171-175
    serial numbers 170
    special tools 170
    top end 183-189
    service precautions 169-170
    servicing in hull 170
  1100 cc
    crankcase and
      crankshaft 246-258
    cylinder. 236-239
    lubrication. 219
    muffler housing 226-233
    piston, piston pin
      and piston rings 239-246
    precautions 221-222
    removal. 222-226
    serial numbers 221
    service precautions 220-221
    servicing in hull 221
    special tools 221
    top end 233-236
  operation. 5-6
  troubleshooting............ 61-75
    starting system 43-47
  tune-up 108-117
Exhaust guide and muffler housing,
  500 cc engine 133-139

F

Fasteners..................... 7-10
Flame arrestor............... 336-346
Flywheel
  and stator plate,
    500 cc models.......... 415-419
  idler gear and stator plate
    1100 cc models......... 423-427
    650, 700, and
      760 cc models......... 419-423

Fuel system
  carburetor 351-361
  flame arrestor 336-346
  intake manifold and
    reed valve 346-349
  meter, RA700, RA700A, RA760,
    RA1100, WB760, WRA650,
    WRA650A, WRA700, WRB650,
    WRB650A, WRB700, WVT700
    and WVT1100 443-446
  petcock 370-381
  precautions 336
  tank 362-370
  troubleshooting............ 57-61
  vent trap system, FX700, RA700,
    RA700A, RA700B, RA760,
    RA1100, WB700, WB700A
    AND WB760 381
Fuse......................... 452

G

Gasket sealant................. 12-13
General information
  basic hand tools 14-18
  bearing replacement 27-30
  clearing a submerged
    water vehicle............ 30-31
  engine, operation 5-6
  fasteners 7-10
  gasket sealant 12-13
  lubricants. 10-12
  mechanic's tips 24-27
  parts replacement 6
  precision measuring tools...... 18-23
  seals....................... 30
  special tools 23-24
  threadlocking compound 13-14
  torque specifications........... 6-7

I

Idler gear, flywheel and stator plate
  650, 700, and
    760 cc models.......... 419-423
  1100 cc models........... 423-427
Ignition system,
  troubleshooting............. 49-57
Intake manifold
  and reed valve............ 346-349
Intermediate shaft and housing
  FX700, SJ650, SJ700, SJ700A,
    WB700, WB700A, WB760,
    WRB650, WRB650A
    and WRB700 273-279
  RA700, RA700A, RA700B,
    RA760, RA1100, WVT700
    AND WVT1100 279-284
  WR500 and WR650 261-267
  WRA650, WRA650A
    AND WRA700 267-273

J

Jet pump
  FX700, SJ650, SJ700, SJ700A,
    WB700, WB700A, WB760,
    WRB650, WRB650A
    and WRB700.......... 319-332
  RA700, RA700A, RA700B, RA760,
    RA1100, WRA650, WRA650A,
    WRA700, WVT700
    and WVT1100.......... 300-319
  troubleshooting............. 77-78
  WR500 and WR650 285-300

L

Lubrication 86-90
  1100 cc engine 219
  500 cc engine 126-127
  650, 700 and 760 cc engines..... 168
  lubricants.................. 10-12
  oil injection service 90-94

M

Maintenance
  10-hour inspection 85
  50 and 100 hour schedule 94-107
  break-in procedure 85
  cooling system flushing 107-108
  operational checklist 80-85
  storage................. 119-120
Mechanic's tips 24-27
Muffler housing
  1100 cc engine 226-233
  650, 700 and
    760 cc engines......... 177-183
  and exhaust guide,
    500 cc engine 133-139
Multifunction meters......... 441-443

N

Nozzle control cable 535-538

O

Oil
  check valves 465-466
  injection service 90-94
  meter, RA700A, RA760,
    RA1100, WVT700
    and WVT1100.......... 446-449
  pump..................... 466
  sub oil tank 465
  tank 454
  tank breather valve 466
  warning indicator, RA700,
    RA700B and WB760 449-450

Operating requirements,
    troubleshooting............. 40-41
Operational checklist 80-85
Overheat warning indicator, RA700,
    RA700A, RA760, RA1100, WB760,
    WRB650, WRB650A, WRB700,
    WVT700 and WVT1100 451-452

P

Parts replacement 6
Piston, piston pin, and piston rings
    650, 700 and
        760 cc engines 192-200
    1100 cc engine 239-246
Precautions
    650, 700 and 760 cc engines 171
    1100 cc engine 221-222
    fuel system 336
Pull cables................... 538-540

R

Reed valve
    and intake manifold 346-349
Ride plate adjustment......... 333-334

S

Sealant, gasket 12-13
Seals 30
Serial numbers
    500 cc engine 128
    650, 700 and 760 cc engines 170
    1100 cc engine 221
Service precautions
    500 cc engine 127-128
    650, 700 and
        760 cc engines 169-170
    1100 cc engine 220-221
Servicing engine in hull
    500 cc 128
    650, 700 and 760 cc 170
    1100 cc 221
Shift cable, WRA650, WRA650A,
    WRA700, WVT700
    and WVT1100 531-535
Special tools, 650, 700
    and 760 cc engines 170
Specifications, torque............. 6-7
Starting system
    electric 405-415
    engine, troubleshooting 43-47

Stator plate
    and flywheel,
        500 cc models........... 415-419
    idler gear and flywheel
        650, 700, and
        760 cc models......... 419-423
        1100 cc models.......... 423-427
Steering
    assembly
        RA700, RA700A, RA700B,
        RA760, RA1100, WB700,
        WB700A, WB760, WR500,
        WR650, WRA650, WRA650A,
        WR700, WRB650, WRB650A,
        WRB700, WVT700
        and WVT1100 481-495
        SJ650, SJ700, SJ700A
        and FX700 495-510
    cable
        RA700, RA700A, RA700B,
        RA760, RA1100, WB700,
        WB700A, WB760, WR500,
        WR650, WRA650, WRA650A,
        WR700, WRB650, WRB650A,
        WRB700, WVT700 and
        WVT1100............ 521-524
        FX700, SJ650, SJ700
        and SJ700A........... 510-521
    choke cable 530-531
    nozzle control cable 535-538
    pull cables
    shift cable, WRA650, WRA650A,
        WRA700, WVT700
        and WVT1100 531-535
    throttle cable
        RA700, RA700A, RA700B,
        RA760, RA1100, WB700,
        WB700A, WB760, WR500,
        WR650, WRA650,
        WRA650A, WR700,
        WRB650, WRB650A,
        WRB700, WVT700
        and WVT1100 525-529
        FX700, SJ650, SJ700
        and SJ700A........... 524-525
    trim control cables, RA700,
        RA700A, RA760
        and RA1100 535
    troubleshooting............... 77
Storage 119-120
Sub oil tank.................... 465
Switches 438-441

T

Threadlocking compound....... 13-14
Throttle cable
    RA700, RA700A, RA700B, RA760,
    RA1100, WB700, WB700A,
    WB760, WR500, WR650,
    WRA650, WRA650A, WR700,
    WRB650, WRB650A,
    WRB700, WVT700
    and WVT1100.......... 525-529
    FX700, SJ650, SJ700
        and SJ700A............ 524-525
Tools
    basic hand 14-18
    precision measuring 18-23
    special 23-24
        500 cc engine 129
        1100 cc engine 221
Torque, specifications 6-7
Trim control cables, RA700,
    RA700A, RA760 and RA1100... 535
Troubleshooting
    charging system 47-49
    emergency................. 41-43
    engine 61-75
    engine, starting system 43-47
    fuel system 57-61
    ignition system 49-57
    jet pump................... 77-78
    operating requirements 40-41
    steering 77
    two-stroke pressure testing 76-77
Tune-up
    carburetor synchronization .. 117-118
    choke synchronization 118-119
    engine 108-117
Two-stroke pressure testing,
    troubleshooting............. 76-77

V

Vent trap system, FX700, RA700,
    RA700A, RA700B, RA760,
    RA1100, WB700,
    WB700A and WB760.......... 381

W

Water box 383-385
Wiring diagrams 545-559

WIRING DIAGRAMS

545

WR500

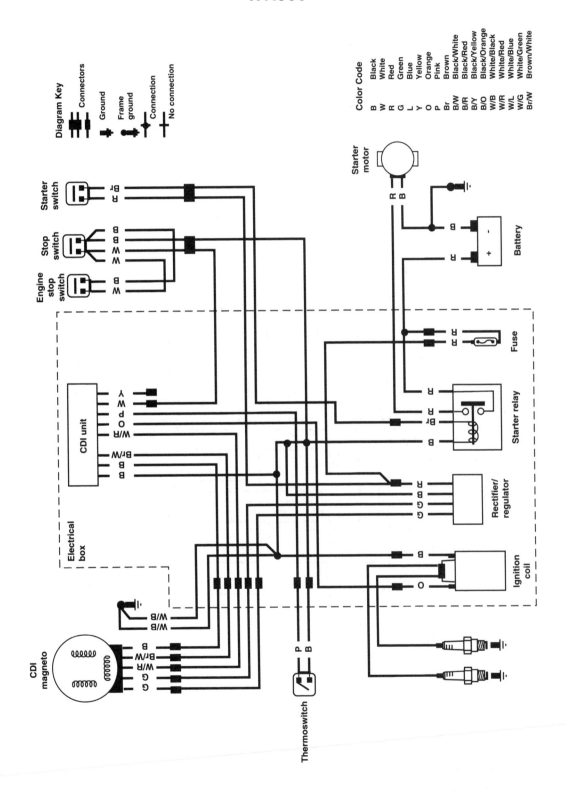

SJ650

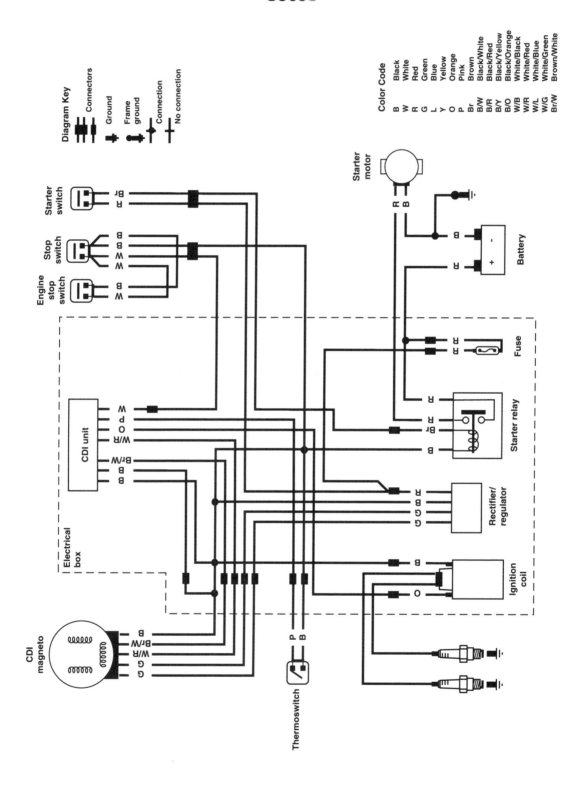

WIRING DIAGRAMS

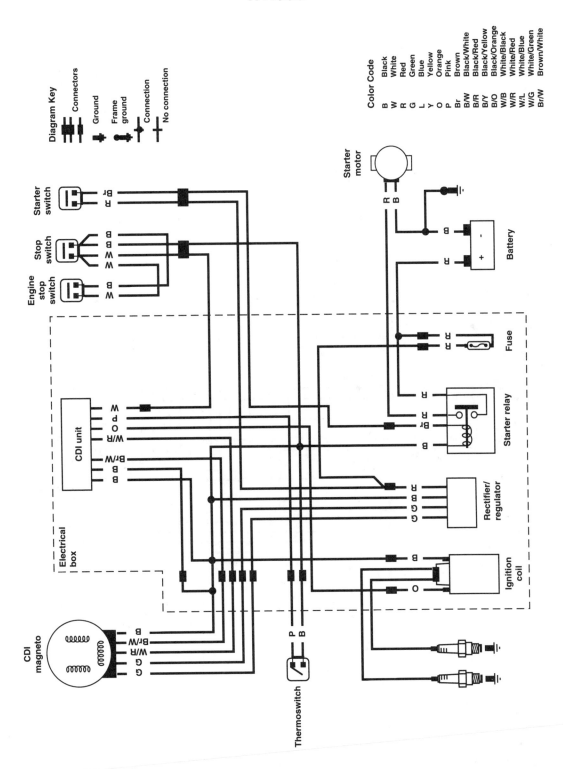

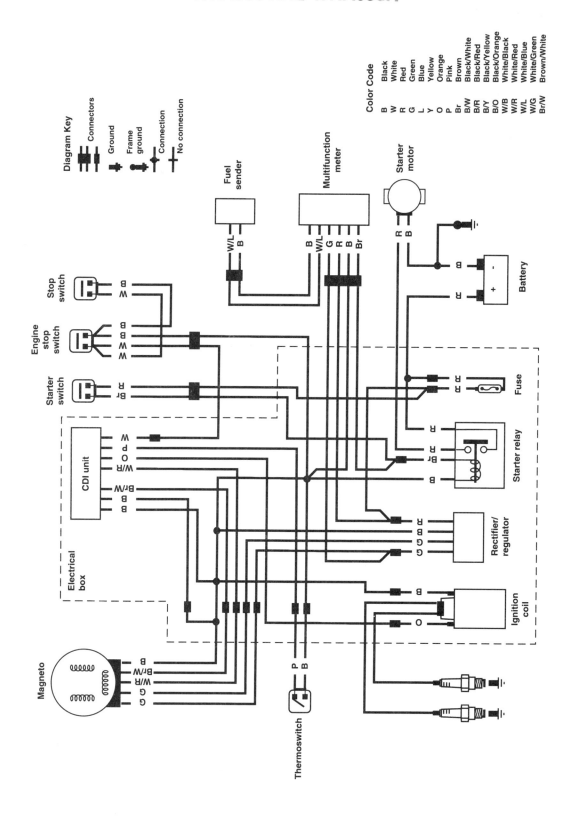

WIRING DIAGRAMS

WRB650 AND WRB650A

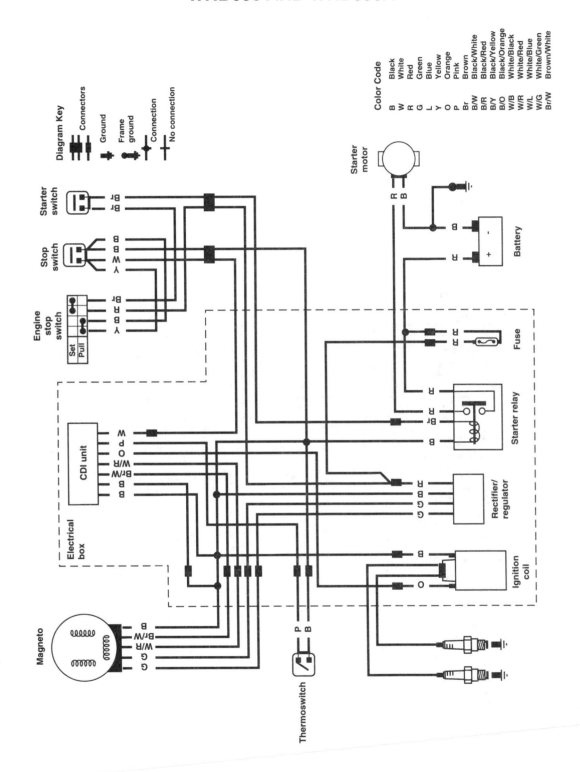

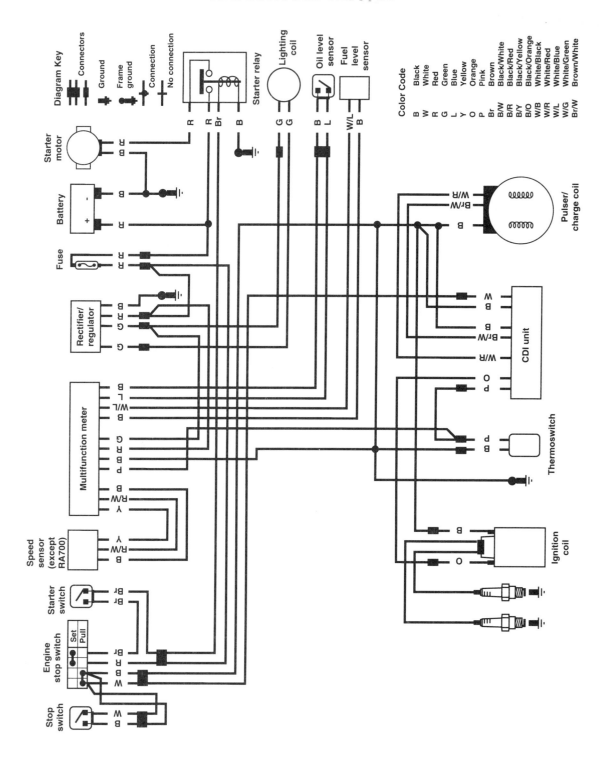

# WIRING DIAGRAMS

RA700 AND RA700B

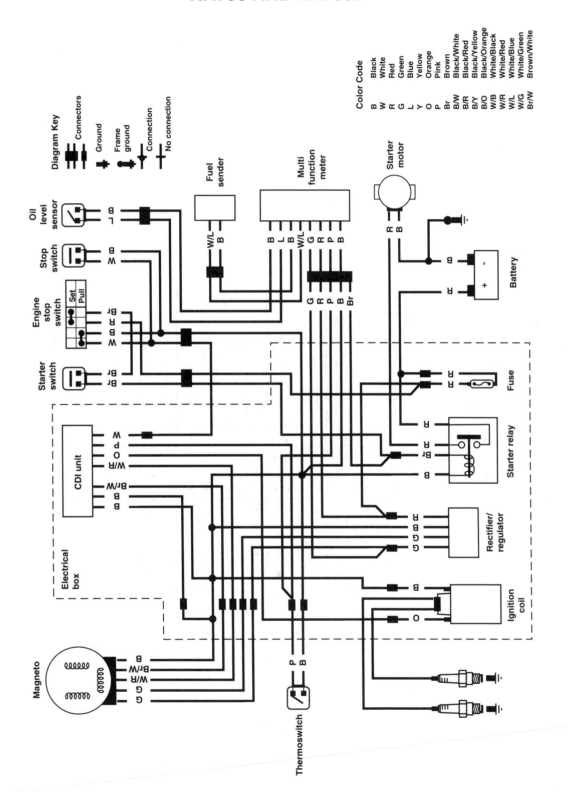

WIRING DIAGRAMS

WB700 AND WB700A

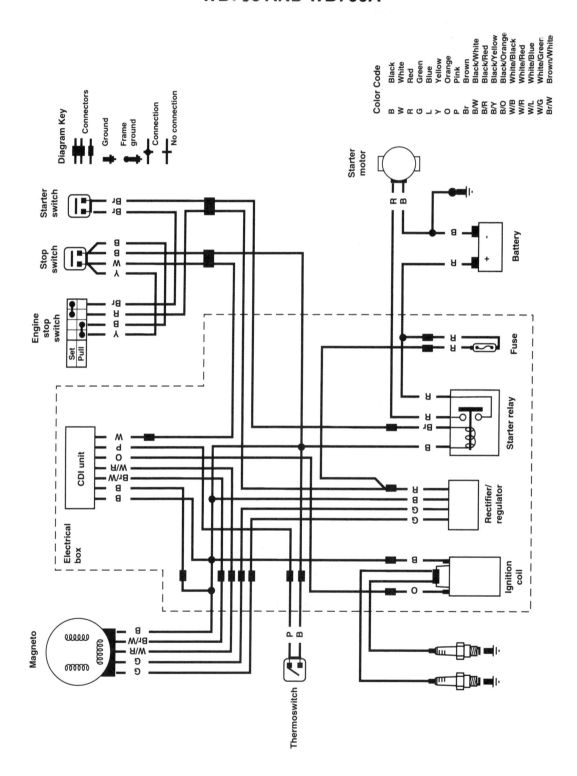

WIRING DIAGRAMS

WRA700

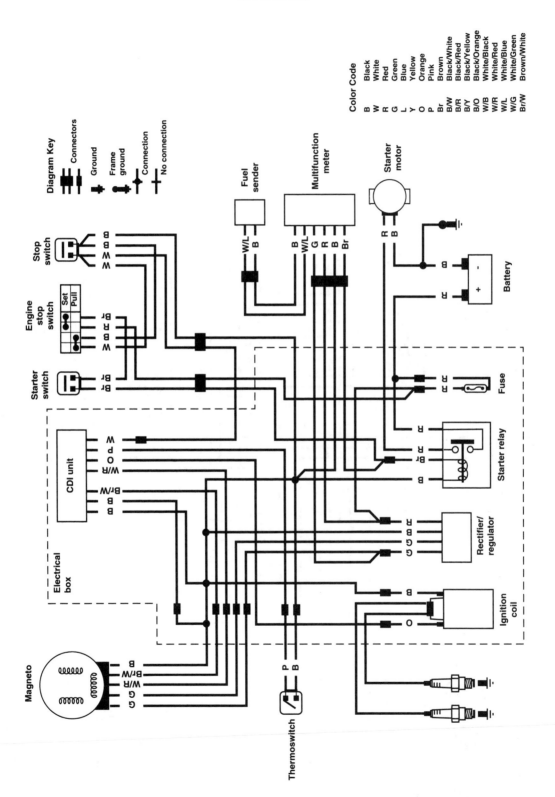

WIRING DIAGRAMS

WRB700

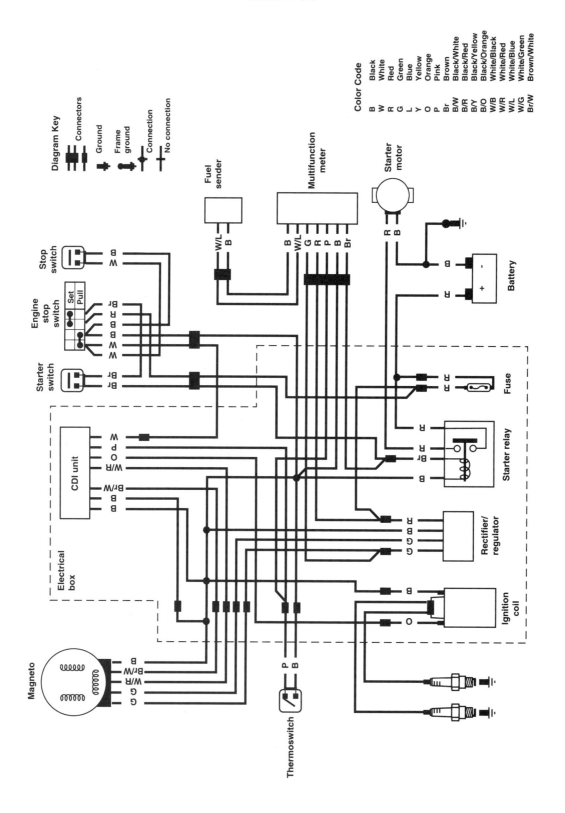

WIRING DIAGRAMS

SJ700 AND FX700

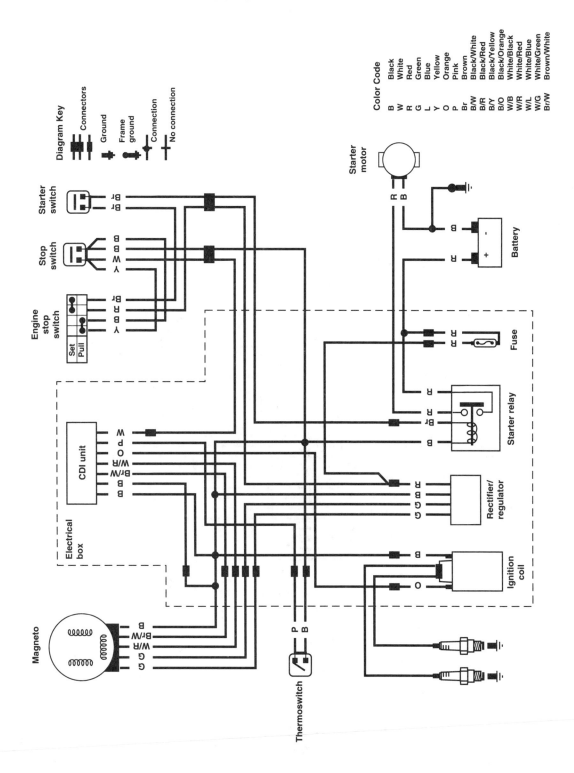

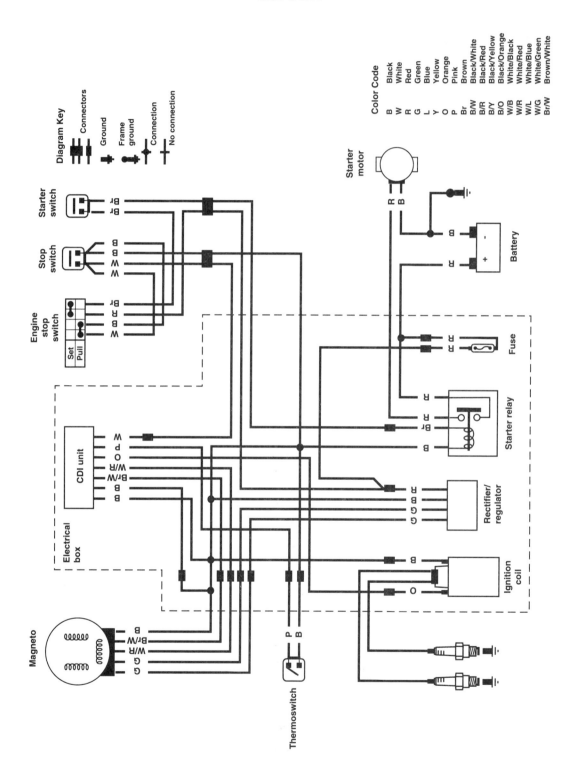

WIRING DIAGRAMS

RA760

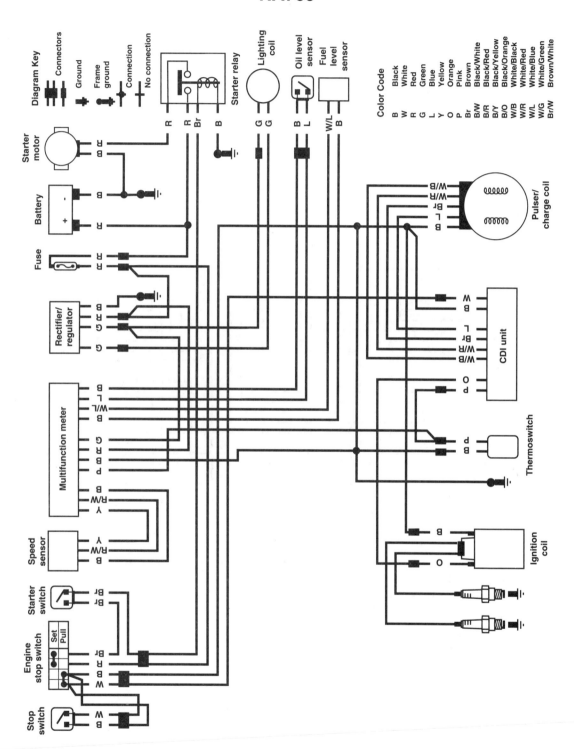

WB760

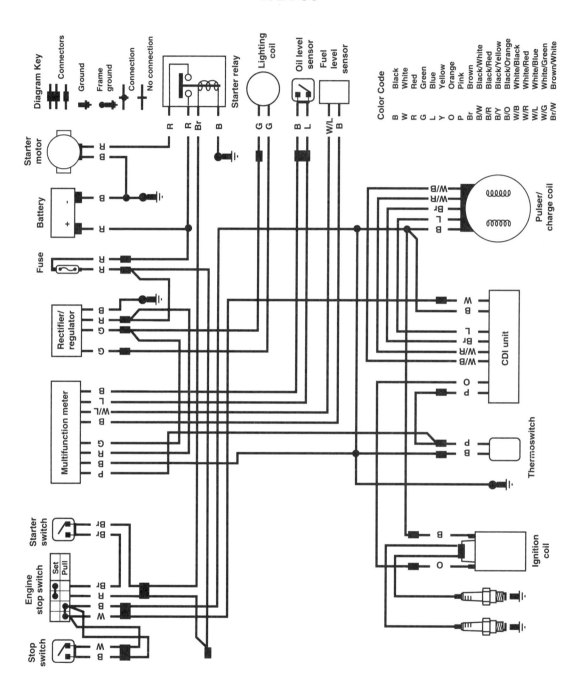

WIRING DIAGRAMS

RA1100 AND WVT1100

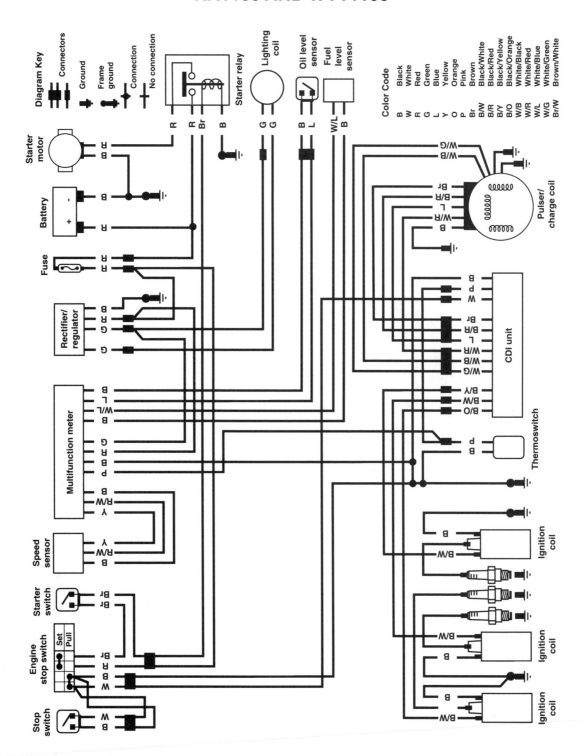

NOTES

NOTES

NOTES

NOTES

MAINTENANCE LOG

Date	Maintenance performed	Engine hours